Ecotechnology and Environmental Issues

Ecotechnology and Environmental Issues

Editors
Professor (Dr.) Arvind Kumar
FLS (London), FASc (Swiss), FISEC, FSESc, FAZ, FZA (Gold Medalist)
Vice Chancellor
S.K.M. University, Dumka – 814 101 (Jharkhand)
&
Er. Kumar Prasun Ramakrishnan
Lecturer,
Department of BCA and Engineering in Maharaja College,
University of Rajasthan,
Jaipur, Rajasthan

Daya Publishing House®
A Division of
Astral International Pvt. Ltd.
New Delhi – 110 002

Reprinted, 2021
ISBN: 9789351240105 (International Ed.)

Published by : **Daya Publishing House®**
A Division of
Astral International Pvt. Ltd.
– ISO 9001:2015 Certified Company –
4736/23, Ansari Road, Darya Ganj
New Delhi-110 002
Ph. 011-43549197, 23278134
E-mail: info@astralint.com
Website: www.astralint.com

Preface

The rising world population coupled with rising standard of living and consumerism appears to be the crux of the problem as this would necessarily cause increasing demand of natural resources and would thus threaten the environment. The products of science and technology no doubt have brought us comfort but whether they made us happier or not is a question especially when environmental degradation, pollution of air, ground or water is taken into consideration. The material possessions do reflect the strata of civilization but the modern life style is unsustainable and damaging to environment because of the greater use of natural resources and energy to satisfy the demands of human beings.

Most of the environmental problems arise due to wrong usage of science and technology for advancement of human beings. However, it is our own responsibility to treat all environment related problems as social problems.

Pollution has become the world wide phenomena with population explosion and mushroom growth of industries. As there is a close link between ecological and technical fields, the environmental protection installations require regular maintenance in the industries. Not only this, the personnel training must periodically be imparted to make the personnel aware of environmental issues. But, unfortunately proper attention is not paid to and, therefore, the industrial effluents are posing toxicological hazards to the environment.

As the environment in which we live is of immediate interest and important to us, there is an urgent need for immediate policy initiatives on environment and appropriate steps for proper eco-management. Therefore, keeping this in view, the present book has been edited to fill this gap.

My special thanks and appreciation go to the scientists whose contributions have enriched this volume. I wish to express my sincere gratitude to Dr. Victor Tigga, Hon'ble Vice Chancellor, S. K. M. University, Dumka who has been a source of constant inspiration. I owe my special thanks to Professor M. C. Dash, Hon'ble former Vice Chancellor of Sambalpur University, Professor N. C. Datta of Calcutta University, Professor K. C. Pandey of Lucknow University, Professor S. K. Konar of Kalyani University,

Professor D. K. Belsare of Bhopal University, Professor U. C. Goswami of Gauhati University, Professor P. C. Mishra of Sambalpur University, Professor P. Natarajan of Kerala University, Professor P. S. Murthy of Bangalore University, Professor Ajit Varma of JNU, Professor A. L. Bhatia of Jaipur University, Professor A. K. Mittal of BHU, Professor S. P. Hosmani of Mysore University, Professor K. Kapoor of Udaipur University, Professor K. B. Reddy of Nagarjuna University, Professor M. Vikram Reddy of Pondicherry University, Professor B. K. Tiwari of NEHU, Professor G. Tripathi of Jodhpur University, Professor R. Ramalingam of Annamalai University, Professor Sharif U. Ahmad of Nagaland University, Professor G. K. Kulkarni of Aurangabad University, Professor S. U. Mehram of Nagpur University, Professor S. K. Battish of PAU, Professor B. M. Sharma of Manipur University, Professor B. D. Joshi of Hardwar University, Professor G. C. Pandey of Faizabad University, Professor K. C. Sharma of Ajmer University, Professor M. Raziuddin of Hazaribag University, Professor U. S. Bagde of Mumbai University, Professor Gurdeep Singh of I. S. M., Dhanbad, Dr. P. K. Goel of Karad, Professor S. P. Roy and Shri Tribhuwan Poddar of Bhagalpur University, Bhagalpur.

I also express my deep sense of gratitude to my parents whose blessings have always prompted me to pursue academic activities deeply. I am also thankful to my sweet wife, Kumari Bimla and my two lovely sons, *Kumar Pallav Shivshankaran* and *Kumar Prasun Ramakrishnan* whose natural smiles extended to me relief all through this tiresome endeavour.

Last but not the least, I am also thankful to Mr. Anil Mittal, Proprietor, Daya Publishing House, New Delhi for taking keen interest in bringing out of this book. Finally, I will always remain a debtor to all my well-wishers for their blessings, without which this book would not have come into existence.

Arvind Kumar

Contents

Chapter 1

Environmental Issues: A Critical Review

Arvind Kumar and Chandan Bohra

Environmental Science Research Unit, Post Graduate Department of Zoology, S.K.M. University, Dumka – 814 101, Jharkhand

Introduction

Nature is fascinating. It embodies the spirit of the creator. The soft green leaf moving gently in the breeze, the drop of dew quivering on the petal of rose, the koel singing hidden in a mango grove, all are so enchanting, so enthralling! No less is the child with its hazel eyes, silken hair, rosy cheeks, toothless mouth and tender skin. Its father, the man, too is very handsome and is perhaps the most unique creation of God, unique in the sense that it is man who has made the world so beautiful. But for him, nature would not have been so enjoyable as it is. He has converted wild forests into parks, gardens, orchards and cities, and deserves all praise for that. But in doing that and in his attempts to make life more and more comfortable, he often destroyed forests thoughtlessly, polluted air and water recklessly and despoiled nature ruthlessly. The inevitable consequence was an ominous deterioration of the environment.

Environmental problems have attracted the attention of a wide cross-section of people all over the world during the last two decades. People are becoming increasingly conscious of a variety of problems like global warming, ozone layer depletion, acid rain, famines, droughts, floods, scarcity of fuel, firewood and fodder, pollution of air and water and problems from hazardous chemicals and radiation which have adverse effects on the environment. No nation in the world has been spared nor any citizen untouched.

India, at the time of independence was among the protest countries with little infrastructure for development. The founding fathers of the nation led by Mahatma Gandhi and Jawaharlal Nehru realised that political independence will have meaning to its citizens only if it helps them to quickly get out of the morass of poverty. Thus has started, in the world of Indira Gandhi, "an enterprise unparalled in human history–the provision of basic needs to one-sixth of mankind within the span of one or two generations." This results, in making the infrastructure of the country stronger and in reversing the trend of poverty and its symptoms, have been commensurate with the efforts. The literacy rate has increased to 52.21 per cent in 1991. Death rate has come down to 4.9 per thousand in 1921 to 10.2 per thousand in 1989. Food production has jumped up by 300 per cent from 50 million tonnes at the time of independence. Every indicator of quality of life has shown betterment.

It was in the early seventies that along with the rest of the countries in the world India realised another disheartening trend. The same efforts which helped bring people above the poverty line also put greater pressure of the natural resources of the country. Irrigation facilities provided also produced salinity in the land. Industrial development provided products but also polluted air and water. Medical facilities improved the health of the people but increased the population, thereby increasing the demand for the products of nature. Poverty is the fundamental cause which makes people over exploit the natural resources of the country like land, forests and water for meeting their basic needs, for employment, for shelter, for fuel and fodder for their cattle. As Indira Gandhi put it in her address to the Stockholm Conference in 1972, "Poverty and need are indeed the greatest polluters."

The birth rate in India has come down from 41.2 per thousand in 1961 to 22.32 per thousand in 2001, but simultaneously the death rate has fallen steeply from 19.2 per thousand in the 1961 to 8.2 per thousand in 2001. According to 2001 census the population of India crossed one billion mark. It is widely recognized that population growth is essentially the result of poverty and lack of education, particularly women's education. Official attitude towards population growth has only a marginal effect. India has to consider population growth as a certainty and make provision for providing the basic minimum needs to an ever-increasing population. India's ongoing population explosion has placed great strain on the country's environment. This rapidly growing population, along with a move toward urbanisation and industrialization, has placed significant pressure on India's infrastructure and its natural resources. Deforestation, soil erosion, water pollution and land degradation continue to worsen and are hindering economic development in rural India, while the rapid industrialization and urbanisation in India's booming metropolises are straining the limits of municipal services and causing serious air pollution problems.

Following the1984 Bhopal disaster in which a toxic leak from the city's Union Carbide Chemical Plant resulted in the deaths of more than 3000 people environmental awareness and activism in India increased significantly. The Environment Protection Act was passed in 1986 creating the Ministry of Environment and Forests (MoEF) and strengthening India's commitment to the environment, which was enshrined in the 42nd amendment to country's constitution in 1976. Under the 1986 Environmental Protection Act, the MoEF is tasked with the overall responsibility for administering and enforcing environmental laws and policies. The MoEF established the importance of integrating environmental strategies into any development plan for the country.

Nevertheless, despite a greater commitment by the Indian Government to protect public health, forests and wildlife, policies geared to develop the country's economy have taken precedence in the last 20 years. While industrial development has contributed significantly to economic growth in India, it has done so at a price to the environment. Not only is industrial pollution increasing public health risks, but abatement efforts also are consuming a significant portion of India's gross domestic product (GDP). As such, one of MoEF's main responsibilities continues to be the reduction of industrial pollution.

From pre-historical to modern time, from primitive landlines to industrial madness, man has been on the move to safeguard his life against the adversities of nature. In order to attain the highest level of development in the present industrial era, he has passed through an evolutionary process.

Now, man controls energy and its various sources. This has increased his material capabilities to a tremendous degree, and he has applied them to a technology which has become more and more diversified and sophisticated, especially in the fields of modern agriculture, industry, urban development, transport and health. Man's action on his environment is especially motivated by the hope of a better life. Human activity and its results have emerged from elements drawn from natural systems within. Man always interferes with the natural cycling of matter by increasing the fluxes between compartments, and modifying the chemical terms of elements, selectively enriching different systems. This development has been concomitant with population explosion as a result the impact of man on his environment has become increasingly detrimental. Because of the diversity and effectiveness of the means at his disposal, the rapidity and the scale of their effects at the level of the contents and even of the planet, have resulted in wide spread contamination and pollution of the environment. It is an established fact that pollution is a global problem. It emerged as an environmental issue in the seventies and since then it has drawn attention of the world for a better environmental management and conservation of natural ecosystems. The two most important processes that effect the global environment are the resource production and waste disposal. A proper balance between these two is necessary for a healthy biosphere. Every human society, be it rural or urban, industrially or technologically advanced, disposes certain byproducts and waste products into the environment, when these disposal exceed the limits of natural scavenging/removal process, they are bound to affect the normal functioning of the ecosystem and consequently they bear an adverse effect on the biota.

The most rapid increase in environmental pollution has certainly taken place in the last 150 years. It is impossible to trace the origin of the history of pollution. But it is generally believed to have developed simultaneously with other human activities. It can be attributed to many interrelated factors, such as population explosion, unplanned urbanisation, deforestation, profit oriented capitalism, technological advancement and industrialization. Mushrooming of industries throughout the world to cater to the needs of the growing population and technology has resulted in the use and production of a large number of chemicals. These chemicals include extremely toxic substances which can affect man and other living bodies in the ecosystem. A recent US study indicates that no toxicological information what so ever is available for the 80 per cent of all the chemicals used in commerce. Not surprisingly these chemicals keep throwing up surprises with their deadly properties.

By constantly altering has environment or ecosystem by such activities as urbanization, industrialization, deforestation, land reclamation, construction of irrigation canals and dams, man has created for himself new health problems. For example, the greatest threat to human health in India today is the ever increasing, unplanned urbanisation, growth of slums and deterioration of environment by way of industrialization. The deterioration of environment is mainly caused due to dumping of solid wastes produced as a byproduct of the industry in the vicinity of industry on the landmasses, discharge of industrial wastewater (effluent) into the aquatic bodies and discharging poisonous gases into the atmosphere polluting the air. By one industry, the air, the water and the land mass of the area gets polluted day by day. After prolonged exposure, the environment becomes totally polluted. As a result diseases at one time thought to be primarily "rural" have acquired serious urban dimensions. The agents of a number of diseases for example malaria and Kala-azar, which were effectively controlled have shown a recrudescence. The reasons for this must be sought in charges in the human ecology. The Bhopal gas tragedy in 1984 highlights the danger of locating industries in urban area. The nuclear disaster in Soviet Russia in April 1986 is another grim reminder of environmental pollution. The recent fire at Hindustan Petroleum Unit at Visakhapatnam is September 1997 was a great ecological disaster.

It is now being increasingly recognized that environmental factors and ecological considerations must be built into the total planning process to prevent degradation of ecosystem.

Environmental Issues

Air Pollution

Industrialisation and urbanisation have resulted in a profound deterioration of India's air quality. India has more than 20 cities with populations of at least 1 million, and some of them–including New Delhi, Mumbai, Chennai and Kolkata–are among the world's most polluted. Urban air quality ranks among the world's worst. Of the 3 million premature deaths in the world that occur each year due to outdoor and indoor pollution, the highest number is assessed to occur in India. Sources of air pollution, India's most severe environmental problem, come in several forms, including vehicular emissions and untreated industrial smoke. Continued urbanisation has exacerbated the problem of rapid industrialization, as more and more people are adversely affected and cities are unable to implement adequate pollution control mechanisms.

One of the most affected cities is New Delhi, where airborne particulate matter (PM) has been registered at levels more than 10 times India's legal limit. Vehicles are the major source of this pollution, with more than three million cars, trucks, buses, taxis and auto rickshaws already on the roads. In Delhi, vehicles contribute about 64 per cent of the total pollutants emitted in the atmosphere (CPCB, 1993–94). With vehicle ownership rising along with population and income, India's efforts to improve urban air quality have focused in this area. In New Delhi, emission limits for gasoline and diesel powered vehicle came into effect in 1991 and 1992, respectively, and the city has prohibited the use of vehicle more than 15 years old. Emissions standards for passengers cars and commercial vehicles were tightened in 2000 at levels equivalent to the Euro-1 standards of the European Union, while the even more stringent Euro-2 standards have been in place for the metropolitan areas of Delhi, Mumbai, Chennai and Kolkata since 2001. Furthermore, the sulphur content of motor fuels sold in the four cities has been restricted to 500 parts per million (ppm) since 2001 in order to be compatible with lighter vehicle emissions standards. Motor fuel sulphur content in all other regions of India has been limited to 2,500 ppm since January, 2000.

India's high concentration of pollution is not due to the absence of a sound environmental regime, however, but to a lack of environmental enforcement at the local level. Regulatory reforms aimed at improving the air pollution problem in cities such as New Delhi have been difficult to implement. In 1998, India's Supreme court issued a ruling requiring all the city's buses to be run on compressed natural gas (CNG) by March 31, 2001. Compliance waste be achieved either by converting existing diesel engines or by replacing the buses themselves. However, only 200 (out of total fleet of 12,000) CNG-fueled buses were available by the initial deadline and public protests, riots and widespread "commuter chaos" ensued as some appearance of some 15,000 taxis and 10,000 buses in the city were banned from use. To ease the transition, the local government changed course and allowed for a gradual phase out of existing bus fleet.

In addition, India's reliance on coal-fired power plants for its electricity generation has undermined source of the vehicular-oriented air quality improvement initiatives. Despite the fact that India is a large coal consumer, its Central Pollution Control Board has been slow to set sulfur dioxide (SO_2) emissions limits for coal-fired power plants, mainly because most of the coal mined in India is low in sulfur content. Coal-fired power plants do not face any nitrogen oxide (NOx) emissions limits either, although thermal power plants fueled by other fossil fuels are subject of particulate matter emission standards.

Due to air pollution, there is an increase in temperature at the rate of 0.09°C per year (Angell, 1994). An increasing amount of CO_2 in the atmosphere due to removal of vegetation has been clearly marked out (Pandya *et al.*, 1977). Concentration of methane in atmosphere has risen rapidly in recent years. The present levels are perhaps twice as high as those in earlier year (Khalil and Rasmussen, 1994). The concentration of toxic pollutants in urban areas have already reached levels that pose risks to human health (Wagner, 1994). Brandon and Homman (1995) have reported that ambient pollution levels exceeding WHO standards in thirty six major Indian cities/towns account for 40,350 premature death, 1,98,05,000 hospital admissions and sickness of requiring medical treatment and 1201 million incidences of minor sickness annually. Parikh *et al.* (1994) estimated human health damage due to air pollution in Natural capital territory, Delhi. Respirable suspended particulate matters (RSPM) have been linked with human morbidity and mortality (Dockey and Papes, 1994; Anderson *et al.*, 1992). Presence of the lead in the environment has led to an increasing awareness and concern of its detrimental effect to ecosystem and human health (WHO, 1977 and Mathur *et al.*, 1997). It is found that toxic metals are associated with fine particulate matter present in the ambient air of Indian urban agglomerations (Negi *et al.*, 1987; Sandasivan and Negi, 1990; Gajghate and Hasan, 1997 and 1999; Gajghate *et al.*, 1997). In the city of Mumbai, India, lead along with toxic elements has been observed to be present in considerable quantities in ambient air (Khandekar *et al.*, 1984 and Asha *et al.*, 2001). Airborne lead in the city of Varanasi has also been reported (Tripathi, 1994). Status of airborne metals in the environment in India and various strategies and approaches for its management has been discussed (Gajghate and Hasan, 1995 and 1996; Gajghate *et al.*, 2003). Air pollution due to store crushers in Jharkhand and its impact on human health has also been studied by Bohra and Kumar (2003).

Thus, we see that environment is much polluted due to air pollution. The environment effects our output of work, our mental alertness, our desire for change and many other attitudes and responses. To ensure his well-being man has to produce an ecological balance between himself and his environment.

Energy Consumption

India's energy consumption is increasing rapidly, from 4.16 quadrillion Btu (quads) in 1980 to 12.8 quads in 2001. This 208 per cent increase is largely the result of India's increasing population and the rapid urbanisation of the country. Higher energy consumption in the industrial, transportation and residential sectors continues to drive India's energy usage upwards at a faster rate even than China, which experienced a 130 per cent increase in energy consumption from 1980 to 2001.

Despite the rapid growth between 1980 and 2001, India's energy consumption is still below that of Germany (14.35 quads), Japan (21.92 quads), China (39.67 quads) and the United States (97.05 quads). In addition, India's per capita energy consumption, which stood at 12.6 million Btu in 2001, is well below most of the rest of Asia and is one of the lowest in the world (although this may be more the result of India's large population rather than a low level of energy consumption. However, the 103 per cent rise in India's per capita consumption between 1980, when per capita energy usage was just 6.2 million Btu and 2001, is more problematic in the long term.

Carbon Emissions

Between 1990 and 2001, India's carbon emissions increased by an astonishing 61 per cent, a rate surpassed only by China's 111 per cent increase during the same time period. The rise in India's carbon emissions has been exacerbated by the low energy efficiency of coal-fired power plants in the country. Many of India's highly polluting coal-fired power plants will have to remains in operation for the next couple of decades because of high capital costs associated with replacing existing coal-fired plants, a scarcity of capital and the long lead time required to introduce advanced coal technologies. As such, India's contribution to world carbon emissions is expected to increase in coming years, with an estimated average annual growth rate between 2001 and 2005 of 3.0 per cent compared to 3.4 per cent in China and 1.5 per cent in the United States. The absolute increase in emissions will partially be a function of the degree to which coals is relied upon as a major energy source. If coal consumption is substituted by oil and natural gas consumption, India's overall carbon emissions would be reduced.

India's per capita carbon emissions are relatively low–at 0.025 metric tons of carbon per person in 2001. However, the country's per capita carbon emissions are expected to increase in the coming years due to the rapid pace of urbanisation, a conversion away from non commercial towards commercial fuels, increase vehicular usage and the continued use of older and more inefficient coal fired plants. In fact, due to fast-paced industrialization, per capita emissions are expected to triple by 2020.

Energy and Carbon Intensity

India's heavy reliance on coal, much of it low-quality, goes a long way towards explaining the country's relatively high carbon intensity level. Indian economic policies such as high import tariffs on high-quality and subsidies on low quality domestic coal also have contributed to increased used of low-quality coal. Although initiatives have taken to encourage the use of higher quality coal, such as reducing the tariffs on imported coal which may help in reducing the country's carbon intensity. The introduction and adoption of technologies to reduce coal consumption and/or improve the efficiency of the coal that is combusted is an important government priority, given that the majority if India's power generation is coal-fired.

India's energy intensity (energy consumption per dollar of GDP) and carbon intensity (carbon emissions per dollar of GDP) is relatively high compared to its neighbours.

Water Pollution

Water pollution is a phenomenon that is characterized by the deterioration of the quality of land water (rivers, lakes, marshes and groundwater) or sea water as a result of various human activities. In a sense, the existence of human beings itself is a source of water pollution. In primitive ages, the purifying capacity of nature was much greater than the rate of water pollution caused by human living. But as time passed, men began to settle near water areas for the convenience of living, then they developed various industries which provided them with the foundation of civilized life. The emergence of various industries coupled with the development of civilization has constantly increased the quantity of wastewater containing pollutants. On the other hand, the increasing concentration of the population has also invited the concentration of pollution sources. As a result, the rate of environmental pollution has exceeded the rate of natural purification. The type of water pollution in a city varies largely depending on the characteristics of the city, life style of the inhabitants, degree of civilization, degree of development of sewage treatment plants or sewerage systems. The modern sewage treatment technique is capable of removing organic pollutants very efficiently, but it is difficult economically to remove nitrogen and phosphorus contents efficiently with existing facilities. For example, only about 30 per cent of the nitrogen and phosphorous contents can be removed through the currently employed two-step treatment method, which mainly applies the activated sludge method, and the remainder of such contents causes the eutrophication of water areas.

Agriculture has been a victim of water pollution in many instances, but sometimes, it is also responsible for polluting water. Water pollution caused by agriculture is mainly an outcome of fertilizers, insecticides and pesticides. According to a survey conducted by the National Environmental Engineering Research Institute, a staggering 70 per cent of the available water in India is polluted, because these water resources have been freely used of the industrial and public wastewater disposal (Basu, 1986). The domestic sewage and industrial effluents contain pollutants that cause harmful effect on receiving water and adverse impact on human health as well as aquatic biota (Kumar, 1996; 2003; Bohra, 2001).

Domestic sewage mostly contain metabolic wastes, various detergent formulations, consumer products and significant amount of trace metals, making them rich in organic matters and nutrients. The organic matters increase the BOD load of the receiving water bodies and render them unsuitable for the survival of the aquatic animals. The organic matters in the domestic sewages are mostly biodegradable and cause problem only when the rate of input into the environment exceeds the rate of degradation or dispersion the addition of nutrients by the way of domestic sewage into the water bodies leads to eutrophication, which brings change in the community metabolism and the biological composition of inhabiting communities and populations. Modern agriculture with its rapid mechanizations and spreading of fertilizers, besides the use of protective treatments such as herbicides, insecticides and fungicides exerts an ever increasing pressure on the natural environment. Some of these chemicals are non-biodegradable, and hence, persist in the ecosystem. These chemicals can be absorbed and concentrated by the living organisms. This phenomenon is known as bioaccumulation and creates various problems, especially when the pollutants are of high toxic nature with high biological half time life.

Industry is the most important source of pollution. Over 75,000 chemicals are in common use today and several thousands of new compounds are being added to the figure each year (Cairns, 1980; Miller, 1984 and Sahu *et al.*, 2003). These chemicals are now an inevitable part of the process of industrialization and mostly they are produced and used by the industry to deliver finished goods. Some of these chemicals, the byproducts and the waste discharges of the industry are released in advertently or accidentally into the environment to such an extent that they really threaten the ecosystem in a global scale.

The emission of pollutants into the environment is a complex phenomenon and cannot be limited to the fixed image of a water pipe spilling out its effluents into a lake or river. In almost all cases, substances discharged into the environment are going to be carried a long way from their source. Atmospheric and hydrologic circulation systems, then disperse them progressively throughout the biosphere. The pollutants act in many ways on the living systems. Acute and/or long term actions are the most obvious ecotoxicological impacts of a given pollutant. It results demo-ecological effects at pollution levels which are displayed through immediate or premature death, reduced reproductive success, reduced growth, and/or increased loss at a juvenile stages. These are ultimately reflected in the lower abundance and perturbed distribution of the exposed populations of sensitive species. The effect of sub-lethal concentrations of some pollutant is by far, the most frequent ecotoxicological problem and it can be far more noxious in the long run for the exposed species resulting in chronic deterioration of the ecosystem. The persistent non-biodegradable pollutant are absorbed by living organisms and are concentrated many times more than the concentration of the surrounding medium. Further these bio-accumulated pollutants are passed from one trophic level to another and exhibit increasing concentration in organisms related to their trophic status resulting in bio-magnification.

Thus, industry is responsible for creating a fantastic array of new chemicals every year, all of which eventually find their way into the environment. For most of these chemicals, not even the chemical formulae are known and much less is known about their acute, chronic or genetic effect on the living organisms. At present, the industry is the focus of attention, the world over, as the most important source of pollution of the environment. Water pollution is an acute problem in Indian rivers and their tributaries especially those flow through the major towns and cities (Trivedy, 1989). Water from such polluted river is abstracted and supplied to the cities for public uses (Jha *et al.*, 1997).

Population Explosion

Even after introduction of birth control devices, we are unable to bring down population growth in a significant way in the third world countries, particularly in India. India's population had reached one billion by the end of 1999 and still it grows at the rate of 1.9 per cent per annum. The fertility rate remains high in Sub-Saharan Africa and parts of South Asia. The United Nations Environment Programme (UNEP) has projected that the world population would be around 8.9 billion in 2050. The worst thing is that the highest birth rate is found in countries suffering from poverty, malnutrition, insecurity and degradation of natural resources (Anon 2000a and b). The alarming fact is that more than 90 per cent of this population growth occurs in the poor and developing world. India imports more than 17 per cent of the world population on 2.4 per cent of the total land area. Instead of declining, population growth triples in between 1951 to 2000 (361 million in 1951 to 1000 million in 1999). Due to rapid growth of population, we are unable to provide the minimum needs to the people such as basic education, potable water for drinking, shelter, sanitation and health, food, rural roads, and other communication facilities. Owing to the alarming rate of population growth, there is a imbalances in the distribution of natural resources. Incidentally 50 per cent of the total land area is under the control of 23 per cent of the rich population. Again 85 per cent of the world income is controlled by 23 per cent of the affluent class. 73 per cent of the population lines in poverty, mostly in Asia, Africa and Latin America. The poor are the hardest hit by the loss of natural resources, pollution and the also cause for damaging environment. The question now is how long the physical availability of natural resources provides support to the ever increasing population? Unless and otherwise the people adopt the policy of "enough to material consumption and desired family size", it is difficult to control the population in poor countries. Another issue worth mentioning here is, whether the population growth in poor countries alone responsible for degrading environment, desertification, soil erosion, deforestation, loss of soil fertility, water logging, salinisation of the soil and spread of diseases?

Urbanisation and Health

Urbanisation and health are other associated problems of population growth in developing countries when people start migrating to urban centres they may cause severe environmental problem in urban areas. Sanitary conditions will be poor, solid waste remains unremoved, food and drinking water will be contaminated, noise level will also be higher. All these environmental problems affect the lives of the urban population. The urban waste disposal dumped in open places fouls the surfaces, groundwater, land and air. Most of the industrial wastage is toxic and hazardous, which is sent by them to the environment, untreated. These unhygienic practices in urban areas spread diseases. One estimation states that 85 per cent of water discharge brings infectious diseases such as diarrhoea, malaria, schistomiasil and gastroenteritis. In addition, urban slums also spread several ailments, namely communicable diseases, malnutrition, mental and physical retardation and behavioural problems.

Deforestation, Biodiversity and Species Extinction

Deforestation has risen due to excessive felling of trees for timber, over grazing, fire and clearance of land for cultivation and pastures. Obviously all these activities are taking place at an alarming pace in the developing countries. Forests are vanishing at a rate of 17 million hectares per year, an area of about half the size of Finland. The issues related to deforestation continue to cause great concern to the world community because the destruction and degradation of forest cause soil erosion and also disturb the water resources. It is estimated that about 6000 million tonnes of top soil with all essential nutrients are flowing into the sea every year (Khan, 1994). This top soil removal has been heavily affecting the tribal economy in addition to siltation problems in reservoirs and dams. Since deforestation disturbs ecological balance and deteriorates quality of life, it was discussed at greater length at the Rio Summit in 1992.

Deforestation is a major cause of loss of biodiversity and species extinction. Some of the life forms have totally vanished while some others are categorized as endangered. About 3120 species of plants and 172 categories of animals of our country are listed in the IUCN's Red List threatened category.

Land Degradation and Desertification

With the increase in population most of the cultivable area had been diverted for non-farm activities. In order to increase agricultural production farmers are practicing various farm technology in their limited farm size. Use of fertilizers and pesticides for instance adversely affect the nature and land degradation becomes unavoidable. Over two million hectares have been degraded, which could be reclaimed. Through appropriate wasteland reclamation measures, which could help to boost agricultural yield and simultaneously reduce unemployment and poverty. Misuse of land, incorrect methods of disposal of solid and liquid waste and increased amount of waste generated through domestic, industrial and agricultural practices greatly cause soil pollution.

Due to over exploitation of natural resources there has been a drastic change in climate in various parts of the globe. The exploitation and climate change synergistically cause desertification. Accentuated degradation and desertification of land leads to loss of soil fertility. In Africa 60 per cent of agricultural land is arid but dry regions in North and South America, Asia, Australia and Europe have been affected by desertification (Zipf, 2001). Soil degradation today is taking place on more than one third of all the land areas on earth, and approximately one billion people are having to struggle with consequences. It is estimated that some 24 billion metric tonnes of fertile soil are permanently lost each year world wide (Zipf, 2001). Desertification leads to migration to urban areas in search of livelihood resulting into growth and over population in urban areas.

Food Production

One third to the world population is malnourished owing to socio-ecological problems. Enormous deaths in Asia and Sub-Saharan African countries can be attributed directly or indirectly to malnutrition and hunger and most of those who die are children. Due to climate changes countries like India will experience reduction in agricultural yield and face a high risk of hunger. Africa will be worst affected, with 18 per cent more people at risk of hunger by the year 2050 (Anon, 1998b). Although green revolution technology helped to increase agricultural food grain production if widened social disparity. Recent studies have shown that yield stagnation has been noticed in major crops across the globe, especially for the cereals from which people get most of their food energy. The world food production loss due to soil degradation is estimated as much as the 17 per cent of the total production from 1945 to 1990 (Anon, 1998a). Now the issue is how to fulfill the food grain requirement of the growing population? Can science and technology solve all our food related problems without degrading the environment?

Global Warming, Greenhouse Effects and Ozone Layer Depletion

Studies have been shown that the globe is getting warmer and warmers by two to six degree celcius in every 25 years. Because of this reason, tropical forest are disappearing and many of the tropical grasslands face the danger of being transformed to desert or temperate grasslands (Anon, 1998b). Since the mean sea level is rising many islands will submerge into the sea. The climatic changes adversely affect agricultural production in developing countries and positively influence in the western cold countries, but the latter will be the greater sufferers of disease, which is not endemic now.

Burning of fossil fuels is the major culprit for the present problem of greenhouse gases. CO_2, CFC, Methane, Nitrous oxide and other gases are major contributing factors for greenhouse gases. A major portion of CO_2 is omitted for vehicles. Rice fields and cattle produce major methane gases. The increasing concentration of Chloro Fluora Carbon (CFC) in the atmosphere causes ozone layer depletion. The CFC is present in refrigeration, air conditioners and aerosol. It is estimated that 15 million tonnes of CFC was expelled into the atmosphere till 1987. A hole was noted in the ozone layer in Antarctica and the Northern hemisphere. The depletion of ozone layer may lead to skin cancer, reduce the resistance power, reduce the power of photosynthesis in plants, reduce fish availability in sea and this may pave the way for acid rain (Rathakrishnan, 2003).

Solution of Environmental Challenges: Checking the Population and Genetic Engineering

Sustained development cannot be achieved without stabilizing the rate of growth of population. It is perceived that the rate of growth in developed countries shows a declining trend whereas in the developing countries like Asia and Africa, the birth rate have increased with a fall in the mortality rate. If unchecked this is going to pose an unimaginable environmental disaster and disaster in socio-economic condition. In India, infant mortality and fertility rates are high due to poverty and illiteracy among females. Heptulla (2001) holds that Total Fertility Rate (TFR) is at 3.4 level (1993) which is far greater than the ideal number at replacement level (2.1 children). Rural total fertility rate is still 3.7 children which is 36 per cent higher than urban TFR of 2.7. Infant Mortality Rate (IMR) is still higher. There is the need to first reduce IMR much before TFR level can be expected to decline. The role of women is of utmost importance in population stabilization (Sahay *et al.*, 2003). It is suggested that the women should be made well educated. This would automatically gear up their mind to have a smaller family besides strict legislation be imposed on every family to cut the family size in 2 to 1 ration.

Sex education *i.e.* lessons on reproductive biology, family planning methodology, irrespective of caste and community, researches on indigenous plants used in family planning should be made effective. Plant based contraceptive should be encouraged. Application of new technologies "green genetic engineering" is a next option.

Sewage Disposal

One of the major threat to environment (water) pollution is sewage which is indiscriminately being thrown in the rivers like the Ganga, Yamuna, Brahmputra, other major and minor rivers, ponds and lakes etc. The methods for sewage treatment as suggested by NEERI could safely be adopted.

In this method sewage is transferred and stored in shallow ponds where due to excess sunshine water evaporates and the organic matters in the shine bring about an algal bloom and growing of bacteria. Thus waste effluent become harmless and can be used for irrigation purposes.

Another methods suggested by NEERI is the preparation of oxidation ponds (ditches) and the sewage is subjected to mechanical aeration which are then transferred to aerated lagoons and then the water can be used in farm houses. Development of other technologies to purify the environment by not allowing harmful discharges into the environment, and to recycle the products for useful preparations have to be developed.

Development of Alternative Source of Energy

Thermal power plants required coal as fuel in huge quantity for burning and these emit large amount of pollutants such as particulate matters (fly ash), SO_2, nitrogen oxides, carbon monoxide along with hydrocarbons.

To meet these adversities solar hydrogen technology has to be brought into picture. Solar cells have to be installed at suitable places. The electricity produced by photoelectric means will then have to be converted to hydrogen using electrolytic equipment. The hydrogen and electricity then could be used to develop agriculture and for industrial purposes. Hydrogen can be transported to various places through pipe lines. It can also be liquefied and enormous amount of hydrogen can be carried by tankers. A small amount of the said hydrogen can be used to desalinate sea water for irrigation in coastal areas and for the electrolytic process itself. The unsalted water can be transported to the solar hydrogen production centre using a parallel pipeline (Bolkow, 1999).

The advantages of solar hydrogen technology are widely accepted. It is the only complete closed system of energy production, and apart from a great deal of water, which then evaporates into the atmosphere, it produces no waste materials. Other established energy systems produce byproducts which are sometimes poisonous but always involve chemical or isotopic changes and they are simply dumped on mother nature lap. In contrast to this, solar hydrogen does not involve the production of any environmentally damaging material. Saving energy is the best way to protect the environment. We must not forget that we can't continue to treat our environment as we have in the past (Sylvester and Hinteredar, 1988). Nuclear energy (with safety standards) could be a second pillar alongside coal for providing a secure energy supply in future. Indian Institute of Science, Bangalore has developed a new type of solar cell for converting sunlight into electricity. The solar cell does not contain silicon rather Copper Indicum Gallium Disclonide (CIGS). CIGS solar cells are resistant to radiation and making it ideal for space application. Use of wind energy has also to be exploited.

Removal of Poverty

Ignorance and elevation of education would in all probability take care of the present environmental assault. Poverty leads to its own form of environmental destruction. Poor people are extremely vulnerable and poverty shortens peoples perspectives. If people cannot afford oil they are bound to cut down trees for firewood. If they are driven because of debt from the farm they lease, they move to agriculturally marginal land or try to earn a living cutting wood or extracting coal illegally. Poverty and population are linked, therefore, removal of poverty, elevation of education and removal of ignorance about environmental conservation are urgently needed.

Mass Awareness

In United Nations Conference on Human Environment held in Stockholm (1972), decision was taken to take appropriate measures for the preservation of the natural resources of the earth which inter alia included the preservation of quality of air and control of air pollution. Earth summit on control of environmental degradation was held at Rio in 1972 under the auspices of the United Nations focussed the world attention on the uncontrollable situations and resolved for concerted multidimensional action plan. Rio conference concerned about the risk to life on earth due to ozone depletion and ultraviolet radiation. So many laws and rules are existing such as Air (Prevention and Control of Pollution) Act, 1981; Water (Prevention and Control) Pollution Act, 1974 etc. but even then large scale industrialization and automobilisation has given a rise to the problem of formidable pollution. Big conferences are of importance of those who are at the helm of affairs but awareness to the grave situation of pollution is hardly known to a common man. It is suggested that mass awareness programme should be brought into action. People's awareness programme should be of the level that used batteries, motor oil, tyres are taken to special collection centres and that medicines, varnish, paints are disposed off in accordance with appropriate regulations. For this waste sorting containers should be made available so that every body takes to an effort of sorting refusable and disposable materials for recycling programme. Recycling laws should be made more effective (Sahay *et al.*, 2000).

Conclusion

Most of the environmental problems arise due to wrong use of science and technology for advancement of human beings. However, it is our own responsibility to treat all environment related problems as social problems and participate ourselves in solving these problems on war footing. Non Governmental Organizations and self-help groups can actively participate in development programs and create awareness among the people about safe drinking water, afforestation, better sanitation, and good health practices. It is the responsibility of the state to provide basic minimum needs to the people without harming the environment. The state should try to adopt a strategy of sustainable and eco-friendly development for all its future economic development. Similarly all citizens should act in a responsible manner while either extracting natural resources or using the output of these natural resources without harming the environment.

References

Anderson, K.R., Avot, E.L., Edward, S.L., Shamoo, D.A., Peng, R.C., Linn, W.S. and Hackny, J.S., 1992. Controlled exposures of volunteers to respirable carbon and sulphuric acid aerosols. *J. Air and Waste Manag. Assoc.*, 42: 771–776.

Angell, J.K., 1994. Global hemisphere and zonal temperature anomalies derived from radiosonde records. In: *Trends 93: A Compendium of Data on Global Change*, (Eds.) T.N. Boden, D.P. Kaiser, R.J. Sepanski and F.W. Stoss), ORNC/CDIAC-65. Carbon Dioxide Information Analysis Centre, Oak Ridge National Lab., Oak Ridge, Tenn. USA, pp. 636–672.

Anon, 1998a. Valuing the global environment: Actions and investments for the twenty first century. *Global Environmental Policy*, pp. 162.

Anon, 1998b. Climate change. Report prepared as background for Buenos Aires Meeting of CCF. Hadley Centre for Climate Prediction and Research.

Anon, 2000. *Global Environment Outlook*. United Nations Environmental Programme, pp. 12.

Basu, B., 1986. Environment protection: A many faceted problem. *Yojana* 30: 4–6.

Bohra, C., 2002. Comparative studies on some aspects of ecobiology of two lentic freshwater ecosystems of Santhal Paragna (Bihar) with special reference to primary productivity. *Ph.D. Thesis*, S.M.K. University, Dumka, pp. 500.

Bohra, C. and Kumar. A., 2003. Air pollution due to stone crushers: A challenge for human health in Jharkhand. In: *Environmental Challenges of the Twenty First Century*, (Ed.) A. Kumar. APH Publ. Corp., New Delhi, pp. 233–237.

Bolkow, L., 1999. A man with vision. *Scale*, 11: 41–43.

Brandon, C. and Homman, K., 1995. The cost of inaction: Valuing the economy wide cost of environmental degradation in India. World Bank, Washington D.C.

Cairns, J.J., 1980. Indicator species by the concept of community structure as an indices of pollution. *Water Resources Bulletin*, 10: 338–347.

CPCB, 1993–94. *Pollution Statistics*. Delhi Central Pollution Control Board, Delhi.

CPCB, 1995. *Implementation Status of the Pollution Control Programme in Major Polluting Industries*. Program objective series PROBES/62/1994–95. Central Pollution Control Board, Delhi.

Dockery, D.W. and Papes, C.A., 1994. Acute respiratory effects of particulate air pollution. *Ann. Rev. Publ. Hlth.*, 15: 107–132.

Gajghate, D.G. Hassan, M.Z., 1995. Status of aerosol with specific reference to toxic trace metals constituents in urban air environment. *J. Chem. and Environ. Sci.*, 4: 67–74.

Gajghate, D.G. and Hassan, M.Z., 1996. Approaches for management of air pollution and control strategy. In: 3rd *Int. Conf. on Env. Planning and Mgt.*, Nagpur, pp. 169–175.

Gajghate, D.G. and Hassan, M.Z., 1997. Lead pollution from gasline powered motor vehicles and abatement strategies. *Ind. J. Env. Protection*, 17: 86–90.

Gajghate, D.G. and Hassan, M.Z., 1999. Ambient lead levels in urban areas. *Bull. Environ. Contam. Toxicol.*, 62: 403–408.

Gajghate, D.G., Thakre, R. and Agrawal, A.L., 1997. Strategic considerations for lead pollution control in Kanpur City. *Ind. J. of Chem. Soc.*, 75: 23–26.

Gajghate, D.G., Jain, S. and Hassan, M.Z., 2003. Investigation on air pollution for fine dust and toxic metals in Kochi city. In: *Environment Pollution and Management*, (Eds.) A. Kumar, C. Bohra and L.K. Singh. APH Publ. Corp., New Delhi, pp. 105–115.

Heptulla, N., 2001. The role of women in population stabilization. *Employ. News*, 26(4): 1–2.

Jha, A.K., Latif, A. and Singh, J.P., 1997. River pollution in India: An overview. *J. Environ. Poll.*, 4: 143–151.

Khalil, M.A.K. and Rasmussan, R.A., 1994. Global CH record derived from six globally distributed locations. In: *Trends 93: A Compendium of Data on Global Change*, (Eds.) T.N. Boden, D.P. Kaiser, R.J. Sepanski and F.W. Stoss), ORNI/CDIAC-65. Carbon Dioxide Information Analysis Centre, Oak Ridge National Lab., Oak Ridge, Tenn. USA, pp. 262–272.

Khan, M.S., 1994. *Current Problems of Indian Economy*. APH, New Delhi, pp. 262.

Khandekar, R.N., Mishra, U.C. and Vohra, K.G., 1984. Environmental lead exposure of an urban Indian population. *Sci. and Env.*, 20: 269–278.

Kumar, A., 1996. Impact of industrial pollution on the population status of endangered Gangetic dolphin (*Platanista gangetica*) in the river Ganga in Bihar. *Poll. Arch. Hydrobiol.*, 43: 469–476.

Kumar, A., 2003. An interaction of human society with environment as a challenge for eco-restoration: A critical review. In: *Environmental Challenge of the Twenty First Century*, (Ed.) A. Kumar. APH Publ. Corp., New Delhi, pp. 1–13.

Mathur, P.K., Gajghate, D.G. and Hassan, M.Z., 1997. Environmental lead and children: A review. *Ind. J. Env. Prot.*, 17: 161–165.

Miller, D.R., 1984. Chemicals in the environment. In: *Effects of Pollutants at the Ecosystem Level*. John Wiley and Sons, Chichester, pp. 7.

Negi, B.S., Sadasivam, S. and Mishra, U.C., 1987. A composition of aerosol and sources in urban area in India. *Atmospheric Environment*, 21: 1259–1266.

Pandya, S.C., Sharma, S.C., Jain, H.K., Pathak, S.J., Paliwal, K.C. and Bhanot, V.M., 1977. The environment and cenchrus grazing lands in Western India: An ecological assessment. Department of Bioscience, Saurashtra University, Rajkot, pp. 451.

Parikh, K., Parikh, J., Murlidharan, J.R. and Hadkel, N., 1994. Economic valuation of air quality degradation in Chembur, Bombay, India. A project under the Metropolitan Environmental Improvement Programme, sponsored by World Bank, IGIDR.

Rathakrishnan, L., 2003. India's environmental problems and issues. In: *Environmental Challenge of the Twenty First Century*, (Ed.) A. Kumar. APH Publ. Corp., New Delhi, pp. 33–38.

Sahay, U., Dan, M.T., Sahay, S., Sah, H.C.P. and Jha, A., 2000. Impact of environmental degradation versus human population. In: *Nat. Sent. Impact Environ. Degradation on Human Population*, R.S.P. College, Jharia (Dhanbad).

Sahay, U., Sahay, H.P., Alka and Verma, A., 2003. Environmental pollution and its challenges. In: *Environmental Challenge of the Twenty First Century*, (Ed.) A. Kumar. APH Publ. Corp., New Delhi, pp. 15–31.

Sahu, A., Panigrahi, M.K., Sahu, S.K. and Panigrahji, A.K., 2003. Aquatic mercury pollution by a chlor-alkali industry and its impact on the ecology of human health in Ganjam area: A case study. In: *Aquatic Ecosystems*, (Ed.) A. Kumar. APH Publ. Corp., New Delhi, pp. 83–111.

Sadasivam, S. and Negi, B.S., 1990. Frequency distribution of total suspended particulate matter and trace elements in air. *Ind. J. Env. Hlth.*, 32: 219–224.

Sylvester, W. and Hintereder, P., 1988. The quest for energy. *Scala*, p. 28–39.

Tripathi, R., 1994. Airborne lead pollution in the city of Varanasi. *Atmospheric Environment*, 28: 2117–2323.

Trivedy, R.K., 1989. *Ecology and Pollution of Indian Rivers*. APH Publ. Corp., New Delhi.

Wagner, E., 1994. Impacts on air pollution in urban areas. *J. Env. Mgt.*, 18: 759–765.

WHO, 1977. *Environmental Health Criteria*. 3, Leads, Geneva.

Zipf, M., 2001. *The Expanding Desert*. Deutchland E4 No. 1 February–March, pp, 20–29.

Chapter 2

Phytoextraction: A Green Technology for Metal Removal

***A. Krishnaveni*[1], *A. Bharani*[1] *and R. Sudhagar*[2]**

[1]*Department of Environmental Science, Tamil Nadu Agricultural University, Coimbatore*
[2]*Sugarcane Research Station, Melalathur, Vellore District*

Introduction

Heavy metals are ubiquitous environmental contaminants in an industrial society. Concerns over the possible health and ecosystem effects of heavy metals in soils have increased in recent years. Soil pollution by metals is essentially different persistence of metals in soil is reportedly much longer than in other compartments of biosphere. Once deposited on to the soil, certain metals such as Pb and Cr may be virtually permanent (Kabata Pendias and Adriano, 1995).

Phytoremediation of heavy metals from the environment serves as an excellent example of the process of plant-facilitated bioremediation and its role in removing environmental stress. Traditionally, when an area becomes contaminated with heavy metals the area must be excavated and the soil removed to a landfill site (Lasat, 2000). This process is extremely expensive and therefore not entirely appealing in light of the recent discoveries regarding phytoremediation (Lasat, 2000). Analysts estimate that the cost of cleaning one hectacre of highly contaminated land at a depth of one meter would range from $600,000 to $3,000,000 depending on the extent of the pollution and the toxicity of the pollutant. The cost of phytoremediation could be as much as 20 times less expensive, making this practice far less prohibitive than previous methods (Lasat, 2000). The ideal type of phytoremediator is a species that creates a large biomass, grows quickly, has an extensive root system, and must be easily cultivated and harvested: (Clemens *et al.*, 2002). The only problem with this criterion is that natural phytoremediators often lack these qualities. Therefore, scientists have been forced to become very creative in the development of effective transgenic phytoremediators.

Toxic Heavy metals

The focus of researchers on the phytoremediation of toxic metals in the environment has been considered an area of major scientific and technological progress and is a subject of contemporary relevance (Table 2.1). The widespread accumulation of heavy metals in soils is becoming a serious problem, as a consequence of industrial activity worldwide. Mining wastes, fertilizers, paper mills and toxic elements from atmospheric emissions, have all contributed to the continuous deposition and resulting accumulation of toxic metals in the environment. The concentrations of toxic metals in polluted soils are often hundreds of times greater than that required to exert a toxic effect on the majority of higher plants. Toxic metals can affect the biosphere for long periods of time and can be reached through the soil layers leading to the contamination of the water table. Consequently, the use of plants contaminated with high levels of heavy metals for food, might pose a serious risk to human and animal health (Wang *et al.*, 2003).

Table 2.1: Social and Economic Dimensions of Toxic Trace Elements, Metalloids and Radionuclides and Areas of Scientific Investigation (Prasad, 2004a)

Area	*Scope for Detailed Investigation*
Ecotoxicology and ecophysiology	Nutrient availability efficiency and deficiency for optimization of crop yield
Environmental chemistry waste management and mine reclamation	Bioremediation and restoration of metal contaminated and polluted ecosystems
Agricultural and nutritional sciences food industry	Plant and soil serve as vital links for supplementing nutrients on sustainable basis
Clinical biochemistry	Usage of radiolabelled antibodies and their disposal
Medicine and pharmacology	Use of plants as a source of certain disorders nutrients for
The atomic energy sector	The nuclear weapon testing (release of mainly ^{14}C, ^{137}Cs, ^{90}Sr and ^{95}Zr) production (release of mainly ^{137}Cs, ^{106}Ru, ^{95}Zr) and nuclear power production

It is also known that some metal contaminated soils are difficult to remediate and such soils are usually excavated and land filled. Sites can also be treated by acid leaching, physical separation of the contaminant or electrochemical processes (Cunningham *et al.*, 1995). Moreover, the costs associated with soil remediation are variable and depend on the properties of the soil, the contaminants, site conditions and the volume of material to be remediated (Cunningham *et al.*, 1995). Considering these limitations, phytoremediation is one innovative approach that offers more environmental benefits and a cost effective alternative.

In phytoextraction and phytomining, accumulated toxic metals in plant tissues are harvested for metal recovery and reuse. Normally, the plants termed hyper accumulators are preferably used, since they have the ability to withstand and build up high concentrations of metals, when compared to other plants. These plants can be processed to recover the metals accumulated during the phytoremediation process (Figure 2.1). Although it is cheaper than the conventional methods, phytoremediation is not an easy technology that consists of simply planting and growing some hyper accumulating plants in the metal polluted area (Alkorta *et al.*, 2004). It is in fact a highly technical strategy, requiring expert project designers with field experience that choose the proper species and cultivars for particular metals and regions. Research carried out with hyper accumulator plant species has mainly focused attention on the physiological mechanisms by which the metal is taken up, transported and sequestrated, but little is know regarding the genetic basis of hyper accumulation

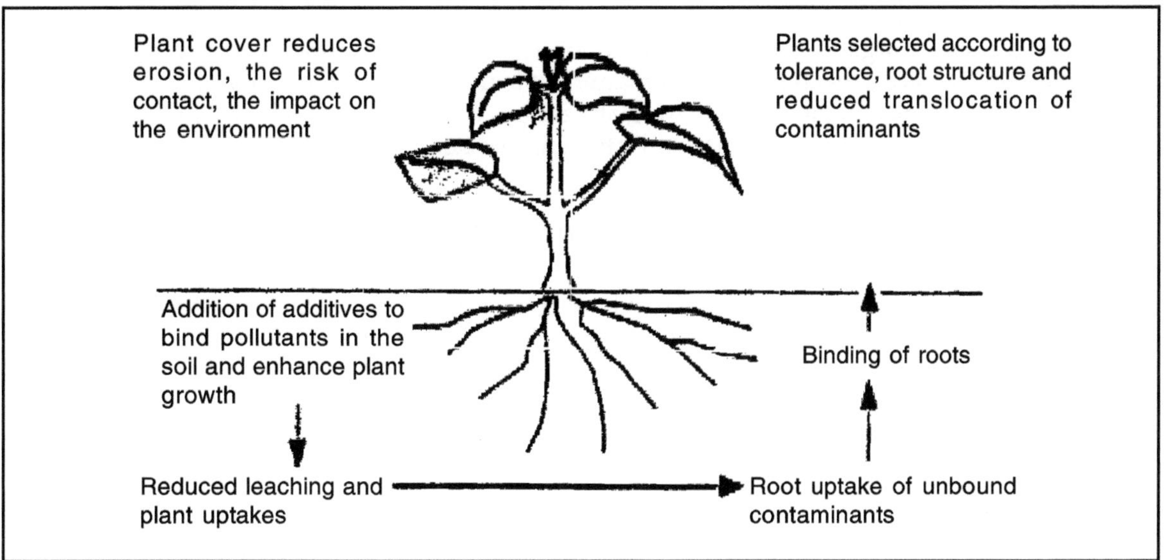

Figure 2.1: Phytoremediation Process

when compared with the genetic basis of metal tolerance. There is evidence for a quantitative genetic variation controlling the ability of these plants to hyper accumulate between and within populations.

Current Phytoremediation

Phytoremediation is the process that introduces plants into an environment and allows them to assimilate the contaminants into their roots and leaves. Such a process has been used to clean up heavy metals, pesticides and xenobiotics (Suresh and Ravishankar, 2004), organic compounds (Newman and Reynolds, 2004), toxic aromatic pollutants (Singh and Jain, 2003) and acid mine drainage (Archer and Caldwell, 2004). Interestingly, phytoremediation was recognized and documented by humans more than 300 years ago, however the scientific study and development of suitable plants was not conducted until the early 1980's (Lasat, 2000).

Phytoremediation is considered to be an environmentally friendly technology, that is a safe and also a cheap way to remove contaminants, in some cases doing the same job as a group of engineers for one tenth of the cost. However, such technology cannot necessarily be effective all of the time or be used in all types of contaminated sites. If the contamination runs too deep, or the concentration of toxic compounds is too high, then plants alone cannot efficiently remediate the soil (Cunningham *et al.*, 1995).

Phytoremediation of Toxic Metals

Phytoremediation is characterized by the use of vegetative species for *in situ* treatment of land areas polluted by a variety of hazardous substances (Sykes *et al.*, 1999). Plants are especially useful in the process of bioremediation because they prevent erosion and leaching which can spread the toxic substances to surrounding areas (United States Environmental Protection Agency, 2001). There are several types of phytoremediation being used today. These include phytoextraction, which relies upon a plant's natural ability to take up certain substances (such as heavy metals) from the environment and sequester them in their cells until the plant can be harvested, phytodegradation, a means by which plants convert organic pollutants into a non-toxic form, phytostabilization, where a plant releases certain chemicals that bind to the contaminant to make it less bioavailable and less mobile in the surrounding environment, and phytovolatilization, a process through which plants extract pollutants from the soil and then convert them into a gas that can be safely released into the atmosphere (Bentjen, 2002).

Transgenic Plants Vs Natural Hyper Accumulators

Some plants and tree species are termed as "hyper accumulators". Plants will extract and store extremely high concentrations (in excess of 100 times greater than non-accumulator species) of metallic elements (Lasat, 2000). Research has shown that these hyper accumulators often do not exclude non-essential metals in the absorption process, thus resulting in plants that can extract high levels (1–2 per cent of their biomass) of pollutants from contaminated soil (Lasat, 2000). It is believed that plants initially developed this ability to hyper accumulate non-essential metallic compounds as a means of protecting themselves from herbivorous predators, who would experience serious toxic side effects from ingestion of the hyper accumulator's foliage (Pollard *et al.*, 1997). Lead is one of the non-essential compounds hyper accumulated by several species of plants including *Thlaspi rotundifolium* and *Brassica juncea.* The difficulty with these models, in terms of performing phytoremediation, is that they grow very slowly and have a very low biomass. Therefore, the bioremediation process would be extremely slow because the rate of bioremediation is directly proportional to growth rate and the total amount of bioremediation is correlated with a plant's total biomass. Because no plant yet discovered meets the ideal criteria of an effective phytoremediator (fast growing, deep and extensive roots, high biomass, easy to harvest, hyper accumulators of a wide range of toxic metals), it is necessary to introduce the hyper accumulating genes into non-accumulator to make the plants better suited as agents in the phytoremediation process (Clemens *et al.*, 2002). The authors of a lead absorption study, Huang and Cunningham, cite corn as a perfect phytoremediator due to its large biomass, fast rate of growth, and the existence of extensive genomic knowledge of this crop. The presence of a body of knowledge regarding corn genetics would be helpful in the effective insertion of DNA that codes for hyper accumulation of the desired contaminant. Other potential model phytoremediators would include various varieties of transgenic trees. Trees are ideal in the remediation of heavy metals because they can withstand higher concentrations of pollutants due to their large biomass, they can accumulate large amounts of the contaminants in their systems because of their size, they can reach a huge area and great depths due to their extensive root systems, they can stabilize an area and prevent erosion, and hence the spread of the contaminant, because of their perennial presence. They can also be easily harvested and removed from the area with minimal risk, effectively taking with them a large quantity of the pollutants that were once present in the soil. The introduction of hyper accumulating genes as well as genetic information that would better prepare these species to deal with diverse climatic conditions, up-regulate this hyper accumulation, or increase the organism's growth rate and overall size–into these model agricultural species would undoubtedly accelerate the remediation process and result in faster removal of environmental stress from the ecosystem. Such metal accumulation and tolerance could be enhanced by over expressing natural or modified genes encoding antioxidant enzymes or those that are involved in the biosynthesis of glutathione and phytochelatins (Figure 2.2).

It is apparent that in order to breed plants with superior phytoremediation potential, it is necessary that these plants have a high rate of biomass production and are sufficiently strong and competitive in the climate they are to be used for phytoremediation (Pilon-Smits and Pilon, 2002). Moreover, plants suitable for phytoremediation should possess the ability to accumulate the target metal in the above

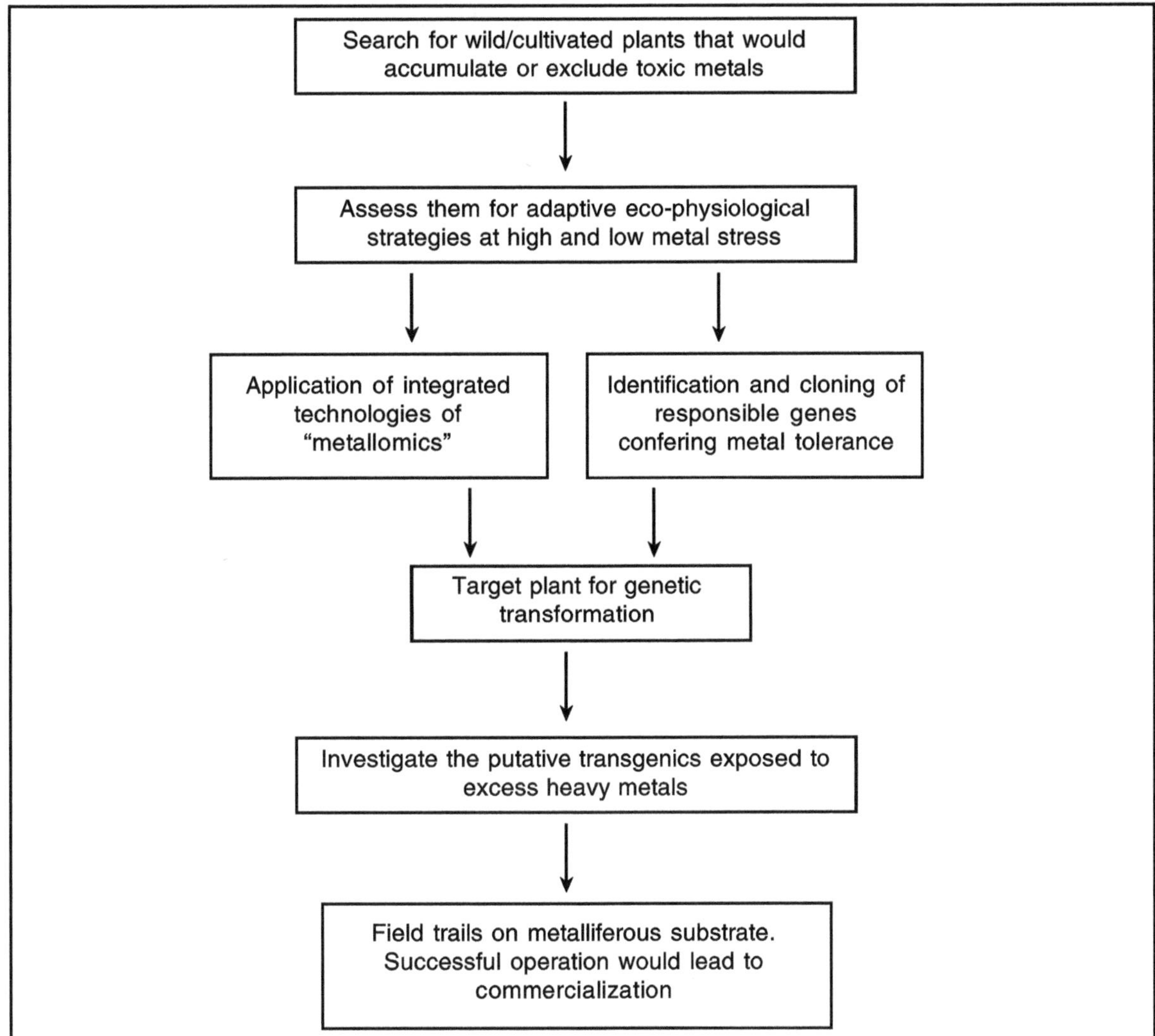

Figure 2.2: The Use of Genetic Engineering to Breed Plants with Superior Phytoremediation Potential (Modified after Karenlampi *et al.*, 2000; Prasad, 2004b).

ground parts, and to tolerate the metal concentration that is to be accumulated (Karenlampi *et al.*, 2000). The over expression of a gene encoding a rate limiting gene product would be expected to lead to a faster overall rate of the pathway and to more efficient phytoremediation (Pilon-Smits and Pilon, 2002). However, it must be considered that metals rarely occur alone in the environment and an adaptive tolerance may be essential for several metals simultaneously (Karenlampi *et al.*, 2000).

Transgenic Plants for Toxic Heavy Metals

S-metabolism, Glutathione and Phytochelatins

Selenium (Se) is essential as a micronutrient for humans and animals but is toxic at medium to high concentrations. Se occurs naturally in soils as selenate and selenite and often as a pollutant, following the industrial use of coal. Se and sulphur (S) have very similar chemical properties and their uptake and assimilation as selenate and sulphate, to incorporation in proteins as selenomethionine and methionine, proceed through common pathways and are catalyzed by the same enzymes. A research group lead by Norman Terry and Elizabeth Pilon-Smits has considered the possibility of over expressing a range of genes encoding key enzymes in sulphur metabolism, with the aim of obtaining selenium tolerant plants.

Indian mustard (*Brassica juncea*) plants over expressing ATP sulphurylase were shown to have higher shoot Se concentrations and enhanced Se tolerance compared to wild type when grown in the presence of selenate in either hydroponic systems or soil (Van Huysen *et al.*, 2004). Interestingly, young transgenic plants that were over expressing ATP sulphurylase, were more tolerant than the wild type to As(III), As(V), Cd, Cu, Hg, and Zn, but less tolerant to Mo and V (Wangeline *et al.*, 2004). Other transgenic Indian mustard plants over expressing cystathionine-O-synthase (CGS) showed a higher Se volatilization rate, lower shoot Se levels, and higher Se tolerance than wild type grown in either soil or hydroponically (Van Huysen *et al.*, 2004).

Astragalus hisulcatus is a native plant that has the capacity to grow on Se containing soils and accumulate Se to high concentrations but it has a slow growth rate. It has been proposed that in *A. bisulcatus* selenocysteine methyl-transferase (SMT) specifically methylates selenocysteine (SeCys) to produce the non-protein amino acid methylselenocysteine MetSeCys, which causes a reduction in the intracellular concentrations of SeCys and selenomethionine (SeMet), thus preventing their incorrect insertion into protein. Indian mustard plants over expressing the *A. bisulcatus* SMT gene, exhibited a greatly increased accumulation of MetSeCys and tolerance to Se compounds, in particular selenite (LeDuc *et al.*, 2004).

Cd is a toxic element that normally occurs in low concentrations in soils, however the concentration can be significantly increased by activities such as zinc mining, iron foundries and the use of sewage sludge as a fertilizer in agriculture (Zhao *et al.*, 2003). Cd may be detoxified in plants by a family of sulphur rich peptides termed phytochelatins (PCs) that are able to bind Cd and some other heavy metals (Cobbett and Goldsbrough, 2002). The peptides are structurally related to glutathione and contain a varying number (normally 2–5) of glutamate and cysteine, linked through the D-carboxyl group of glutamate. Phytochelatin synthase (PCS) has been characterized as a specific O-glutamyl cysteine dipeptidyl transpeptidase (EC 2.3.2.15) (Vatamaniuk *et al.*, 2004), which carries out the conversion of glutathione to PCs, and has been shown to be activated by Cd.

Various attempts have been made to increase the formation of PCs, by over expressing genes encoding enzymes that could stimulate the synthesis of cysteine and glutathione. Cysteine synthase catalyses the last step in the assimilation of sulphate into the amino acid. Transgenic tobacco plants over-expressing cysteine synthase in the cytosol, had elevated concentrations of PCs, were more tolerant to Cd, but did not accumulate the metal in the leaves (Harada *et al.*, 2001). In contrast, Dominguez-Solis *et al.* (2004) over expressed cysteine synthase in the cytosol of *A. thaliana.* One transgenic line was shown to be particularly resistant to Cd and to accumulate high concentrations in the leaves mainly in the trichomes. Transgenic tobacco plants over expressing cysteine synthase in either the cytosol or chloroplasts were more tolerant to metals such as Cd, Se and Ni. F_1 plants with expression in both the cytosol and chloroplast exhibited a higher tolerance than the other transgenic lines and accumulated Cd in the shoots (Kawashima *et al.*, 2004).

Sudhagar (2003) investigated the Cd tolerance limit for maize crop. The cadmium phytoremediation capacity of maize in breds and the genetics of cadmium tolerance were investigated. Fifty one maize genotypes with more divergence were utilized for the study. The cadmium critical concentration was fixed by studying the shoot and root development pattern of four randomly selected experimental genotypes. The fifty percent reduction in shoot and root length were considered and critical concentration was fixed through SPSS analysis. All the genotypes were screened for cadmium tolerance at this critical concentration. Germination parameters, morphology and growth and varied biochemical parameters were utilized for isolating cadmium tolerant genotypes. The genotypes were scrutinized for the activity of H^+ ATPase, nitrate reductase, superoxide dismutase, catalase, ascorbate peroxidase, monodehydroascorbate reductse, dehydroascorbate reductase, glutathione reductase and phytochelaytin sulfhydrill. Reduction in growth and biochemical parameters was observed invariably across the test genotypes. The inbreds with less and no reduction in morphology, growth traits and heavy metal induced enzymes were treated as tolerant genotypes. The genotypes with sharp decline in growth and related biochemical parameters under cadmium-imposed condition were considered as cadmium tolerant. The selected cadmium susceptible and tolerant inbreds were further test verified by isoenzyme and protein analysis. In the tolerant genotypes one additional catalase, peroxidase and poly phenol oxidase isoenzyme loci were observed. The tolerant genotypes manifested enhanced expression of 52 KDa protein that was absent or less expressed in susceptible genotypes. The tolerant and susceptible genotypes were crossed in line and tester mating design to obtain the genetic information of cadmium tolerance and other cadmium induced biometrical traits.

Indian mustard over expressing the *E. coli gshI* gene encoding O-glutamylcysteine synthetase (O-ECS) exhibited increased tolerance to Cd and had higher concentrations of PCs and D-GluCys, compared to the wild-type. When tested in a hydroponic system, O-ECS plants grew better and accumulated Cd in the shoots (Zhu *et al.*, 1999b). Transgenic Indian mustard containing the *E. coli gshII* gene encoding glutathione synthetase (GS) accumulated significantly more Cd than the wild type in the shoot and the plants showed enhanced tolerance to Cd at both the seedling and mature plant stages (Zhu *et al.*, 1999a). Zhu *et al.* (1999b) concluded from the results obtained with the two types of transgenic plants, that under normal conditions that O-ECS limits the rate of glutathione and PC synthesis, but that in the presence of Cd O-ECS is activated and that GS becomes rate limiting. When grown in contaminated soil, the thiol-overproducing transgenic O-ECS and GS Indian mustard plants showed enhanced phytoextraction capacity, yet still produced the same shoot biomass as the wild type. As a result, the total shoot metal accumulation of the O-ECS and OS transgenic were 1.5-fold higher for Cd, and 1.5- to 2-fold higher for Zn. Furthermore, the D-ECS transgenic accumulated 2.4 to 3-fold more Cr, Cu, and Pb, relative to the wild type (Bennett *et al.*, 2003).

Following the positive results obtained with over expressing the enzymes involved in cysteine and glutathione synthesis, it was expected that over expressing PCS would be even more successful, however this did not prove to be the case for Cd toxicity. When PCS was over expressed in *A. thaliana*, the transgenic plants were shown to be hypersensitive to Cd and Zn but not to Cu (Lee *et al.*, 2003b). In a similar series of experiments, over expressing PCS in *A. thaliana*, Li *et al.* (2004) demonstrated that the transgenic plants were highly resistant to arsenic (As), accumulating 20–100 times more biomass on 250 and 300 μM arsenate, than the wild type. However, again the PCS plants were hypersensitive to Cd, despite the fact that they contained 2–6 fold higher concentrations of PCs. However, transgenic tobacco *Nicotiana glauca* expressing wheat PCS showed a slight increase in tolerance to Cd (Gisbert *et al.*, 2003). These somewhat surprising results were discussed in detail by Li *et al.* (2004) and they concluded that "much is still to be learned about the processing of cadmium in Cd-PC peptide complexes."

Metallothioneins

Metallothioneins (MTs) are low molecular mass cysteine rich proteins that were originally isolated as Cu, Cd and Zn binding proteins in mammals. There is now good evidence that four categories of these proteins occur in plants, which are encoded by at least seven genes in *A. thaliana* (Cobbett and Goldsbrough, 2002). A wide range of MT genes from various sources have been over expressed in plants, these include human, mouse, Chinese hamster and yeast. There was some variation in the range of Cd tolerance obtained. Metal uptake and accumulation in the shoots was not markedly altered, ranging from 70 per cent less to 30 per cent more than the wild type controls.

When a pea (*Pisum sativum*) MT gene *PsMTA* was expressed in *A. thaliana,* more Cu (several-fold in some plants) accumulated in,the roots of transformed than of control plants (Evans *et al.*, 1992). Similarly when a type 2 MT gene, *tyMT,* cloned from Cattail (*Typha latifolia*), a wetland plant with constitutional tolerance, was introduced into *A. thaliana,* the transgenic plant showed an increased tolerance to both Cu^{2+} and Cd^{2+}. The *A. thaliana* metallothionein proteins AtMT2a and AtMT3 were introduced as fluorescent protein-fused fonTIs into the guard cells of *Vicia faba.* The MTs protected guard cell chloroplasts from degradation upon exposure to Cd, by reducing the presence of reactive oxygen species. It was concluded that the Cd stays bound to the MT in the cytoplasm and is not sequestered into the vacuole, as occurs when Cd is detoxified by PCs (Lee *et al.*, 2004).

Transporters

There has been considerable interest in the possibility of manipulating transporters within plants in order to achieve different objectives, these include exclusion of a toxic metal ion, transporting the metal into the apoplastic space and transporting the metal into the vacuole where it would be less likely to exert a toxic effect (Tong *et al.*, 2004). In this article we will only cover a little of the extensive work that has been carried out.

Lee *et al.* (2003a) transformed *A. thaliana* plants with the *E. coli* gene, *ZntA,* which encodes a Pb(II)/Cd(II)/Zn(II) transporter. The *ZntA*-transgenic plants grew better than the wild-type plant in Pb, Cd and Zn-containing medium. The shoots of the transgenic plants exhibited decreased Pb and Cd contents, whereas transgenic protoplasts showed lower accumulation of Cd and faster release of pre-loaded Cd than wild-type protoplasts. Lee *et al.* (2003a) proposed that the *ZntA*-transgenic plants excluded the metal ions at the cellular level by pumping them from the plasma membrane to the extracellular space.

The over expression of the *AtHMA4* gene, which encodes a *A. thaliana* P-1B-ATPase Zn and Cd transporter, improved the growth of both primary and secondary roots in the presence of toxic concentrations of Zn, Cd and Co. A determination of metal content demonstrated that the over expressing lines, when exposed to toxic concentrations of Zn or Cd, translocated these metals at a greater extent to the shoot, compared to the control plants, an important criteria for phytoremediation. In contrast, the metal level was found to be rather similar in roots, indicating that the metal uptake by the roots compensated for the increased metal translocation to the shoot (Verret *et al.*, 2004).

The vacuole is generally considered to be the main storage site for metals in yeast and plant cells and there is evidence that phytochelatin-metal complexes are pumped into the vacuole. The best characterized of the known vacuolar transporters and channels involved in metal tolerance is YCF1 from *Saccharomyces cerevisiae.* YCF1 is a MgATP-energized glutathione S-conjugate transporter responsible for vacuolar sequestration of compounds after their S-conjugation with glutathione. Song *et al.* (2003) overexpressed the *YCF1* gene in *A. thaliana* and the YCF1 proteins were found to be associated with the tonoplast and the plasma membrane. The vacuoles of the YCF1-transgenic plants exhibited a 4-fold higher rate of glutathione-Cd uptake than those of wild-type plants, indicating that expression of *YCF1* strongly increases Cd transport activity. The transgenic plants showed improved resistance to both Cd and Pb and elevated metal content, characteristics desirable for phytoremediation.

Other transporter proteins that could be of value include: the *A. thaliana* antiporter CAX2 (Hirschi *et al.*, 2000), LCT1, a nonspecific transporter for Ca^{2+}, Cd^{2+}, Na^{+} and K^{+} (Antosiewicz and Hennig, 2004), the *Thlaspi caerulescens* heavy metal ATPase, TcHMA4 (Papoyan and Kochian, 2004), a novel family of cysteine rich membrane proteins that mediate Cd resistance in *A. thaliana* (Song *et al.*, 2004) and AtMRP3, an ABC transporter (Bovet *et al.*, 2005).

It can be seen from the above account of work that has been already carried out, that there are a large number of possibilities to improve the ability of plants to withstand the stresses of high concentrations of toxic metals and also to accumulate them. Additional work not mentioned includes over expressing genes encoding: ACC deaminase (Grichko *et al.*, 2000), mercuric ion reductase, arsenate reductase aldose/aldehyde reductase (Hegedüs *et al.*, 2004) and enzymes of histidine biosynthesis (Wycisk *et al.*, 2004).

Clearly there may be some opportunity to combine some of these genes in fast growing species, but more importantly the plants need to be tested under carefully regulated environmental conditions.

Natural Hyper Accumulators

Natural metal hyper accumulators plants can accumulate and tolerate greater metal concentrations in shoots than those usually found in non-accumulators, without visible symptoms. According to Baker and Brooks (1989), the minimum threshold tissue concentrations for Co, Cu, Cr, Pb or Ni hyper accumulators should be 0.1 per cent dry weight, while for Zn or Mn the threshold is 1 per cent. Over 400 hyper accumulator plants have been reported and include members of the Asteraceae, Brassicaceae, Caryophyllaceae, Cyperaceae, Fabaceae, Flacourtiaceae, Lamiaceae, Poaceae, Violaceae, and Euphorbiaceae. The Brassicaceae is a very important group when heavy metal accumulation is concerned, with several species being able to hyper accumulate more than one metal (Prasad and Freitas, 2003).

Pistia stratiotes was examined in the presence of several distinct heavy metals separately, in order to determine the ability of the plant to tolerate and accumulate the metals tested and for eventual use for phytoremediation of wastewater or natural water bodies polluted with these heavy metals (Odjegba and Fasidi, 2004). *P. stratiotes* exhibited different patterns of response to Ag, Cd, Cr, Cu, Hg, Ni, Pb and Zn, and although concentrations as high as 5 mM resulted in distinct levels of growth inhibition and biomass production, all the elements accumulated at high concentrations mainly in the root system. Furthermore, this plant species exhibited the highest tolerance index to Zn and the lowest to Hg (Odjegba and Fasidi, 2004).

Spartina plants have been shown to be 3-fold more tolerant to Hg than tobacco plants, due to an ability to absorb organic Hg and transform it into an inorganic form (Hg^{+}, Hg^{2+}). The inorganic Hg then accumulates in the underground parts of the plants and is transferred back to the soil by diffusion and permeation, indicating that this species may be used in the phytoremediation of a Hg polluted environment (Tian *et al.*, 2004). According to Bennicelli *et al.* (2004), the water hyperaccumulator fern

Azolla caroliniana Willd. (Azollaceae) has the capacity to purify waters polluted by Hg and Cr by accumulation of these heavy metals in its tissues.

Helianthus annuus has been shown to concentrate Pb in the leaf and stem indicating that it has the prerequisites of a hyperaccumulator plant that could be used in the restoration of abandoned mines and factories sites contaminated with elevated Pb levels in the soil (Boonyapookana *et al.*, 2005). In an almost similar manner, *Hemidesmus indicus* has also been shown to be a Pb hyperaccumulating plant species, but the heavy metal was mainly accumulated in roots and shoots (Chandra *et al.*, 2005), exhibiting a high potential for use in the phytoremediation of industrial areas contaminated with this metal. The hyperaccumulator *Sesbania drummondii* is a plant species that has been shown to predominantly accumulate Pb as lead acetate in roots and leaves, although lead sulfate and sulfide have also been detected in leaves, whereas lead sulfide was detected in root samples (Sharma *et al.*, 2004). These results have indicated that *S. drummondii* is able to biotransform lead nitrate in the nutrient solution to lead acetate and sulfate in its tissues and the complexation with acetate and sulfate may be a lead detoxification strategy in this plant species (Sharma *et al.*, 2004).

Studies related to As accumulation in *Lemna gibba* have demonstrated that this plant species can be a preliminary bioindicator for As transfer from substrate to plants, indicating its use to monitor the transfer of arsenic from lower to higher trophic levels in abandoned mine sites. Moreover, *L. gibba* can be used for As phytoremediation of mine tailing waters because of its high accumulation capacity (Mkandawire and Dudel, 2005). *Pteris vittata* can hyperaccumulate As from naturally contaminated soils, but may be suitable for phytoremediation only in the moderately contaminated soils (Caille *et al.*, 2004). In addition to *P. vittata, P. cretica, P. longifolia* and *P. umbrosa* are also able to hyperaccumulate As to a similar extent (Zhao *et al.*, 2002).

Broadhurst *et al.* (2004) have developed commercially viable phytoremediation/phytomining technologies employing *Alyssum* Ni-hyperaccumulator species, where the majority of Ni is stored either in the leaf epidennal cell vacuoles, or in the basal portions *of* the numerous stellate trichomes. The metal concentration in the trichome basal compartment was the highest ever reported for healthy vascular plant tissue, approximately 15–20 per cent dry weight (Broadhurst *et al.*, 2004).

Solanum nigrum and *Conyza canadensis* have not only been shown to accumulate high concentration of Cd, but also to be tolerant to the combined action of Cd, Pb, Cu and Zn. The hyperaccumulator *Thlaspi caerulescens* J.&C. Presl. has a high Cdaccumulating capability, acquiring Cd from the same soil pools as non-accumulating species, by very efficient mechanisms (Schwartz *et al.*, 2003). Thus, *T. caerulescens* may be used as a tool to efficiently reduce the availability of Cd in soils, providing appropriate populations are used (Schwartz *et al.*, 2003). Moreover, *Arabis gemmifera* is a hyperaccumulator of Cd and Zn, with phytoextraction capacities almost equal to *Thlaspi caerulescens* (Kubota and Takenaka, 2003) and plants of the mined ecotype of *Sedum alfredii* have a greater ability to tolerate, transport, and accumulate Cd, when compared to the non-mined ecotype.

Understanding the mechanisms of rhizosphere interaction, uptake, transport and sequestration of metals in hyperaccumulator plants will lead to designing novel transgenic plants with improved remediation traits (Eapen and Souza, 2005). Moreover, the selection and testing of multiple hyperaccumulator plants could enhance the rate of phytoremediation, making this process a successful one for bioremediation of environmental contamination (Suresh and Ravishankar, 2004).

Advantages and Limitations of Phytoremediation

Several processes can limit the performance of plants in phytoremediation, such as the availability of the toxic metal ions in the soil for uptake by plant roots, the rate of uptake of the contaminants by plant roots, their translocation from roots to shoots and the extent of tolerance, or the rate of chemical transformation into less toxic, possibly volatile compounds (Prasad, 2003) (Table 2.2).

Table 2.2: Advantages and Limitations of Some of the Sub-processes of Phytoremediation (Prasad, 2004b)

Advantage	*Limitation*
Phytoextraction	
The plant must be able to produce abundant biomass in short time. *e.g.* in greenhouse experiments, gold was harvested from plants (Anderson *et al.*, 1998)	Metal hyperaccumulators are generally slow-growing and bioproductivity is rather small and shallow root systems. Phytomass after process must be disposed off properly
Phytostabilization	
It circumvents the removal of soil, low cost and is less disruptive and enhances ecosystem restoration/revegetation	Often requires extensive fertilization or soil modification using amendments, long-term maintenance is needed to prevent leaching
Phytovolatilization	
Contaminant/pollutant will be transformed in to less-toxic forms. *e.g.* elemental mercury and dimethyl selenite gas. Atmospheric processes such as photochemical degradation for rapid decontamination/transformation	The contaminant or a hazardous metabolite might accumulate in vegetation and be passed or in later products such as fruit or lumber. Low levels of metabolites have been found in plant tissue.
Phytofiltration/Rhizofiltration	
It can be either *in situ* (floating rafts on ponds) or *ex situ* (an engineered tank system); terrestrial or aquatic	pH of the medium to be monitored continually for optimizing uptake of metals, chemical speciation and interactions of all species in the influent need be understood; functions like a bioreactor and intensive maintenance is needed

Although the use of transgenic plants as phytoremediators is showing increasing potential for soils contaminated with toxic metals, possible risks should be considered, including-the uncontrolled spread of the transgenic plants due to interbreeding with populations of wild relatives. Moreover, the transformation of metals into forms more bioavailable could increase the exposure of wildlife and humans to metals. However, the risk of metal accumulation in plant shoots that can be ingested by the wildlife, can be minimized by the reduction of the growth period and thus exposure of the transgenic plants. Furthermore, it is important that in the case of harvesting crop plants normally used to feed livestock and humans, the translocation of the toxic elements to the seeds is avoided at all costs.

Full utilization of plant resources after they have been used for phytoremediation is an unsolved problem. Therefore the testing of the plants used in phytoremediation is necessary and may support their continuous use in contaminated soils. For instance, *Elsholtzia argyi* flowers are used as fragrances and antiseptics due to the perfume ingredient and antibacterial components existing in their essential oils, and the analysis of these plants in Pb/Zn mined area, where they normally occur revealed that they can be safely exploited (Peng and Yang, 2005).

While phytoremediation processes hold a great promise as a way to remediate contaminated soils, there are disadvantages and limitations that must be carefully considered.

Advantages of Phytoextraction

1. Soil stabilization
2. Aesthetically pleasing

3. Cost effective remedial alternative
4. Applicability to wide range of contaminants of varying concentration. These are subsequently rendered harmless or removed, with relatively low amounts of wastes being generated. Where metals are concerned, the contaminants can be recycled from the residues
5. Minimal environmental disturbance
6. Elimination of secondary air or water borne wastes

Limitations to Phytoextraction

1. Limited regulatory acceptance as a viable treatment alternative
2. Often long duration before there is clean up to acceptable threshold levels
3. Potential contamination of the food chain if organisms graze on the vegetation
4. Difficulty in establishing and maintaining vegetation in media containing toxic wastes
5. Potential for off-site migration and leaf transportation of metals to surface when using tree species
6. To minimize dispersal of seeds or pollen ensuring that metals are not mobilized down the soil profile
7. Preventing herbivores accumulating dangerous amounts of elements

To Overcome the Limitations

1. The plants need to be harvested before they flower (*i.e.*) during a period of maximum biomass and element yield
2. Growth of non-crop or non-palatable species which might reduce the chances of biomagnification.

Future Needs

1. Appropriate agronomic practices such as planting density, animal management, weed control and harvest methods to increase the biomass of hyperaccumulators to enhance phytoextraction of metals
2. Use of appropriate soil amendments to increase the bioavailability of metals to hyperaccumulators.
3. Use of growth modifying hormones such as auxins and cytokinin as other compounds of accumulator plants with the aim of maximizing removal. Increased production of hyperaccumulators by tissue culture techniques
4. Use of plant symbionts like bacteria or fungi (VAM) to enhance accumulation. Identification of gene or genes responsible for hyperaccumulation and insert them into a high biomass plant. None of the genes involved has been identified and cloned.
5. To identify the interactions of nutrients which affect the uptake of metals in the shoot.
6. Elaboration of methods for the extraction of metals in hyperaccumulators.

Conclusions

It is clear that phytoremediation is not type, but hope as there are many examples that had the potential for improving environmental quality. Plants are very flexible and react quickly to environmental stress, but from a plant physiology point of view, what is the meaning of phytoremediation? Many plants are around us, but who will screen them? Exploiting the natural biodiversity is thus an important issue with choice of appropriate species for phytoremediation. Why should we go transgenic plants when natural variability is not yet explored? The economic utilization of crops becomes more important than their heavy metal uptake ability or the time demand of phytoremediation.

It is evident that phytoremediation has benefits to restore balance to a stressed environment, but it is important to proceed with caution. The study and use of genetic modifications must be performed in order to determine the true costs and benefits of this technology to the ecosystem as a whole, before it is to be applied to a larger scale. Progress in the field of molecular genetics, will allow the analysis of metal hyperaccumulator plants and should provide new insights into metabolic detoxification processes and identify tolerance genes, thus providing considerable more information about the genomes of these model organisms.

Advances in other methods involving analysis of "omics" technologies could further reveal the non-targeted identification of all gene products in a specific biological sample, which could be followed by a refined analysis of quantitative dynamics in biological systems. The genomics can accelerate the discovery of genes that confer key traits, allowing their modification. In addition, metabolomics can provide biochemical and physiological knowledge about network organization in plants subject to toxic metal stress, providing a much more detail understanding of the molecular basis of hyperaccumulation.

Approaches allowing recombination hotspots to be highlighted will further aid plant breeding efforts. Many signalling pathways and proteins can contribute to the cellular stress response, and the identification of the key ones within the stress response network is essential. The development of DNA and RNA micro array chip technologies in systematic genome mapping, sequencing, functioning and experimentation may allow the identification and genotyping of mutations and polymorphisms, allowing better insight into structure-function interaction of genome complexity under toxic metal stress.

Molecular genetics approaches such as insertion mutagenesis involving populations of T-DNA, can be used to identify genes involved in hyper-accumulation or transposon tagged plants screened to identify mutants impaired in the ability to accumulate metals (Pollard *et al.,* 2002). Recently, considerable progress has been made in identifying plant genes encoding metal ion transporters with important functions in cation transport and homeostasis (Papoyan and Kochian, 2004; Weber *et al.,* 2004). Modem molecular techniques, bioinformatics and computational techniques are effective tools for detailed structure-function genome analysis. It is clear that both fundamental and applied research must be carried out in association.

Phytoremediation technology is still in its early development stage and full scale applications are still limited. The results already obtained have indicated that the plants are effective and could be used in toxic metal remediation. Although it appears to be common sense among scientists, engineers, and regulators about the more widespread future use of this technique, it is important that public awareness about this technology is considered and clear and precise information is made available to the general public to enhance its acceptability as a global sustainable technology to be widely used.

References

Alkorta, I., Hemández-Allica, J., Becerril, J.M., Amezaga, I., Albizu, I. and Garbisu, C., 2004. Recent findings on the phytoremediation of soils contaminated with environmentally toxic heavy metals and metalloids such as zinc, cadmium, lead and arsenic. *Rev. Environ. Sci. Biotechnol.,* 3: 71–90.

Antosiewicz, D.M. and Hennig, J., 2004. Overexpression of *LCT1* in tobacco enhances the protective action of calcium against cadmium toxicity. *Environ. Pollut.*, 129: 237–245.

Archer, M.J.G. and Caldwell, R.A., 2004. Response of six Australian plant species to heavy metal contamination at an abandoned mine site. *Water Air Soil Poll.*, 157: 257–267.

Baker, A.J.M. and Brooks, R.R., 1989. Terrestrial higher plants which hyperaccumulate metallic elements: A review of their distribution, ecology and phytochemistry. *Biorecovery*, 1: 81–126.

Bennett, L.E., Burkhead, J.L., Hale, K.L., Terry, N., Pilon, M. and Pilon-Smits, E.A.H., 2003. Analysis of transgenic Indian mustard plants for phytoremediation of metal-contaminated mine tailings. *J. Env. Qual.*, 32: 432–440.

Bennicelli, R., Stepniewska, Z., Banach, A., Szajnocha, K. and Ostrowski, J., 2004. The ability of *Azolla caroliniana* to remove heavy metals [Hg(II), Cr(III), Cr(VI)] from municipal wastewater. *Chemosphere*, 55: 141–146.

Bentjen, Steve, 1998. *Bioremediation and Phytoremediation Glossary.*

Boonyapookana, B., Parkplan, P., Techapinyawat, S., DeLaune, R.D. and Jugsujinda, A., 2005. Phytoaccumulation of lead by sunflower (*Helianthus annuus*), tobacco (*Nicotiana tabacum*), and vetiver (*Vetiveria zizanioides*). *J. Environ. Sci. Heal.*, A40: 117–137.

Bovet, L., Feller, U. and Martinoia, E., 2005. Possible involvement of plant ABC transporters in cadmium detoxification: A cDNA sub-microarray approach. *Environ Int.*, 31: 263–267.

Broadhurst, C.L., Chaney, R.L., Angle, J.S., Maugel, T.K., Erbe, E.F. and Murphy, C.A., 2004. Simultaneous hyperaccumulation of nickel, manganese, and calcium in *Alyssum* leaf trichomes. *Environ. Sci. Technol.*, 38: 5797–5802.

Caille, N., Swanwick, S., Zhao, F.J. and McGrath, S.P., 2004. Arsenic hyperaccumulation by *Pteris vittata* from arsenic contaminated soils and the effect of liming and phosphate fertilisation. *Environ. Pollut.*, 132: 113–120.

Chandra Sekhar, K., Kamala, C.T., Chary, N.S., Balaram, V. and Garcia, G., 2005. Potential of *Hemidesmus indicus* for phytoextraction of lead from industrially contaminated soils. *Chemosphere*, 58: 507–514.

Clemens, Stephan, Palmgren, Micheal, G. and Kramer, Ute, 2002. A long way ahead: Understanding and engineering plant metal accumulation. *Trends in Plant Science*, 7(7): 309–314.

Cobbett, C. and Goldsbrough, P., 2002. Phytochelatins and metallothioneins: Roles in heavy metal detoxification and homeostasis. *Ann. Rev. Plant Biol.*, 53: 159–182.

Cunningham, S.D., Berti, W.R., Huang, J.W., 1995. Phytoremediation of contaminated soils. *TIBTECH*, 13: 393–397.

Dominguez-Solis, J.R., Lopez-Martin, M.C., Ager, F.J., Ynsa, M.D., Romero, L.C. and Gotor, C., 2004. Increased cysteine availability is essential for cadmium tolerance and accumulation in *Arabidopsis thaliana. Plant Biotechnol. J.*, 2: 469–476.

Eapen, S. and D'Souza, S.F., 2005. Prospects of genetic engineering of plants for phytoremediation of toxic metals. *Biotech. Adv.*, 23: 97–114.

Evans, K.M., Gatehouse, J.A., Lindsay, W.P., Shi, J., Tommey, A.M. and Robinson, N.J., 1992. Expression of the pea metallothionein-like gene *PsMTA* in *Escherichia coli* and *Arabidopsis thaliana* and analysis of trace metal ion accumulation: implications for PsMTA function. *Plant Mol. Biol.*, 20: 1019–1028.

Gisbert, C., Ros, R., De Haro, A., Walker, D.J., Bernal, M.P., Serrano, R. and Navarro-Avino, J., 2003. A plant genetically modified that accumulates Pb is especially promising for phytoremediation. *Biochem. Biophys. Res. Commun.*, 303: 440–445.

Grichko, V.P., Filby, B. and Glick, B.R., 2000. Increased ability of transgenic plants expressing the bacterial enzyme ACC deaminase to accumulate Cd, Co, Cu, NI, Pb, and Zn. *J. Biotechnol.*, 81: 45–53.

Harada, E., Choi, Y.E., Tsuchisaka, A., Obata, H. and Sano, H., 2001. Transgenic tobacco plants expressing a rice cysteine synthase gene are tolerant to toxic levels of cadmium. *J. Plant Physiol.*, 158: 655–661.

Hegedüs, A., Erdei, S., Janda, T., Tóth, E., Horváth, G. and Dudits, D., 2004. Transgenic tobacco plants overproducing alfalfa aldose/aldehyde reductase show higher tolerance to low temperature and cadmium stress. *Plant Sci.*, 166: 1329–1333.

Hirschi, E.D., Korenkey, V.D., Wilganewski, N.I. and Wagner, G.I., 2000. Expression of *Arabidopsis* CAX2 in tobacco: Altered metal accumulation and increased manganese tolerance. *Plant Physiol.*, 124: 128–133.

Jansen, S., Broadley, M.R., Robbrecht, E. and Smets, E., 2002. Aluminum hyperaccumulation in angiosperms: A review of its phylogenetic significance. *Bot. Rev.*, 68: 235–269.

Kabata Pendias, A. and Adriano, D.D., 1995. *Trace Metals in Soil Amendments and Environmental Quality.* (Eds.) J.E. Recheigh. Lewis Publishers, New York, pp. 139–167.

Karenlampi, S., Schat, H., Vangronsveld, J., Verkleij, J.A.C., van der Lelie, D., Mergeay, M. and Tervahauta, A.I., 2000. Genetic engineering in the improvement of plants for phytoremediation of metal polluted soils. *Environ. Pollut.*, 107: 225–231.

Kawashima, C.G., Noji, M., Nakamura, M., Ogra, Y., Suzuki, K.T. and Saito, K., 2004. Heavy metal tolerance of transgenic tobacco plants over-expressing cysteine synthase. *Biotechnol. Lett.*, 26: 153–157.

Kubota, H. and Takenaka, C., 2003. *Arabis gemmifera* is a hyperaccumulator of Cd and Zn. *Int. J. Phytoremediation*, 5: 197–120.

Lasat, M.M., 2000. Phytoextraction of metals from contaminated sites: A critical review of plant/soil/ metal interaction and assessment of pertinent agronomic issues. *J. Hazard Substance Res.*, 2: 1–25.

LeDuc, D.L., Tarun, A.S., Montes-Bayon, M., Meija, J., Malit, M.F., Wu, C.P., AbdelSamie, M., Chiang, C.Y., Tagmount, A., DeSouza, M., Neuhierl, B., Bock, A., Caruso, J. and Terry, N., 2004. Overexpression of selenocysteine methyltransferase in arabidopsis and Indian mustard increases selenium tolerance and accumulation. *Plant Physiol.*, 135: 377–383.

Lee, J., Bae, H., Jeong, J., Lee, J.Y., Yang, Y.Y., Hwang, I., Martinoia, E. and Lee, Y., 2003a. Functional expression of a bacterial heavy metal transporter in *Arabidopsis* enhances resistance to and decrease uptake of heavy metals. *Plant Physiol.*, 133: 589–596.

Lee, J., Shim, D., Song, W.Y., Hwang, I. and Lee, Y., 2004. Arabidopsis metallothioneins 2a and 3 enhance resistance to cadmium when expressed in *Vida laba* guard cells. *Plant Mol. Biol.*, 54: 805–815.

Lee, S., Moon, J.S., Ko, T.S., Petros, D., Goldsbrough, P.B. and Korban, S.S., 2003b. Overexpression of *Arabidopsis* phytochelatin synthase paradoxically leads to hypersensitivity to cadmium stress. *Plant Physiol.*, 131: 656–663.

Li, Y., Dhankher, O.P., Carreira, L., Lee, D., Chen, A., Schroeder, J.I., Balish, R.S. and Meagher, R.B., 2004. Over-expression of phytochelatin synthase in *Arabidopsis* leads to enhanced arsenic tolerance and cadmium hypersensitivity. *Plant Cell Physiol.*, 45: 1787–1797.

Mkandawire, M. and Dudel, E.O., 2005. Accumulation of arsenic in *Lemna gibba* L. (duckweed) in tailing waters of two abandoned uranium mining sites in Saxony, Germany. *Sci. Total Environ.*, 336: 81–89.

Newman, L.A. and Reynolds, C.M., 2004. Phytodegradation of organic compounds. *Curr. Opin. Biotech.*, 15: 225–230.

Odjegba, V.J. and Fasidi, I.O., 2004. Accumulation of trace elements by *Pistia stratiotes*: Implications for phytoremediation. *Ecotoxicology*, 13: 637–646.

Papoyan, A. and Kochian, L.V., 2004. Identification of *Thlaspi caerulescens* genes that may be involved in heavy metal hyperaccumulation and tolerance. Characterization of a novel heavy metal transporting ATPase. *Plant Physiol.*, 136: 3814–3823.

Peng, H.Y. and Yang, X.E., 2005. Volatile constituents in the flowers of *Elsholtzia argyi* and their variation: A possible utilization of plant resources after phytoremediation. *J Zhejiang Univ. Sci.*, 6: 91–95.

Pilon-Smits, E. and Pilon, M., 2002. Phytoremediation of metals using transgenic plants. *Crit. Rev. Plant Sci.*, 21: 439–456.

Poliard, A.J., Powell, K.D., Harper, F.A. and Smith, J.A.C., 2002. The genetic basis of metal hyperaccumulation in plants. *Crit. Rev. Plant Sci.*, 21: 539–566.

Prasad, M.N.V., 2004. Phytoremediation of metals in the environment for sustainable development. *Proceedings of the Indian National Science Academy*, 70:71–98.

Prasad, M.N.V., 2003. Phytoremediation of metal-polluted ecosystems: Hype for commercialization. *Russ. J. Plant Physiol.*, 50: 686–700.

Prasad, M.N.V. and Freitas, H., 2003. Metal hyperaccumulation in plants: Biodiversity prospecting for phytoremediation technology. *Electronic Journal of Biotechnology* 6: 275–321 (Online electronic journal).

Sharma, N.C., Gardea-Torresdey, J.L., Parsons, J. and Sahi, S.V., 2004. Chemical speciation and cellular deposition of lead in *Sesbania drummondii. Environ. Toxicol. Chem.*, 23: 2068–2073.

Schwartz, C., Echevarria, G. and Morel, J.L., 2003. Phytoextraction of cadmium with *Thlaspi caerulescens. Plant Soil*, 24: 27–35.

Singh, O.V., and Jain, R.K., 2003. Phytoremediation of toxic aromatic pollutants from soil. *Appl. Microbiol. Biotech.*, 63: 128–135.

Song, W-Y., Sohn, E.J., Martinoia, E., Lee, Y.J., Yang, Y.Y., Jasinski, M., Forestier, C., Hwang, I. and Lee, Y., 2003. Engineering tolerance and accumulation of lead and cadmium in transgenic plants. *Nature Biotech*, 21: 914–919.

Sudhagar, R., 2003. Cadmium tolerance in maize (*Zea mays* L.): A genetic and biochemical perspective. *Ph.D. Thesis*, Tamil Nadu Agricultural University, Coimbatore.

Suresh, B. and Ravishankar, G.A., 2004. Phytoremediation: A novel and promising approach for environmental clean-up. *Crit. Rev. Biotechnol.*, 24: 97–124.

Sykes, Marguerite, Yang, Vina, Blankenburg, Julie, and AbuBakr, Said, 1999. Biotechnology: Working with nature to improve forest resources and products. *International Environmental Conference*, p. 631–637.

Tian, J.L, Zhu, H.T., Yang, Y.A. and He, Y.K., 2004. Organic mercury tolerance, absorption and transformation in Spartina plants. Zhi Wu Sheng Li Yu Fen Zi Sheng Wu Xue Xue Bao, 30: 577–582.

Tong, Y.P., Kneer, R. and Zhu, Y.G., 2004. Vacuolar compartmentalization: A second-generation approach to engineering plants for phytoremediation. *Trends Plant Sci.*, 9: 7–9.

United States Environmental Protection Agency, 2001. *A Citizen's Guide to Phytoremediation.*

Van Huysen, T., Terry, N. and Pilon-Smits, E.A.H., 2004. Exploring the selenium phytoremediation potential of transgenic Indian mustard overexpressing ATP sulfurylase or cystathionine-O-synthase. *Int. J. Phytoremed.*, 6: 111–118.

Vatamaniuk, O.K., Mari, S., Lang, A., Chalasani, S., Demkiv, La and Rea, P.A., 2004. Phytochelatin synthase: A dipeptidyltransferase that undergoes multisite acylation with gamma-glutamylcysteine during catalysis–Stoichiometric and site-directed mutagenic analysis of *Arabidopsis thaliana* PCS1-catalyzed phytochelatin synthesis. *J. Biol. Chem.*, 279: 22449–22460.

Verret, F., Gravot, A., Auroy, P., Leonhardt, N., David, P., Nussaume, L., Vavasseur, A. and Richaud, P., 2004. Overexpression of AtHMA4 enhances root-to-shoot translocation of zinc and cadmium and plant metal tolerance. *FEBS Lett.*, 576: 306–312.

Wang, Q.R., Cui, Y.S., Liu, X.M., Dong, Y.T. and Christie, P., 2003. Soil contamination and uptake of heavy metals at polluted sites in China. *J. Environ. Sci. Heal.*, A38: 823–838.

Wycisk, K., Kim, E.J., Schroeder, J.I., Kramer, U., 2004. Enhancing the first enzymatic step in the histidine biosynthesis pathway increases the free histidine pool and nickel tolerance in *Arabidopsis thaliana. FEBS Lett.*, 578: 128–134.

Zhang, Y.W., Tam, N.F.Y. and Wong, Y.S., 2004. Cloning and characterization of type 2 metallotltionein-like gene from a wetland plant, *Typha latifolia. Plant Sci.*, 167: 869–877.

Zhao, F.J., Dunham, S.J. and McGrath, S.P., 2002. Arsenic hyperaccumulation by different fern species. *New Phytol.*, 15: 27–31.

Zhu, Y.L., Pilon-Smits, E.A.H., Jouanin, L. and Terry, N., 1999a. Over expression of glutathione synthetase in Indian mustard enhances cadmium accumulation and tolerance. *Plant Physiol.*, 119: 73–79.

Zhu, Y.L., Pilon-Smits, E.A.H., Tarun, A.S., Neber, S.U., Jouanin, L., Terry, N., 1999b. Cadmium tolerance and accumulation in Indian mustard is enhanced by over expressing gamm-aglutamylcysteine synthetase. *Plant Physiol.*, 121: 1169–1178.

Chapter 3

Effect of Gamma Rays on Multiple Shoot Induction from Stem Explants of *Cucumis melo* var. *Utilissimus*

M. Venkateshwarlu

Department of Botany, Kakatiya University, Warangal – 506 009, Andhra Pradesh

ABSTRACT

Mutagenic effectiveness is a measure of the frequency of mutations induced by a unit dose of mutagen while mutagenic efficiency gives an idea of the proportion of mutations in relation to other associated undesirable biological effects such as lethality and sterility induced by the mutagen (Konaz *et al.*, 1965). The high effectiveness was recorded in 10kR and 15kR. For any successful mutation breeding programme, selection of efficient and effective mutagenesis is very essential to recover high frequency of desirable mutations. The present study reveals on effectiveness, efficiency and synergistic effect of gamma rays in *Cucumis melo* var. *Utilissimus*. The greatest advantage of *in vitro* systems is the case of treating great number of cells with mutagenic agents, which has provided more effective and rapid results than conventional techniques. King (1984) reported mutagenesis in *in vitro* as a n important field for crop improvement. Shoot multiplication was obtained fro shoot apices of *Nigeri* when cultured on MS medium supplemented with 1.0 to 3.0 mg/l BAP (Ahmed and Pande, 1988).

Keywords: *Gamma rays, Multiple shoot, Stem explants, Mutagenic agents.*

Introduction

This variety resembles cucumber and is used as a vegetable. The fruits are slender and elongated, the length varying from a few inches to about 3 ft. They are pale or dark green in colour, smooth or ridged, with soft downy hairs covering the skin when tender. Seeds are smaller than those of the musk melon. The primary aim of this study has been to gain some knowledge about the genotypic differences for callus initiation and high frequency plant regeneration from long term callus cultures of *Cucumis melo* var. *Utilissimus*. This variety is cultivated both as a hot weather crop and as a rainy season crop. It grows on any kind of soil, but thrives best on well-manured rich loamy soils with abundant water supply. The seeds are small and edible, and are used in confectionery. A refreshing drink is prepared from ground kernels. In comparison to lactose for different straw lactose 3 per cent of lemon grass straw (580 gram) was proved to be less effective to others. The similar findings were also reported by Singh (2005). Si estimation was carried out according to the method described by Ma, *et al.* (2002). Ahamed John (1991) found that there was a decrease in mutagenic efficiency with increase in mutagenic dose. Gamma rays was found to be generally efficient and effective mutagen (Gautam *et al.*, 1991), in *Vigna mungo.* The new insights into the complex role played by silicon in imparting resistance to rice blast disease (Rodrigues *et al.*, 2003).

Material and Methods

The stem raised from control seeds could produce only callus on MS with different supplements which was regenerated into a single shoot. Well filled undamaged and uniform sized seeds were hand picked from the seed lot and equilibrated to the moisture content of 12 per cent. For each dose of physical mutagen, and random sample of 100 seeds were treated in *Cucumis melo* var *Utilissimus* variety. The dry seeds of *Cucumis melo* (1, 2, 3, 4, 5, 10kR and 15kR). The seeds were irradiated at different doses starting from 10kR to 15kR with an interval of 10kR. The recommended agronomic practices and plant protection measures were followed uniformly for all treatment stem explants. In the present studies, the induction of multiple shoots were reported from stem explants after gamma rays treatment. The seeds were removed, and the cotyledons were surface sterilized with 2 ml Teepol, in 98 ml sterilized distilled water for 2 minutes, and then rinsed twice with distilled water; later stem were sterilized in 0.1 per cent mercuric chloride for 5–7 minutes, followed by rinsing four times in sterile distilled water. Filter paper bridges, placed in the test tubes containing distilled water, were closed with cotton plugs and were sterilized by autoclaving. Later, the seeds were kept for germination on the filter paper bridges. After 8 days of germination the resulted seedlings were irradiated. Stem explants were excised aseptically and were inoculated on the MS based medium supplemented by kinetin or BAP at concentrations ranging from 0.5 to 5 mg/l. cultures were incubated under 10 h fluorescent light at 25±2°C temperature.

Results and Discussion

Mutagen (gamma rays) treated stem explants were tested for their effectiveness. The lower doses induced callus and greening of callus in stem cultures (Plate 3.1, Figure 1). The doses 4 kR or 5 kR induced high frequency of multiple shoots (Plate 3.1, Figures 3 and 4). The higher doses 10, 15 and 20 kR induced excessive callus growth and browning of callus. The maximum number of shoots obtained at dose of 4 and 5 kR and they ranged from 12–16 from stem cultures. The higher dose of 20 kR reduced the number of shoots and only greening of callus (Plate 3.1, Figure 2) and doses beyond that totally suppressed the formation of shoots from stem explants (Tables 3.1 and 3.2). With the increase in the level of BAP (2.0 to 4.0 mg/l). The maximum number of shoots on the explants were observed at 4.0 mg/l BAP or 5.0 mg/l kinetin, but at higher levels of BAP or kinetin, the formation of callus had taken place and the number of shoots per explant was reduced. The exact cause for the multiple shoot induction by gamma rays is not known, but it may by possibly due to the stimulation of endogenous hormones by the physical mutagen (gamma rays) which in turn induces multiple shoots.

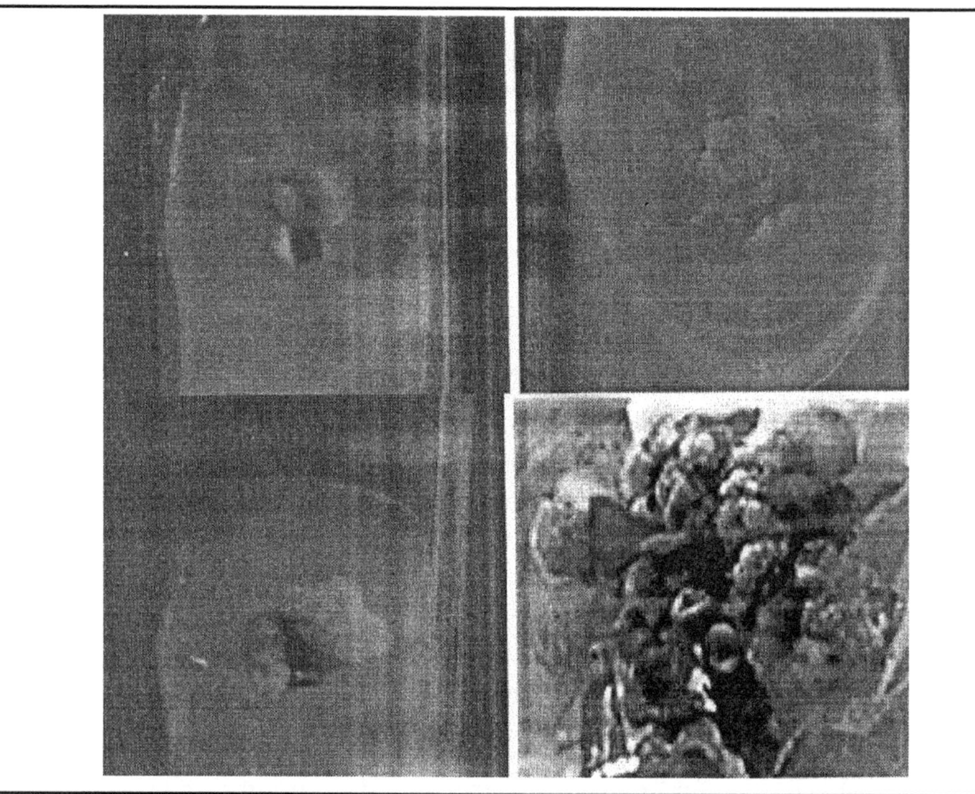

Plate 3.1: Induction from Stem Explants of *Cucumis melo* var *Utilissimus*

Table 3.1: Multiple Shoot Induction from Stem Explants of *Cucumis melo* var. *Utilissimus*

Per cent of Growth Regulators (mg/)	*Stem*	
	Mean Number of Shoots per Shoot Tip	*Per cent of Callus Production*
MS + 1.0 BAP + 1.0 NAA	10.2±2.1	42
MS + 2.0 BAP + 1.0 NAA	9.2±2.3	30
MS + 3.0 BAP + 1.0 NAA	6.2±1.5	35
MS + 4.0 BAP + 1.0 NAA	1.4±0.2	30
MS + 5.0 BAP + 1.0 NAA	6.5±1.2	26
MS + 2.0 BAP + 1.0 L-glutamic acid	12.3±1.0	20
MS + 3.0 BAP + 1.0 L-glutamic acid	15.4±0.5	20
MS + 4.0 BAP + 1.0 L-glutamic acid	14.5±0.4	14
MS + 10 per cent CM + 0.5 BAP	11.2±2.1	12
MS + 15 per cent CM + 0.5 BAP	11.3±2.3	8
MS + 20 per cent CM + 0.5 BAP	10.5±1.4	5

Table 3.2: Effect of Gamma Rays on Stem Explants of *Cucumis melo* var. *Utilissimus*

Dose of Irradiation (kR)	*Stem (No. of. shoots)*	*Callus Growth*
Control	3–4	Normal callus
1	2–3	Callus growth
2	2–4	Better growth
3	3–6	Better growth
4	4–6	Better greening of callus
5	2–3	Growth in abundance
10	4–5	Better growth
15	3–4	Better growth

The stem treated with 1, 2, 3, 4, 5, 10, 15 and 20 kR irradiation produced a number of multiple shoots. 4 or 5 kR is most potent in inducing most number of multiple shoots. The isolated *in vitro* raised shoots of 1–2 cm long, rooted profusely on MS medium with BAP (2 mg/l) + NAA (1 mg/l) within 15 days resulting in the formation of complete plantlets.

Conclusion

The stem explants transferred to a fresh medium containing the some concentration of growth regulators, again resulted in the formation of multiple shoots. In the present study the effectiveness of different doses by gamma rays was assessed by their mutagen dose and the efficiency was assessed on four different biological parameters. The efficiency of mutagenic agent not only depends on the biological system but also of physical damage, chromosomal aberration and sterility induced in addition to mutation.

References

Ahamed John, S., 1991. Mutation studies in black gram [*Vigna Mungo* (L) Hepper], *Ph.D., Thesis,* Bharathidasan University, Thiruchirapalli.

Ahmed and Pande, 1988. Shoot multiplication was obtained from hypocotyls and cotyledon explants of Niger. *Bionature,* 8: 95.

Bottino, P.I., 1975. The potential of genetic manipulation in plant cell cultures for plant breeding. *Rad Bot.,* 15: 1–16.

Gautam, A.S., Sood, K.C and Richaria, A.C., 1991. Mutagenic effectiveness and efficiency of gamma rays, EMS and their synergistic effects in black gram *Vigna mungo L. Cytologia,* 57: 85–89.

King, P.J., 1984. Mutagenesis in cultured cells. In: *Cell Culture and Somatic Cell Genetics, Vol. 1: Laboratory Procedure and their Application* (Ed) L.K. Vasil. Academic Press, Orlando, pp. 547–551.

Ma, J.F., Higashitani, A. Sato, K. and Takeda, K., 2002. Genotypic variation in silicon concentration of barley grains. *Plant and Soil,* 249(2): 383–387.

Murashige, J., 1974. Plant propagation through tissue culture. *Annu. Rev. Plant Physiol.,* 25: 135–166.

Rodrigues, F.A., Benhamou, N., Datnoff, L.E., Jones, J.B. and Belanger, R.R., 2003. Ultrastructure and cytological aspects of silicon mediated rice blast resistance. Phytopathology, 93: 535–546.

Singh, Rochica, 2005. Studies on some larger fungi of Faizabad with reference to their eco-physiological characteristics. *Ph.D. Thesis,* Dr. R.M.L. Avadh University, Faizabad, pp. 114–115.

Chapter 4

Particulate Transport of Trace Metals in a Riverine Ecosystem–Karamana River, South Kerala, India

***P.R. Jayaraman*[1], *C.G. Radhika*[2], *T. Ganga Devi*[3] *and T. Vasudevan Nayar*[4]**

[1]*Department of Botany, Government College for Women, Thiruvananthapuram, India*
[2]*Department of Botany, University College, Thiruvananthapuram, India*
[3]*Principal (Retired), Government College for Women, Thiruvananthapuram, India*
[4]*Scientific Officer (Retired), Division of Marine Chemistry, Department of Aquatic Biology and Fisheries, University of Kerala*

ABSTRACT

Suspended Particulate Matter (SPM) and trace metal fluxes in Karamana River, the major source of drinking water for Thiruvananthapuram, the capital city of Kerala, South India, were monitored for a period of one year from February 1998 to January 1999. Six trace metals, *viz.*, copper, lead, cadmium, zinc, manganese and iron were estimated in the particulate matter using Atomic Absorption Spectroscopy. The total particulate flux (15.68 mg L^{-1}) in the river was very low as compared to Indian and world averages. Significant spatial and temporal variations were evident in the total flux of SPM and associated trace metals to varying extents. Co-transportation of trace elements in different combinations during different seasons was noticed. SPM associated transport of cadmium (155.96 µg g^{-1}), lead (464.01 µg g^{-1}) and iron. (56,633 µg g^{-1}) remained high, whereas that of copper (212.91 µg g^{-1}), zinc (623.58 µg g^{-1}) and manganese (957.13 µg g^{-1}) was found low in comparison with standard values as suggested by Aston and Chester (1976). Particulate flux of all the six metals was found higher than the standard values for average shale, upper continental crust, average mud and average values for Indian rivers, while a slightly lower value was noticed in the case of manganese when compared to world river SPM and average river particulate. Particulate load of cadmium was very high as compared to the tentative threshold values, while the values of particulate cadmium, lead and iron were found low compared to metal content in world river borne detritus. Leaching of heavy metals into the lotic ecosystem resulting from plantations along with other anthropogenic activities have been highlighted. Toxic metals cadmium and lead were detected in obvious quantities indicating the association of these elements in the river water. In general, the levels of these trace metals have not reached any alarming level yet, while cadmium and lead pose great concern.

Keywords: *Fluvial systems, Suspended particulate matter, Biogeochemical cycles, Trace metals.*

Introduction

Suspended sediments play a significant role in the biological and biogeochemical cycling of trace elements in fluvial systems (Horowitz, 1996). Riverine flora ranging from microscopic phytoplankton to angiosperms depend largely on river borne particulates for their nutrients. The allochthonous materials in the river are rapidly adsorbed and transported by particulates (Walling and Webb, 1985) and therefore Suspended Particulate Matter (SPM) reflects the effect of pollution/deterioration in a given area. Hence, spatial and temporal variations of trace metals in Suspended Particulate Matter (SPM) in fluvial environments have been recognized as an area, which demands serious attention.

Suspended particulate matter in aquatic systems according to Burton (1976) can be considered as suspended matter retained by a 0.45 µm filter. A large proportion of the fluviatile transport of matter occurs in the form of suspended material with mean grain sizes of 2–63 µm (silt grain size) as opined by Forstner and Wittman (1983). Particulate matter is an extremely important, but poorly studied substrate for the transportation of trace metals in fluvial systems and particulate transport represents a major pathway in the biogeochemical cycling of trace contaminants (Allen, 1979). Sediments and suspended matter in aquatic systems absorb and concentrate comtaminants such as trace metals. Rivers are the important transporting agents for continental material (both natural as well as anthropogenic) to the oceans that happen mostly in particulate phase amounting to 13 billion tons (Milliman and Meade, 1983). Studies on the chemical composition of particulates have been found to contribute much to the field of geochemical prospecting in river beds (Hawkes and Webb, 1962).

Although a number of studies have been carried out to assess the role of particulate matter in riverine transport in India (Subramanian, 1979; Borole *et al.*, 1982; Subramanian *et al.*, 1985; Subramanian *et al.*, 1987; Biksham and Subramanian 1988; Alagarsamy and Zhang, 2005; Jain and Sharma, 2006) no serious attempt in this regard has yet been made on rivers in Kerala but for the studies by Paul and Pillai (1983) and Shibu *et al.* (1990) on the lower reaches of Periyar and Muvattupuzha rivers. A systematic study on the flux and dynamics of trace metals in lotic water bodies is important to understand their transport and role in biogeochemical processes as well as in detecting their sources. Hence, the present study was undertaken to evaluate the occurrence and abundance as well as the spatial and temporal variations of selected trace metals in the waters of Karamana River, which is the main source of drinking water for Thiruvananthapuram, the capital city of Kerala.

Karamama River originates from Chemmungi Mottai, a peak in Western Ghats and after flowing for about 68 km through the Peppara Wildlife Sanctuary, plantations, rural and urban areas, it debauches into the Lakshadweep Sea at Panathura, through a barmouth which remains closed during summer. The river along with its tributaries lies between 8°21′ and 8°42′N and 76°52′ and 77° 15′E, within the administrative boundaries of Thiruvananthapuram District, South Kerala. 10 stations were identified along the river for the collection of samples (Figure 4.1). Station 1 was located down to Peppara dam in the upper reaches of the river under the influence of land runoff from thick forests and

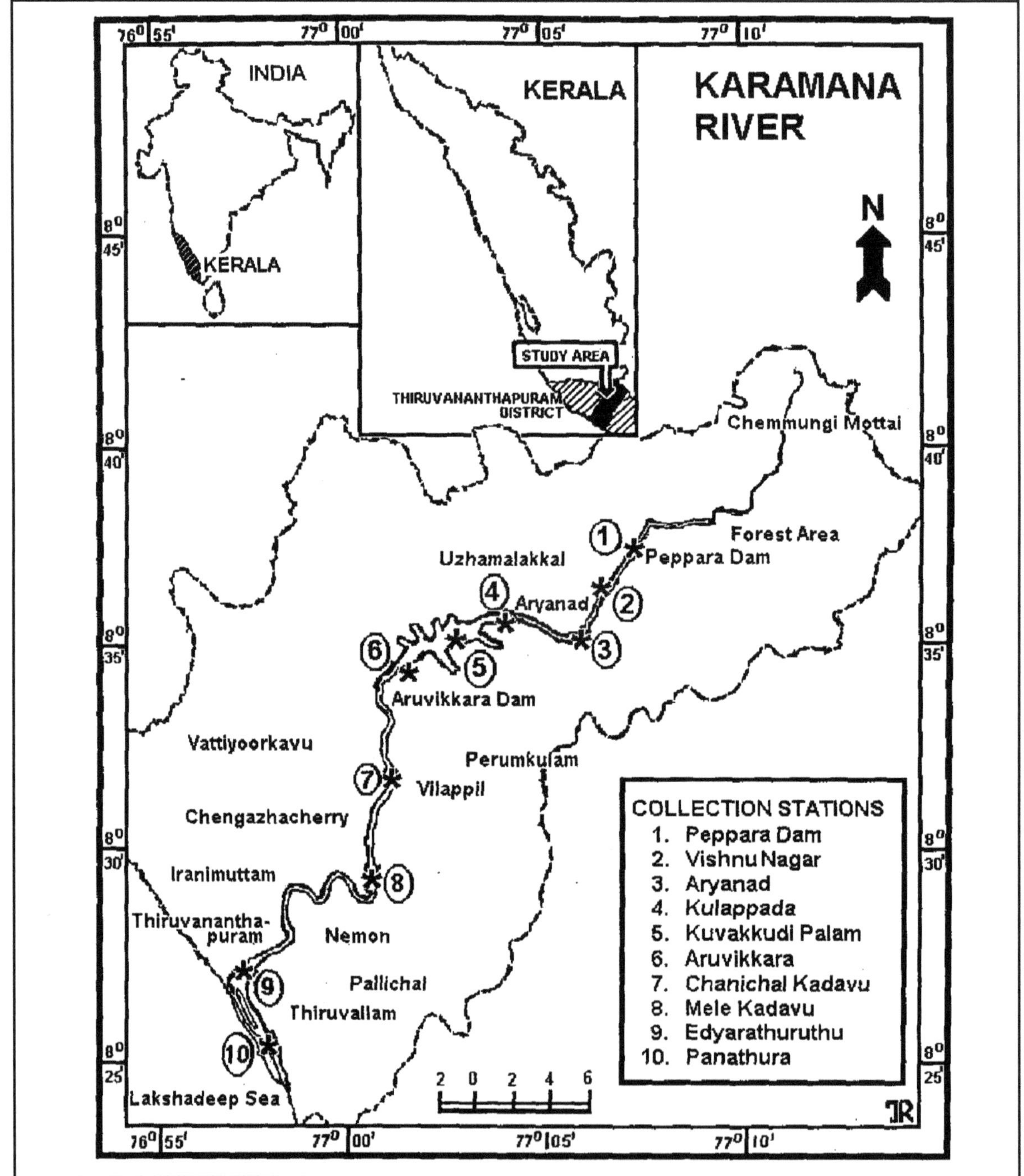

Figure 4.1: Map of Karamana River Showing Sampling Sites

discharge from the dam. Station 2 was identified at Vishnu Nagar, a housing colony that receives discharges from plantations and agricultural lands and wastes from nearby hutments. Station 3 was fixed at Aryanad Township influenced by domestic effluents. Station 4 was at Kulappada, where a ferry is being operated across the deep waters of the river. Station 5 was set at Kuvakkudi Palam (bridge) at the upper reaches of Aruvikkara dam characterized by turbid and stagnant water throughout the year. Station 6 was established down to Aruvikkara dam floodgate, where the river flows very fast through rocky terrain. Station 7 was placed at Chanichal Kadavu, a bathing ghat with more or less stagnant water that is under the influence of sand-mining. Station 8 was at Mele Kadavu, another bathing Ghat, where the river is shallow and slow flowing and under the influence of a granite quarry nearby. Station 9 was located at Edyarathuruthu where the Parvathyputhenar loaded with effluents from Thiruvananthapuram city joins the river. Besides, the station has estuarine features and is under the influence of retting of coconut husk and sand-mining. Station 10 was at Panathura, where the barmouth remains closed during summer and sea water enters the river through tidal effect during rainy season.

Materials and Methods

Water samples were collected in acid-cleaned, non-reactive plastic containers once a month from the 10 stations along the Karamana River for a period of one year from February 1998 to January 1999. Known volumes of water samples were filtered to collect the particulate matter on acid-treated, dried and pre-weighed 0.45 micron Millipore filter papers using an all-glass Millipore filtration unit. Filter papers were then oven dried and weighed to find out the particulate load. The particulate matter was acid digested following standard procedures (APHA, 1992) and subjected to Atomic absorption spectrophotometric analysis for the estimation of six trace metals *viz.*, copper, cadmium, lead, zinc, manganese and iron and values were expressed in $\mu g\ g^{-1}$ (A.A.S. Perkin Elmer Model 2380). Monthly data were pooled and expressed on seasonal basis as pre-monsoon (February to May), monsoon (June to September) and post-monsoon (October to January). Annual average transport of SPM and metals was computed based on the annual discharge of water by the Karamana River, *viz.*, 1324 million m^3 (Kerala State Land Use Board, 1995). The results were compared with standard values for average shale (Turekian and Wedepohl, 1961), the tentative threshold values (Aston and Thornton, 1977), geochemistry of average river borne detritus (Aston and Chester, 1976), average river particulate (Martin and Meybeck, 1979), average mud and upper continental crust (Taylor and McLennan, 1985) world river SPM (Martin and Windom, 1991) and also with known reports on particulate metal content in rivers in India and abroad. Enrichment Factors (EF) were computed for each metal in the suspended particulate matter at Stations 2 to 10 that are apparently exposed to anthropogenic inputs by weighing the respective concentrations against background values at Station 1 located in the comparatively pristine forest environment. The results were subjected to regression analysis and ANOVA test to evaluate the relationships among the various factors involved.

Results and Discussion

The data on the season-wise distribution of SPM and associated heavy metal concentrations in the Karamana River are presented in Figures 4.2 to 4.8. Table 4.1 compares the SPM content and annual flux of SPM in the Karamana River with that of other rivers. Table 4.2 summarizes geochemistry of SPM from the Karamana River compared with other rivers of India and abroad, other water bodies and standard values for rocks and sediments. Metal concentration in SPM of the Karamana River in comparison with that of the tentative threshold values as suggested by Aston and Thornton (1977) is presented in Table 4.3. Enrichment Factors (EF) computed for each metal in the SPM at Stations 2 to 10 are depicted in Table 4.4.

Table 4.1: SPM Content and Annual Flux of SPM in the Karamana River Water Compared to Indian and World Rivers

Name of River	*SPM (mg L^{-1})*	*Annual Flux (10^9 kg yr^{-1})*
Ganges	1250.0[r]	460.10[su2]
Brahmaputra	1370.0[r]	711.20[su2]
Godavari	122.0[r]	16.20[su2]
Krishna	130.0[r]	8.50[su2]
Narmada	10.0[r] (1154[b])	6.20[su2]
Tapti	13.0[r] (445[b])	2.70[su2]
Cauvery	48.0[r]	0.71[su2]
Netravati	54.0[sh]	1.40[kib]
Gurpur	52.0[sh]	0.10[kib]
Mahanadi	93.8-596[ry]	–
Krishna	600[sul]	–
Godavari	2000[sul]	–
Mulki-Pavanje	38[k]	–
Danube	20[g]	–
Global flux	500[h]	12000–13000[mm]
US rivers	400[t]	–
Karamana (present study)	15.68	0.02

h: Holeman (1968), t: Turekian (1969), r: Rao (1975), sul: Subramanian (1979), b: Borole *et al.* (1982), mm: Milliman and Meade (1983), ry: Ray *et al.* (1984), su2: Subramanian *et al.* (1985), kib: Karnataka Irrigation Department (1986), k: Karbassi (1989), sh: Shankar and Manjunatha (1997), g: Guieu *et al.* (1998).

Table 4.2: Geochemistry of Particulates from the Karamana River Compared with Other Indian and Foreign Rivers, Other Water-bodies, Rocks and Sediments

Sample Description	*Elements (g g^{-1})*			*Fe (%)*		
	Cu	*Cd*	*Pb*	*Zn*	*Mn*	*Fe*
World river SPM[mw]	100	1.20	35	250	1050	4.80
Average river particulate[mm]	100	1.00	150	350	1050	4.80
World river–borne detritus[ac]	2500	18.00	280	950	4300	2.20
Average shale[tw]	45	0.30	20	95	850	4.67
Upper continental crust[tm]	25	0.098	20	71	600	3.50
Average mud[tm]	50	–	20	85	850	5.10
Indian average[su]	28	–	–	16	605	2.90
Ganga river SPM[sa]	75	–	–	–	1000	7.97

Contd...

Table 4.2–Contd...

Sample Description	*Elements (g g^{-1})*			*Fe (%)*		
	Cu	*Cd*	*Pb*	*Zn*	*Mn*	*Fe*
Ganga river SPM[az]	252	–	37	643	3450	9.00
Brahmaputra river SPM[az]	108	–	–	916	522	10.90
Ganga and Brahmaputra rivers SPM[su]	19	–	–	46	522	2.53
South West coast rivers of India SPM[mj]	71	–	–	90	724	6.50
Krishna[az]	220	–	45	270	2540	13.20
Cauvery[az]	60	–	40	500	1300	6.20
Netravathi river SPM[sh]	75	–	43	95	884	6.40
Gurpur river SPM[sh]	70	–	57	86	546	6.59
Tapti river SPM[b]	139	–	–	157	1075	7.10
Tapti river SPM[az]	136	–	–	157	1304	7.90
Mulki–Pavanje river SPM[k]	88	–	55	324	884	7.50
Narmada river SPM[b]	133	–	–	143	1182	7.40
Narmada river SPM[az]	128	–	–	143	1125	7.60
Periyar and Muvattupuzha rivers and Cochin estuary SPM[sb]	20–300	8–60	–	60–1550	–	–
Karapad creek, Tuticorin (polluted) SPM[ch]	0.9–34.5	0.4–4.5	5.89–28.1	21.3–110	–	0.93–3.56
Major American rivers SPM[c]	–	–	–	–	1500	4.41
Mississipi[az]	38.5	–	39	193	1260	4.43
Amazon[az]	266	–	–	–	1033	5.55
Yellow river SPM[l]	33	–	35	75	800	3.20
Yangtze river SPM[l]	71	–	87	115	960	5.40
Huanghe river SPM[az]	26.7	–	16.5	69.8	767	3.72
Danube river SPM[g]	115	1.1	84	248	1704	3.60
Patuxent river (USA) SPM[r]	17–40	0.2–7.3	–	–	123–230	
Lena river (Russia) SPM[mg]	28	–	23	143	–	
Wisconsin rivers (USA) SPM[sop]	22.9	0.7	36.2	–	128	–
St. Lawrence[az]	40.2	–	36.2	342.7	1311	5.07
Orinoco[az]	60.8	–	–	76.5	588	7.40
Zaire[az]	100	–	220	300	1400	7.10
Taylor Creek, southern Nigeria[oo]	–	5.9	19	83	90	0.12
Karamana River SPM (present study)	212.91	155.96	464.01	623.58	957.13	5.66

tw: Turekian and Wedepohl (1961), ac: Aston and Chester (1976), mm: Martin and Meybeck (1979), sa: Sarin *et al.* (1979), b: Borole (1980), l: Li *et al.* (1984), su: Subramanian *et al.* (1985), tm: Taylor and McLennan (1985). k: Karbassi (1989), sb: Shibu *et al.* (1990), mw: Martin and Windom (1991), mg: Martin *et al.* (1993), c: Canfield (1997), sh: Shankar and Manjunatha (1997), g: Guieu *et al.* (1998), mj: Manjunatha *et al.* (1998), sop: Shafer *et al.* (1999), r: Riedel *et al.* (2000), ch: Chandrasekhar (2001), az: Alagarsamy and Zhang (2005), oo: Okafor and Opuene (2006).

Table 4.3: Metal Concentration in SPM of the Karamana River in Comparison with that of the Tentative Threshold Values (The concentration level in stream sediments at which it is likely that associated waters may, on occasion, exceed the highest desirable levels (HDL) as suggested by Aston and Thornton, 1977)

Elements	*HDL (g L^{-1})*	*Tentative Threshold Value*	*Metal in SPM of the Karamana River*
Cu	50	1000 g g^{-1}	212.91 g g^{-1}
Cd	10	10 g g^{-1}	155.96 g g^{-1}
Pb	50	500 g g^{-1}	464.01 g g^{-1}
Zn	5000	2000 g g^{-1}	623.58 g g^{-1}
Mn	500	1000 g g^{-1}	957.13 g g^{-1}
Fe	100	6 per cent	5.66 per cent

Table 4.4: Enrichment Factors (EF) Computed for Trace Metals in SPM at Stations 2 to 10 Against the Background Values at Station 1

Trace Metals	*Stations*								
	2	*3*	*4*	*5*	*6*	*7*	*8*	*9*	*10*
Cu	0.76	0.66	1.10	0.73	0.85	0.62	0.50	0.64	0.58
Cd	1.20	0.93	0.95	0.76	0.91	0.95	1.05	0.93	0.96
Pb	1.12	0.95	1.82	0.55	1.35	0.35	0.71	0.41	1.42
Zn	0.97	0.79	0.83	0.92	0.92	0.81	0.73	0.68	0.47
Mn	1.89	1.59	1.47	1.92	2.14	2.00	1.58	1.42	0.75
Fe	1.14	1.21	0.92	0.82	0.79	0.66	0.71	0.57	0.60

Transport of Suspended Particulate Matter (SPM)

Seasonal variations in SPM flux in the river is depicted in Figure 4.2. The maximum monthly SPM load was observed during May (24.91 mg L^{-1}), may be due to the impact of the first summer showers and the minimum was during February and March (8.50 mg L^{-1} and 9.95 mg L^{-1} respectively) characterized by slow flow rate of the river in summer. The maximum yearly SPM flux was registered at Stations 9 (20.87 mg L^{-1}), 7 (20.84 mg L^{-1}) and 5 (19.63 mg L^{-1}) owing to plankton growth within the dam at Stations 5 and the combined effect of plankton growth and sand mining at Stations 7 and 9. The impact of tidal fluxes might also have been a contributing factor at Station 9. The minimum annual SPM load exhibited at Stations 1 and 2 (10.43 mg L^{-1}) may be attributed to the settling of SPM in the reservoir beyond Station 1.

Based on the annual discharge of water, the Karamana River was found to transport an annual average load of 15.68 mg L^{-1} SPM that was lower by a factor of 32 than the global average and by a factor of 25.5 than the average for US rivers. The SPM load was also much lower compared to Indian rivers the Ganges, Brahmaputra, Godavari, Krishna, Cauvery, Netravati, Gurpur, Mahanadi and Mulki-Pavanje but slightly higher than that of rivers Narmada and Tapti (Table 4.1). It could evidently be a positive index of the drinking quality of water and at the same time it may also indicate the poor status of the river as a potential transporter of nutrients and toxic elements. The river was found to contribute an yearly load of 20,786.8 metric tons of SPM to the Lakshadweep Sea that was much lower than the annual flux of SPM by other Indian rivers.

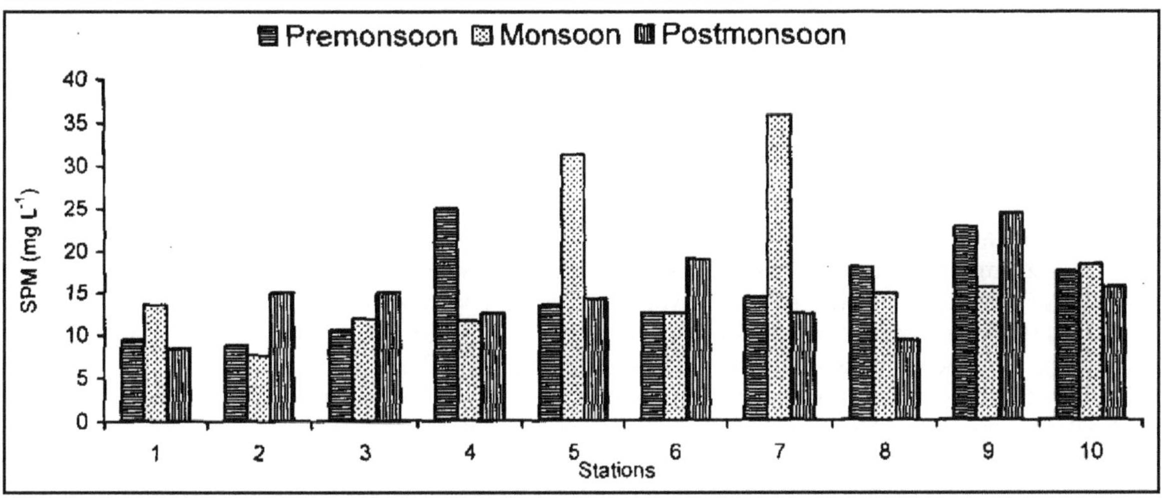

Figure 4.2: Seasonal Distribution of SPM in Water

SPM transport in the river did not exhibit any significant season-wise and station-wise variations, but month-wise variations were significant (ANOVA: $df = 11, F = 2.378, P \leq 0.012$) indicating that the SPM flux in the river may be independent of spatial and temporal environmental attributes other than month-wise variations. Short residence time due to fast flow rate may account for the absence of significant station-wise variations. Total SPM load showed weak, yet significant negative correlations with particulate metals iron ($r = -0.322, y = 21.184 - 0.0001x, p \leq 0.0003, n = 120$), manganese ($r = -0.332, Y = 20.091 - 0.005x, P \leq 0.0002, n = 120$), zinc ($r = -0.393, y = 24.424 - 0.140x, P \leq 0.00001, n = 120$), cadmium ($r = -0.326, Y = 18.432 - 0.018x, P \leq 0.0003, n = 120$) and lead ($r = -0.219, Y = 17.274 - 0.0003x, P \leq 0.02, n = 120$). The apparent negative relationships may be due to the normal component of small sized metal-rich particulates occasionally getting mixed up with large sized metal poor sediment fractions by resuspension and advection resulting in higher rates of particulate flux. The inverse relationship between particle size and metal content is well documented (Förstner and Wittmann, 1979; Salomons and Förstner, 1984; Horowitz, 1988; Combest, 1991; Verma *et al.*, 1993: Bertin and Bourg, 1995; Mohan and Hosetti, 1997). No correlation was evident in the case of copper suggesting an organic association of particulate copper as reported elsewhere (Hasle and Abdullah, 1981; Shibu *et al.*, 1990; Pardo *et al.*, 1990). Significant correlations were not evident in the relationship of SPM with hydrographical parameters (Jayaraman *et al.*, 2003) but for the weak, yet significant positive correlations with nitrate-nitrogen ($r = 0.305, Y = 11.601 + 22.460x, P \leq 0.0007, n = 120$) and silicate-silicon ($r = 0.244, Y = 9.620 + 1.706x, p \leq 0.007, n = 120$) and it may be due to the dynamic nature of the lotic system. The weak correlation with nitrate–nitrogen and silicate–silicon may be attributed to terrigenous land run-off preceding particulate flux.

Particulate Transport of Trace Metals

Colloidal surfaces are greater in fluvial particulates as compared to bottom sediments (Hart, 1982) and hence particulates may have a greater role in physico-chemical sorption reactions. Sedimentation of enriched particulate matter is a potentially important mechanism by which sediments may concentrate trace metals.

It is well established that transport of elements in rivers, except that of mercury, is through solid phase rather than in dissolved form (Blachford and Ongley, 1984). The ability to adsorb elements onto SPM depends on surface charges and it was investigated by several workers (Gibbs, 1973; Aston and Chester, 1973; Bowares and Huang, 1987).

Copper

Most copper minerals are of low solubility as it is strongly immobilized in the crystal lattices of mineral micelles (Burton, 1976) and hence it is transported dominantly or exclusively by SPM. Only 5.1 per cent of total copper flux happens in dissolved form as reported for Amazon and Yukon (Gibbs, 1973) and only 30 per cent is leachable into dissolved phase (Duursma, 1976).

Seasonal distribution of particulate copper in the river is presented in Figure 4.3. Analysis of the monthly data at various stations showing the highest value of SPM associated copper during July (479.29 µg g^{-1}) may be due to the influence of monsoon rains, whereas the lowest was during December (59.41 µg g^{-1}) owing to the reduced rate of water discharge. The maximum copper content at Stations 1, 2, 3, 7, 8 and 9 were experienced during monsoon, while at Stations 4, 5, 6 and 10 the maximum were during pre-monsoon. The pre-monsoon and monsoon seasonal maxima may suggest the dominant role of particulate adsorption and terrigenous inputs respectively. The maximum annual flux of particulate copper was at Station 4 (315.25 µg g^{-1}), whereas Station 8 (141.73 µg g^{-1}) recorded the least. A slightly higher rate was observed at Station 4 followed by Stations 1, 6 and 2 which may suggest the role of pesticides and fungicides used in rubber plantations.

The annual average rate of particulate copper transport was found to be 212.91µg g^{-1} with an annual load of 4.43 metric tons. The annual average rate was higher by a factor of two than that of world river SPM and average river particulate, twelve times lower than that of world river-borne detritus, five times higher than that of baseline levels in average shale, more than eight times higher than that of upper continental crust and four times higher than that of average mud. It was also much higher in comparison with Indian average and of most Indian rivers and other aquatic systems (Table 4.2). Copper content in SPM was five times lower than that of the tentative threshold value of 1000 µg g^{-1} at which the related water may exceed the highest desirable level of 50 µg L^{-1} (Table 4.3). The seasonal and annual average station-wise distribution was without any pattern or trend that ruled out the possibility of any point source for particulate copper. The Enrichment Factor (EF) computed for particulate copper on an annual cycle remained less than unity and did not indicate any appreciable degree of enrichment and this may be due to removal under the impact of terrigenous leachates

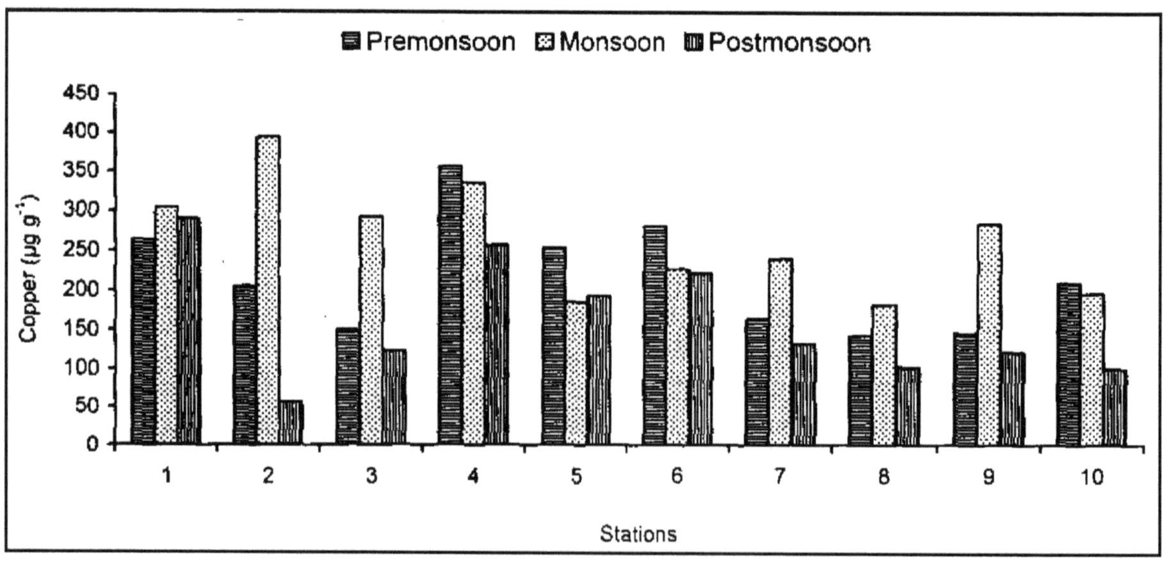

Figure 4.3: Seasonal Distribution of Particulate Copper

(Table 4.4). Hence, it is reasonable to state that the water of the Karamana River did not pose any threatening level of pollution with respect to particulate copper.

The distribution of SPM-associated copper showed statistically significant temporal variations (ANOVA: month-wise –$df = 11$, $F = 8.560$, $p \leq$ 2E-10; season-wise $df = 2$, $F = 7.814$, $p \leq 0.004$), while significant station-wise variations were evident in season-wise distribution alone (ANOVA: $df = 9$, $F = 2.555$, $P \leq 0.04$) reflecting the extent of the influence of spatial and temporal environmental variables. Particulate copper showed low, but significant positive correlations with other metals, *viz.*, zinc ($r = 0.439$, $y = 55.510 + 0.253x$, $p \leq$ 1E-06, $n = 120$), cadmium ($r = 0.420$, $y = 156.031 + 0.366x$, $P \leq$ 1.76E-06, $n = 120$) and lead ($r = 0.181$, $y = 192.261 + 0.045x$, $p \leq 05$, $n = 120$), but such affinities were not evident in the case of iron and manganese. Copper in SPM of the river may at least partially be associated with zinc, cadmium and lead suggesting a mineralogical and/or anthropogenic origin. A sympathetic association of copper with organic matter in the sediments of the Karamana River was substantiated by its positive correlation with SOC, carbohydrates, proteins and lipids (Jayaraman, 2003) suggesting an organic association of particulate copper as reported elsewhere (Hasle and Abdullah, 1981; Shibu *et al.*, 1990; Pardo *et al.*, 1990). Hence, it is reasonable to hypothesize that copper content in the SPM of the Karamana River may be partly anthropogenic (as copper fungicides applied in the plantations and farmlands in the catchment area) and partly mineralogical in origin. Regression analysis revealed that hydrological parameters did not exert any significant influence on SPM-associated copper transport in the river.

Cadmium

Seasonal pattern of particulate flux of cadmium is depicted in Figure 4.4. Month-wise data revealed the maximum particulate load of cadmium during March (548.03 µg g^{-1}) and the minimum during December (0.12 µg g^{-1}). Season-wise analysis showed a uniform pre-monsoon > monsoon > post-monsoon pattern at all the 10 stations that clearly indicated a pre-monsoon enrichment of cadmium by adsorption onto SPM and a subsequent desorption into dissolved phase under the impact of terrigenous bioorganic leaching agents as opined by Gardiner (1974) and Duursma (1976) during monsoon and post-monsoon.

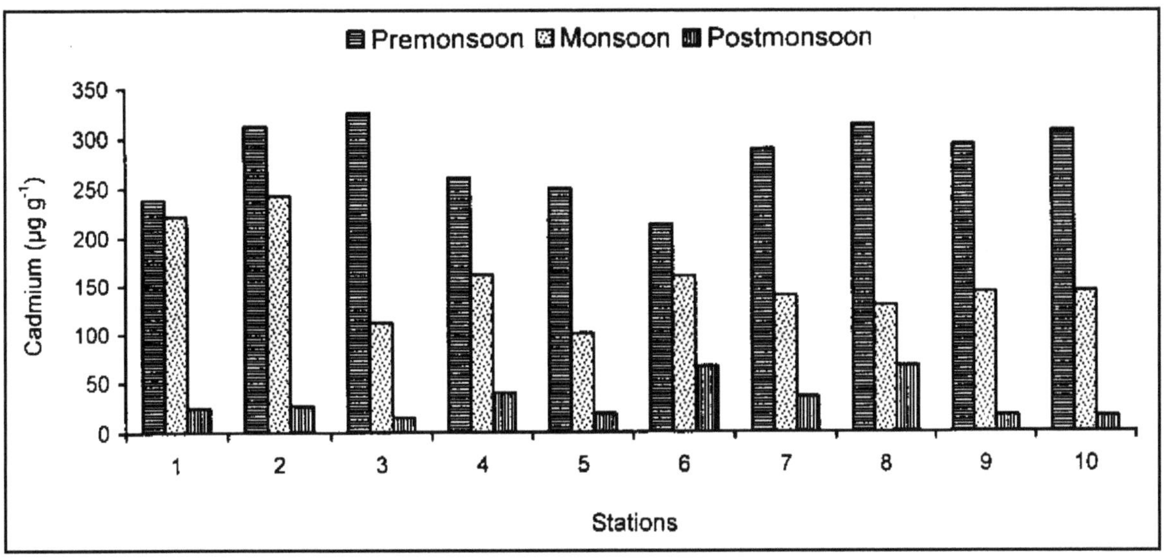

Figure 4.4: Seasonal Distribution of Particulate Cadmium

An annual average load of 155.96 µg g^{-1} with an yearly load of 3.24 metric tons of particulate flux of cadmium was observed in the present study. The annual particulate cadmium load estimated was 130 times higher than that of world river SPM, 156 times greater than average value for river particulate; approximately 9 times greater than the average value for river borne detritus, 520 times greater than the baseline level for average shale and 1591 times higher than that for upper continental crust (Table 4.2). The values were also much higher in comparison with most of the reported cases. Cadmium content in SPM was more than 15 times as high as the tentative threshold value of 10 µg g^{-1} at which the related water may exceed the highest desirable level of 10 µg L^{-1} (Table 4.3). All these findings are indicative of high level of particulate cadmium in the river. Enrichment Factor (EF) computed for particulate cadmium revealed slight enrichment at a few stations during pre-monsoon and post-monsoon, however on an annual basis there was no discernable incidence of enrichment, which is likely to be under the influence of terrigenous leachates (Table 4.4).

The distribution of cadmium in the riverine particulates showed significant month-wise ($df = 11$, $F = 90.551, p \leq$ E-46) and season-wise (ANOVA: $df = 2, F = 107.111$, $p \leq$ 1E-10) variations, but station-wise variations were not significant during all the three seasons that indicated non-point particulate cadmium input into the lotic ecosystem. Cadmium content in SPM correlated significantly, but with small coefficients to other metal elements such as iron ($r = 0.193, y = 94.673 + 0.001x, p \leq 0.04, n = 120$), manganese ($r = 0.273, y = 88.812 + 0.070x, p \leq 0.003, n = 120$), zinc ($r = 0.299, y = 36.462 + 0.193x, p \leq 0.001, n = 120$), lead ($r = 0.251, y = 122.430 + 0.072x, p \leq 0.006, n = 120$) and copper ($r = 0.420, y = 52.824 + 0.483x, p \leq$ 1.7E-06, $n = 120$) indicating very low level of affinity. It also showed significant yet weak correlation with BOD ($r = 0.322, y = 90.489 + 2.173x, p \leq 0.0003, n = 120$), which may be an indirect indication of organic association. It exhibited small coefficients of negative correlation with nitrate-nitrogen ($r = -0.309, y = 232.873 - 422.017x, p \leq 0.0006, n = 120$) and silicate-silicon ($r = -0.369, y = 323.68 - 46.068x, p \leq$ 4E-05, $n = 120$), may be due to a negative relationship with land runoff from agricultural areas well reflecting the seasonal behaviour of particulate cadmium. The absence of strong and clear correlations experienced in the present study may indicate the possibility of several types of geochemical support phases and or the chance for mixed associations that are in close agreement with that of the findings by Bertin and Bourg (1995) in Lot River system in France. Significant relationships not evident in the case of other hydrographical parameters may be due to the short residence time and fast flow rate.

Although the particulate level of cadmium was very high, the present data clearly indicated that the amount of cadmium available for biological cycling in the Karamana River has not reached any alarming level due to the very low particulate matter content in the river. But it deserves serious attention as a potential contaminant in the years to come.

Lead

The seasonal distribution profile of particulate lead in the Karamana River is presented in Figure 4.5. Data on SPM-associated lead revealed the maximum flux during March (1142.48 µg g^{-1}) followed by January (762.92 µg g^{-1}) and February (693.46 µg g^{-1}) that may indicate pre-monsoon accumulation due to phytoplankton growth. Minimum month-wise values observed during September (82.52 µg g^{-1}) and October (136.61 µg g^{-1}) may be under the impact of flood borne organic leaching agents.

The annual average rate of particulate lead transport was found to be 464.01 µg g^{-1} with an yearly flux of 9.64 metric tons that was about 13.25 times greater than that of world river SPM, 3.09 times higher than that of average river particulate, 1.66 times greater than the world average lead content for riverine detritus and 23 times greater than baseline levels in average shale, upper continental crust

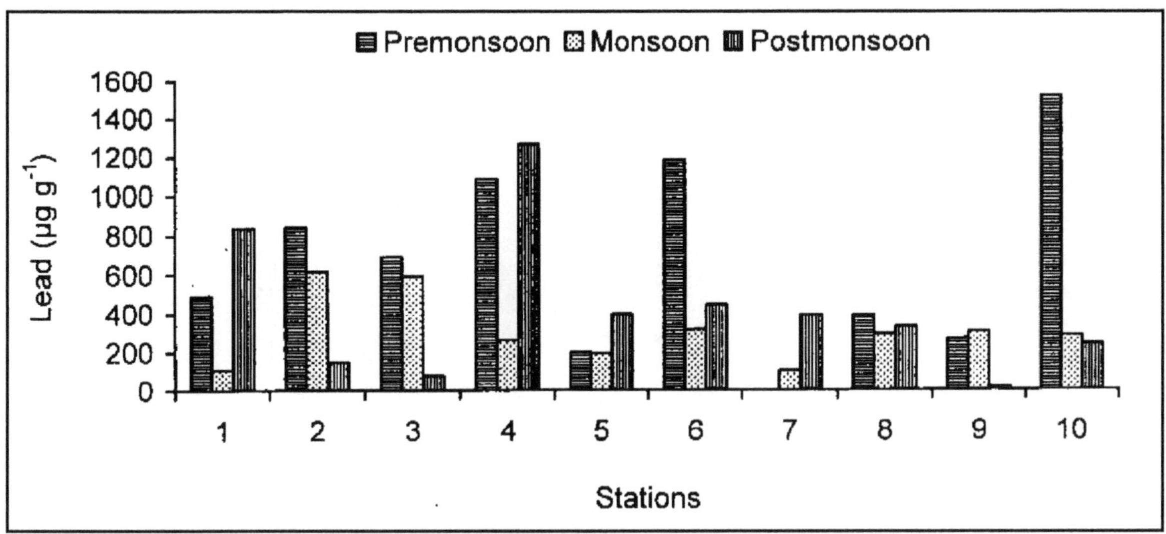

Figure 4.5: Seasonal Distribution of Particulate Lead

and average mud. It was also much higher in comparison with most of Indian and world rivers (Table 4.2). The particulate load of lead was slightly higher than that of the tentative threshold value (500 µg g^{-1}) corresponding to the highest desirable level of 50 µg L^{-1} (Table 4.3). Enrichment Factor (EF) computed for SPM-associated lead revealed mild enrichment at various stations especially during pre-monsoon and monsoon indicating the input of terrigenous particulates (Table 4.4).

Significant month-wise, season-wise and station-wise variations were not evident in the particulate transport of lead in the Karamana River suggesting non-point source, possibly air borne particulates from automobile exhausts from Thiruvananthapuram metropolitan area that is brought in by surface runoff as recorded elsewhere (Knoppers *et al.*, 1990; Bertin and Bourg, 1995). Significant, yet weak positive correlations were evident in the relationship of lead with iron ($r = 0.217, y = 223.765 + 0.004x, p \leq 0.02, n = 120$), copper ($r = 0.0.181, y = 309.036 + 0.0.728x, p \leq 0.05, n = 120$) and cadmium ($r = 0.251, y = 337.606 + 0.0.876, p \leq 0.006, n = 120$) indicating weak mineralogical affinity with these metals. The results were also indicative of involvement of several forms of geochemical support phases and/or mixed associations (Bertin and Bourg, 1995). Hydrographical parameters except nitrate nitrogen ($r = -0.238, y = 671.38–1134.53x, p \leq 0.01, n = 120$) were found to be less influential on the disposition of sedimentary lead owing to the dynamic nature of the river. Hence, it seems reasonable to suggest an anthropogenic input of SPM-associated lead into the riverine environment rather than a mineralogical origin. Transport of particulate lead from automobile exhaust through surface runoff might be a significant factor in this respect.

Although particulate lead concentration was higher than the standard levels for SPM, as the total flux of particulate matter is much less in the river water it can be concluded that lead concentration has not reached any alarming level in the river yet. However, the possibility of reaching higher levels due to anthropogenic influences in the near future can never be overlooked.

Zinc

Seasonal distribution profile of SPM associated Zinc at Stations 1 to 10 is presented in Figure 4.6. Month-wise data showed a maximum load of particulate zinc during January (1050. 60 µg g^{-1}) and the

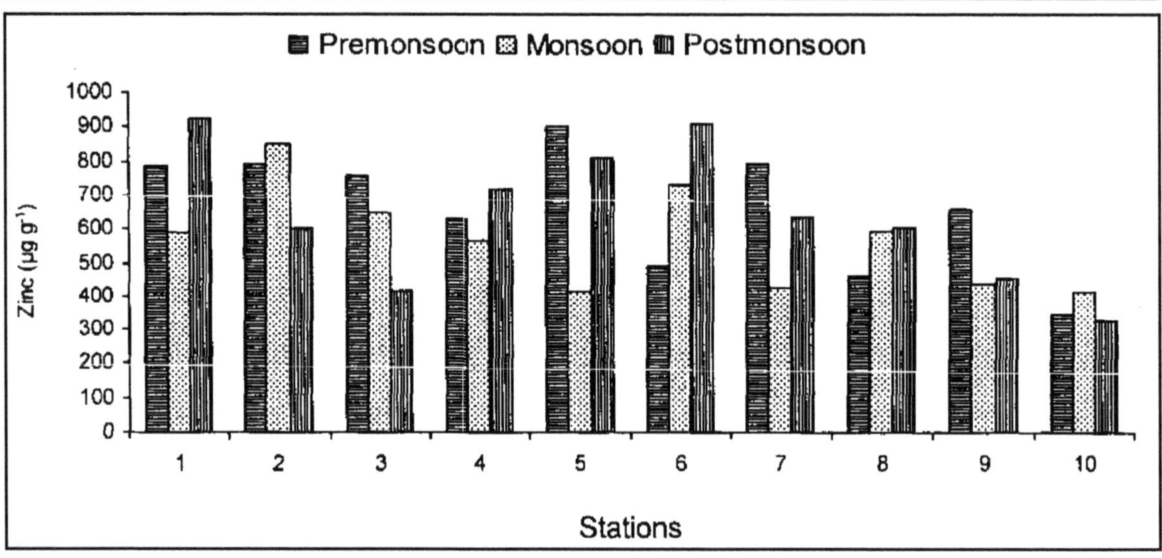

Figure 4.6: Seasonal Distribution of Particulate Zinc

minimum during October (378.43 µg g^{-1}). Particulate flux of zinc was generally low at Station 10 with a Station 9 > Station 10 pattern owing to salinity dependent desorption as suggested by Evans and Cutshall (1973).

Particulate transport of zinc in the river was found to be at a rate of 623.58 µg g^{-1} with an annual load of 12.96 metric tons. The average rate was about 2.5 times greater than that of world river SPM, 4.16 times higher than that of average river particulate, 2/3rd of the world river-borne detritus, 6.6 times greater than zinc content in average shale, 8.8 times higher than that of upper continental crust, 7.3 times greater than that of average mud, while it remained 39 times higher than the Indian average for riverine SPM. It was generally higher in comparison with most of the Indian and world rivers, but within the range of values recorded for Periyar and Muvattupuzha Rivers and Cochin estuary (Table 4.2). Besides, it was about 3.2 times lower than the tentative threshold value of 2000 µg g^{-1} at which the related water may exceed the highest desirable level of 5000 µg L^{-1} (Table 4.3). Enrichment Factor (EF) computed for SPM-associated zinc did not indicate any appreciable extent of enrichment in particulate phase as the values mainly remained below unity which is suggestive of mobilization into dissolved phase under the influence of river-borne leachates (Table 4.4).

SPM-associated zinc in the riverine system showed significant month-wise (ANOVA: $df = 11$, $F = 4.167$, $p \leq 1E\text{-}05$) and station-wise (ANOVA: $df = 9$, $F = 2.235$, $p \leq 0.03$) variations in its distribution, but seasonal variations were not significant reflecting the extent of the influence of spatial and temporal environmental variables. Regression analysis showed that particulate zinc flux in the river had significant positive correlations with iron ($r = 0.236$, $y = 510.497 + 00.002x$, $p \leq 0.0005$, $n = 120$), manganese ($r = 0.360$, $y = 489.867 + 0.139x$, $p \leq 0.0001$, $n = 120$), copper ($r = 0.439$, $y = 460.985 + 0.764x$, $p \leq 5.25E\text{-}07$, $n = 120$) and cadmium ($r = 0.289$, $y = 555.457 + 0.438x$, $p \leq 0.001$, $n = 120$), however the coefficients were low. The relationship with iron and manganese is consistent with earlier observations on the scavenging properties of iron and manganese oxides in metal ion accumulation. The relationship with copper and cadmium may suggest mineralogical association. It showed negative correlation with low coefficients to hydrographical parameters, *viz.*, COD ($r = -0.213$, $Y = 652.703 - 0.445x$,

$p \leq 0.02$, $n = 120$), nitrite-nitrogen ($r = -0.260$, $y = 675.224 - 1990.623x$, $p \leq 0.004$, $n = 120$), nitrate-nitrogen ($r = -0.268$, $y = 725.066 - 555.202x$, $p \leq 0.003$, $n = 120$) and hardness ($r = -0.209$, $y = 644.832 - 0.267x$, $p \leq 0.02$, $n = 120$) indicating their significant roles in particulate zinc transport. Other water parameters were without any significant relationships. Results of correlation studies revealed a dominant mineralogical association of particulate zinc.

Although the SPM associated zinc content in the river was generally higher than that of most of the reported cases, considering the low rate of total particulate flux in the river it can reasonably be concluded that there is no immediate threat of contamination by particulate zinc.

Manganese

Seasonal variation in SPM associated manganese in the river is depicted in Figure 4.7. Monthly data on particulate manganese content at various stations showed the highest value during April (1392.74 µg g^{-1}) and the lowest during December (374.62 µg g^{-1}).

Particulate manganese in the Karamana River was found to be mobilized at a rate of 957.13 µg g^{-1} with an annual transport of 19.89 metric tons. The observed mean value of SPM associated manganese was slightly lower than that of world river SPM and average river particulate, about 4.5 times lower than the average for world river borne detritus and higher than that of average shale, upper continental crust, average mud and average for Indian rivers. The particulate manganese load was higher than or within the range of the reported values for many of the Indian and world rivers. However, the values remained lower than that for major American rivers and Danube River (Table 4.2). Manganese content in SPM was slightly lower than the tentative threshold value of 1000 µg g^{-1} at which the related water may exceed the highest desirable level of 500 µg L^{-1} (Table 4.3). Enrichment Factor (EF) computed for manganese revealing incidence of enrichment at various stations may be due to adsorption on hydroxy coatings on sediment grains (Gibbs, 1977; Wilbur and Hunter, 1979; Tessier *et al.*, 1982; Brook and Moore, 1988) in the conducive environment during pre-monsoon. The slight enrichment recorded during monsoon and post-monsoon may be under the impact of floodwater.

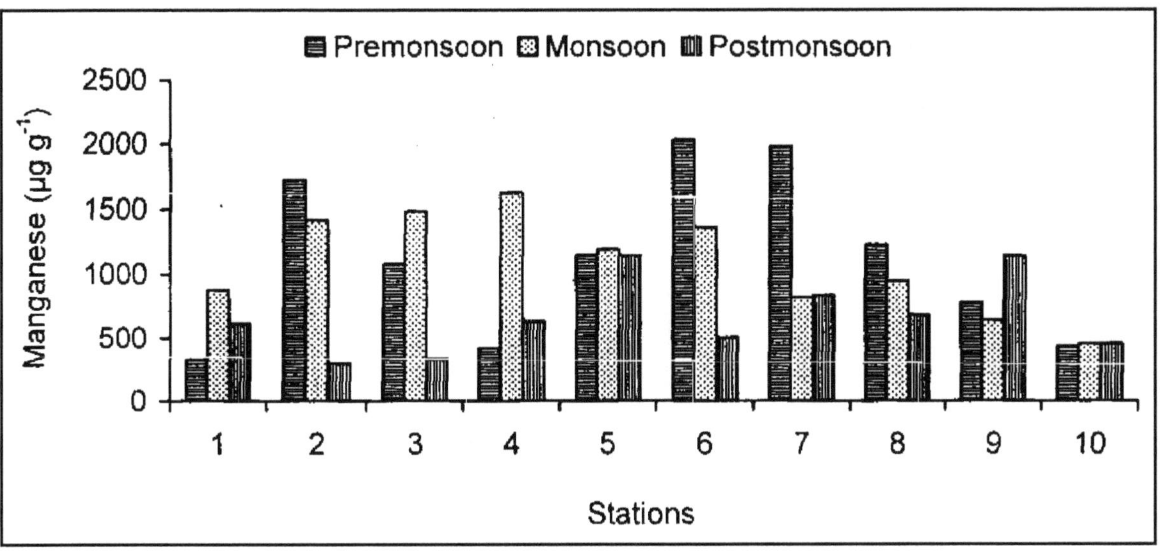

Figure 4.7: Seasonal Distribution of Particulate Manganese

In the present study, particulate manganese showed statistically significant month-wise variations (ANOVA: $df = 9$, $F = 2.427$, $P \leq 0.01$) but season-wise as well as station-wise variations were not significant reflecting the limited influence of spatial and temporal environmental variables. Particulate manganese showed significant positive correlations with low coefficients to various factors such as iron ($r = 0.262$, $Y = 634.171 + 0.006x$, $p \leq 0.004$, $n = 120$), zinc ($r = 0.360$, $y = 377.782 + 0.929x$, $p \leq 5.3\text{E-}05$, $n = 120$), cadmium ($r = 0.273$, $y = 791.52 + 0.064x$, $p \leq 0.0003$, $n = 120$), pH ($r = 0.189$, $y = -2357.86 + 482.39x$, $p \leq 0.04$, $n = 120$) and silicate-silicon ($r = -0.199$, $y = 1317.95 - 101.114x$, $p \leq 0.03$, $n = 120$). The apparent weak correlations with iron, zinc and cadmium suggested the possibility of several types of geochemical support and/or mixed associations (Bertin and Bourg, 1995). The weak relationships with pH and silicate-silicon showed at least a partial involvement of these hydrographical parameters in the mobilization of particulate manganese. The other metals and hydrographical parameters did not show any significant relationships. Results of regression analysis suggested a dominantly mineralogical association of SPM associated manganese.

Considering the very low level of SPM flux in the river, it could reasonably be concluded that for the time being there is no immediate threat of contamination by particulate manganese in the Karamana River.

Iron

Seasonal variations in SPM associated iron flux in the Karamana River is depicted in Figure 4.8. Month-wise data revealed maximum particulate iron content during March (111,136 µg g^{-1}) and minimum during February (36,375 µg g^{-1}). On an annual basis, maximum SPM-associated iron was registered at Station 3 (81,566 µg g^{-1}) followed by Stations 2 (76,945 µg g^{-1}) and 1 (67,223 µg g^{-1}), whereas the minimum was recorded at Station 9 (36,180 µg g^{-1}) and at Station 10 (40,067 µg g^{-1}) showing a decreasing trend towards downstream reflecting the increasing influence of organic leaching agents (Duursma, 1976) and salinity dependent desorption at the last two stations (Evans and Cutshall, 1973).

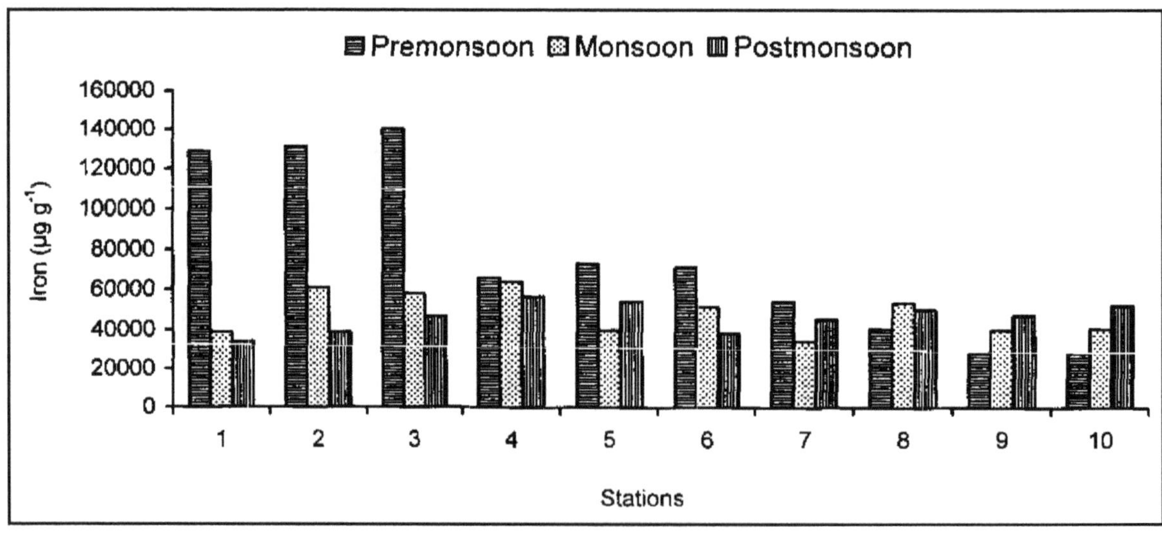

Figure 4.8: Seasonal Distribution of Particulate Iron

Particulate iron flux in the river was found to be at a rate of 56,633 µg g^{-1} (5.66 per cent) with an annual load of 1177.22 metric tons. The average rate of particulate iron flux was found higher in comparison with world river SPM, world river-borne detritus, average river particulate, upper continental crust, average mud, average shale, and the average for Indian rivers. The values generally remained higher in comparison with most of the rivers abroad, but lower than those of the rivers in the south west coast of India and River Ganga (Table 4.2). Particulate iron content in the Karamana River was slightly lower when compared to the tentative threshold value (6 per cent) at which the related water may exceed the highest desirable level of 100µg L^{-1} (Table 4. 3). Enrichment Factor (EF) computed for SPM-associated iron revealed slight enrichment at various stations during monsoon and post-monsoon, however on an annual basis there was no discernable incidence of enrichment, except at Stations 2 and 3 (Table 4.4). Terrigenous organic leachates might have played a significant role in this respect.

Statistically significant station-wise (ANOVA: $df = 9, F = 2.519, p \leq 0.01$), month-wise (ANOVA: $df = 11, F = 4.563, p \leq 1E\text{-}05$) and season-wise variations (ANOVA: $df = 2, F = 4.251, p \leq 0.03$) were evident in the distribution of SPM associated iron in the riverine environment reflecting the influence of temporal variables supplemented by local disturbances such as sand mining and illegal prospecting for gemstones. Particulate iron showed weak yet significant positive correlations with lead ($r = 0.217$, $y = 51502.51 + 11.128x, p \leq 0.02, n = 120$), cadmium ($r = 0.193, y = 51291.42 + 34.511x, p \leq 0.03, n = 120$), zinc ($r = 0.236, y = 39242.27 + 27.941x, p \leq 0.01, n = 120$) and manganese ($r = 0.262, y = 45180.74 + 11.999x, p \leq 0.004, n = 120$) indicating weak mineralogical affinity with these metals or the involvement of several forms of geochemical support phases and or mixed associations as suggested by Bertin and Bourg (1995). Among the various hydrographical parameters studied, significant relationships were evident only in the case of nitrate-nitrogen ($r = -0.207, y = 61524.5 - 18734.7x, p \leq 0.03, n = 120$). Hydrographical parameters may have only a limited influence on the disposition of SPM-associated iron owing to the dynamic nature of the river. Results of correlation studies suggested a dominantly mineralogical association of particulate iron.

Although the SPM load of iron was higher than world standards considering the very low level of SPM flux in the river, it can very well be pointed out that there is no immediate threat of contamination by particulate iron in the Karamana River.

Conclusions

The present study provides a relatively comprehensive set of measurements of the concentrations of selected particulate trace metals in the Karamana River. The flux of total SPM was very low in the river as compared to reference standards and world averages. The spatial and temporal variations noticed in the composition of SPM may be due to the influence of seasonal changes and local conditions. Co-transportation of trace elements in different combinations during different seasons was noticed. Particulate flux of all the six metals was found higher than that of standard values for average shale and Indian rivers, whereas a slightly lower value was noticed in the case of manganese in comparison with world river SPM and average river particulate, while the particulate flux of copper, zinc and manganese remained lower in comparison with world river-borne detritus. The particulate load of manganese was slightly higher and that of cadmium was much higher as compared to the tentative threshold values. Possibility of leaching of heavy metals into the lotic ecosystem originating from the plantations along with other anthropogenic activities can not be overlooked. Toxic metals, *viz.*, cadmium and lead detected in significant quantities pose great concern as they cause deleterious effects on biological systems.

Acknowledgement

The authors are thankful to the University of Kerala and Principal, University College, Thiruvananthapuram, Kerala for the facilities provided.

References

Alagarsamy, R. and Zhang, J., 2005. Comparative studies on trace metal geochemistry in Indian and Chinese rivers. *Current Sci.*, 89: 299–309.

Allen, R.J., 1979. Sediment related fluvial transmission of contaminants: Some advances in 1979. Scientific Series No. 107. *Environment Canada*, Ottawa, pp. 24.

APHA (American Public Health Association), 1992. *Standard Methods for the Examination of Water and Wastewater*, 18th edn. APHA, AWWA, WPCF, pp. 1134.

Aston, S.R. and Chester, R., 1976. Estuarine sedimentary process. In: *Estuarine Chemistry*, (Eds.) Burton, J.D. and Liss, P.S. Academic Press Inc. (London) Ltd., pp. 37–91.

Aston, S.R. and Thornton, I., 1977. Regional geochemical data in relation to seasonal variations in water quality. *Sci. Total Environ.*, 7: 247–260.

Bertin, C. and Bourg, A.C.M., 1995. Trends in the heavy metal content (Cd, Pb, Zn) in the drainage basin of smelting activities. *Wat. Res.*, 29: 1729–1736.

Biksham, G. and Subramanian, V., 1988. Elemental composition of Godavari sediments (Central and southern Indian subcontinent). *Chem. Geo.*, 70: 275–286.

Blachford, D.P. and Ongley, E.D., 1984. Biogeochemical pathways of phosphorus, heavy metals and organochlorine residues in the Bow and Oldman rivers, Alberta. Inland Water Directorate (Canada) Scientific Series No. 138, pp. 1980–81:

Borole, D.V., 1980. Radiometric and trace elemental investigations on Indian estuaries and adjacent seas. *Ph.D. Thesis* (Unpubl.), Gujarat University, pp. 155.

Borole, D.V., Sarin, M.M. and Somayajulu, B.L.K., 1982. Composition of Narmada and Tapti estuarine particles and adjacent Arabian Sea sediments. *Indian J. Mar. Sci.*, 11: 51–62.

Bowares, A.R. and Huang, C.P., 1987. Role of Fe (III) in metal complex absorption by hydrous solids. *Wat. Res.*, 21: 757–764.

Brook, E.J. and Moore, J.N., 1988. Particle size and chemical control of As, Cd, Cu, Fe, Mn, Ni, Pb and Zn in bed sediments from the Clark Fork River, Montana (U.S.A.). *Sci. Total Environ.*, 76: 247–266.

Burton, J.D., 1976. Basic properties and processes in estuarine chemistry. In: *Estuarine Chemistry*, (Eds.) Burton, J.D. and Liss, P.S. Academic Press Inc. (London) Ltd., pp. 1–36.

Canfield, D.E., 1997. The geochemistry of river particulates from the continental USA: Major elements. *Geochem. et Cosmochim. Acta*, 61: 3349–3365.

Chandrasekhar, N., 2001. Trace elements in the suspended sediments of salt marsh area of Karapad creek, Tuticorin. *Indian J. Environ. and Ecoplan.*, 5: 81–86.

Combest, K.B., 1991. Trace metals in sediment: Spatial trends and sorption processes. *Wat. Res. Bul.*, 27: 19–28.

Duursma, E.K., 1976. Radioactive tracers in estuarine chemical studies. In: *Estuarine Chemistry*, (Eds.) Burton, J.D. and Liss, P.S. Academic Press Inc. (London) Ltd., pp. 159–181.

Evans, D.W. and Cutshall, N.H., 1973. Effects of ocean water on the soluble-suspended distributions of Columbia River radionucleids. In: *Radioactive Contamination of the Marine Environment*. International Atomic Energy Agency, Vienna, pp. 125–140.

Förstner, U. and Wittman, G.T.W., 1983. *Metal Pollution in the Aquatic Environment*, 2nd Revised Edn. Springer-Verlag, Haeidelberg, pp. 486.

Gardiner, J., 1974. The chemistry of cadmium in natural water–I. A study of cadmium complex function using the cadmium specific-ion electrode. *Wat. Res.*, 8: 23–30.

Gibbs, R.J., 1973. Mechanisms of trace metal transport in rivers. *Science*, 180: 71–73.

Gibbs, R.J., 1977. Transport phases of transition metals in the Amazon and Yukon Rivers. *Geol. Soc. Am. Bull.*, 88: 829–843.

Guieu, C., Martin, J.M., Tankere, S.P.C., Mousty, F., Trincherini, P., Bazot, M. and Dai, M.H., 1998. The trace metal geochemistry in the Danube river and western Black sea. *Estuar. Coast. Shelf Sci.*, 47: 471–485.

Hart, B.T., 1982. Uptake of trace metals by sediments and suspended particulates: A review. *Hydrobiologia*, 91: 299–313.

Hasle, J.R. and Abdullah, M.I., 1981. Analytical fraction of dissolved copper, lead and cadmium in coastal seawater. *Mar. Chem.*, 10: 487–503.

Hawkes, H.E. and Webb, J.S. 1962. *Geochemistry in Mineral Exploration*. Harper and Row, New York, pp. 415.

Holeman, J.N., 1968. The sediment yield of major rivers of the world. *Wat. Resour. Res.*, 4: 737–747.

Horovitz, A., 1988. A review of physical and chemical partitioning of inorganic constituents in sediments. In: *The Role of Sediments in the Chemistry of Aquatic Systems. Proc. Sediment Chemistry Workshop*, (Eds.) Horowitz, A. J. and W.L. Bradford. US Geological Survey, pp. 969.

Horowitz, A.J., 1996. Spatial and temporal variations in suspended sediment and associated trace elements-requirement for sampling, data interpretation and determination of annual mass transport. In: *Suspended Particulate Matter in Rivers and Estuaries, Proceedings of an International Symposium*, (Eds.) Kausch, H. and Michaeils, W. Stuttgart, FRG. Schweizerbart' Sche Verlagsbuchhandlung. 47: 515–536.

Jain, C.K. and Sharma, M.K., 2006. Heavy metal transport in Hindon river basin, India. *Environ. Monit. Assmt.*, 112: 255–270.

Jayaraman, P.R., 2003. Impact of environmental factors on the macro flora of a lotic ecosystem–Karam ana River: A case study. *Ph.D. Thesis*, Department of Aquatic Biology and Fisheries, Kerala University, Thiruvananthapuram, pp. 405.

Jayaraman, P.R., Gangadevi, T. and Nayar, T.V., 2003. Water quality studies on Karamana River, Thiruvananthapuram District, South Kerala, India. *Poll. Res.*, 22: 89–100.

Karbassi, A.R., 1989. Geochemical and magnetic studies of marine, estuarine and riverine sediments near Mulki (Karnataka), India. *Ph.D. Thesis*, (Unpubl.), Mangalore University, Mangalore, pp. 196.

Karnataka Irrigation Department, 1986. River discharge data. Unpubl. No. 5. Gayuging Sub-division, Shimoga, India.

Kerala State Land Use Board, Thiruvananthapuram, 1995. *Land Resources of Kerala*, pp. 150.

Knoppers, B.A., Lacerda, L.D. and Pachineelam, S.R., 1990. Nutrients, heavy metals and organic micropollutants in an eutrophic Brazilian lagoon. *Mar. Poll. Bull.*, 21: 381–384.

Li, Y.H., Teraoka, H., Yang, T.S and Chen, J.S., 1984. The elemental composition of suspended particles from the Yellow and Yangtze rivers. *Geochim. Cosmochim. Acta*, 48: 1561–1564.

Manjunatha, B.R., Yeats, P.A., Smith, J.N., Shankar, R, Narayana, A.C. and Prakash, T.N., 1998. Accumulation of heavy metals in sediments of marine environments along the southwest coast of India. *Extended Synopses, International Symposium on Marine Pollution, Monaco*, International Atomic Energy Agency, pp. 93–94.

Martin, J.M. and Meybeck, M., 1979. Elemental mass balance of materials carried by major world rivers. *Mar. Chem.*, 7: 173–206.

Martin, J.M., Guan, D.M., Elbaz-Poulichet, F., Thomas, A.J. and Gordeev, V.V., 1993. Preliminary assessment of the distribution of some trace elements (As, Cd, Cu, Fe, Ni, Pb, and Zn) in a pristine aquatic environment: The Lina River Estuary (Russia). *Mar. Chem.*, 43: 185–199.

Martin, J.M. and Windom, H.L., 1991. Present and future roles of ocean margins in regulating marine biogeochemical cycles of trace elements. In: *Ocean Margin Processes in Global Change*, (Eds.) Mantoura, R.F.C., Martin, J.M. and Wollost, R. Wiley, New York, pp. 45–67.

Milliman, J.D. and Meade, R.H., 1983. Worldwide delivery of river sediments to the oceans. *J. Geology*, 19: 1–22.

Mohan, B.S. and Hosetti, B.B., 1997. Potential phytotoxicity of lead and cadmium to *Lemna minor* grown in sewage stabilization ponds. *Environ. Poll.*, 98(2): 233–238.

Okafor, E.C. and Opuene, K., 2006. Correlations, partitioning and bioaccumulation of trace metals between different segments of Taylor Creek, southern Nigeria. *Int. J. Env. Sci. and Tech.*, 3: 381–389.

Pardo, R., Barrado, E., Perez, L. and. Vega, M., 1990. Determination and speciation of heavy metals in the sediments of the Pisuerga River. *Water Res.*, 24: 373–379.

Paul, A.C. and Pillai, K.C., 1983. Trace metals in a tropical river environment Distribution. *Water, Air, Soil Poll.*, 19: 63–73.

Rao, K.L., 1975. *India's Water Wealth*. Oxford University Press, Delhi, pp. 262.

Ray, S.B., Mohanti, M. and Somayajulu, B.L.K., 1984. Suspended matter, major cations and dissolved silicon in the estuarine waters of the Mahanadi River. *India J. Hydrol.*, 69: 183–196.

Riedel, G.F., Williams, S.A., Riedel, S.G., Gilmour, C.C. and Sanders, J.G., 2000. Temporal and spatial patterns of trace elements in the Pantuxent River: A whole watershed approach. *Estuaries*, 23: 521–535.

Salomons, W. and Förstner, U., 1984. *Metals in the Hydrocycle*. Springer-Verlag, New York, pp. 349.

Sarin, M.M., Borole, D.V. and Krishnaswamy, S., 1979. Geochemistry of sediments from the Bay of Bengal. *Proc. Ind. Acad. Sci.*, 88: 131–154.

Shafer, M.M., Overdier, J.T., Phillips, H., Webb, D., Sullivan, J.R. and Armstrong, D.E., 1999. Trace metal levels and partitioning in Wisconsin rivers. *Water, Air and Soil Poll.*, 110: 273–311.

Shankar, R. and Manjunatha, B.R., 1997. Geochemistry and isotopic studies of riverine, estuarine and marine environments of the western India: A review. In: *Proc. First Regional GEOSAS Workshop on Quaternary Geology of South Asia*, Anna University, Madras, pp. 195–220.

Shibu, M.P., Balachand, A.N. and Nambisan, P.N.K., 1990. Trace metal speciation in a Tropical estuary-significance of environmental factors. *Sci. Total Environ.*, 97/98: 267–282.

Subramanian, V., 1979. Chemical and suspended sediment characteristics of rivers of India. *J. Hydrol.*, 44: 37–35.

Subramanian, V., Biksham, G. and Ramesh, R., 1987. Environmental geology of peninsular river basins of India. *J. Geol. Soc. Ind.*, 30: 393–401.

Subramanian, V., Van' T Dack, L. and Van Grieken, R., 1985. Chemical composition of river sediments from the Indian sub-continent. *Chem. Geol.*, 48: 271–279.

Taylor, S.R. and McLennan, S.M., 1985. *The Continental Crust: Its Composition and Evolution.* Blackwells, pp. 312.

Tessier, A., Campbell, P.G.C. and Bisson, M., 1982. Particulate trace metal speciation in stream sediments and relationships with grain size: Implications for geochemical exploration. *J. Geochem. Explor.*, 16: 77–104.

Turekian, K.K. and Wedepohl, K.H., 1961. Distribution of the elements in some major units of earth's crust. *Bull. Geol. Sve. Amer.*, 72: 175–191.

Turekian, K.K., 1969. The oceans, streams and atmosphere. In: *Handbook of Geochemistry*, Vol. 1, (Ed.) K.H. Wedepohl. Springer-Verlag, Berlin, pp. 297–323.

Verma, D.D., Rao, K.S.R., Rao, A.T. and Desari, M.R., 1993. Trace element geochemistry of clay fractions and bulk sediments from Vamsadhsra River basin, east coast of India. *Indian J. Mar. Sci.*, 22: 247–251.

Walling, D.E. and Webb, B.W., 1985. Estimating the discharges of contaminants to costal waters by rivers: Some cautionary comments. *Mar. Poll. Bull.*, 16: 488–492.

Wilbur, W.G. and Hunter, J.V., 1979. Impact of urbanization on the distribution of heavy metals in bottom sediments of Saddle River. *Water Resour. Bull.*, 15: 790–800.

Chapter 5

Minerals Identification in PVC Pipe Scale Deposits in Salem District, Tamil Nadu Using FT-IR Spectroscopy and SEM-EDX

G. Velraj and V. Velmurugan

Department of Physics, Periyar University, Salem – 636 011, Tamil Nadu

ABSTRACT

An attempt has been made in the present work to identify the minerals which causes scale deposits in Poly Vinyl Chloride (PVC) pipe using Fourier Transform Infrared spectroscopy (FT-IR) and Scanning Electron Microscopy (SEM) with Energy Dispersive X-ray fluorescence spectrometer (EDX). The scale deposits have been collected from the agricultural bore well of Chinnatirupathi, Salem District, Tamil Nadu. The FT-IR analysis has been carried out for the three layers of scale deposits. From FT-IR analysis, the presence of Barium sulphate, Dibasic phosphate, Bicarbonate, Nitrosyl halides, Calcium carbonate have been observed. The samples were also analyzed using microscopic technique with elemental analysis (SEM with EDX). Using SEM method, Calcium carbonate and Barium sulphate minerals were identified. In EDX analysis, the weight in percentage of various elements were identified and the comparative study of these elements in all the three layers of the scale deposits in the PVC pipe are also discussed.

Keywords: *Scale deposits, SEM, EDX, FT-IR.*

Introduction

When untreated water flows through tubes and pipes, calcium carbonate as well as other compounds may form on the tube walls. This process is referred to as scalling. Scaling rates depend strongly on water composition, with hard water most problematic. The precipitation and deposition of scale on equipment surfaces are influenced by various factors, including feed and recirculating water chemistry, pH, temperature, flow velocity, heat exchanger metallurgy and types of additive used in the treatment program (Zahid Amjad *et al.*, 2005). Scale occurs when inorganic salts, such as Calcium carbonate, Calcium sulphate and Barium sulphate precipitate from water to form a deposit on system surfaces.

Several mechanisms for scale formation have been hypothesized, but the main one is based on exceeding solubility limits of the scale components. The initiating setup, called nucleation, involves adsorption of the scale components onto a surface. As the concentration of ions in the bulk phase of the solution increases, the adsorption process continues until ion clusters begin to form on the surface. The ion clusters eventually grow to a critical size and become stable enough to remain on the surface, usually at imperfections or rough locations on the surface itself (Bruce Sithole, 2002).

Scale deposits that have formed beyond control by any or a combination of these mechanisms require removal by mechanical and chemical. The scale deposits in PVC pipe taken for the present work has been separated in to the three different layers and are named as outer (SDOL) middle (SDML) and inner (SDIL) layers. Many spectroscopic techniques are available to identify scale deposits and scaling species in water samples (Elizabeth Y. Anthony *et al.*, 1984). In the present study, three techniques such as FT-IR and SEM with EDX have been carried out in the samples of scale deposits in the PVC pipe.

Materials and Methods

To investigate scale deposits, the sample was collected from the groundwater flow system in bore well of Chinnatirupathi, Salem District, Tamil Nadu, India. The study area is situated in the northern part of Salem district adjoining Dharmapuri district of Tamil Nadu. It covers an area of 188 km^2 and lies between North latitudes 11°45′–12° and East longitude 78°–78°15′. The average rainfall in the area is 1147mm for the period of 10 years (1995–2004). A high quality infrared spectrum was obtained using Fourier Transform Infrared spectroscopy (FT-IR). The preparation of scale deposits specimen for infrared measurements is the alkali halide (KBr) pelleting technique. The sample is powdered and finely grind using agate mortar, and the powdered sample is mixed with KBr at the ratio 1:20 by weight and pressed into a pellet. The FT-IR spectra of the three samples were recorded on a BRUKER, IFS 66V spectrometer.

In SEM with EDX technique, samples have to be prepared carefully to withstand the vacuum inside the microscope. SEM samples are coated with a very thin layer of gold by a machine called a sputter coater. The sample is placed inside the vacuum column through an air tight door. Finally the sample scan position with the resulting signal, an image can be formed. The samples should be examined with a stereomicroscope, overall homogeneity and any material adhering to the sample. The choice of the specific method for sample preparation depends on the size, nature and condition of the specimen, as well as the analytical request. It may be necessary to use multiple preparation methods in order to analyze all sample characteristics.

Results and Discussion

The sample of scale deposit consists of the three different layers named as SDOL, SDIL and SDML. The FT-IR spectra of the sample were presented in Figure 5.1 and their tentative vibrational assignments are given in the Table 5.1. Using the FT-IR technique, the following tentative assignments were carried out.

Table 5.1: FT-IR Absorption Frequencies and Relative Intensities of Scale Deposit Samples with Tentative Vibrational Assignments

SDOL		*SDML*		*SDIL*		*Tentative Vibrational Assignments*
Frequency (cm^{-1})	*Relative Intensity*	*Frequency (cm^{-1})*	*Relative Intensity*	*Frequency (cm^{-1})*	*Relative Intensity*	
3413	VW	3428	VW	3425	VW	Barium Sulphate
2876	VW	–	–	–	–	Dibasic Phosphate
2516	VW	2517	VW	2517	VW	Dicarbonate
1799	W	–	–	–	–	Nitrosyl Halides
1423	VS	1423	VS	1423	VS	Calcite material
875	S	875	S	875	S	Calcite material
713	W	714	VW	714	VW	Calcite material

VW: Very weak; W: Weak; S: Strong; VS: Very strong.

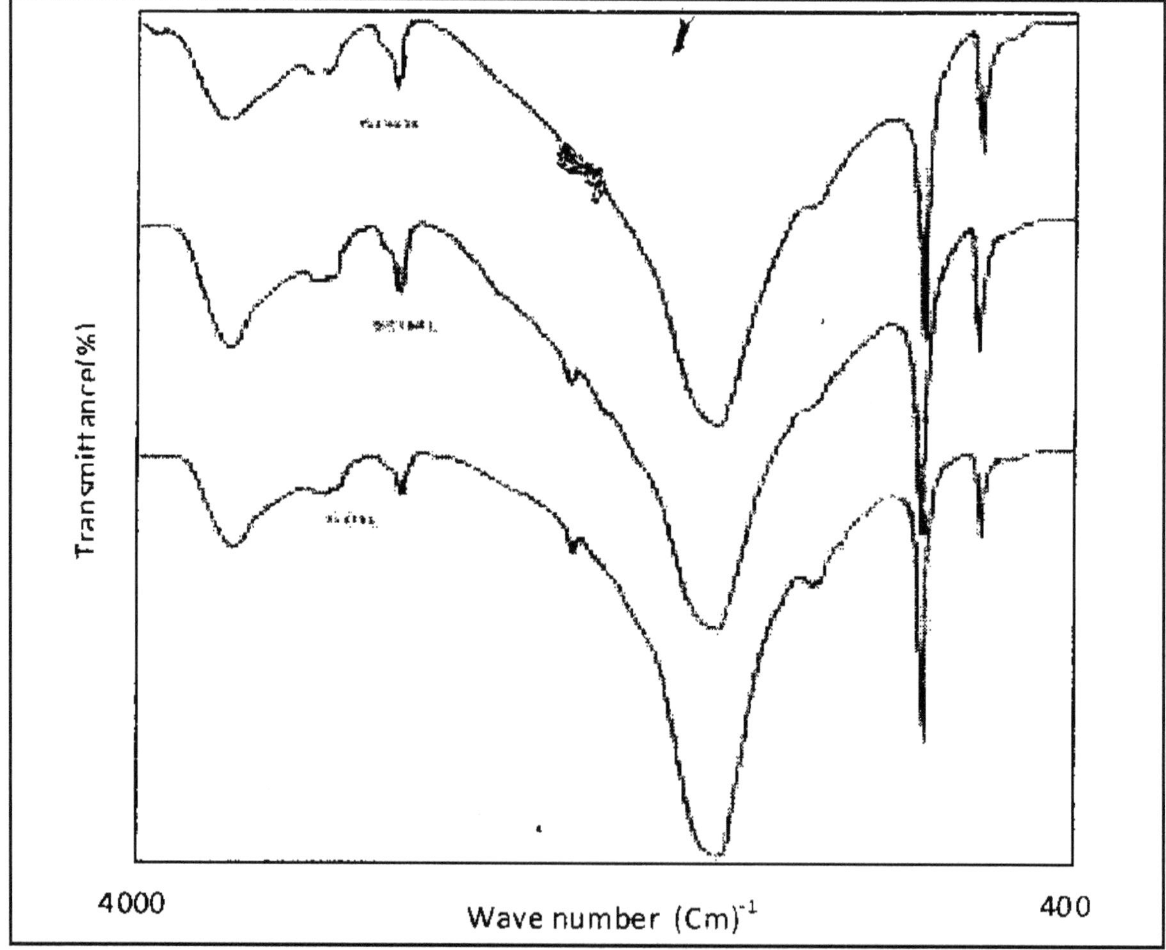

Figure 5.1: FT-IR Spectra of Three Layers in Scale Deposits

FT-IR Spectral Analysis

According to (George Socrates, 2001), a strong broad band in the region 1495–1410 cm^{-1}, a medium to strong band in the region 885–870 cm^{-1} and weak band at 715 cm^{-1} have been suggested as due to the presence of the functional group calcium carbonate. In a similar way, (Kuriyavur *et al.*, 2000) have suggested a strong band at 1440, 873 and 710 cm^{-1} is due to the presence of Calcium carbonate. According to them, a very strong band of absorption around 1423 cm^{-1}, a strong band of adsorption at 875 cm^{-1} and a very weak adsorption band around 713 cm^{-1} indicates Calcium carbonate present in the sample taken for the present study.

A strong band in the region 1850–1790 cm^{-1} is assigned to the functional group of Nitrosyl halide (George Socrates, 2001). According to him a weak adsorption band at 1799 cm^{-1} in SDOL layer indicates the presence of Nitrosyl halide. According to (George Socrates, 2001), a broad band in the region 3300–2000 cm^{-1} have been assigned to bicarbonate. A well ordered bicarbonate have been identified in all the three layers namely SDOL, SDML and SDIL taken for the present work with a very weak adsorption band at 2516, 2517 and 2517 cm^{-1} respectively. A weak with broadband in the region 2900–2750 cm^{-1} has been assigned to Dibasic phosphate by (George Socrates, 2001) in the scale deposits. A very weak band at 2876 cm^{-1} in SDOL layer may be due to the presence of Disbasic phosphate.

The medium with broadband near 3430 cm^{-1} had been reported as due to the presence of Barium sulphate (George Socrates, 2001). A very weak adsorption band at 3413, 3428 and 3425 cm^{-1} in the layers namely SDOL, SDML and SDIL respectively indicates the presence of barium sulphate in all the three layers of the sample taken for the work.

From the FT-IR spectra of the three layers of PVC pipe namely SDOL, SDML and SDIL recorded in the mid IR region one can observe the presence of the complexes Barium sulphate, Dibasic phosphat ebicarbonate, Nitrosyl halides and Calcium carbonate. Among these three layers the outermost layer contains more complex salts formation since it deposited at the beginning. If goes on into the middle and inner layers which deposited later having other remaining complexes. Inorder to study the quantitative analysis of these three layers an attempt has been made to analyze the semi quantification of minerals present in the samples by SEM with EDX.

SEM Analysis

Using the SEM method, morphology of minerals were identified. Plates 5.1 to 5.3 represents the scanning electron microscope photographs, and corresponding energy dispersive spectra of minerals of the scale deposits are given in the Figures 5.2–5.4. SEM and image analysis can be used to define particle shape. Particle shape can be documented with a microphotograph. The image analysis unit has the capability to generate an array of chords across the particle to arithmetically define the shape of the particle. (Arumugam, 2000) has suggested that the orthorhombic shape of the crystal is due to barium sulphate present in the mineral. According to (Sakthivel *et al.*, 2006) the cubic or entangled rhombohedra shape is due to the presence of calcium carbonate in the sample. The presence of calcium carbonate mineral in the three layers SDOL, SDML and SDIL have been identified from the microphotographs plates numbered 5.1, 5.2 and 5.3 respectively and Barium sulphate, which is orthorhombic in crystal shape is also identified from the morphology of the layers SDML and SDIL corresponding to the plates numbered 5.2 and 5.3.

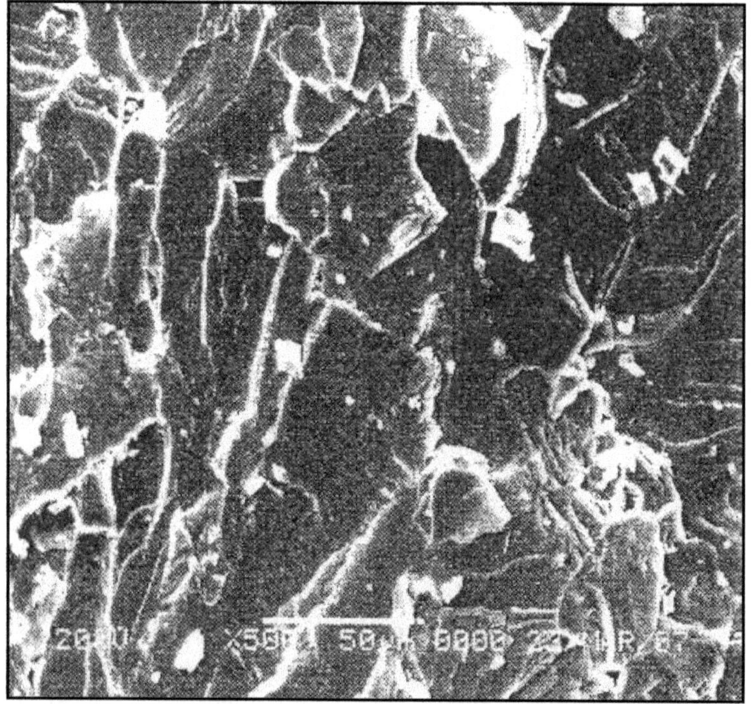

Plate 5.1: Microphotograph of the Sample SDOL Using SEM

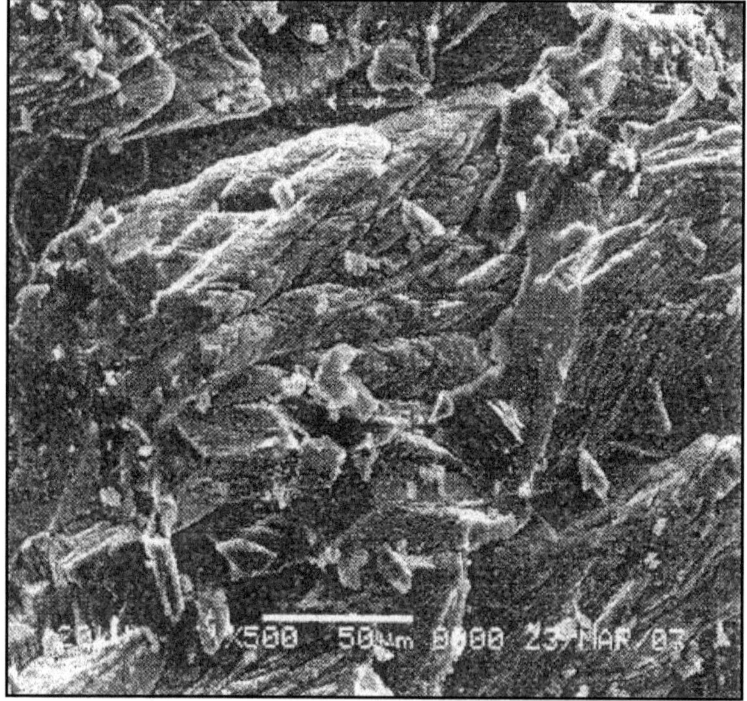

Plate 5.2: Microphotograph of the Sample SDML Using SEM

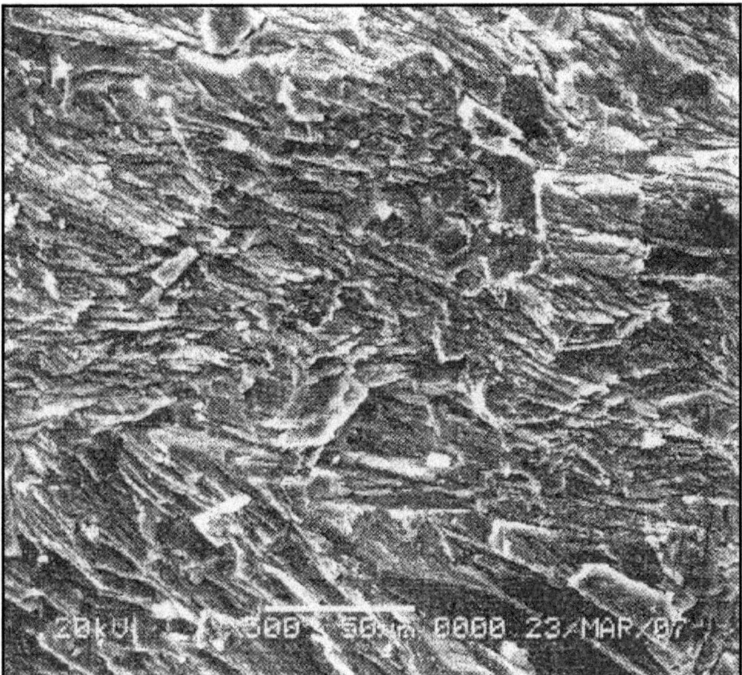

Plate 5.3: Microphotograph of the Sample SDIL Using SEM

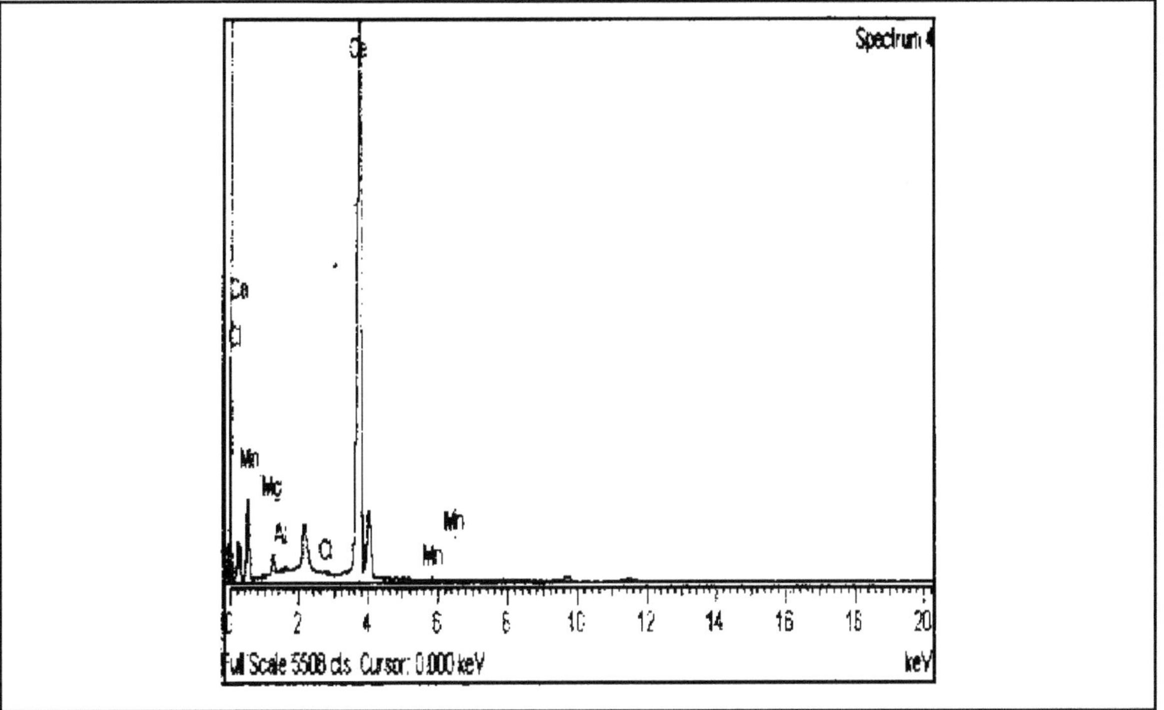

Figure 5.2: The EDX Spectrum of the Sample SDOL

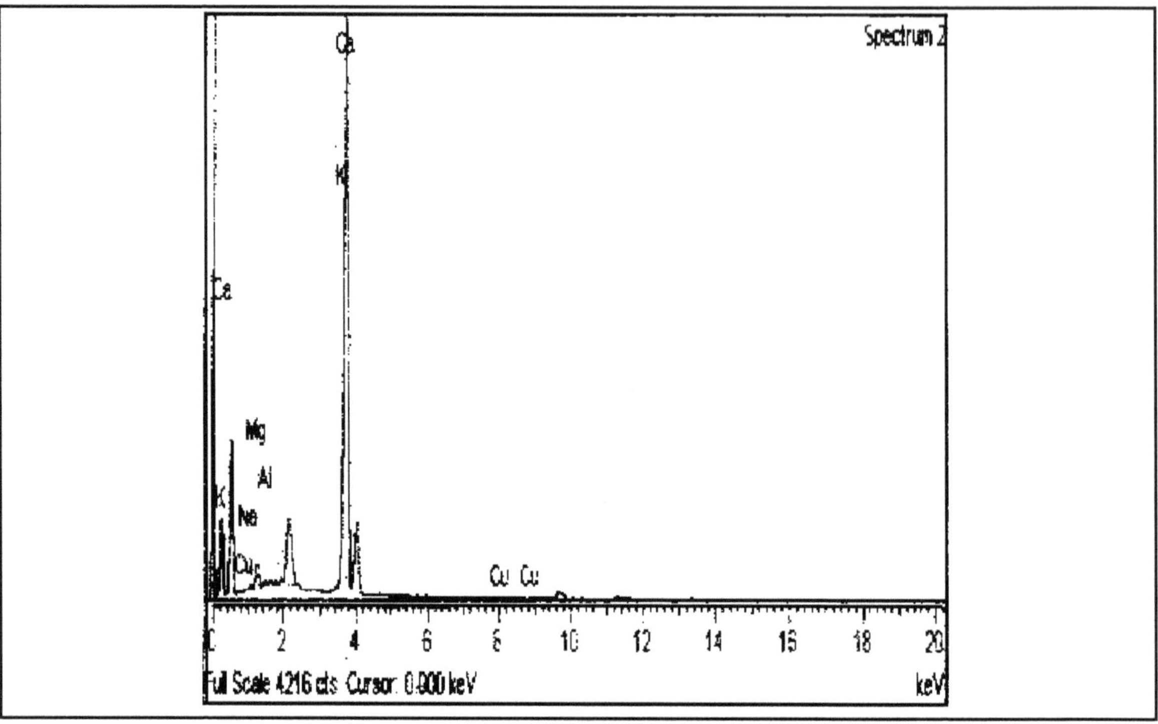

Figure 5.3: The EDX Spectrum of the Sample SDML

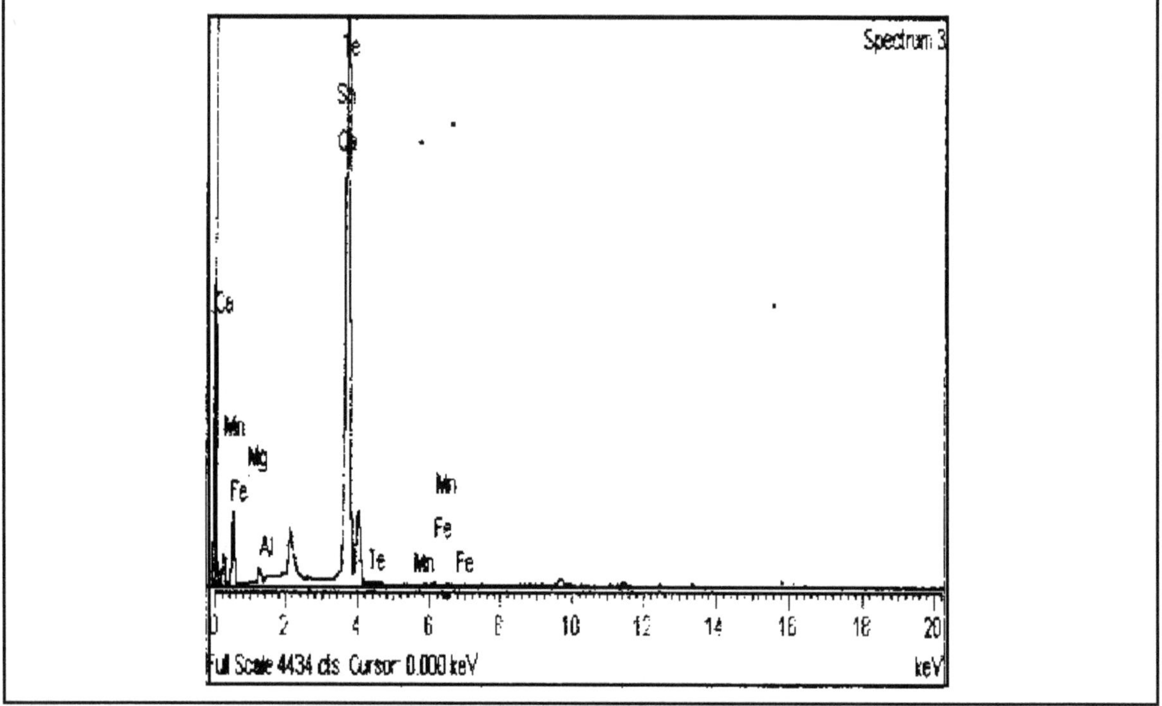

Figure 5.4: The EDX Spectrum of the Sample SDIL

EDX Analysis

From the EDX analysis, the weight in percentage of various elements were identified in the sample of scale deposits and are tabulated in Table 5.2 and the relative concentration of calcium, magnesium and aluminum which are the major three elements present in all the three layers of scale deposits graphically represented in the Figures 5.5–5.7 respectively. From the EDX analysis one can infer that the samples contains the elements of sodium (Na), magnesium (Mg), aluminum (Al), calcium (Ca), copper (Cu), chlorine (Cl), manganese (Mn), iron (Fe), tin (Sn), and tellurium (Te) in different percentages. From the Table 5.2, the elements sodium and copper are present in the SDML layer only and their percentage are 0.18, 0.34 respectively. The percentage of magnesium is 3.19, 4.26 and 2.32 in all the three layers SDOL, SDML and SDIL respectively. The aluminum and calcium are present in all the three layers (SDOL, SDML and SDIL) and the percentage of aluminum is 0.17, 0.75 and 0.53 and that of calcium is 95.99, 94.46 and 89.48 respectively. The element chlorine is present in the SDOL layer in percentage of 0.07. Manganese element is present in SDOL and SDIL layers and their percentages are 0.04 and 0.03 respectively. The percentage of iron, tin and tellurium are 0.16, 2.33 and 5.15 and it is present only in the SDIL layer. One can observe from the above analysis that the element calcium is in more percentage when compared to the other elements and its percentage of composition is above 89 per cent.

Tao Chen *et al.* (2005) suggested that, scaling of metallic or insulating walls in contact with water supersaturated with respect to calcium carbonate causes scale deposits. The elements of magnesium,

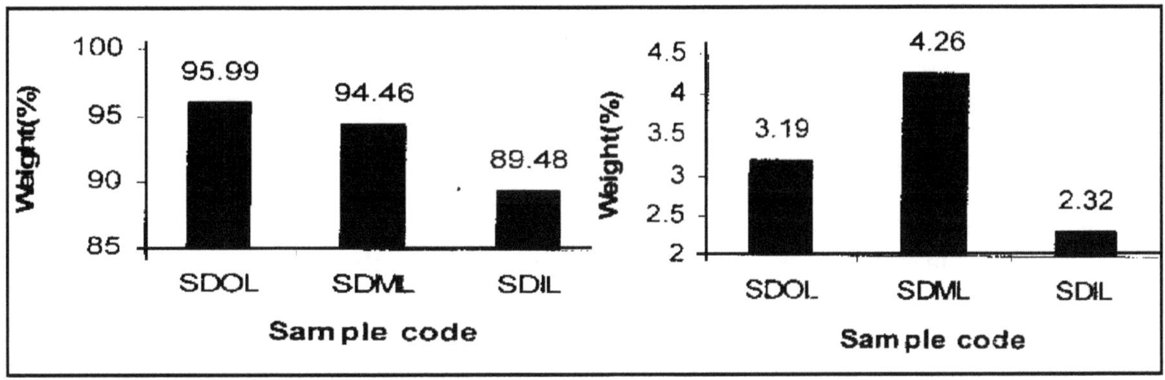

Figure 5.5: Relative Concentration of Calcium Present in the Samples

Figure 5.6: Relative Concentration of Magnesium Present in the Samples

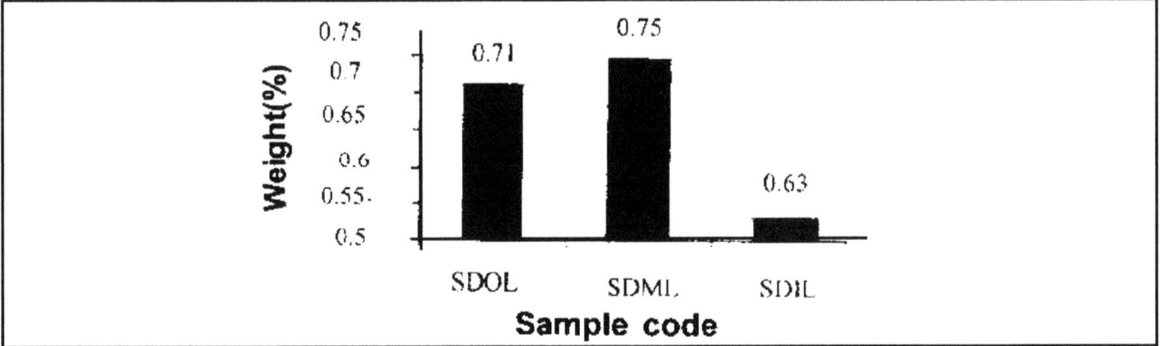

Figure 5.7: Relative Concentration of Aluminium Present in the Samples

tin and tellurium are present below the level of 6 per cent. The other elements of sodium, aluminum, copper, chlorine, manganese and iron are presents in small amount (1 per cent) of the sample. From the Figures 5.5 to 5.7 and Table 5.2 one can infer that the concentration of calcium is exponentially decreasing from the outer (SDOL) layer to the inner (SDIL) layer. But in the case of magnesium and aluminium greatest concentration is in SDML layer, greater in SDOL layer and least in SDIL layer.

Table 5.2: Elemental Concentration of Scale Deposit Sample

Elements	*SDOL Weight (%)*	*SDML Weight (%)*	*SDIL Weight (%)*
Na	–	0.18	–
Mg	3.19	4.26	2.32
Al	0.71	0.75	0.53
Ca	95.99	94.46	89.48
Cu	–	0.34	–
Cl	0.07	–	–
Mn	0.04	–	0.03
Fe	–	–	0.16
Sn	–	–	2.33
Te	–	–	5.15

Conclusion

From the present work, the following conclusions can be drawn. From FT-IR analysis which have been carried out for all the three layers of the scale deposits, one can observe the presence of Barium sulphate, Dibasic phosphate, Bicarbonate, Nitrosyl halides, Calcium carbonate. Using the SEM analysis, Calcium carbonate and Barium sulphate minerals were identified. In EDX analysis, the elements calcium, aluminum and magnesium are present in all the three layers with various levels. But calcium element is present in more present percentage when compared to other elements. FT-IR results also reflect with SEM with EDS analysis. Therefore it is concluded that the presence of Calcium carbonate is the major factor for the formation of scale deposits in the PVC pipe. PVC pipe taken from the study area has been affected by scale deposits due to the presence of more percentage of Calcium carbonate.

Acknowledgement

The authors acknowledge with thanks to the Director, SAIF, IIT Madras, Chennai, for his help to record the FT-IR spectra in time and the staff-in-charge CSIL Lab, Annamalai University, Annamalainagar helped to record SEM with EDS to the samples.

References

Arumugam, M., 2000. *Material Science*. Anuradha Agencies, Chennai, p. 43.

Bruce Sithole, 2002. Scale deposit problems in pulp and paper mills. *African Pulp and Paper Week.*

Chen, Tao, Neville, Anne and Yuan, Mingdong, 2005. Calcium carbonate scale formation assessing the initial stages of precipitation and deposition. *Journal of Petroleum Science and Engineering*, 46: 185–194.

Elizabeth, Y., Anthony, James T., Reynolds and Beane, Richard E., 1984. Identification of daughter minerals in fluid inclusions using scanning electron microscopy and energy dispersive analysis. *American Mineralogist*, 69: 1053–1057.

Kuriyavar, Satish I., Vetrivel, Rajappan, Hedge, Sooryakant G., Arumugamangalam, V. Ramaswamy, Chakrabarty, Debojit and Mahapatra, Samiran, 2000. Insights into the formation of hydroxyl ions in calcium carbonate: Temperature dependent FTIR and molecular modelling studies. *Journal of Material Chemistry*, 10: 1835–1840.

Shakthivel, P. and Vasudevan, T., 2006. Acrylic acid-diphenylamine sulphonic acid copolymer threshold inhibitor for sulphate and carbonate scales in cooling water system. *Desalination* 197: 179–189.

Socrates, George, 2001. *Infrared and Raman Characteristic Group Frequencies*, 3rd edn. John Wiley and Sons Ltd., New York, p. 276 –291.

Zahid, Amjad and Zuhl, Robert W., Noveon, Inc. and John F. Zibrida, Zibex, Inc., 2005. *Factors Influencing the Precipitation of Calcium–Inhibitor Salts in Industrial Water Treatment Industry*, 12.

Chapter 6

Hydrodynamic Investigations in Hybrid Rotating and Reciprocating Perforated Plate Bubble Column

S. Dhanasekaran* and V. Vijayagopal

Mass Transfer Laboratory, Department of Chemical Engineering, FEAT, Annamalai University, Annamalai Nagar – 608 002, Tamil Nadu

ABSTRACT

This chapter reports an experimental and theoretical investigation carried out to evaluate the gas holdup, bubble diameter and interfacial area for a gas-liquid mixture in a novel Hybrid Rotating and Reciprocating Perforated Plate Bubble Column. Counter current air-water system is used in this study. The response of this hybrid column is found to be similar to that of reciprocating plate column showing mixer-settler, transition and emulsion regions. This chapter results the effect of agitation level, superficial gas velocity, superficial liquid velocity, perforation diameter and plate spacing on gas holdup. The gas holdup is found to be least at an agitation level of 1.5 cm/s in the transition region between mixer-settler and emulsion regions. The size of the gas bubbles is found to have a direct relation with the agitation level. Effect of agitation level and superficial gas velocity on interfacial area are also analyzed and found to be significant. Correlations have been developed to predict the gas holdup, bubble diameter and interfacial area and found to be significant from the statistical analysis. The values predicted by the developed correlations concur with the experimental values.

Keywords: *Hybrid rotating and reciprocating perforated plate bubble column, Reciprocating plate column, Gas-liquid contact, Gas holdup, Bubble diameter, Interfacial area.*

---

* Corresponding Author; E-mail: sudha040506@yahoo.co.in.

Introduction

In order to accelerate mass transfer between gas and liquid phases, reactors in which the gas is dispersed as fine bubbles into the liquid are often used. Such reactors are particularly important in the process industries for absorption and fermentation. The absorption of sparingly soluble gases as in aerobic fermentations is commonly carried out in stirred tank reactors, bubble columns and mechanically agitated gas-liquid reactors. The stirred tank reactors (Archis *et al.*, 2002) normally provide highly efficient contact by means of an external agitation but a single such reactor normally provides no more than one equilibrium stage. Bubble columns operated counter currently offer the possibility of multistage performance. An additional advantage of a tall bubble column is the increased solubility of gas at the base of the column (liquid exit). This is due to the higher hydrostatic pressure in accordance with Henry's law. Axial mixing in the liquid phase of the bubble columns generally can not be ignored; the axial dispersion coefficients in the absence of internals increase approximately as the 4/3 power of column diameter (Baird *et al.*, 1975). Thus as the scale of the column is increased, the advantages of the counter current operation (relative to a stirred tank absorber) is reduced (Ramo Rao *et al.*, 1988). Axial mixing in bubble columns is largely due to the radial non-uniformity in two phase flow pattern. It can be reduced by careful vertical alignment of the column (Tinge *et al.*, 1986) by inserting baffles, packing or partitions and by agitation whether rotary or reciprocatary. The insertion of baffles and the provision of agitation also have additional beneficial effects of increasing the interfacial area. It is due to increased holdup and decreased bubble size (Mersmann *et al.*, 1988; Maruoka *et al.*, 1989; Midoux *et al.*, 1984, and Brauer, 1988). If an absorption process is accompanied by a fast chemical reaction, the available interfacial area per unit volume will be the main design parameter since it is much more affected directly by the gas holdup and agitation level (Veljkovic *et al.*, 1988).

In order to increase the overall mass transfer rate between the phases, it is necessary to increase the liquid film mass transfer coefficient and/or interfacial area. This is connected with extra cost usually as external energy input. It is advisable that an increase in the mass transfer rate can be achieved in more suitable way by increasing interfacial area rather than liquid film mass transfer coefficient. An advancement of bubble column is Karr's Reciprocating Plate Column (KRPC). It is a type of multiphase reactor in which agitation is provided by a set of perforated plates moving up and down through a central shaft in the column. By introducing a small amount of external energy through the reciprocating plates, uniform gas dispersion is created and a frequent renewal of the interfacial area is resulted along with increased interfacial mass transfer. The mam advantages of this reciprocating plate column are large gas holdup (Mustapha *et al.*, 1993), nearly uniform bubble size, large interfacial area (Yang *et al.*, 1986) and reduced axial mixing. Although the studies on reciprocating plate columns (Mustapha *et al.*, 1993; Brauer *et al.*, 1979; Ramo Rao *et al.*, 1983, 1986, and 1988 Veljkovic *et al.*, 1983, 1989; Yang, 1986) and its application are quite abundant as a gas-liquid contactor, there has been considerable general interest in the effects of rotary and reciprocatary flow on gas-liquid mass transfer.

The novelty of this work lies in combining the effects of stirred tank reactor and the reciprocating plate column. This resulted in Hybrid Rotating and Reciprocating Perforated Plate Bubble Column. At first it is necessary to study the hydrodynamics of this column. This knowledge is very important for the mechanical design of this contacting system and for the calculation of the energy necessary for the desired level of agitation. The objective of the present investigation is to study the hydrodynamic characteristics namely, gas holdup, bubble diameter and interfacial area for this hybrid column.

Experimental

Setup

The experimental setup and its ancillary connections are shown schematically in Figure 6.1. The column is 10.0 cm in internal diameter and 120 cm in height flanged acrylic column. The plate stack height is 100 cm, extending from a position of 10 cm above the base of the column. The plate is of 0.2 cm thickness plastic sheet with circular perforation. The plates are mounted on a 1.0 cm diameter acrylic drive shaft with equal spacing. Copper sleeves are used to the arrangement the plate spacing. The plate stack is driven in vertical and horizontal hybrid motion by the bevel gear arrangement, attached

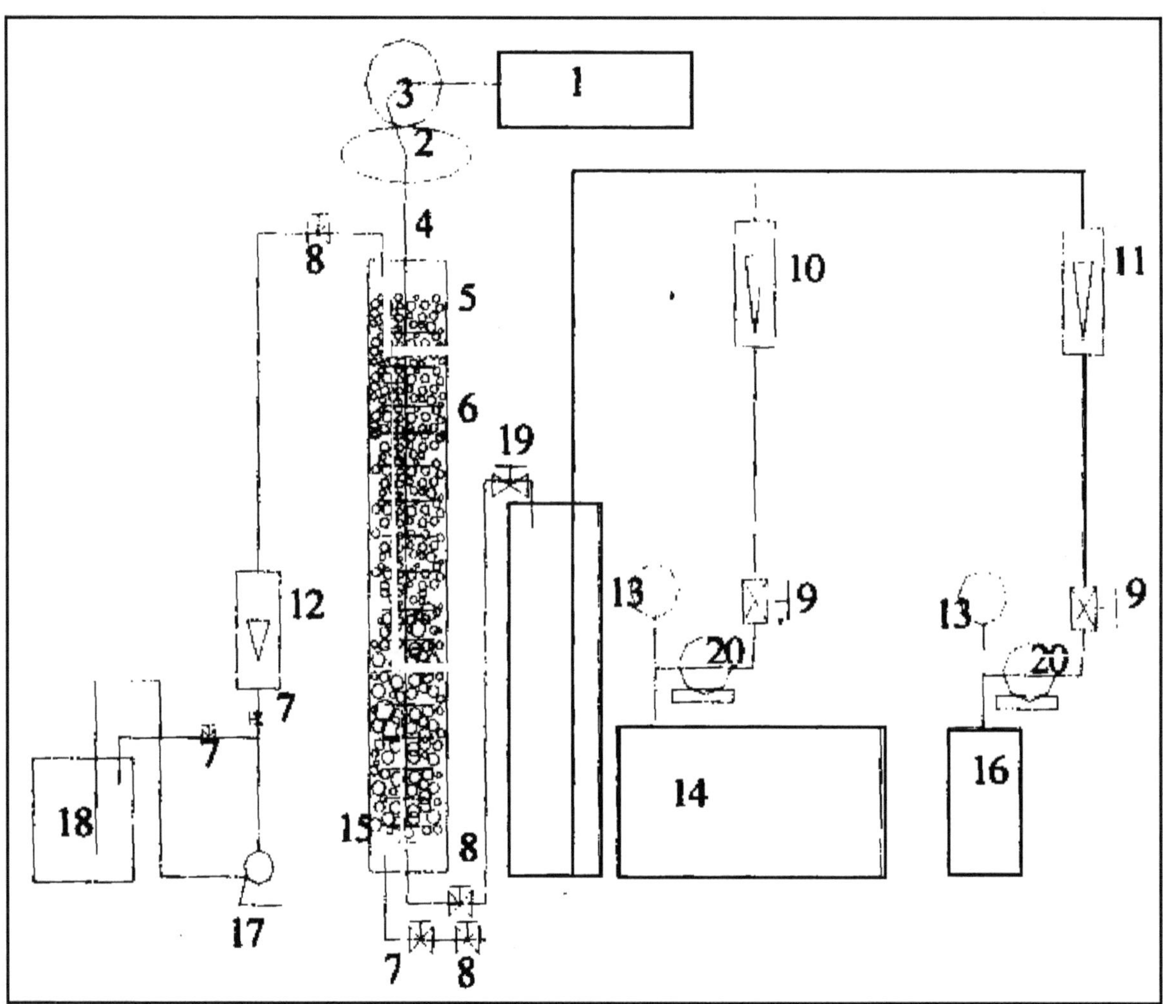

Figure 6.1: Experimental Setup: Hybrid Rotating and Reciprocating Perforated Plate Bubble Column

1: DC motor; 2: Bevel gear arrangement; 3: Connecting rod; 4: Central rod; 5:Column; 6: Perforated plates; 7: Globe valve; 8: Solenoid valve; 9: Needle valve; 10: Air Rota meter; 11: Gas Rota meter; 12: Water Rota meter; 13: Pressure gauge; 14: Air compressor; 15: Air distributor; 16: Gas storage tank; 17: Liquid pump; 18: Water recirculation tank; 19: Non return valve; 20: Pressure regulator.

to a variable speed DC motor M. A frequency controller FC is used to control the speed. The amplitude A (Half of the full reciprocation stroke) could be adjusted up to 1 cm. The drive frequency could be maintained up to 3.5 sec^{-1}. The water is supplied from a tank to the top of the plate stack through a globe valve and rotameter. Two solenoid valves in the water line one at the inlet and the other at the outlet permits shut-off during holdup measurements. A globe valve permits adjustment of the water outlet to control the interface level IL at the top of the column. Air is fed to the column from the air compressor through the pressure regulator, surge tank and rotameter. The surge tank is used to avoid the pressure fluctuations. The air line is also provided with a solenoid valve which enables to measure holdups. Air leaves the top of the column at atmospheric pressure.

Procedure

Average gas holdup is measured by running the column in steady state under counter current condition at the desired settings of flow rates and agitation level. Then the three solenoid valves water inlet, water outlet and air inlet are simultaneously shut-off. The drop in interface level was measured and is used to calculate the volume holdup. The column section is being photographed. The mean bubble sizes were obtained from photographs and used to calculate the effective interfacial area.

Result and Discussion

Gas Holdup

Gas holdup is an important parameter for the design and the selection of the operation conditions of the column. In addition, gas holdup has a direct effect on the overall mass transfer coefficient and on the power consumption. The fractional gas holdup is measured by noting the dispersion heights in the presence and absence of gas as per the following equation (1)

$$\varepsilon_G = \frac{H - H_o}{H} \tag{1}$$

Effect of Agitation Level on Gas Holdup

Figures 6.2a to 6.2e results the effect of agitation level Af (product of amplitude and plate stack frequency) on gas holdup for two phase flow condition used in this investigation. For a given plate spacing and superficial velocities of each phases, holdup decreases slowly up to a level of 1.5 cm/s. Beyond this level, holdup increases significantly with increasing agitation level. In reciprocating plate bubble column, Ramo Rao *et al.* (1988) reported that gas holdup increases slowly with agitation till a level of about 6 cm/s is reached; beyond this level, holdup increases significantly with agitation. Miyanami *et al.* (1978) reported that the gas holdup hardly increases until Af exceeds a certain vibrating speed Af_c (about 2 cm/s) for the air-water system, and then at higher speeds the holdup increases almost linearly with Af. In reciprocating plate column for air water system, Ramo Rao *et al.* (1983) reported that at low vibrating speeds the dispersed phase holdup decreased with an increase in the vibrating speeding in the mixer-settler (up to 6 cm/s) and reached a minimum in the transition region (about 6 cm/s); thereafter holdup increased with an increase in the vibrating speed (above 6 cm/s) in emulsion region. Mustapha Lounes *et al.* (1993) reported that the gas holdup decreases up to the agitation level of approximately 5 to 7 cm/s in the reciprocating plate gas liquid column. The same was reported by Veljkovic *et al.* (1986) in reciprocating plate columns for gas-liquid contacting system. In the present study the transition occurred at an agitation level of 1.5 cm/s as compared to 6 cm/s which was reported by Ramo Rao *et al.* (1983) and then increases. This value ensures that the attainment

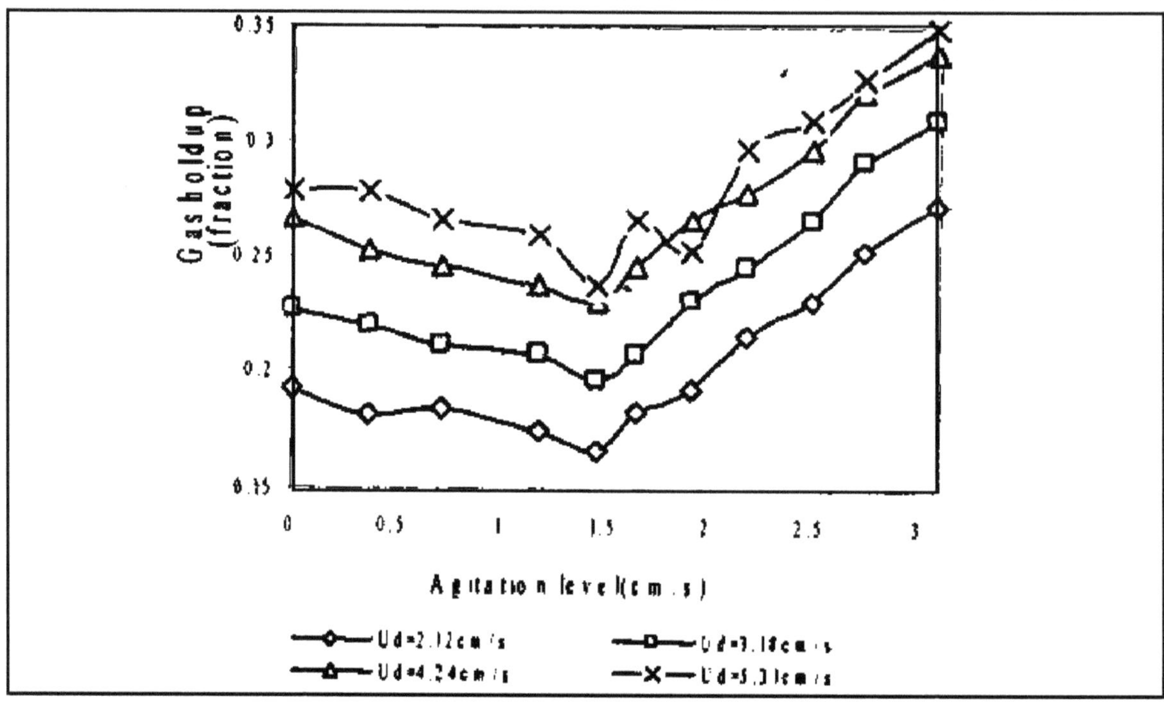

Figure 6.2a: Effect of Agitation Level on Gas Holdup
U_c: 0.14cm/s; d_p: 0.8cm; S_p: 4.5cm; n_p: 10.

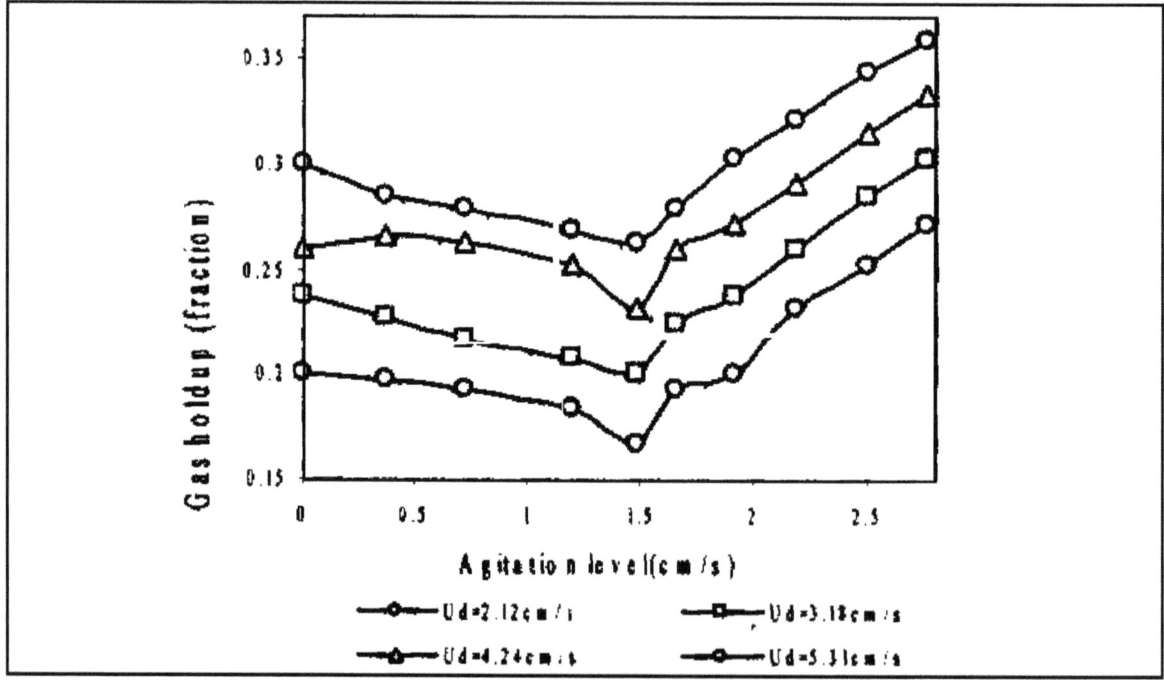

Figure 6.2b: Effect of Agitation Level on Gas Holdup
U_c: 0.18cm/s; d_p: 0.8cm; S_p: 4.5cm; n_p: 10.

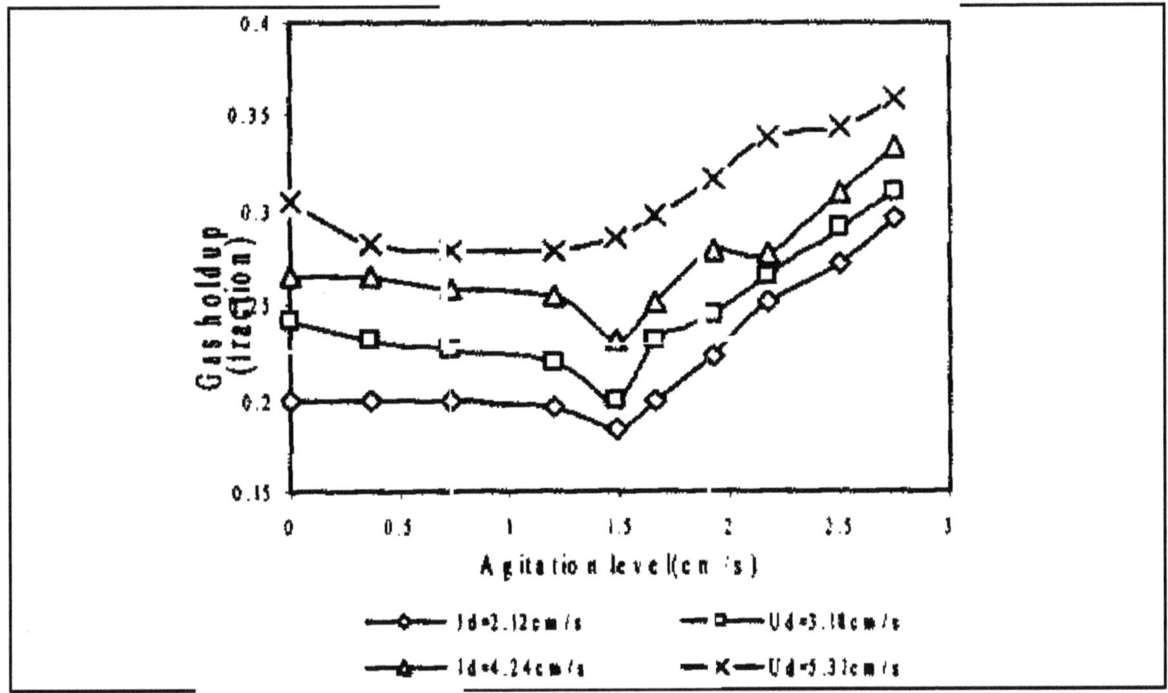

Figure 6.2c: Effect of Agitation Level on Gas Holdup

U_c: 0.21cm/s; d_p: 0.8cm; S_p: 4.5cm; n_p: 10.

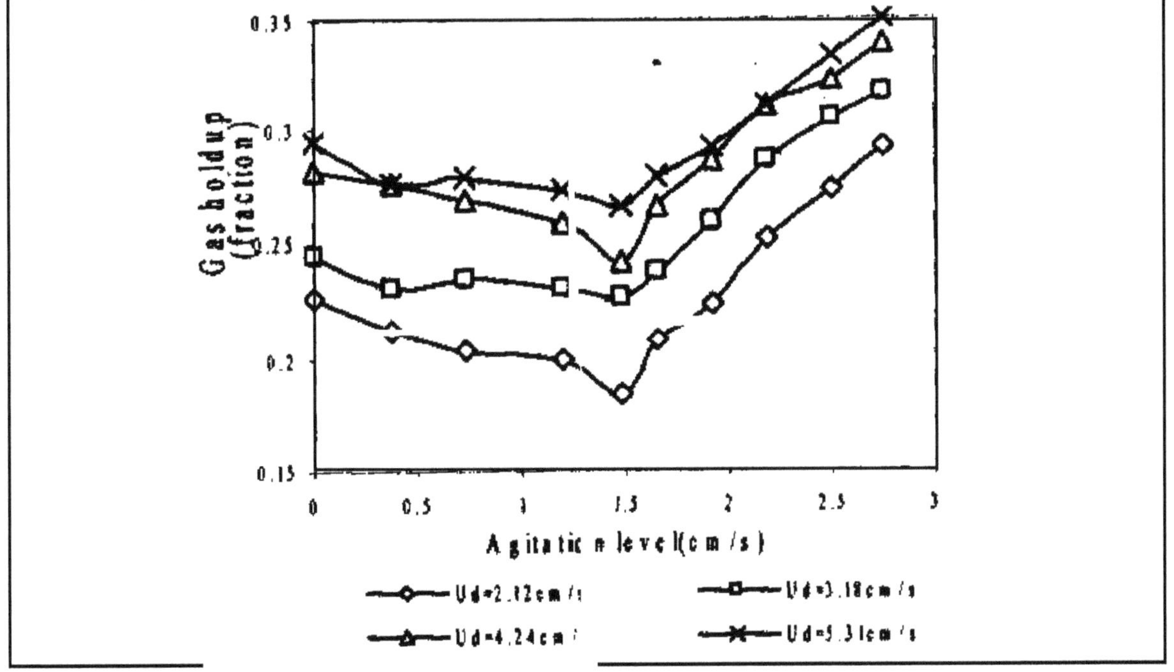

Figure 6.2d: Effect of Agitation Level on Gas Holdup

U_c: 0.25cm/s; d_p: 0.8cm; S_p: 4.5cm; n_p: 10.

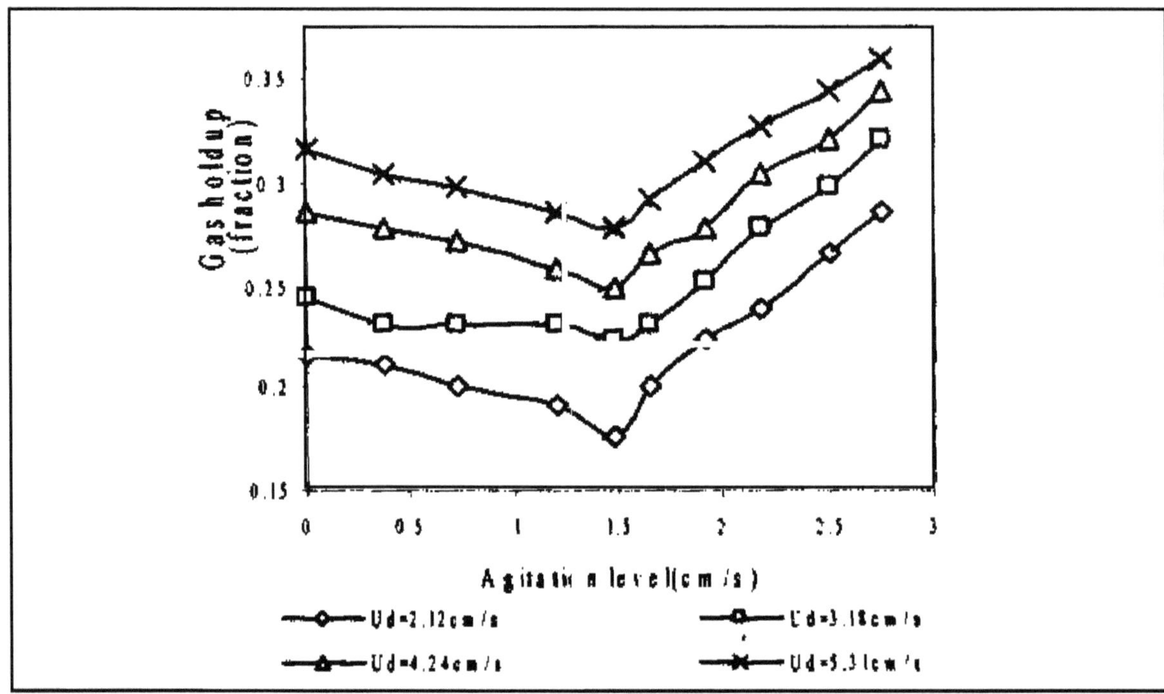

Figure 6.2e: Effect of Agitation Level on Gas Holdup
U_c: 0.25cm/s; d_p: 0.8cm; S_p: 4.5cm; n_p: 10.

of minimum level falls early when compared to the values obtained by other researchers. The lower level of agitation obtained in this study is mainly due to the introduction of rotational effect along with the reciprocation in this column. This rotational effect hinders the axial movement of the bubbles. The bubbles move in a rotational manner along the axis and these results in increased residence time. This increased residence time results more stagnation gas in the column. Observation showed that at low agitation levels, the gas bubbles moves faster from plate to plate through the perforation in the upward direction. While increasing the agitation level, bubbles retain in the upper space between the plates. During the upward movement of the plate stock, the down coming liquid flow hinders the upward movement of the clustered bubbles around the plates. It increases the stagnation of gas within the column and results an increased gas holdup. During the successive downward stroke, the bubbles clustered at the bottom plates get uniformly dispersed into the liquid. The space occupied by the gas dispersion between the plates, especially near the bottom surface of the plates increases gradually from the bottom plate to the top. Visual observation ensures the same. This is the clear evidence for the complete intimate contact between the gas and liquid. It indicates that introduction of rotational effect increases the dispersion of gas into the liquid phase at the top of the column where the real counter current operation exists. This enhances the rate of mass transfer of the gas into the liquid.

Effect of Superficial Gas Velocity on Gas Holdup

It is interesting to note from Figure 6.3a to 6.3e that the effect of superficial gas velocity on the holdup is very strong. The holdup increases almost linearly with superficial gas velocity when other

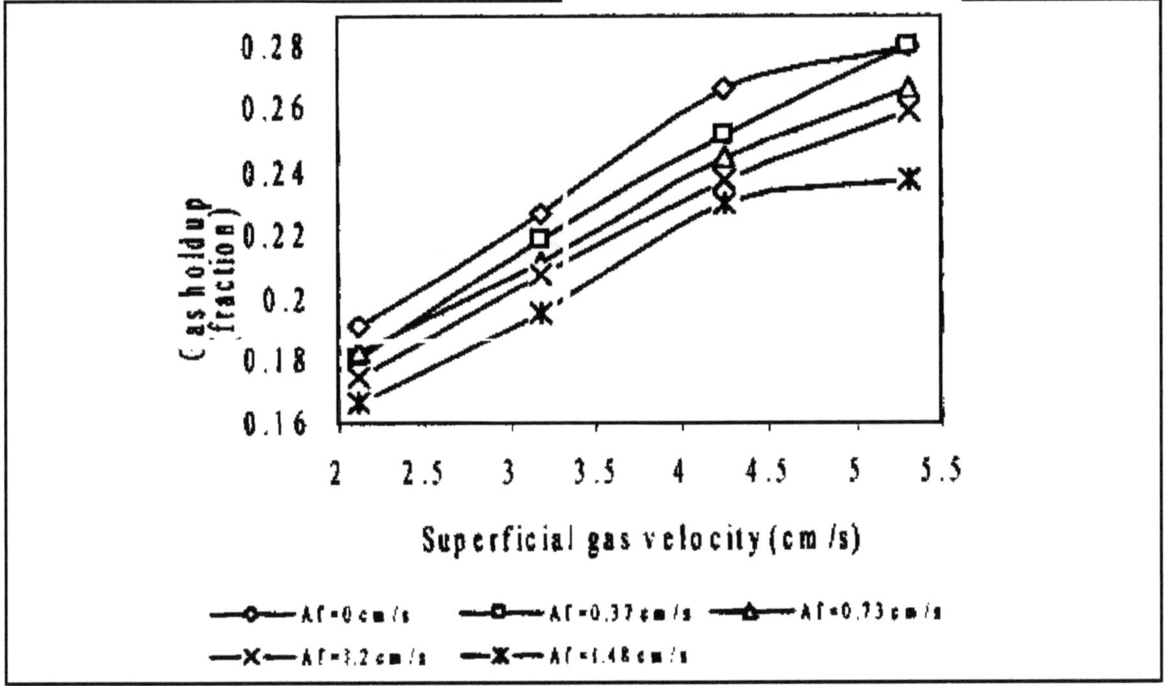

Figure 6.3a: Effect of Superficial Gas Velocity on Gas Holdup
U_c: 0.14cm/s; d_p: 0.8cm; S_p: 4.5cm; n_p: 10.

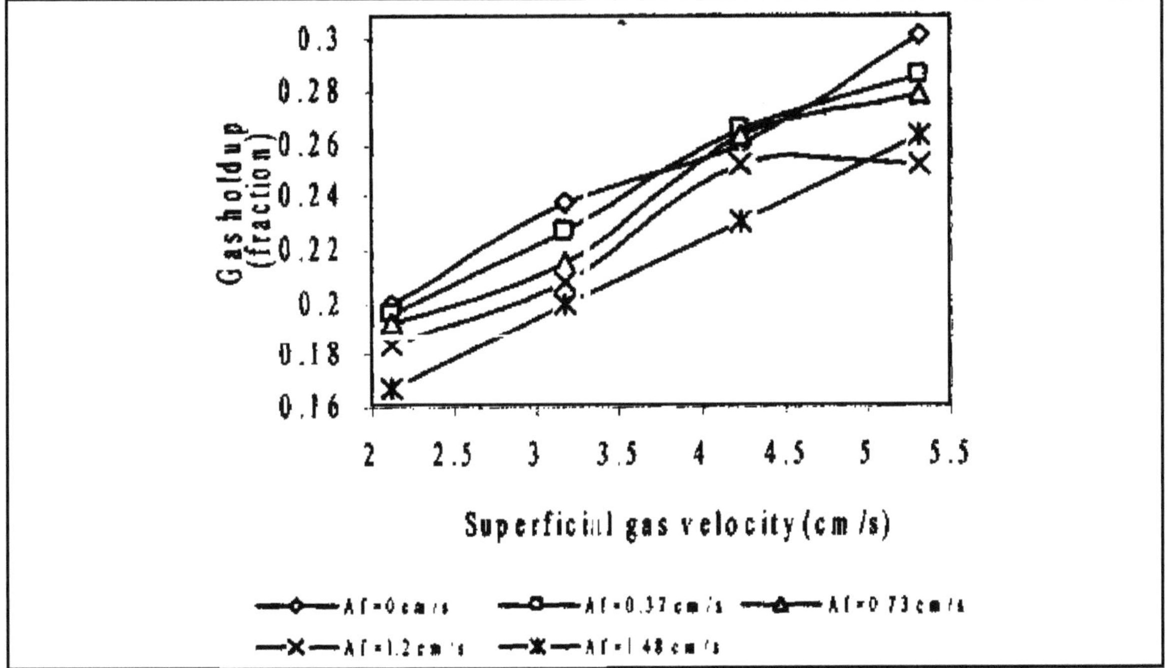

Figure 6.3b: Effect of Superficial Gas Velocity on Gas Holdup
U_c: 0.18cm/s; d_p: 0.8cm; S_p: 4.5cm; n_p: 10.

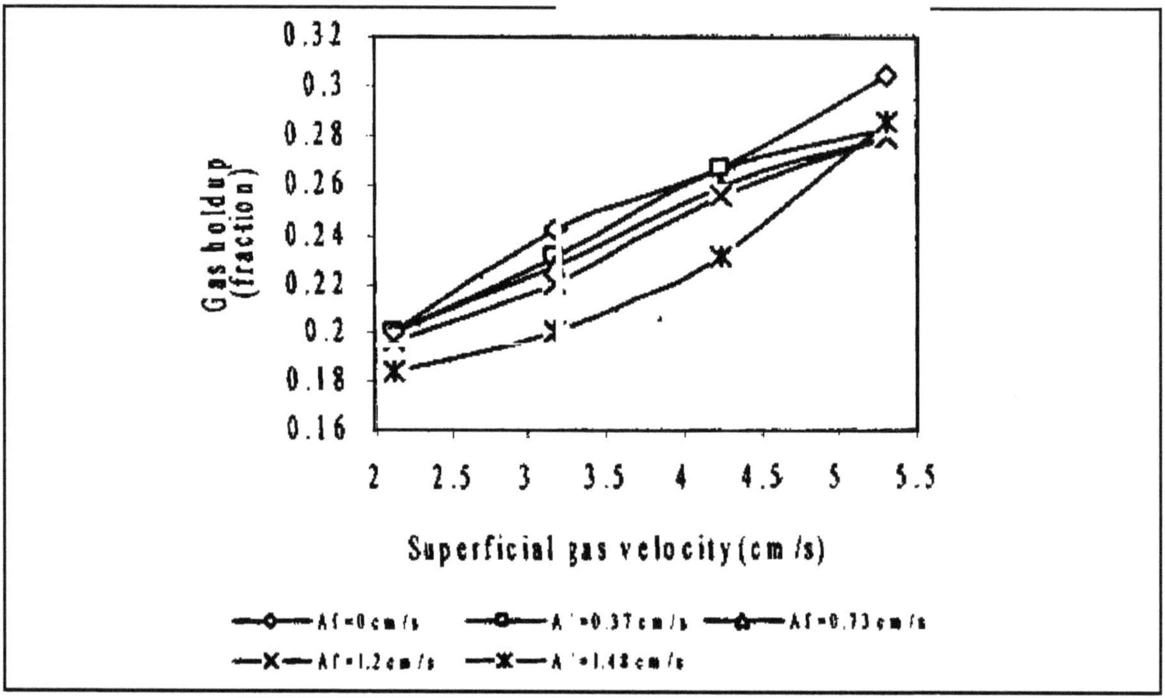

Figure 6.3c: Effect of Superficial Gas Velocity on Gas Holdup

U_c: 0.21cm/s; d_p: 0.8cm; S_p: 4.5cm; n_p: 10.

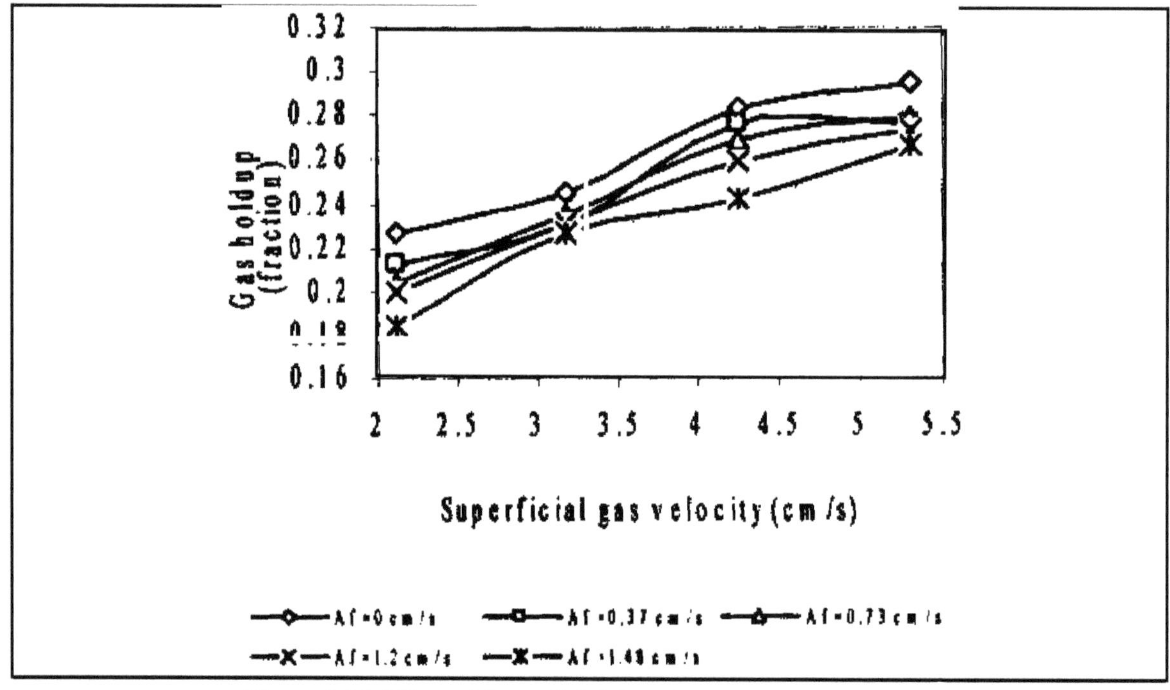

Figure 6.3d: Effect of Superficial Gas Velocity on Gas Holdup

U_c: 0.25cm/s; d_p: 0.8cm; S_p: 4.5cm; n_p: 10.

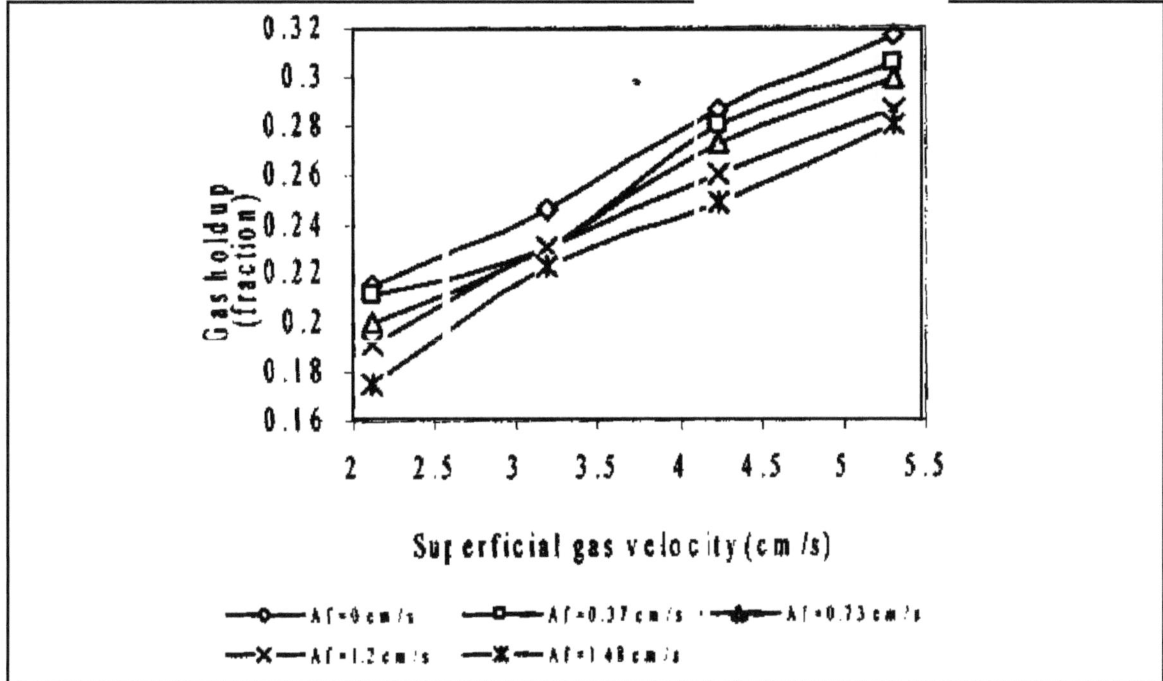

Figure 6.3e: Effect of Superficial Gas Velocity on Gas Holdup
U_c: 0.28cm/s; d_p: 0.8cm; S_p: 4.5cm; n_p: 10.

parameters kept constant. The slope of the curve gradually reduces and reaches almost constant beyond certain values of superficial gas velocities. This similar effect was observed and reported in modified multi-stage bubble column scrubber by Bhic *et al.* (2002). In the initial stages *i.e.*, at the minimum gas velocities, visual observation shows that less number of bubbles were formed. The bubble sizes are abruptly large, non-uniform and unstable. Therefore bubble breaking is less and bubble movement is very slow. An increase in superficial gas velocity results in the formation of a large number of gas bubbles and seems to be move faster. While continuously increasing the superficial gas velocity, the bubble breaking is drastically increased thus decreasing the bubble size and enhancing uniform gas bubble distribution.

Effect of Superficial Liquid Velocity on Gas Holdup

From the Figures 6.4a to 6.4d, it is observed that the effect of liquid flow velocity is less significant than that of gas velocity. It is seen from the above figures that with increasing liquid velocities the gas holdup increases slightly. This slight increment in holdup might be due to the uniform distribution of the gas bubbles and clustering behaviours of the gas bubbles at higher liquid down flow.

Effect of Plate Spacing on Gas Holdup

Figure 6.5 show that effect of plate spacing as an independent variable. The closer plate spacing provides more number of plates in the stack. The mean bubble size would decrease with an increasing number of plates that results increased gas holdup.

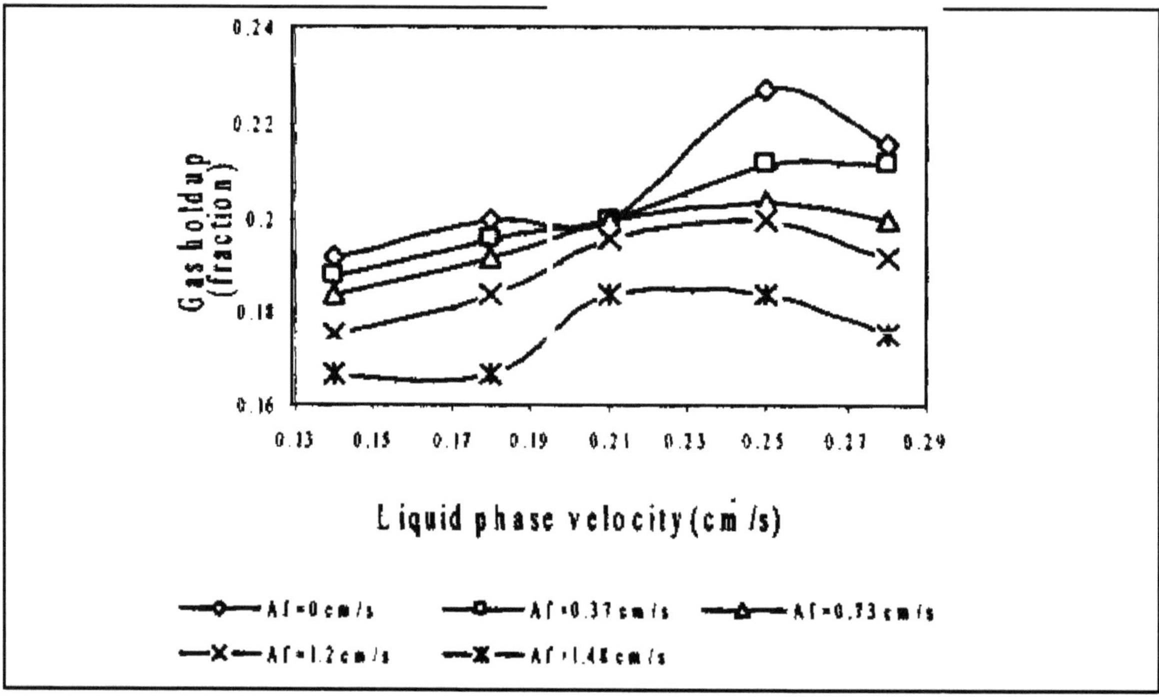

Figure 6.4a: Effect of Superficial Liquid Velocity on Gas Holdup
U_d: 2.12cm/s; d_p: 0.8cm; S_p: 4.5cm; n_p: 10.

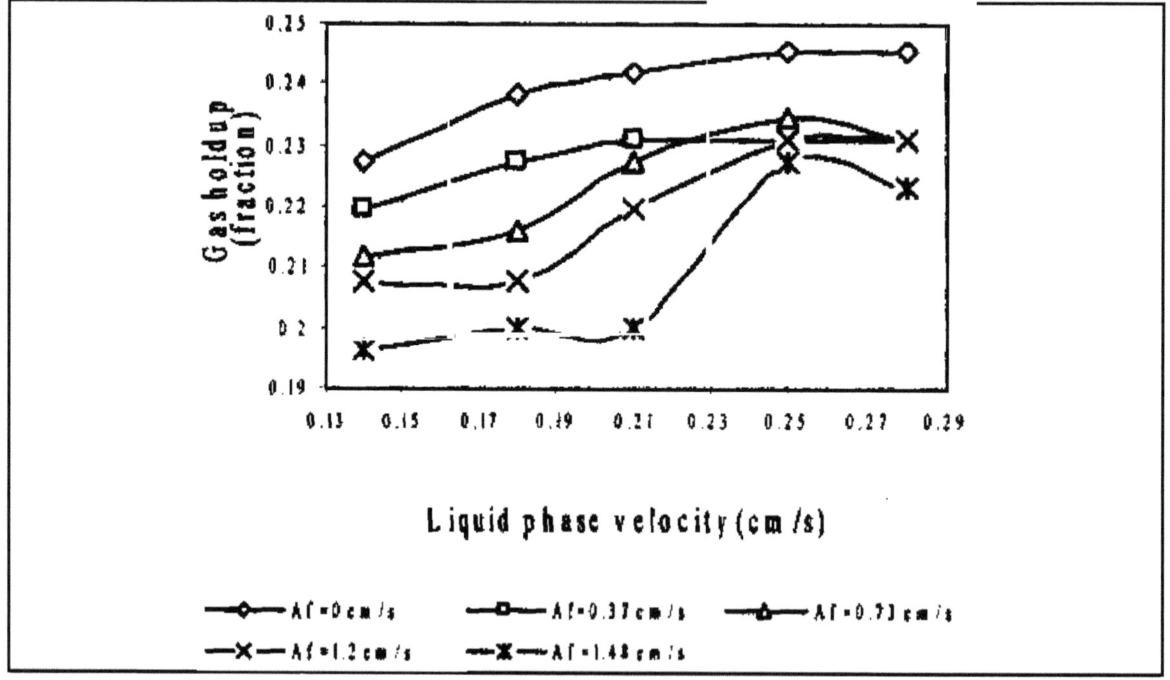

Figure 6.4b: Effect of Superficial Liquid Velocity on Gas Holdup
U_d: 3.18cm/s; d_p: 0.8cm; S_p: 4.5cm; n_p: 10.

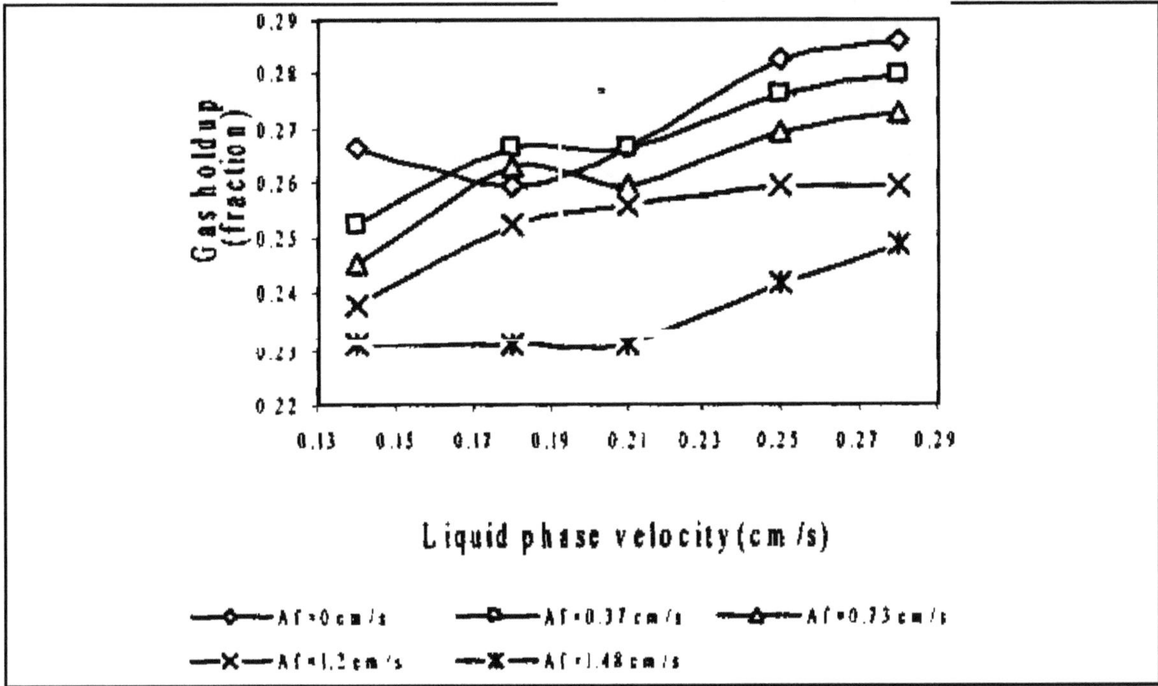

Figure 6.4c: Effect of Superficial Liquid Velocity on Gas Holdup
U_d: 4.24cm/s; d_p: 0.8cm; S_p: 4.5cm; n_p: 10.

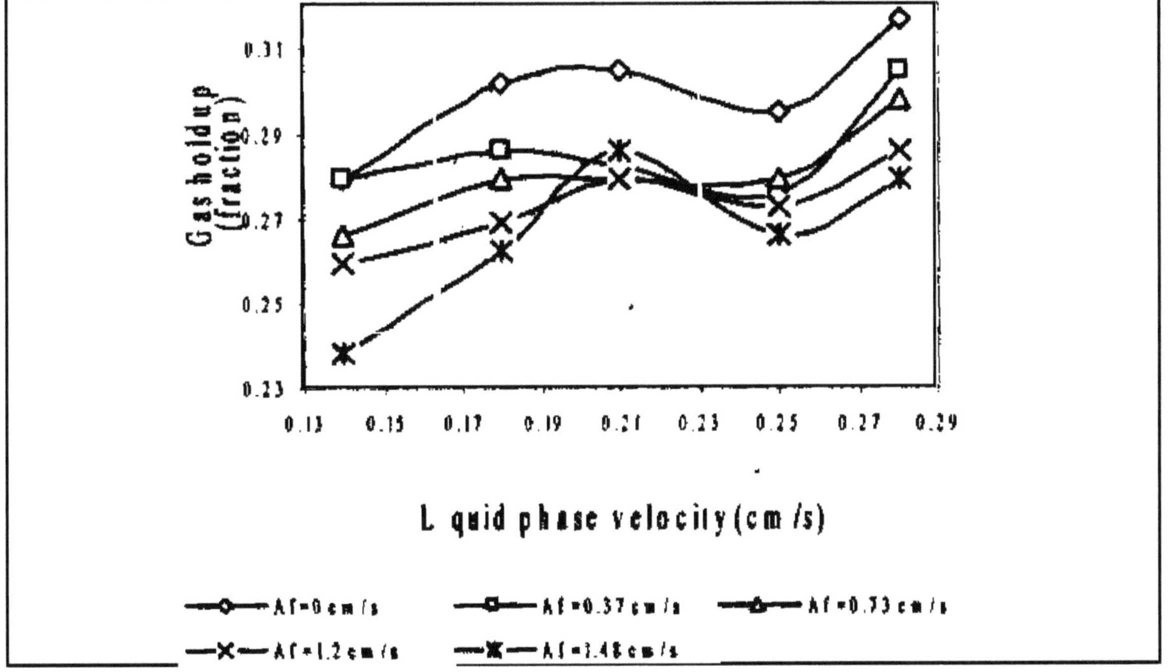

Figure 6.4d: Effect of Superficial Liquid Velocity on Gas Holdup
U_d: 5.31cm/s; d_p: 0.8cm; S_p: 4.5cm; n_p: 10.

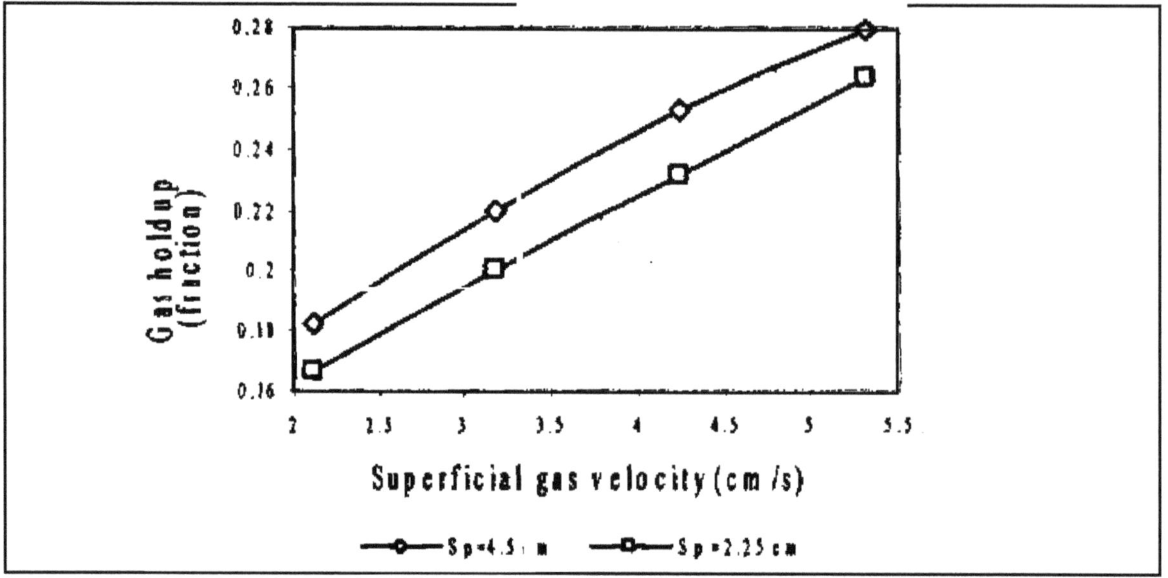

Figure 6.5: Effect of Plate Spacing on Gas Holdup

U_c: 0.28cm/s; Af: 0.37cm/s; d_p: 0.8cm

Correlations for Gas Holdup

In this present investigation the performance of the novel hybrid rotating and reciprocating perforated plate bubble column has been investigated. Flow conditions: superficial gas and liquid velocities, fluid properties: densities of gas and liquid, liquid phase viscosity and surface tension of the liquid, geometrical properties: plate perforation diameter and plate spacing and hybrid property: agitation level has been found to have strong effect on the gas holdup. Therefore if a theoretical relationship exists between the gas holdup and the above said parameters then the gas holdup may be written as follows

$$\varepsilon_G \quad f_1(d_p, A_f, \ \rho, U_d, S_p, \sigma, \mu_L, g) \tag{2}$$

By using the dimensionless analysis, an equation for gas holdup can be written as

$$\varepsilon_G = f_2 \left[\frac{U_d}{Af}\right]^{b1} \left[\frac{\sigma}{d_p.Af^2.\Delta\rho}\right]^{b2} \left[\frac{\mu_L}{d_p.Af^2.\Delta\rho}\right]^{b3} \left[\frac{d_p g}{Af^2}\right]^{b4} \tag{3}$$

To establish the functional relationship between EG and the various dimensionless groups in equation (3), regression analysis has been used to evaluate the constant and coefficients of the equation. This yielded a regression coefficient of 0.954 and a standard deviation of percentage error 5.86 for the equation towards from the mixer-settler region to the transition region is

$$\varepsilon_G \quad 0.0608 \ \frac{U_d}{Af}^{0.4} \ \frac{\sigma}{d_p.Af^2. \ \rho}^{0.005} \ \frac{\mu_L}{d_p.Af^2. \ \rho}^{0.22} \ \frac{d_p g}{Af^2}^{0.01} \tag{4}$$

The regression coefficient is found to be 0.993 and a standard deviation of percentage error is 1.7005 for the equation towards from transition region to the emulsion region is

$$\varepsilon_G = 5.6\times10^4\left[\frac{U_d}{Af}\right]^2\left[\frac{\sigma}{d_p.Af^2.\Delta\rho}\right]^{-2}\left[\frac{\mu_L}{d_p.Af^2.\Delta\rho}\right]^{1.5} \tag{5}$$

The values of gas holdup predicted by equation (4) and (5) have been plotted against the experimental values in Figure 6.6. In addition to that, to test the acceptability of the correlations, various statistical tests have been carried out (statistical details have been discussed in Appendix-I). It is pointed out from the statistical analysis that correlation 4 is significant with 90 per cent confidence level and correlation 5 is significant with 99 per cent confidence level.

Bubble Diameter

The mean bubble diameter is calculated using a Zeiss particle size counter. From the photographs the bubbles are counted approximately with respect to their sizes and the mean bubble diameter was calculated from the following relationship

$$d_B = \frac{\sum_{i=1}^{n} n_i d_i^3}{\sum_{i=1}^{n} n_i d_i^2} \tag{6}$$

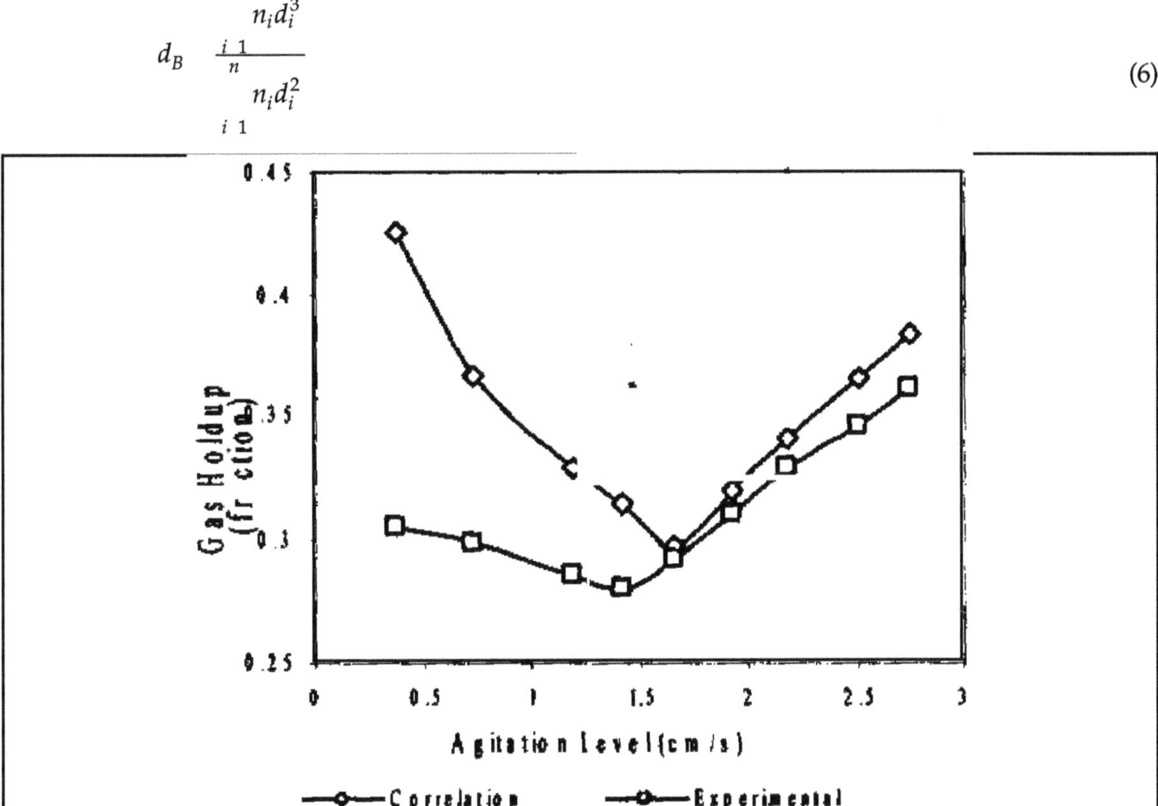

Figure 6.6: Comparison of Dispersed Phase Holdup Predicted Theoretically with the Developed Correlation

U_d: 5.31cm/s; U_c: 2.8cm/s; d_p: 0.8cm; S_p: 4.5cm; n_p: 10.

In absence of agitation the bubble size distribution is non-uniform. The size ranges from 6 to 15 mm approximately. Even though small sized bubbles are formed, it is disturbed due to coalescence. The agitation level has a strong effect on the gas bubble distribution. The gas and liquid phase flow rates had less significant effect on bubble sizes. This similar effect was found previously for gas liquid system by Ramo Rao *et al.* (1988) in counter current reciprocating plate bubble column.

Correlation for Bubble Diameter

The following equation expresses the functional relationship between the bubble diameter and other variables in terms of volume of the bubble

$$V_B = f_3(d_p, U_d, \Delta\rho, S_p, Af) \tag{7}$$

The following dimensionless correlation, theoretically relates the variables such as agitation level, superficial gas velocity, plate spacing and perforation diameter with the mean bubble diameter in terms of volume of the bubble which is assumed to a spherical

$$\frac{V_B}{d_p^3} = f_4\left(\frac{S_p}{d_p} \times \frac{Af}{U_d}\right)^{b1} \tag{8}$$

To establish the functional relationship between bubble diameter and dimensionless group in equation 8, regression analysis has been used to evaluate the constant and coefficient of the equation. It yielded a regression coefficient of 0.907 and a standard deviation of percentage error 18 for the equation given below

$$\frac{V_B}{d_p^3} = 0.0748\left(\frac{S_p.Af}{d_p.U_d}\right)^{-0.51} \tag{9}$$

As per the statistical analysis, the developed correlation is significant with 90 per cent confidence level. Since the shape of the bubble is assumed to be spherical, after rearrangements, the above equation becomes,

$$d_B = 0.5228 d_p \left(\frac{S_p Af}{d_p U_d}\right)^{-0.17} \tag{10}$$

Figure 6.7 shows the comparison of bubble diameter predicated experimentally with correlation values.

Interfacial Area

Physical method is employed for the determination of interfacial area. This method is based on measuring the gas holdup and bubble diameter and does not require a detailed description except the picture taking technique. The emulsion that is gas bubbles is photographed through the transparent walls of the column. The count of bubbles taken from the photographs permit's the determination of the distribution of bubble diameter. This measurement yields an average valve for the interfacial area from the following experimental relationship,

$$a = \frac{6\varepsilon_G}{d_B} \tag{11}$$

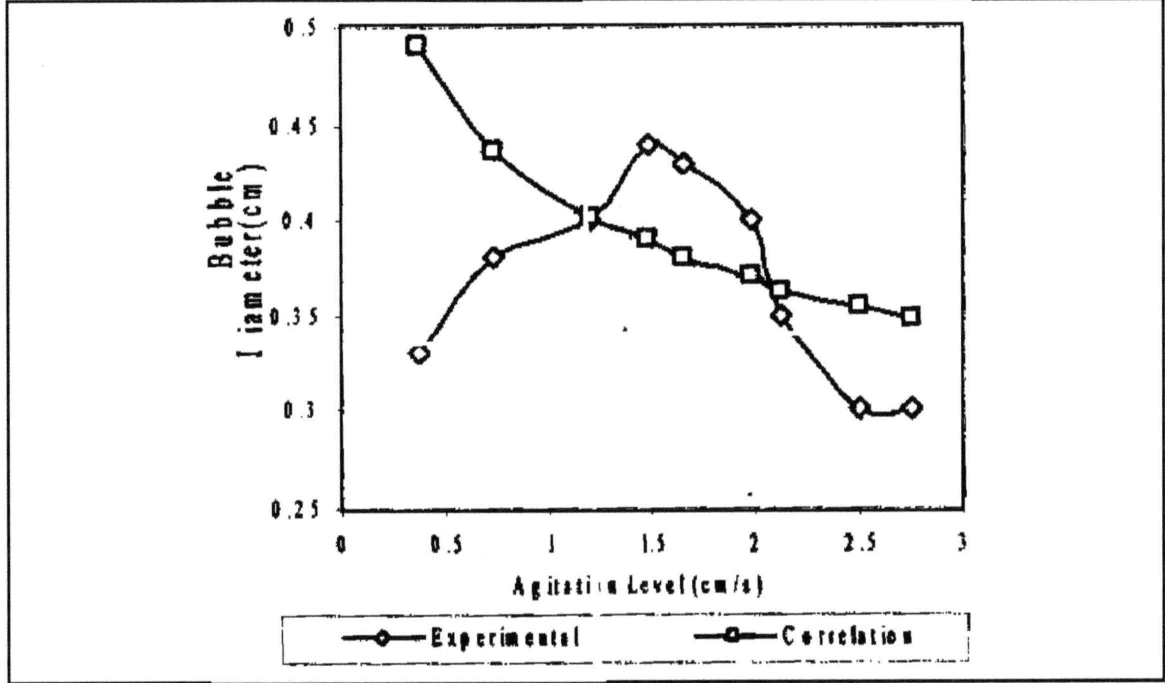

Figure 6.7: Comparison of Bubble Diameter Predicted Theoretically with the Developed Correlation U_d: 5.31cm/s; U_c: 2.8cm/s; d_p: 0.8cm; S_p: 4.5cm; n_p: 10.

The influence of the operating conditions on the interfacial area can be seen in the Figures 6.8a and 6.8b. The idea of the minimum agitation level as in the case of gas holdup can be utilized again. With increase of agitation level, the interfacial area increases slowly up to the transition region and then it increases significantly with increasing agitation level. This means that the interfacial area will begin to increase when the dispersion of the gas bubbles becomes effective, *i.e.* when the flow regime is changed from that of a mixer settler to an emulsion regime. The effect of superficial gas velocity is also found to be significant on interfacial area. It increases with increasing superficial gas velocity.

Correlation for Interfacial Area

The specific interfacial area in this hybrid column can be expected to depend on the operating conditions (agitation level, superficial gas velocity), geometrical characteristics (plate spacing, perforation diameter of the plates) and physical properties of the phases (density and surface tension of the liquid phase). The following equation expresses the functional relationship between the interfacial area and other variables

$$a = f_5(\varepsilon_p, Af, \rho_L, \sigma_L, U_d) \tag{12}$$

The following dimensionless correlation, theoretically relates the variables such as gas holdup, agitation level, superficial gas velocity, density and surface tension of the liquid phase with the interfacial area

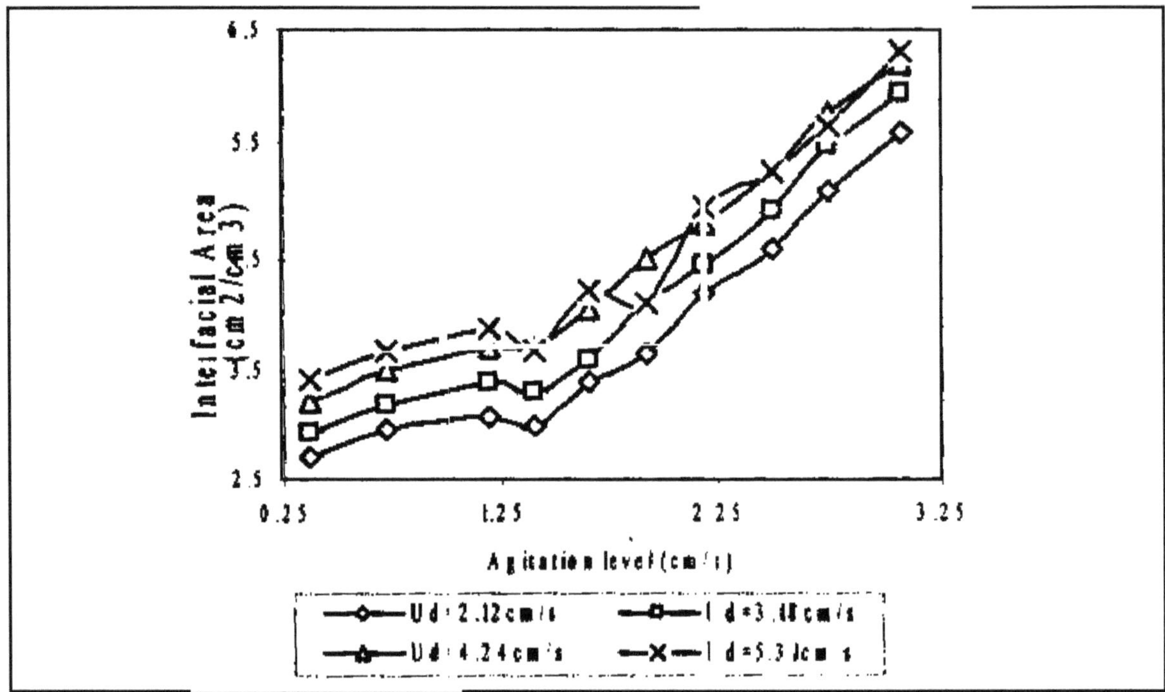

Figure 6.8a: Effect of Agitation Level on Interfacial Area

U_c: 0.14cm/s; d_p: 0.8cm; S_p: 4.5cm; n_p: 10.

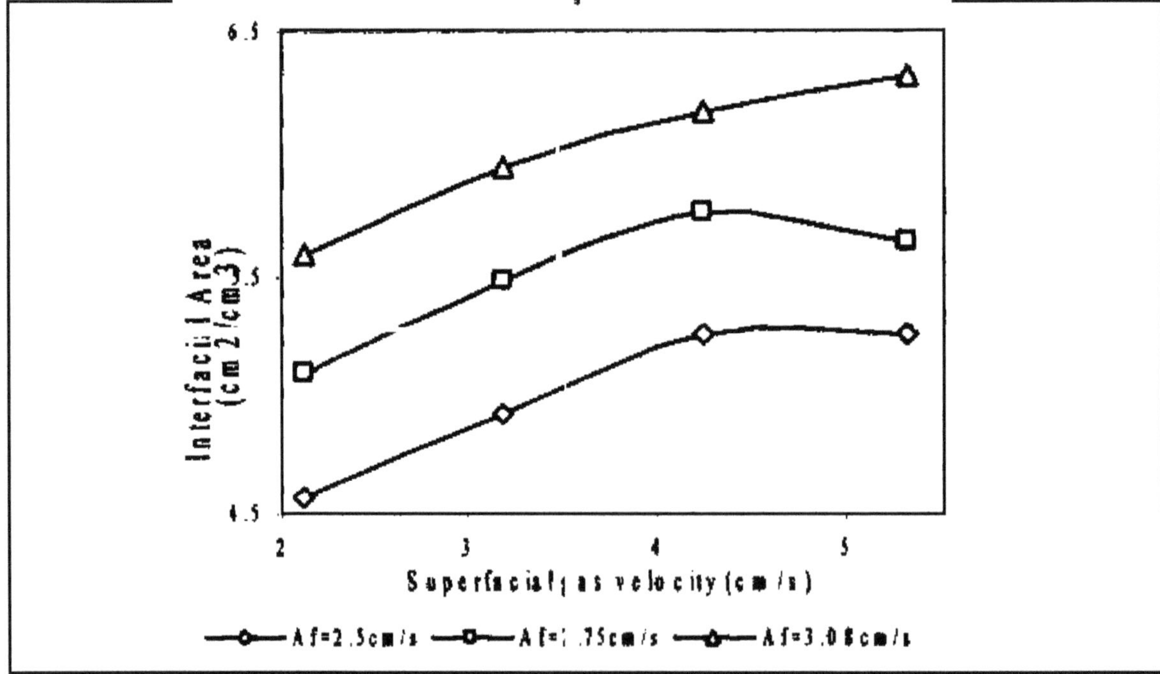

Figure 6.8b: Effect of Superficial Gas Velocity on Interfacial Area

U_c: 0.14cm/s; d_p: 0.8cm; S_p: 4.5cm; n_p: 10.

$$a = f_6\left(\frac{Af^2\rho_L\varepsilon_G}{\sigma_L}\right)^{b1}\left(\frac{U_d}{Af}\right)^{b2} \tag{13}$$

To establish the functional relationship between interfacial area and dimensionless group in equation (13), regression analysis has been used to evaluate the constant and coefficient of the equation. It yielded a regression coefficient of 0.94 and a standard deviation of percentage error 2.1 for the equation given below

$$a \quad 11.76 \quad \frac{Af_2\rho_L\varepsilon_G}{\sigma_L}^{0.2357} \quad \frac{U_d}{Af}^{0.1247} \tag{14}$$

This correlation is significant with 95 per cent confidence level as per the statistical analysis. Figure 6.8c shows the comparison of interfacial area predicated experimentally with correlation values.

Conclusion

This investigation is concerned with the characteristic study of gas holdup, bubble diameter and interfacial area in a novel Hybrid Rotating and Reciprocating Perforated Plate Bubble Column. It is shown that the gas holdup has a strong function of agitation level. First it goes through a level of minimum before increasing steadily with an increase in agitation level. This minimum agitation level falls least when compared with the other investigation done in the reciprocating plate column. This is

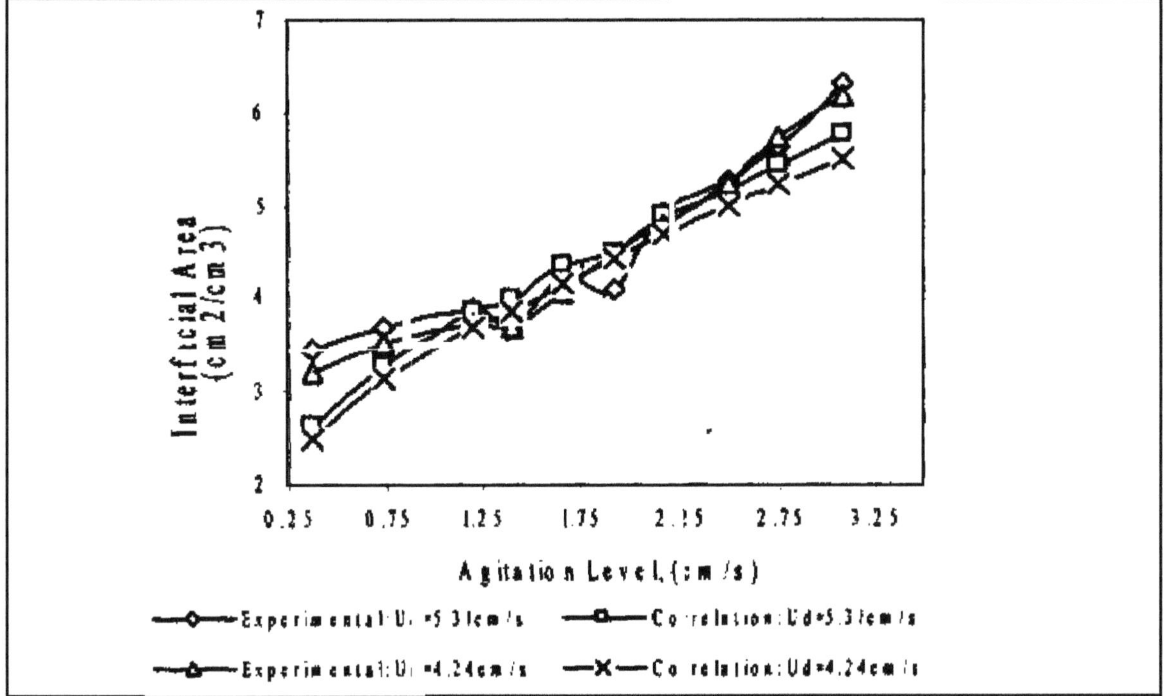

Figure 6.8c: Comparison of Interfacial Area Predicted Theoretically with the Developed Correlation U_c: 2.8cm/s; d_p: 0.8cm; S_p: 4.5cm; n_p: 10.

mainly due to the introduction of hybrid effect. The gas holdup increases with increasing superficial gas velocity but the liquid phase velocity has less effect. The closer plate spacing shows a strong effect on the gas holdup. The size of the gas bubbles is found to have a direct relation with the agitation level. With increase of agitation level, the interfacial area increases slowly up to the transition region and then it increases significantly with increasing agitation level. The effect of superficial gas velocity is also found to be significant on interfacial area. Correlations have been developed for the determination of gas holdup, bubble diameter and interfacial area of this Hybrid Rotating and Reciprocating Perforated Plate Bubble Column. The correlations have been tested statistically and found to be of high significance.

Nomenclature

a: Interfacial area, cm^2/cm^3
b1, b2, b3, b4: Coefficients of the dimensionless equation
A: Amplitude of reciprocation, cm
Af: Agitation level, cm/s
d_i: Diameter of individual bubble, cm
d_p: Diameter of perforation in the plate, cm
$\overline{E}$: Mean error
f: Frequency, s^{-1}
f1, f2: Coefficients of the dimensionless equation
g: Acceleration due to gravity, cm/s^2
H: Height of the gassed column, cm
H_o: Height of the ungassed column, cm
n: Number of variables in the least square method
n_p: Number of the plates
S_p: Space between the plates, cm
s: Standard deviation of error
t: t-distribution
U_d: Superficial gas velocity, cm/s
U_c: Superficial liquid velocity, cm/s

Greek Symbols

σ: Surface of the liquid phase, gm/s^2
ε: Gas holdup, fraction
$\Delta\rho$: Density difference between liquid and gas phase, gm/cm^3
μ_L: Liquid phase viscosity, gm/cm.s

Appendix

Statistical Analysis of the Correlation (*Equation-4*)

The following statistical parameters have been calculated to test the significance of the correlation.

Sum of the square of deviation removed by regression:

$\Sigma c^2 = 1.165 \times 10^{-2}$

Sum of the square of deviation:

$\Sigma y^2 = 1.279 \times 10^{-2}$

Residual Sum of square of deviation:

$\Sigma R^2 = \Sigma y^2 - \Sigma c^2 = 1.142 \times 10^{-3}$

Correlation coefficient:

$$r = \sqrt{\frac{\Sigma c^2}{\Sigma y^2}} = 0.954$$

Variance of estimate:

$$s^2(Y) = \frac{\Sigma R^2}{N-K-1} = 1.142 \times 10^{-3}$$

The diagonal Gaussian multipliers for the correlation are:

$C11 = 1.09521 \times 10^4$

$C22 = 0.35880 \times 10^4$

$C33 = 0.60250 \times 10^4$

$C44 = 0.17681 \times 10^4$

Variance of regression coefficients:

$S^2\,(b1) = s^2(Y) \times C11 = 0.02036$

$S^2\,(b2) = s^2(Y) \times C22 = 0.00669$

$S^2\,(b3) = s^2(Y) \times C33 = 0.0112$

$S^2\,(b4) = s^2(Y) \times C44 = 0.032975$

The *t* value for 0.05 probability level and 5 degree of freedom as obtained from statistical table is 2.571. Therefore at 95 per cent confidence level, the regression coefficients are:

$b1 = a \pm 2.571\sqrt{S^2(a)} = -0.4344 \pm 0.3669 = [(-0.0675) + (-0.8013) \div 2 = -0.4344$

$b2 = b \pm 2.571\sqrt{S^2(b)} = 0.005564 \pm 0.21029 = (0.21585) + (-0.20472) \div 2 = 0.0056$

$b3 = c \pm 2.571\sqrt{S^2(c)} = 0.216226 \pm 0.27209 = [(0.4883)+(-0.05586) \div 2 = 0.216228$

$b4 = d \pm 2.571\sqrt{S^2(d)} = -0.01086 \pm 0.46687 = [(0.4560)+(-0.4\,777) \div 2 = -0.01086$

Constant of the regression correlation:

$Ln\,(f_2) = Y - bl \times X1 - b2 \times X2 - b3 \times X3 - b4 \times X4 = -2.80059;\ f_2 = 0.0608$

Mean Square: Regression $= 5.825 \times 10^{-3}$

Mean Square: Residual $= 5.710 \times 10^{-4}$

$$F = \frac{5.825 \times 10^{-3}}{5.710 \times 10^{-3}} = 10.202$$

Significance of the Correlation by F Test

For checking the significance of the correlation, the Null hypothesis is used. The F value for the regression is 10.202. The table value of F for (2, 3) degree of freedom at 10 per cent probability level obtained from statistical table is 5.46. Since the calculated value of F is greater than the table value, H0 is rejected. It means that the developed correlation 4 is highly significant with 90 per cent confidence level.

Equation 5 is also analyzed statistically for significance in same manner. The F valve for the regression is 268.334. The table value of F for (1, 4) degree of freedom at 1 per cent probability level obtained from statistical table is 21.20. Since the calculated value of F is greater than the table value, H0 is rejected. It means that the developed correlation 5 is highly significant with 99 per cent confident level.

References

Yawalkar, Archis A., Pangarkar, Vishwas G. and Beenackers, Anthony A.C.M., 2002. Gas hold up in stirred tank reactors. *Can. J. Chem. Eng.*, 80: 158–166.

Baird, M.H.I. and Rice, R.G., 1975. Axial dispersion in large unbaffled columns. *Chem. Eng. J.*, 9: 171–174.

Meikap, Bhim C., Kundu, Gautam and Biswas, Manindra N., 2002. Prediction of dispersed phase holdup in a modified multi-stage bubble column scrubber. *Can. J. Chem. Eng.*, 80: 306–312.

Brauer, H. and Sucker, D., 1979. Biological wastewater treatment in a high efficiency reactor. *Ger. Chem. Eng.*, 2: 77–86.

Brauer, H., 1988. Mass transfer machines: The road to better design of equipment in the processing industries. *Int. Chem. Eng.*, 28(2): 207–219.

Lee, Dong-Hyun, Grace, John R. and Epstein, Norman, 2000. Gas holdup for gas velocities in a gas-liquid concurrent upward-flow system. *Can. J. Chem. Eng.*, 78: 1006–1010.

Lounes, M.J., Thibault, Audet, J., and Leduy, A., 1995. Mass transfer in a reciprocating plate bioreactor. *Bioprocess Eng.*, 13: 1–11.

Maruoka, Y. and Brauer, H., 1989. Fluid dynamics and mass transfer in a multistage rotating-disk reactor. *Int. Chem. Eng.*, 29(4): 577–615.

Mersmann, A., Voit, H. and Zeppenfeld, R., 1988. Do we need mass transfer machines. *Int. Chem. Eng.*, 28(1): 1– 13.

Midoux, N. and Charpentier, J.-C., 1984. Mechanically agitated gas-liquid reactors, Part 1: Hydrodynamics. *Int. Chem. Eng.*, 24(2): 49–287.

Midoux, N. and Charpentier, J.-C., 1984. Mechanically agitated gas-liquid reactors, Part 2: Interfacial area. *Int. Chem. Eng.*, 24(2): 452–481.

Miyanami, K., Tojo, K., Minami, I. and Yano, T., 1978. Gas-liquid mass transfer in a vibrating disk column. *Chem. Eng. Sci.*, 33: 601–608.

Mustapha, Lounes and Thibault, Jules, 1993. Hydrodynamics and power consumption of a reciprocating plate gas-liquid column. *Can. J. Chem. Eng.*, 71: 497–506.

Rama Rao, N.V., Srinivas, N.S. and Varma, Y.B.G., 1983. Dispersed phase holdup and drop size distribution in reciprocating plate column. *Can. J. Chem. Eng.*, 61: 168–177.

Rama Rao, N.V. and Baird, M.H.I., 1986. Gas liquid pressure drop studies in a reciprocating plate column. *Can. J. Chem. Eng.*, 64: 42–47.

Rama Rao, N.V. and Baird, M.H.I., 1988. Characteristics of a countercurrent reciprocating plate bubble column. I. Holdup, pressure drop and bubble Diameter. *Can. J. Chem. Eng.*, 66: 211–221.

Tinge, J.T. and Drinkenburg, A.A.H., 1986. The influence of slight departures from vertical alignment on liquid dispersion and gas holdup in a bubble column. *Chem. Eng. Sci.*, 41: 165–169.

Veljkovic, V. and Skala, D., 1986. Hydrodynamic investigations of gas-liquid contacting in a reciprocating plate column. *Can. J. Chem. Eng.*, 64: 906–914.

Veljkovic, V. and Skala, D., 1988. Mass transfer characteristics in a gas-liquid reciprocating plate column. II. Interfacial area. *Can. J. Chem. Eng.*, 66: 200–210.

Veljkovic, V. and Skala, D., 1989. On the possibility of general prediction of gas holdup in reciprocating plate column. *Can. J. Chem. Eng.*, 67: 1015–1018.

Yang, N.S., Shen, Z.Q., Chen, B.H. and McMillan, A.F., 1986. Pressure drop, gas holdup and interfacial area for gas-liquid contact in Karr columns. *Ind. Eng. Chem. Proc. Des. Dev.*, 25: 660–664.

Chapter 7

Amylase and Protease Production by *Pleurotus ostreatus* and *Laccaria fraterna* under Submerged and Solid-state Fermentations

***K. Balaraju*[1]**, *J. Joel Gnanadoss*[2]*, *S. Arokiyaraj*[2], *P. Agastian*[2] *and S. Ignacimuthu*[1]**

[1]*Entomology Research Institute, Loyola College, Chennai – 600 034*
[2]*Department of Plant Biology and Biotechnology, School of Life Sciences, Loyola College, Chennai – 600 034, India*

ABSTRACT

Studies were carried out on the production of amylase and protease enzymes using agro-waste materials such as Rice bran, Paddy straw, Saw dust and Wheat bran by *Pleurotus ostreatus* and *Laccaria fraterna* (White-rot basidiomycetes) in submerged fermentation (SmF) and solid-state fermentation (SSF). Protease and amylase were produced by the test fungi. Enzymes were assayed both in solid and submerged fermentation. Enzyme production was higher in SSF than in submerged culture. Higher level of protease activity was seen in solid substrate cultivation. *Laccaria fraterna* showed higher protease activity (1760 units/mg protein/min) than *P. ostreatus* in submerged cultivation (1600 units/mg protein/min). In SSF Rice bran and wheat bran supported more protease

* Corresponding Author; E-mail: joel@loyolacollege.edu; Phone: +91 9444365288.
** E-mail: k_balaraju1980@yahoo.co.in.

activities (908 units/mg protein/min). Results indicated that these two fungi can be utilized for large scale production of amylase and protease using cost effective technology.

Keywords: *Laccaria fraterna, Pleurotus ostreatus, Amylase, Protease, Solid-state and Submerged fermentation.*

Introduction

SSF holds tremendous potential for the production of enzymes. Agro-industrial residues or wastes are generally considered as suitable substrates for the production of enzymes in this process (Pandey *et al.*, 1999, 2000; Pandey, 2003), which can be of special interest in those processes where the crude fermented product may be used directly as the enzyme source (Tengerdy, 1998).

SSF reproduces the natural microbiological processes like composting and ensiling. In industrial applications this natural process can be utilized in a controlled way to produce a desired product which also offers numerous advantages for the production of bulk chemicals and enzymes (Hesseltine, 1977; Pandey, 1999; Soccol, 1994). The ability of filamentous fungi to secrete large amounts of extra cellular proteins makes them well suited for the production of industrial enzymes (Agger, 2001). Starch-degrading amylolytic enzymes are of great significance in biotechnological applications ranging from food, fermentation, textile to paper industries (Un *et al.*, 1997; Pandey *et al.*, 2000) and amylases can be derived from several sources such as plants, animals and microbes (Kathiresan 2006).

Acid protease production of *Mucor bacilliformis* was investigated under solid-state fermentation conditions as a model of recovery and purification of a protein from a solid culture medium (Fernandez, 1998). Fibrinolytic protease activity was detected in *Pleurotus ostreatus* which is very much useful as chemotheraphy reagent for cardiovascular disease caused blood clot (Hye-Seon Choi, 1998). Alkaline protease is an important additive used to improve the decontamination effect in the detergent industry. Until now, alkaline protease has been produced mostly by submerged fermentation (SMF) (Hongzhang, 2006). Characteristically *Pleurotus ostreatus* is an edible and wood rotting fungus of great biotechnological interest, not only for its ability to grow on numerous agricultural residues to produce mushrooms of high quality, but also produces some proteins and enzymes of industrial importance. These properties significantly increase its commercial value. *L. fraterna,* though not edible, is an important ectomycorrhizal fungi which can also grow on agro-wastes.

The present study investigates the effect of submerged and solid-state cultivation techniques using various agro wastes for growth, and protease and amylase production of *Laccaria fraterna* and *Pleurotus ostreatus.*

Materials and Methods

Test Organisms

Laccaria fraterna and *Pleurotus ostreatus* used in the study were obtained from Tamil Nadu Agriculture University, Coimbatore, South India. They were maintained on potato-dextrose agar (Hi-Media, Mumbai) and were subcultured fortnightly and stored at 25°C.

Raw Materials

Agro-waste materials such as Rice bran, Saw dust, Paddy straw and Wheat bran were obtained from processing unit and Marketing Corporation, Chennai, India. They were further mechanically dehydrated until the moisture content of dried materials reached 4±1 per cent, which generally took

3–4 hours. Dried materials were packed in polythene pouches for further studies (Vinod Kumar Joshi, 2006).

Submerged Fermentation

Liquid culture was grown in Erlenmeyer flasks (150 ml, Borosil) containing 30 mL of the medium. Mycelium was cultured in Nitrogen limiting medium: Glucose 15 g; Malt extract 0.4 g; $MnCl_2$ 0.3 g; $FeSO_4.7H_2O$ 0.004g; $MgSO_4. 7H_2O$ 0.04g; 3′3 dimethyl butyric acid 20 mM; Distilled water 1000 mL.

Calculations were done using the following formula:

$$\text{Units/mL enzyme} = \frac{(\text{r A530nm/min Test} - \text{r A530nm/min Blank})\ (\text{df})}{(0.001)(0.5)}$$

df: Dilution factor, rA: Relative absorbency, 0.001: The change in A530nm/minute per unit of enzyme at pH 6.5 at 30°C in a 3 mL reaction mixture, 0.5: Volume (in milliliter) of enzyme used

$$\text{Units/mg solid} = \frac{\text{Units/mL enzyme}}{\text{mg solid/mL enzyme}}$$

Solid-State Fermentation

Agro-wastes such as wheat bran, rice bran, paddy straw and saw dust were used as substrates. Paddy straw was cut into small pieces of one inch in size. The substrates were separately transferred into Erlenmeyer flasks and distilled water was poured in all the flasks and was soaked over night. The next day, water was drained off, and care was taken to retain sufficient moisture (about 70 per cent) for the mycelium to grow. The flasks were then sterilized in an autoclave. After cooling, all the flasks were inoculated with the fungal isolates under aseptic condition. The flasks were incubated for the complete growth of mycelium.

After 7, 14 and 21 days of incubation the flasks were taken and Phosphate buffer (pH-7) was added under aseptic condition to the bottles containing different agricultural materials. It was mixed well with glass rod and some more buffer solution was added to give a final volume of 50 mL. The flasks were kept on the shaker for 30 minutes maintained at 120 rpm to extract the enzyme. The flasks were kept in cold room overnight. Later, the contents were filtered with a muslin cloth, and centrifuged for 20 minutes at 4000 rpm. The clear supernatant was collected in clean containers. The crude enzyme source was then estimated for amylase and protease.

Estimation of Amylase

Amylase was estimated by Nelson's method. The amount of reducing sugar was estimated by measuring absorbance at 595 nm and the result was expressed as the amount of reducing sugar released per mg weight.

Estimation of Protease

Protease activity of culture broth and solid extract was determined by a modified Anson's method (Yang and Huang, 1994). Microorganisms produce protease inductively in the presence of protein substrate casein. Casein is a milk protein usually substituted as a nitrogen source for microbial growth. After a sufficient period of incubation, fungal mycelium is removed and the culture filtrate is assayed for enzyme activity, and readings were taken at 650 nm and the enzyme activity was expressed per ml of filtrate.

Results and Discussion

Amylase Activity of *Laccaria fraterna* and *Pleurotus ostreatus* Under Submerged Solid-state Condition

Amylase Activity Under Submerged Condition

In the present investigation *Pleurotus ostreatus* and *Laccaria fraterna* showed 240 and 290 units/ml of amylase activity under submerged culture condition. On 14th day there was an increase in activity in *Pleurotus ostreatus* grown culture but with *Laccaria fraterna* only 50 units/ml was observed (Figures 7.1 and 7.2). Starch and glucose were the best carbon sources for amylase production whereas maximum mycelial biomass was recorded in the medium containing sucrose as the sole carbon source. The maximum mycelial biomass and amylase activity were recorded at 30°C and pH 6 (Amadioha 1998). *Bacillus subtilis* (ATCC 21556) was used to produce alpha-amylase, using potato peel as a substrate. To enhance the production of this enzyme, two nonionic synthetic surfactants were used, Tween 80 and Tween 20. One anionic surfactant, SDS at concentrations of 0.05 per cent or 0.10 per cent (v/w) and a biosurfactant produced by *Bacillus subtilis* (ATCC 21332), known as surfactin, at concentrations of 0.003 per cent, 0.007 per cent, 0.013 per cent or 0.03 per cent (w/w) were also used. The results have shown that surfactants significantly increased the production of alpha-amylase. Tween 80 at

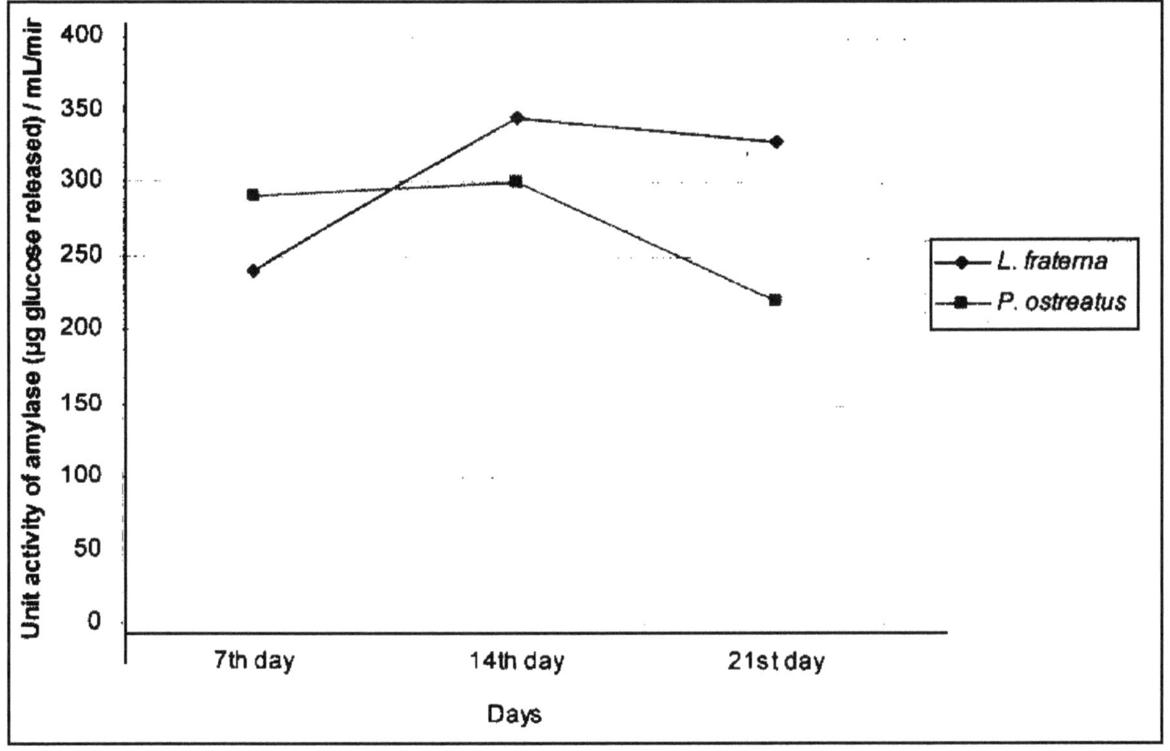

Figure 7.1: Amylase Activity of *L. fraterna* and *P. ostreatus* in Submerged Fermentation

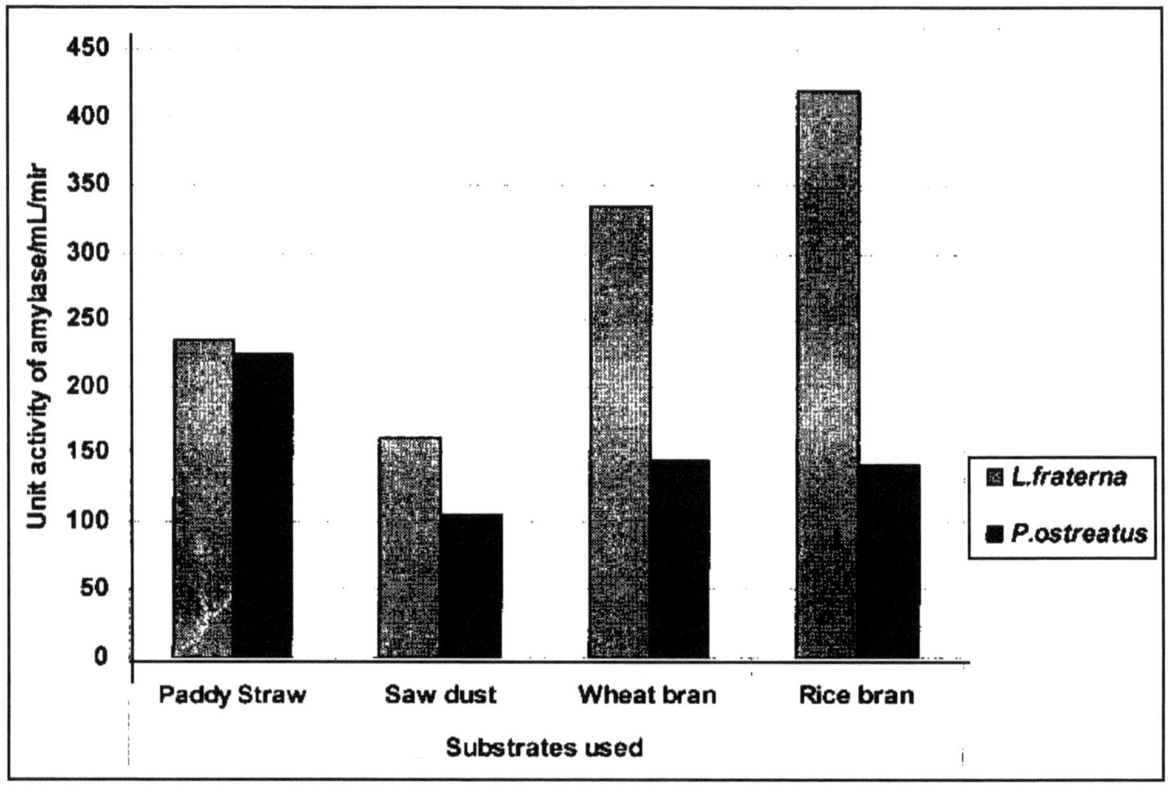

Figure 7.2: Amylase Activity of *L. fraterna* and *P. ostreatus* in Solid State Fermentation with Different Types of Substrates on Different Days of Incubation

0.1 per cent and surfactin at 0.013 per cent provided the highest enzyme activity compared with the control (Goes and Sheppard 1999).

Amylase Activity Under Solid-state Condition

In solid substrate cultivation Rice bran (420 units/ml) and wheat bran (335 units/ml) showed higher activity in case of *Laccaria fraterna.* It was 162 and 165 units/ml in wheat bran and rice bran, respectively. In *Pleurotus ostreatus* except paddy straw (225 units/m), rice bran, wheat bran and saw dust showed only moderate activities (Figure 7.2). The solid-state fermentation technique was carried out using various agro industrial wastes from different sources to produce fungal alpha-amylase economically by a locally isolated strain of *A. niger*, F-21. A high level of enzyme activity was obtained after 48 hours using potato peels as a substrate with an initial pH 6.0 or using whey permeate level of 50 per cent as a moistening agent and urea as a cheap nitrogen source at 400 mg NIL moistening agent. The obtained enzyme has an activity of 620 U/g fermented substrate. The crude enzyme was heat stable when incubated at 70°C for 39 hours retaining its maximum activity. Crude enzyme was applied for potato starch digestion using different slurries concentrations up to 30 per cent giving about 96 per cent digestion (Fadel, 2000).

Protease Activity of *Laccaria fraterna* and *Pleurotus ostreatus* Under Submerged Solid-state Condition

Protease Activity Under Submerged Condition

The test fungi exhibited protease activity both in submerged and solid-state fermentations. The *L. fraterna* showed higher activity (1760 units/ml) than *P. ostreatus* in submerged cultivation (1600 units/ml) on 7th day itself and gradual decrease in activity was seen when the days of incubation increased. Rice bran and wheat bran (908 units/ml) supported more protease activities in solid substrate cultivation (Figures 7.3 and 7.4).

Protease activity was well pronounced in the test fungi *L. fraterna* and *P. ostreatus* (2160 units/ml and 2320 units/ml) respectively. Many *Pleurotus* spp. produce different types of proteases. *Pleurotus ostreatus* produced a medicinally important fibrinolytic protease, which is proposed as medicine for removing blood clots. Protease activity was also assayed in solid-state fermentation. According to Tunga, (1999), 10 per cent ethanol with 3 per cent glycerol was the best solvent for protease extraction, and the soaking time was 2 hours at 30°C. It was further observed that double wash of fermented biomass yielded almost total enzyme in the leachate. In submerged culture, inspite of protein supplementation in the medium, there was lesser protease activity than solid-state cultivation. Higher protease production in solid substrate than in the medium was also reported with *Phaenaerochaetae chrysosporium* and *Phlebia radiata* (Cabaleiro *et al.*, 2002). Protease is involved not only in fructification of mushrooms but also has many applications in food and leather industries. Protease affects the

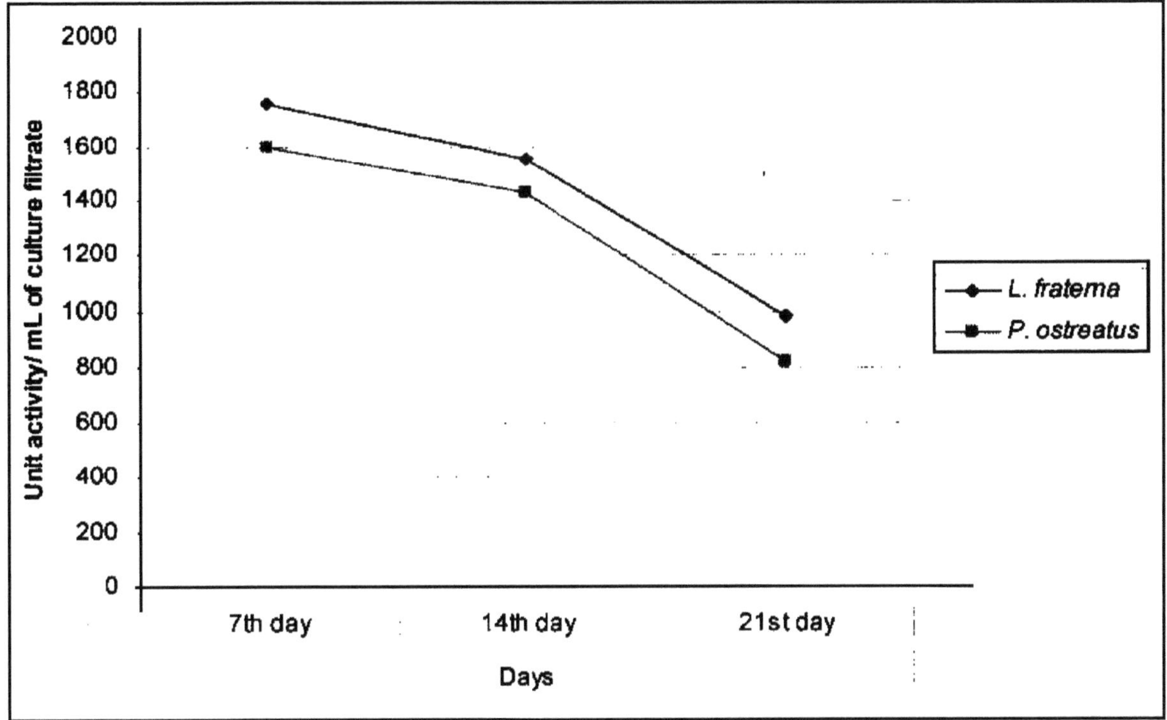

Figure 7.3: Protease Activity of *L. fraterna* and *P. ostreatus* Under Submerged Fermentation

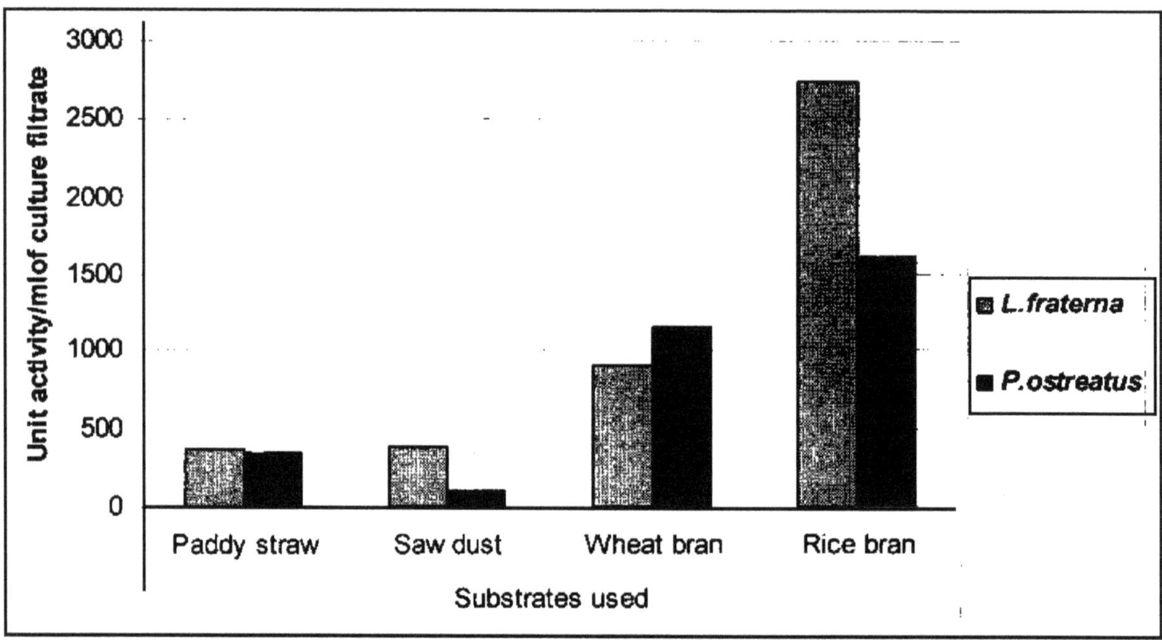

Figure 7.4: Protease Activity of *L. fraterna* and *P. ostreatus* from SSF

lignolytic enzymes; hence it is necessary to know in which state protease is secreted and when lignin peroxidase is secreted so as to optimize applications of these for degradation of xenobiotic compounds. *Rhizopus oryzae* is grown on wheat bran for the production of alkaline protease. The production of this enzyme through solid-state fermentation process was attempted (Tunga, 1999).

Protease Activity Under Solid-state Condition

Protein rich rice bran, wheat bran and paddy straw mixture supported high levels of protease activity in solid state fermentation without any protein supplement. Paddy straw is rich in cellulose and contains very less amount of proteins. The low protein content would not have been sufficient to induce protease production in *Pleurotus ostreatus* and *Laccaria fraterna* during SSF. The results of the study clearly indicate that cheap source of raw materials such as paddy straw; saw dust, wheat bran and rice bran used can induce protease production in SSF. The method developed is cost effective. Further studies are however required to optimize conditions for large scale production of protease under SSF.

Matsushima *et al.* (1981) indicated that the pH drop in the acid protease production might be because of the accumulation of organic and the residue of sulphate ion in the utilization of ammonium sulphate. The pH lowering would be prevented when urea is used as the nitrogen source (Raimbault and Alazard, 1980; Yang, 1988). A strain of *Aspergillus fumigatus* B 149 isolated produced alkaline protease on wheat bran and pigeon pea residue. Amylase and protease are two industrially important enzymes and assayed in the test fungi. These are degrading enzymes and are commercially isolated using protoplast released from plant cell. Usually enzymes will be detected ten days after inoculation or at the on set of fruiting which may be reason for the earlier activity in the seventh day (Kaviyarasan and Natarajan, 1997).

From the above study, it is concluded that in *L. fraterna* showed more of amylase and protease activity than *P. ostreatus*, and under both solid-state and submerged condition in rice bran and wheat bran.

References

Agger, T., Spohr, A.B. and Nielsen J., 2001. α-Amylase production in high cell density submerged cultivation of *Aspergillus oryzae* and *A. nidulans*. *Appl. Microbial. Biotechnol.*, 55: 81–84.

Amadioha, A.C., 1998. Effects of cultural condition on the growth and amylolytic enzyme production by *Rhizophus oryzae*. Acta phytopathologica et. *Entomologica Hungarica*, 33: 1–2.

Cabaleiro, D.R., Rodriguez, S., Sanroman, A. and Longo, M.A., 2002. Comparison between the protease production ability of ligninolytic fungi cultivated in solid state media. *Proc. Biochem.*, 37: 1017–1023.

Fadel, M., 2000. Economic utilization of agro-industrial wastes through solid state fermentation by *Aspergillus niger* F-21 for alpha amylase production. *Egyptian Journal of Microbiology*, 35: 173–189.

Fernandez, Lahore, H.M., Fraile, E.R. and Cascone, 1998. Acid protease recovery from a solid-state fermentation system. *Journal of Biotechnology*, 62: 83–93.

Goes, A.P. and Sheppard, J.D., 1999. Effect of surfactants on α-amylase production in a solid substrate fermentation process. *Journal of Chemical Technology and Biotechnology*, 74: 709–712.

Hesseltine, C.W., 1977. *Solid-state Fermentation, Part 1: Process Biochemistry*, 12: 24–27.

Hongzhang, C.W., Hui, Z. Aijun and Zuohu, L., 2006. Alkaline protease production by solid-state fermentation on polyurethane foam. *Chem. Biochem. Eng.*, 20(1): 93–97.

Hye-Seon, Choi and Hymn-Hee, Skin, 1998. Purification and partial characterization of a fibrolytic protease in *Pleurotus ostreatus*. *Mycologia*, 90(4): 674–679.

Joshi, Vinod Kumar, Parmar, Mukesh and Rana, Neerja S., 2006. Pectin esterase production from apple pomace in solid-state and submerged fermentations. *Food Technol. Biotechnol.*, 44(2): 253–256.

Kathiresan, K. and Manivannan, S., 2006. Amylase production by *Penicillium fellutanum* isolated from mangrove rhizosphere soil. *African Journal of Biotechnology*, 5(10): 829–832.

Kaviyarasan, V. and Natarajan, K., 1997 Changes in extra cellular enzyme activities during growth and fruiting of *Pleurotus cornucopiae* var. *citrinopileatus*. In: *Advances in Mushroom Biology and Production*, (Eds.) Rai, R.D., Dhar, B.L. and Verma, R.N. National Research Centre for Mushroom, Solan, pp. 309–320.

Lin, L.L., Hsu, W.H. and Chu, W.S., 1997. A gene encoding for amylase from thermophilic *Bacillus sp.*, strain TS-23 and its expression in *Escherichia coli*. *J. Appl. Microbiol.*, 82: 325–334.

Matsushima, K.M., Hayakawa, M. Ito and Stimada, K., 1981. Ecological study on the proteolytic enzymes of micro-organisms. Part III. Features of the proteolytic enzymes system of hyper-acid-productive and non-acid productive fungi. *J. Gen. Appl. Microbiol.*, 27: 423–426.

Pandey, A., 2003. Solid-state fermentation. *Biochem. Eng. J.*, 13: 81–84.

Pandey, A., Selvakumar, P., Soccol, C.R. and Nigam, P., 1999. Solid-state fermentation for the production of industrial enzymes. *Current Science*, 77: 149–162.

Pandey, A., Soccol, C.R. and Mitchell, D., 2000. New developments in solid state fermentation. Part I. bioprocesses and applications. *Process Biochem.*, 35: 1153–1169.

Raimbault, M. and Alazard, D., 1980. Culture method to study fungi growth in solid fermentation. *Europ. J. Appl. Microbiol. Biotechnol.*, 9:199–209.

Soccol, C.R., Iloki, I., Marin, B. and Raimbault, M., 1994. Comparative production of α-amylase, glucoamylase and protein enrichment of raw and cooked cassava by *Rhizopus* strains in submerged and solid-state fermentations. *Journal of Food Science and Technology*, 31: 320–323.

Tengerdy, R.P., 1998. *Advances in Biotechnology*, (Ed.) A. Pandey. Educational Publishers and Distributors, New Delhi, p. 13–16.

Tunga, R., Banerjee, R. and Bhattacharya, B.C., 1999. Some studies on optimization of extraction process for protease production in SSF. *Bioprocess Engineering*, 20: 485–489.

Yang, S.S. and. Huang, C.I., 1994. Protease production by amylolytic fungi in solid-state fermentation. *J. Chin. Agric. Chem. Soc.*, 32: 589–601.

Yang, S.S., 1988. Protein enrichment of sweet potato residue with amylolytic yeast by solid-state fermentation. *Biotechnol. Bioeng.*, 32: 886–890.

Chapter 8

Studies on the Distribution of Photosynthetic Pigments in a Lotic Ecosystem, Vamanapuram River, Kerala, South India

I. Mini, C.G. Radhika and T. Ganga Devi*

Department of Botany, University College, Thiruvananthapuram – 695 034, Kerala

ABSTRACT

Water sampled from ten Stations along the Vamanapuram river at monthly intervals for a period of one year from February 1999 to January 2000 was analysed for photosynthetic pigments such as chlorophyll a, b, c, carotenoids and phaeopigments. The monthwise distribution of chlorophyll *a* (bdl to 30.94 mg m^{-3}), chlorophyll *b* (bdl to 57.19 mg m^{-3}) chlorophyll *c* (bdl to 86.57 mg m^{-3}), Carotenoids (bdl to 10.87 mg m^{-3}) and Phaeopigments/phaeophytin (bdl to 32.47 mg m^{-3}) exhibited spatial as well as temporal variations. The study revealed comparatively low plankton population in terms of pigments due to the swift movement of water coupled with reduced residence time of plankton in a lotic system as compared to a lentic environment. The ratio among various photosynthetic pigments (>1) indicated a healthy crop of plankton in the river with significant variation in species composition. High concentration of chlorophyll *c* may be suggestive of abundant diatom population. ANOVA substantiated significant spatial and temporal variation noticed in the distribution of pigments.

Keywords: *Chlorophyll, Phaeopigments, Lotic system, Plankton.*

---

* Former Principal, Government College for Women, Thiruvananthapuram – 695 014, Kerala.

Introduction

Estimation of chlorophyll is important for water resource management as well as marine and fresh water research. The identification and quantification of photosynthetic pigments in water provide information on the distribution and abundance of phytoplankton in aquatic ecosystems. It denotes the diversity of planktonic flora, productivity of an aquatic system and water quality as well. It is an effective tool in the identification of eutrophic waters as frequent water blooms of eutrophic waters results in unpleasant odour, taste and toxins that adversely affect water quality. The standing crop of plankton indicates the availability of food for animals at the primary stage (Sarupriya and Bhargava, 1997). Chlorophyll and Carotenoids occur either singly or in combination have been successfully used for chemosynthetic identification of plankton in oceanic waters (Jeffrey *et al.*, 1975; Gieseks and Kraay, 1983; Barlow *et al.*, 1993).

The estimation of pigments in marine and estuarine systems were reported by various authors (Devassy and Bhattathitri, 1981; Sasammal *et al.*, 1985; Nittala and Rao, 1989; Satyanarayana *et al.*, 1994; Vijayakumaran *et al.*, 1996; Jiyalalram *et al.*, 1997; Nair *et al.*, 1998; Sarupriya and Bhargava, 1998; Redekar and Waugh, 2000; Rajasegar *et al.*, 2000). Studies on pigments constitution of fresh water environments are very few such as on Chilka Lake (Raman *et al.*, 1990), Periyar (Joy *et al.*, 1990), Moosi River (Nirmala kumari *et al.*, 1991; Rani and Venkateswaralu, 1999), Karamana River (Jayaraman, 2002) and Vellayani Lake (Radhika, 2005). The present investigation is aimed at analyzing the phytoplankton population in terms of pigment constitution to draw conclusions about the diversity as well as spatial and temporal variation.

Materials and Methods

The Vamanapuram River with a length of 88 km and a drainage area of 687 sq.km. and lying between longitudes 76°44–77°12′E and latitudes 8°49′N, originates from Chemmungi mottai in Western ghats flows southwards through Ponmudi and Palode, then meanders westwards towards Vamanapuram and drains into Arabian Sea (Lakshadweep Sea) through Muthalapozhi. Ten Stations along the river were identified for the collection of samples. Station 1–Kallar means stone river which flows through a bed of stone is situated under thick cover of vegetation and water is cold and clear. Station 2–Anapara is located where a major tributary Chittar joins the main river and is wider and less disturbed by anthropogenic activities. Station 3–Chettachal characterized by sandy river bed with layers of submerged rock and water is pumped to nearby places. Station 4–Palode where the river bed is more or less sandy with huge rocks in between and a temporary check dam is constructed across the river. Station 5–Aruvippuram is about 40.5 km from the bar mouth. Sand mining is a regular feature at this station. Station 6–Vamanapuram is situated about 27.5 km from the bar mouth. Anthropogenic pressure is more at this station. Station 7–Poovanpara is about 12 km away from the barmouth where the river takes a broader flow and water is used for bathing, washing and cleaning vehicles. Station 8–Kollampuzhakadavu located at a distance of 2 km down from the seventh station and river water is saline due to the intrusion of sea water. Station 9–Pulimootilkadavu is an estuarine environment with abundant coconut retting and coir manufacturing units and sand mining is a regular feature. Station 10–Muthalapozhi is the location where a sand bar separates the river from the Arabian sea. The sand bar opens during floods the resulted in the intrusion of sea water in to the river.

Water samples were collected at monthly interval from ten stations along the Vamanapuram river for a period of one year from February 1999 to January 2000. The chlorophylls, carotenoids and phaeopigments were estimated by the method of Strickland and Parsons (1977) for which the water samples were filtered through 0.45 μm Millipore membrane filter in a Millipore vacuum filtration unit and extracted with 90 per cent acetone and kept overnight at 4°C, centrifuged and read spectrophotometrically. Phaeopigments were analysed after acidification with hydrochloric acid. The quantity of pigments was calculated using SCOR/UNESCO (1966) and expressed in mg m^{-3}. The period of study was divided in to three seasons–pre-monsoon (February to May), monsoon (June to September) and post-monsoon (October to January) depending on the intensity and duration of rain fall in order to explain the variability of each parameter. The results were subjected to statistical analysis (ANOVA) to elucidate the spatial and temporal variation of the parameters under investigation.

Results and Discussion

The month-wise distribution of chlorophyll *a* varied from below detectable level (bdl) to 30.94 mg m^{-3} with an annual average between 0.95 mg m^{-3} at Station 1 to 5.69 mg m^{-3} at Station 10. The seasonal values (Figure 8.1) varied from 0.26 mg m^{-3} to 7.63 mg m^{-3}, 0.15 mg m^{-3} to 3.72 and 0.84 mg m^{-3} to 8.72 mg m^{-3} during pre-monsoon, monsoon and post-monsoon respectively. The concentration of chlorophyll *a* remained high at Station 10 irrespective of the seasons. The maximum concentration of chlorophyll *a* was registered during pre-monsoon (Stations 3, 6, 9 and 10) and post-monsoon (Stations 2, 4, 5, 7 and 8) and minimum at all stations during monsoon except at Station 1 where the minimum was during pre-monsoon. Maximum chlorophyll *a* at majority of the stations encountered during pre-monsoon may be as a result of high plankton density due to conducive atmosphere (low flow rate, high temperature, insinuation of light etc). The production of phytoplankton is temperature dependent and is accelerated with increase in temperature between 25 to 40°C (Joy *et al.*, 1990) and upwelling of water currents coupled with clear skies with enhanced irradiance (Zingone *et al.*, 1995). Chlorophyll *a* plays a pivotal role and is an effective measure in the primary productivity of any aquatic system (Rani and Venkateswarlu, 1999). It is the universal pigment participating directly in the photochemical act. The chlorophyll *a* content encountered in the present investigation is in unison with the reports of Rani and Venkateswarlu (1999) in River Moosi (12.4 µg L^{-1} during June to 112.0 µg L^{-1} during March) and Rajasegar *et al.* (2000) in Zuari estuary (0.04–20.72 mg m^{-3}), and inferred that minimum and maximum concentrations were observed during monsoon and summer respectively. Comparatively

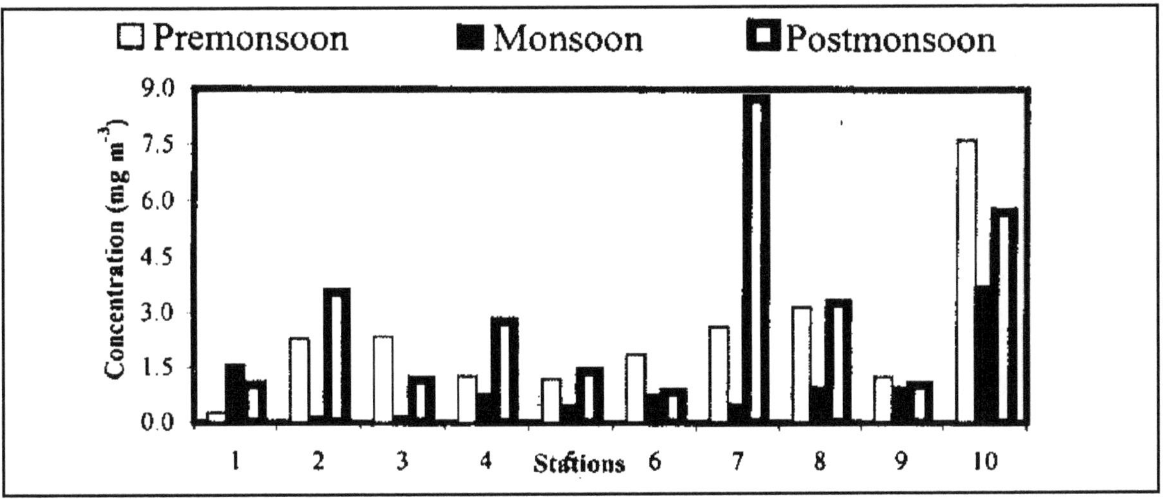

Figure 8.1: Seasonal Distribution of Chlorophyll *a*

higher values were documented by Sasamal *et al.* (1985) in coastal waters of north western Bay of Bengal (3.95–5.55 mg m^{-3}), Joy *et al.* (1990) in Periyar river (0.20–54.48 mg m^{-3}), Satyanarayana *et al.* (1994) in VIsakhapatnam harbour (1.65–2.85 mg m^{-3}), Jiyalalram *et al.* (1997) in Uran, Maharashtra (2.1–11.8 mg m^{-3}) and Nair *et al.* (1998) in Vasishti estuary (3.1 mg m^{-3}). Comparatively lower values were reported in marine environments by Sarojini and Subbarangiah (2000) in the coastal waters of Bay of Bengal along northern Andhra Pradesh (0.001 mg m^{-3}–0.090 mg m^{-3}) and Bhargava *et al.* (1978) in Arabian sea (0.023–0.578 mg m^{-3}) Raman *et al.* (1990) suggested that high concentration of chlorophyll *a* is often associated with high population density of phytoplankton. The variation noticed in the chlorophyll *a* content in the present investigation may be due to the fact that the streams are subjected to fluctuating conditions and are not characterised by any definite plankton species peculiar to it as suggested by Palmer (1980). The distribution of Chlorophyll *a* exhibited statistically significant month-wise (d.f. = 11, F = 2.527217, $p \leq 0.007$) and season-wise (d.f. = 2, F = 4.510191, $p \leq 0.025$) whereas spatial variation was insignificant.

The month-wise distribution of chlorophyll *b* ranged from bdl to 57.19 mg m^{-3} and annual average fluctuated between 0.25 mg m^{-3} to 5.60 mg m^{-3} at Station 5 and 7 respectively. The seasonal average varied from (Figure 8.2) 0.22 mg m^{-3} to 2.63 mg m^{-3}, 0.12 to 1.95 and bdl to 15.58 mg m^{-3} during pre-monsoon, monsoon and post-monsoon respectively. They are the dominant pigments of Chlorophyta, Charophyta and Euglenophyta. In green algae chlorophyll *a* and *b* make up about two third of the total pigments in the chloroplast. Chlorophyll *b* may also act to stabilise and protect chlorophyll *a* in the algal cells. The high chlorophyll b content during July, August and January at various stations might have reflected in the maximum values noticed during monsoon and post-monsoon seasons indicative of high productivity and abundance of chlorophycean and euglenoid algae in the surface waters. The high nutrient enrichment associated with the southwest and northeast monsoonal rains may account for the above observation. The swift flow of water at Stations 1 and 3 may be the reason for the decrease in the chlorophyll *b* content as compared to the rest of the stations. Joy *et al.* (1990) analysed the chlorophyll *b* content in Periyar (0.20 to 28.04 m^{-3}) which is in agreement with the values obtained in the present study. Higher values were noticed at all stations of the Vamanapuram river as compared

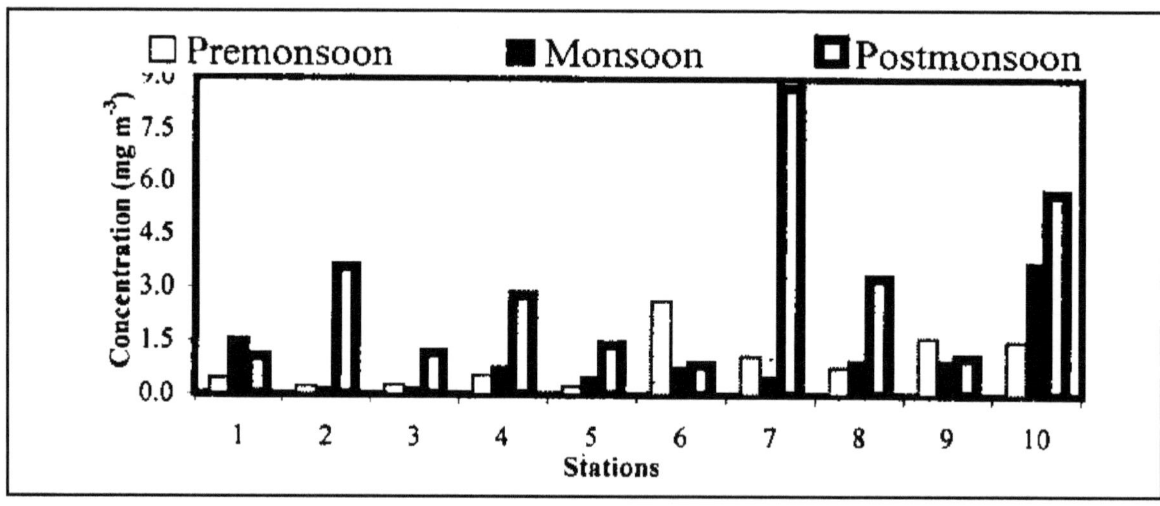

Figure 8.2: Seasonal Distribution of Chlorophyll *b*

to the values in Visakhapatnam harbour (bdl to 2.72 mg m^{-3} where a significant correlation was observed with primary production (Vijayakumaran *et al.*, 1996) and coastal waters of Bay of Bengal (0.001 and 0.540 mg m^{-3}) with a mean value of 0.14 mg m^{-3} (Sarojini and Subbarangaiah, 2000). Significant statistical variation in the distribution of chlorophyll *b* was lacking either spatially or temporally in the Vamanapuram river.

Chlorophyll *c* is common in certain categories of phytoplankton such as diatoms, dinoflagellates and brown algae. Month-wise values of chlorophyll *c* ranged from bdl to 86.57 mg m^{-3} at station 7 during January and an annual average between 1.38 mg m^{-3} (Station 3) to 13.69 mg m^{-3} (Station 7). The seasonal data (Figure 8.3) varied from 0.96 to 3.91 mg m^{-3}, 1.77 to 15.23 mg m^{-3} and 0.88 to 31.61 mg m^{-3} during pre-monsoon, monsoon and post-monsoon respectively. Chlorophyll *c* content recorded high values during pre-monsoon (Stations 6, 9 and 10), monsoon (Stations 1, 3, 4 and 5) and post-monsoon (2, 7 and 8). Minimum concentration of chlorophyll *c* was exhibited at Stations 3, 4, 5 and 6 located in the fresh water stretch of the river during post-monsoon. There is remarkable increase in the chlorophyll c content at Stations 7 and 8 during post-monsoon. The concentration of chlorophyll c remained high as compared to chlorophyll *b* which is indicative of the abundance of diatoms in the river water. Devassy and Bhattathiri (1981) suggested that the coastal waters rich in nutrients can sustain highest population of diatoms. Joy *et al.* (1990) reported chlorophyll *c* content (0.11 to 22.26 mg m^{-3}) in Periyar which corroborates with the present study. The chlorophyll *c* content noticed at various stations of the Vamanapuram river was comparatively higher than the reported values of Sarojini and Subbarangaiah (2000) in Bay of Bengal (0.001 to 0.144 mg m^{-3}) and Vijayakumaran *et al.* (1996) in Visakhapatnam harbour (0 to 3.77 mg m^{-3}). Significant statistical variation was absent either spatially or temporally in the month-wise or season-wise distribution of chlorophyll *c*.

Carotenoids exhibited monthly values ranging between bdl to 10.87 mg m^{-3} at Station 10. The annual average values remained low (0.25 mg m^{-3}) at Station 1 and high (4.10 mg m^{-3}) at Station 10. Season-wise data (Figure 8.4) showed values ranging from 0.41 to 7.43 mg m^{-3}, 0.03 to 2.24 mg m^{-3} and 0.04 to 2.65 mg m^{-3} during pre-monsoon, monsoon and post-monsoon respectively. Chloroplasts of

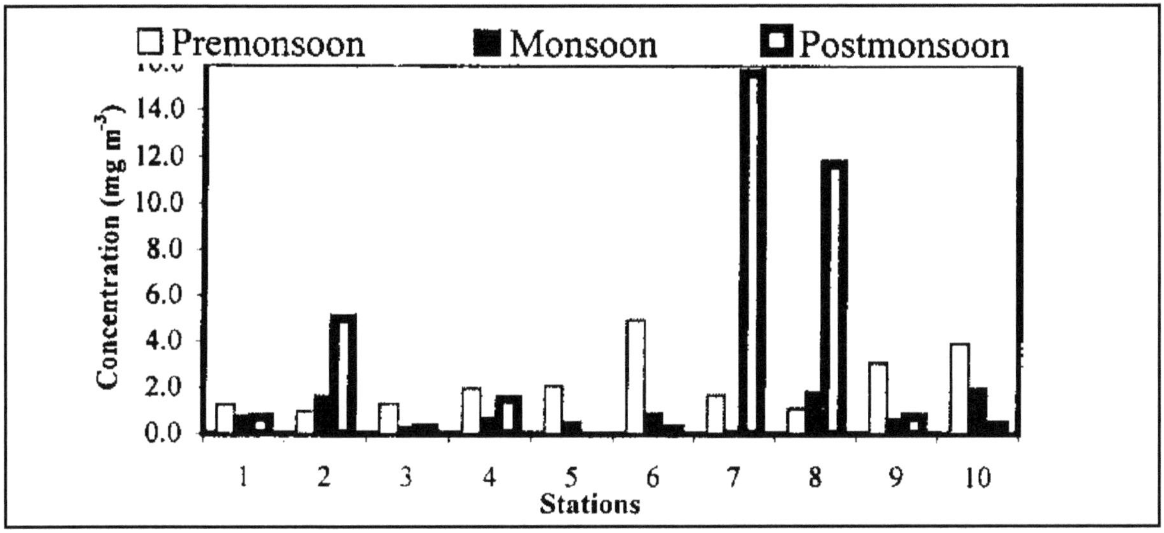

Figure 8.3: Seasonal Distribution of Chlorophyll *c*

photosynthetic organisms contain bewildering variety of carotenoids, a group of yellow, orange and brown pigments whose principal role is that of extending the spectral region of harvesting light with subsequent energy and transfer to chlorophyll. A possible subsidiary role of carotenoids is that of protection of chlorophyll at high light intensities. The carotenoid content was high during pre-monsoon at all stations of the Vamanapuram river. The concentration of carotenoid was high at Station 10 irrespective of the seasons as compared to the rest of the stations along the entire stretch of the river. The minimum carotenoid content was registered during monsoon (Stations 8, 9 and 10) in the saltwater stretch of the river while it was low either during monsoon (Stations 3, 4 and 5) or post-monsoon (Stations 1, 2, 6 and 7) in the fresh water stretch of the river. The distribution of carotenoid content in the river water followed a general pattern of pre-monsoon > post-monsoon > monsoon at Stations 8, 9 and 10 located downstream. The low carotenoid content encountered during monsoon and post-monsoon may be attributed to the heavy rainfall coupled with high flow rate reducing the retention of plankton in the river water. Vijayakumaran *et al.* (1996) reported high carotenoid content while assessing the surface productivity of Visakhapatnam during summer remained in unison with the results of the present investigation. The distribution of carotenoid in the entire stretch of the Vamanapuram river exhibited definite spatial (d.f. = 9, F = 6.886147, $p \leq 0.0001$) and monthwise (d.f. = 11, F = 15. 82982, $p \leq 0.0001$) variations while the season-wise variation remained insignificant.

The month-wise data on Phaeopigments/phaeophytin ranged from bdl to 32.47 mg m^{-3} with annual average values between 0.37 mg m^{-3} and 3.52 mg m^{-3} at Stations 1 and 10 respectively. The season-wise data varied (Figure 8.5) from 0.37 to 2.39 mg m^{-3}, bdl to 4.14 mg m-3 and 0.12 to 9.74 mg m^{-3} during pre-monsoon, monsoon and post-monsoon respectively. Chlorophylls tend to loose their magnesium metal under mild acid conditions, forming a characteristic pigment called phaeopigments or phaeophytin in natural aquatic environments. Phaeopigments have long been recognised as degradation products formed during isolation of chlorophyll from plant tissues. The concentration of phaeopigments is mainly governed by the physicochemical as well as biotic conditions leading to the degradation of chlorophyll. The growth of phytoplankton, cell death, grazing by herbivores and the

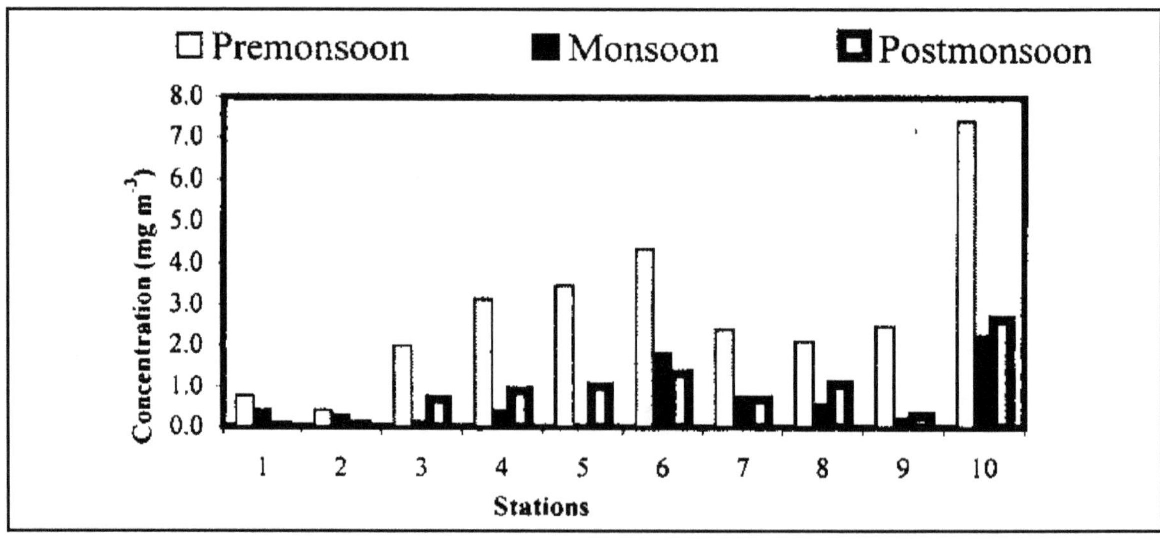

Figure 8.4: Seasonal Distribution of Chlorophyll Carotenoids

faecal pellets of zooplankton may alter the pheophytin content. Moreover, the quantification of phaeophytin is an effective measure to describe the physiological state of plankters and the extent of degradation of chlorophyll in the aquatic ecosystem. The high phaeophytin content was exhibited during monsoon or post-monsoon (Stations 1 and 6) at majority of the stations located in the fresh water stretch of the Vamanapuram river may be due to the combined effect of high productivity along with zooplankton grazing. Yashel *et al.* (1998) opined that phytoplankton grazing accounted for the significant elevation in the pheophytin content while analysing the phytoplankton distribution and grazing near coral reefs. The concentration of phaeophytin (1.0–3.5 mg m^{-3}) in the oil terminal at Uran in Maharashra (Jiyalalram *et al.*, 1997) which is lower than the values observed in the present investigation. The high concentration of phaeopigments at the estuarine stations during pre-monsoon may be attributed to the reduced flow of water as well as characteristic and conducive hydrological conditions and low values in the surface waters may be due to its efficient settling onto the sediments. Lorenzen and Welschmeyer (1983) indicated that the phaeopigments produced by herbivores are in turn packaged into pellets that exhibit short residence time in water column due to their rapid sinking rates. Nakajima (1973) opined that the phaeopigment content indicated the physiological state of phytoplankton and was governed mainly by the intensity of light and grazing pressure.

The ratios relating chlorophyll *a* : *b*, *a* : *c*, *a* : carotenoids and *a* : phaeophytin (Table 8.1) indicated the physiological state of plankton. The seasonal data of chlorophyll *a* : *b* varied from 0.59 to 10.41, 0.09 to 2.09, bdl to 14.04 during pre-monsoon, monsoon and post-monsoon respectively. Seasonal fluctuations revealed the occurrence of a composite group of plankton in Vamanapuram river. The seasonal average values chlorophyll *a* : *c* ranged from 0.21 to 2.94, 0.03 to 1.82 and 0.10 to 2.91 during pre-monsoon, monsoon and post-monsoon respectively. The chlorophyll *a* to carotenoid registered a maximum of 62.09 and a minimum of 0.34 at Stations 2 and 1 and seasonal data fluctuated from 0.34 to 5.55, 0.43 to 16.00 and 0.63 to 62.09 during pre-monsoon, monsoon and post-monsoon respectively. Schlueter *et al.* (1997) opined that wide fluctuations in the chlorophyll *a* to carotenoid ratio may be due to low phosphorous and nitrogen content and is in agreement with the reports of hydrological parameters in the Vamanapuram river water (Mini *et al.*, 2003). The seasonal average of chlorophyll *a*

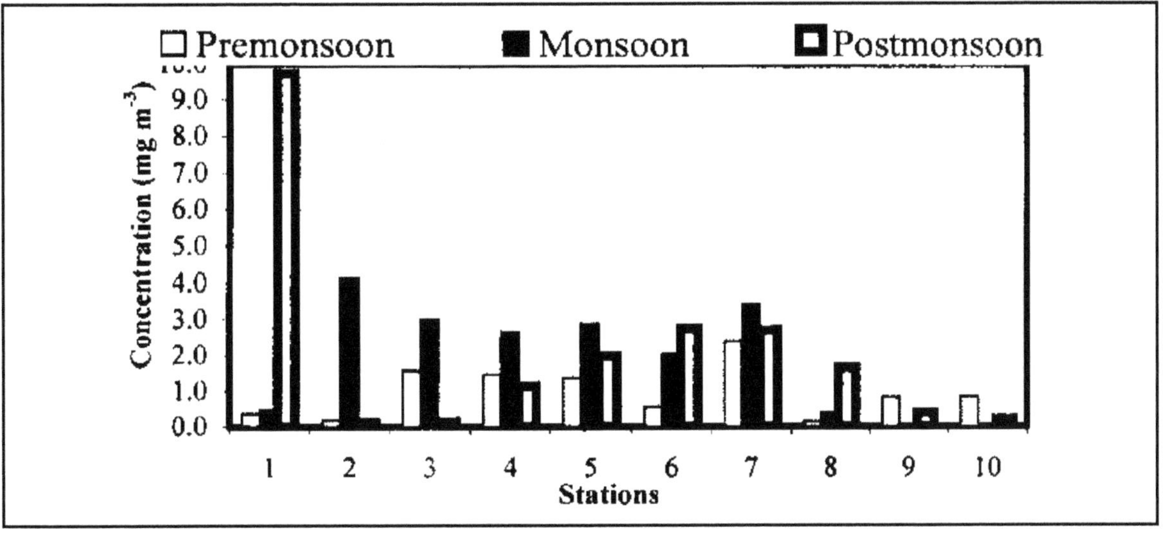

Figure 8.5: Seasonal Distribution of Chlorophyll Phaeopigments

to phaeopigments varied from 0.71 to 22.61, 0 to 3.34 and 0.11 to 30.04 during pre-monsoon, monsoon and post-monsoon respectively. For healthy crop the values may be greater than unity. Millie *et al.* (1993) opined that although the pigments vary within taxon or between taxa, the abundance of diagnostic pigments generally reflects the major distribution of phytoplankton to the division or class level. The ratios fluctuate within each station and among stations may be as a result of the lotic environment, physicochemical characteristics of the water body, diversity in the plankton population and physiological state of plankters.

Table 8.1: Seasonal Variation of chl *a* : chl *b*, chl *a* : chl *c*, chl *a* : ctd, chl *a* : Phaeopigments in Water

Station	*Season*	*chl a/chl b*	*chl a/chl c*	*chl a/ctd*	*chl a/phaeo*
	A	0.59	0.21	0.34	0.71
1	B	2.09	0.43	3.83	3.34
	C	1.45	0.52	29.71	0.11
	A	10.41	2.39	5.55	12.85
2	B	0.09	0.08	0.51	0.04
	C	0.72	0.29	62.09	30.04
	A	8.74	1.85	1.19	1.52
3	B	0.54	0.08	1.21	0.05
	C	4.68	1.33	1.71	8.35
	A	2.36	0.65	0.41	0.88
4	B	1.16	0.35	1.89	0.30
	C	1.88	1.98	3.13	2.45
	A	5.05	0.58	0.35	0.89
5	B	0.92	0.13	16.00	0.17
	C	2.74	1.71	1.39	0.71
	A	0.71	0.38	0.43	3.40
6	B	0.87	0.44	0.43	0.38
	C	3.61	4.00	0.63	0.31
	A	2.40	1.58	1.10	1.10
7	B	4.17	0.03	0.67	0.15
	C	0.56	0.36	12.62	3.25
	A	4.21	2.94	1.51	22.61
8	B	0.53	0.35	1.69	2.57
	C	0.28	0.10	3.08	2.01
	A	0.78	0.40	0.51	1.53
9	B	1.52	0.42	4.36	0.96
	C	1.39	0.40	3.71	2.71
	A	5.10	1.95	1.03	9.07
10	B	1.91	1.82	1.66	5.17
	C	14.04	2.91	2.16	26.47

A: Pre-monsoon; B: Monsoon; C: Post-monsoon.

Conclusion

The study revealed comparatively low plankton population in terms of photosynthetic pigments in the lotic ecosystem due to swift movement of water reducing the residence time of plankton unlike lentic water bodies. Seasonal fluctuations in chlorophyll *a* to *b* ratio revealed the occurrence of a composite group of plankton in Vamanapuram river. The concentration of various photosynthetic pigments and their ratios (>1) is suggestive of the presence of a healthy crop of plankton in the Vamanapuram river. The diatom population may be abundant, represented by the high chlorophyll *c* content as compared to the rest of the pigments under investigation. Wide fluctuations in the *a* to carotenoid ratio may be due to low phosphorous and nitrogen content in the river water. In general, the physicochemical factors influenced production and distribution of photosynthetic pigments which in turn reflected the pattern of spatial and temporal distribution of phytoplankton communities in the river.

Acknowledgement

The authors are highly indebted to The Principal, University college, Thiruvananthapuram for the facilities provided. One of the authors expresses sincere gratitude to UGC for financial assistance.

References

Barlow, R.G., Mantoura, R.F.C., Gough, M.A. and Fheman, T.W., 1993. Pigment signatures of the phytoplankton composition in the northeastern Atlantic during the 1990 spring bloom. *Deep Sea Res.*, 40: 459–477.

Devassy, V.P. and Bhattathiri, P.M.A., 1981. Distribution of phytoplankton and chlorophyll *a* around Little Andaman Island. *Ind. J. Mar. Sci.*, 10: 248–252.

Gieskes, W.W.C. and Kraay, G.W., 1983. Unknown chlorophyll *a* derivatives in the North Sea and the tropical Atlantic Ocean revealed by HPLC analysis. *Limnol. Oceanogr.*, 28: 757–766.

Jayaraman, P.R., 2002. Impact of environmental factors on the macroflora of a lotic ecosystem Karamana river: A case study. *Ph.D. Thesis*. Aquatic Biology and Fisheries, University of Kerala, pp. 413.

Jeffrey, S.W., Sielicki, M. and Haxo, F.T., 1975. Chloroplast pigment patterns in dinoflagellates. *J. Phycol.*, 11: 374–384.

Joy, C.M., Balakrishnan, K.P. and Ammini Joseph, 1990. Effect of Industrial discharges on the ecology of phytoplankton in the river Periyar (India). *Water. Res.*, 24: 787–796.

Jiyalalram, M., Desai, B.N. and Abidi, S.A.H., 1984. Diurnal variations of phytoplankton pigments and population in near shore waters of Thai (Maharashtra). *Mahasagar*, 17: 41–47.

Lorenzen, C.J. and Welschmeyer, M.A., 1983. The *in situ* sinking rates of herbivore faecal pellets. *J. Plankton. Res.*, 5: 929–933.

Mini, I., Radhika, C.G. and Ganga Devi, T., 2003. Hydrological studies on a lotic ecosystem Vamanapuram river, Thiruvananthapuram, Kerala, South India. *Poll. Res.*, 23(4): 617–626.

Nair, V.R., Mustaffa, S., Mehta, P., Govindan, K., Ram, M.J. and Gajbhiye, S.N., 1998. Biological characteristics of the Vasishti estuary, Maharashtra (West coast of India). *Ind. J. Mar. Sci.*, 27(3–4): 310–316.

Nakajima, K., 1973. *Mem. Fac. Fish.*, Hokkaido. Univ., 20: 1.

Nittala, S. Sharma and Rao, M.U., 1989. Alkali and alkaline earth metals in surface sediments of Bhimunipatnam–Amalpuram, Central East coast of India (Bay of Bengal). *Ind.* J. *Mar. Sci.*, 28: 375–379.

Palmer, C.M., 1980. *Algae and Water Pollution*. Castle House Publications Ltd., p. 119.

Radhika, C.G., 2005. Limnological studies of a tropical fresh water lake (Vellayani Lake) with special reference to associated flora. *Ph.D. Thesis*, University of Kerala.

Rajasegar, M., Srinivasan, M. and Rajaram, R., 2000. Phytoplankton diversity with shrimp farm development in Vellar estuary, S. India.

Raman, A.V., Satyanarayana, K., Adiseshadri and Prakash, K.P., 1990. Phytoplankton characteristics of the Chilka lake: A brackish water lagoon along the east coast of India. *Ind. J. Mar. Sci.*, 19: 274–277.

Rani, S.P. and Venkateswarlu, V., 1999. Study of biochemical aspect of chlorophyll *a* in the river Moosi. *Ecol. Env. and Cons.*, 5(1): 77–78.

Sarojini, Y. and Subbarangaiah, 2000. Spatial variation in distribution of chlorophyll and phytoplankton in coastal waters of Bay of Bengal along northern Andhra Pradesh. *Seaweed Res.*, 22(1 and 2): 93–100.

Sarupriya, J.S. and Bhargava, R.M.S., 1997. Seasonal distribution of chlorophyll *a* in the exclusive economic zone (EEZ) of India. *Ind. J. Mar. Sci.*, 27: 292–297.

Sasamal, R., Panigrahy, R.C. and Sahu, B.K., 1985. Distribution of photosynthetic pigments and particulate organic carbon in coastal waters of northwestern Bay of Bengal. *Ind. J. Mar. Sci.*, 14: 167–168.

Satyanarayana, D., Sahu, S.D. and Panigrahy, P.K., 1994a. Distribution of phytoplankton pigments and particulate organic carbon in the coastal waters of Visakhapatnam (Bay of Bengal). *Ind. J. Mar. Sci.*, 23: 47–51.

Schlueter, J., Riemann, B. and Soendegaard, M., 1997. Nutrient limitation in relation to phytoplankton and chlorophyll *a* to carotenoid ratios in fresh water mesocosms. *J. Plankton Research*, 19: 891–906.

SCOR/UNESCO, 1966. *Monographs on Oceanographic Methodology*. Publ. UNESCO.

Strickland, J.D.H. and Parsons, T.R., 1977. *A Practical Handbook of Seawater Analysis*, 2nd edn. Bulletin of the Fisheries Research Board of Canada, 310 pp.

Redekar, P.D. and Waugh, A.B., 2000. Studies on fouling diatoms from the Zuari estuary, Goa (West coast of India). *Seaweed Res.*, 22(1 and 2): 113–119.

Yashel, D., Vaulot, D. and Genin, A., 1998. Phytoplankton distribution and grazing near coral reefs. *Limnol. Oceanogr.*, 43(4): 551–563.

Zingone, A., Casotti, M.R., d'Akala, M. Scardi and Marino, D., 1995. St. Martin's Summer: The case of an autumn phytoplankton bloom in the Gulf of Naplis (Mediterranean Sea). *J. Plankton Res.*, 17: 375–393.

Chapter 9

In vitro Selection of Secondary Metabolites of Medicinal Plants

M. Venkateshwarlu

Department of Botany, Kakatiya University,
Warangal – 506 009, Andhra Pradesh

ABSTRACT

Ayurvedic medicinal plants are rich in secondary metabolites, which are potential sources of drugs and approximately 25 per cent of, prescribed medicines (Farnsworth and Bingel, 1977). It has been estimated that about 1000 Ayurvedic remedies, prepared from around 750 plants are being used at present. Plant metabolites are major sources of pharmaceuticals, food additives, fragrances and pesticides. It is estimated that the market potential for herbal drugs in the western world alone could be from US $ 4.9 billion to 47 billion by the year 2000 if the AIDS epidemic continues unchecked. Nearly 75–80 per cent of the world population depends upon crude plant drug preparations to tackle their health problems. Ayurveda is an ancient Indian system of medicine. On the global basis, at least 130 drugs, all of single chemical entities have been extracted from higher plants, and have been care fully standardized, and their efficacy and safety for a suggested application demonstrated. In several cases considerable research has also been carried out on their mechanism of action.

Introduction

The secondary metabolites often have complex stereo structure with many chiral centres, which may be essential for biological activity, many of these, cannot be synthesized economically on a commercial basis (Farnsworth and Morris, 1976). Besides these plants collected as minor forest produce, show a wide disparity in their values, due to lack of information on their life cycles; maturity and regeneration times, all of which change the quality of active chemical ingredients present (Balick, 1994). The continuous and non-organized exploitation has resulted in many plants becoming rare and some even become extinct. There are fluctuations in the concentrations and quantities of secondary metabolites in field grown plants as the biosynthesis of secondary metabolites, although controlled genetically, is affected strongly by environmental influence. They are therefore extracted from plants cultivated in fields or growing in wild stands. It has been reported that nearly 95 per cent of the plants used traditionally as ingredients in crude drugs are collected from forests and other natural resources (Lozoya, 1994).

Table 9.1: Some Medicinal Plant Crude Drugs which have Received Pharmacological Clinical Support for their Therapeutic Claims

Sl.No.	Botanical Name	Active Component
1.	*Acorus calamus*	Calamus oil with asarone
2.	*Adathaoda zeylanfca*	Vasicine
3.	*Andrographis paniculata*	Andrographolide
4.	*Asparagus racemoses*	Shatavarin I
5.	*Azadirachta indica*	Gedunin
6.	*Bacopa monneri*	Baccosides
7.	*Boerhaavia diffusa*	Puranavine
8.	*Butea frondosa*	Palasonin
9.	*Centella asiatica*	Asiaticosides
10.	*Curcuma longa*	Curcumin
11.	*Halarrhena antidysentrica*	Conessine
12.	*Picrorhiza kurroa*	Picroside, Kutocide
13.	*Psoralea corylifolia*	Psoralen, Bakuchiol
14.	*Swertia chirata*	Ciratin, Ophelic acid

Materials and Methods

Recently variation in MS media composition and selection of high yielding cell line in the callus have been used to get higher yield of drugs. Changing the media composition has been found to bring about changes in the biosynthetic ability of the cultured cells. Various production media have been developed that remarkably increased the yield of certain products, *viz.* ajamalcine in *Catharanthus roseustisue* culture (Knobloch *et al.*, 1982), diosgenin in *Dioscorea species* (Tal and Goldberg 1982) and shikonin in *Lithospermum eryhrorhizon* (Fujita *et al.*, 1981). Plant tissue and cell cultures consist of a heterogeneous cell population and it is now possible to select and clone high drug producing cell lines/strins. Using the cell selection method, the production of ubiquinone was augmented 10 times to nearly 15 times in the case *Coleus blumei* (Matsumoto *et al.*, 1982). The important energy source used in the media was sucrose. Increased concentrations of sucrose have been favourable for the synthesis of shikonin in cell cultures of *Lithospermum erythrorhizon*, diosgenin in *Dioscorea* ad anthroquinone production in cell cultures of *Gallium mullungo* (Wilson and Balague 1985).

Results and Discussion

Cell culture technology has been applied to a number of medicinal plants to obtain pharmaceutically important drugs (Table 9.2). But the results have not been encouraging, as the yield is too low to be commercially feasible.

Only a few products such as shikonin and ginseng biomass are manufactured at a larger scale. However, the strategies such as biotransformation, cell permeabilization, immobilization, elicitation and hairy root culture indicate the likelihood of many products to reach commercial level.

Table 9.2: Secondary Metabolites Produced *in vitro* in Medicinal Plants

Name of the Plant	*Uses*	*Active Principle/ Secondary Metabolites*	*Techniques Used*
Cephaelis epecacuanha	Emetic, amoebic dysentery expectorant	Emetine, cephaeline	Cell immobilization
Datura candida	Narcotic, antispasmodic	Scopolamine, hyoscyamine	Hairy root culture
Datura stramonium	Narcotic, antispasmodic	Hyoscyamine and scopolamine	Hairy root culture
Morinda citrifolia	Cathartic, febrifuge	Anthraquinone	Cell suspension culture
Papaver somnifera	Narcotic, sedative, antispasmodic nypnotic, emetic	Codeine, morphine papaverine sanquinarine	Cell culture
Solanum eleagnifolium	Cardiac stimulant	Solasodie	Batch suspension culture
Solanum paludosum	Cardiac stimulant	Solamargine	Multiple shoot
Taxus baccata	Anti-cancer	Toxol	Cell suspension culture

Conclusion

Some of the other compounds produced are anti-cancer agents toxol and camptothecin, obtained trom *Taxus bervifolia* and an indigenous tree *Nothapodytes foetida,* the anti-malarial drug artemisinin, anti-HIV agent castano spermine, forskolin, the anti platelet activity compounds from *Coleus forskohli.* Considering the production cost, the tissue culture of medicinal plants is not a solution to the continuous supply of drugs in large quantities at a cheaper rate. To date, only three compounds have been produced commercially. These are shikonin from cell cultures of *Lithospermum erythrorhizon,* barberine, an alkaloid possessing anti-septic activity from *Coptis japonica* and ginsengosides used in gastric disorders from cell cultures of *Panax ginseng.* Production of secondary metabolites of Medicinal plants is still in the experimental level. A concerted effort is needed to make the results obtained already, to be of commercial use.

References

Balick, M.J., 1994. *Ethnobotany and the Search for New Drugs.* John Wiley and Sons, Chicester, pp. 4–18.

Fujita, Y., Hara, Y., Suga, C. and Marimoto, T., 1981. Production of shikonin derivatives by cell suspension cultures of *Lithospermum erythrorhizon.* II. A new medium for the production of shikonin derivatives. *Plant Cell Rep.,* 1: 61–63.

Knobloch, K.H., Bast, G. and Berlin, J., 1982. Medium- and light-induced formation of serventine and anthocyanins in cell suspension cultures of *Catharanthus roseus. Phytochemistry,* 21: 591–594.

Lozoya, X., 1994. *Ethnobotany and the Search for New Drugs.* John Wiley and Sons, Chicester, pp. 130–140.

Matsumoto, T., Ikeda, T., Okimura, C., Obi, V., Kiisaki, T. and Noguchi, M., 1982. Production of uniquinone 10 (by UQ-10) by UQ highly producing strains selected by a cell cloning technique. In: *Plant Tissue Culture,* (Ed.) A. Fujiwara. Japan. Assoc. Plant Tissue Cult., Tokyo, pp. 275–277.

Tal, B. and Goldberg, I., 1982. Growth and diosgenin production by *Dioscorea deltoidea* cells in batch and continuous cultures. *Plant Med.,* 44: 107–110.

Wilson, G. and Balague, C., 1985. Biosynthesis of anthraquinone by cells of *Galium mollugo* L. grown in chemostat with limiting sucrose or phosphate. *J. Exp. Bot.,* 36: 485–493.

Chapter 10

DGPS: Analysis of Differential Correction Patterns

D. Vijayalakshmi[1]*, *I.V. Murali Krishna*[2] *and E.G. Rajan*[3]

[1]*Associate Professor, VNR Vignana Jyothi Institute of Engineering and Technology, Hyderabad*

[2]*Professor in Centre for Spatial Information Technology, JNTU, Hyderabad*

[3]*Professor in Department of ECE, Vasavi College of Engineering, Hyderabad*

ABSTRACT

Differentially correcting GPS data by post processing uses a base GPS receiver that logs positions at a known location and a rover GPS receiver that collects positions in the field. Most of these errors can be "Calculated out" of the measurements by the process called differential correction. The specific objective of the study is to estimate the accuracies involved with DGPS analysis in post processing mode. The errors are reported using the factor 1.96σ. The deviation values (Co-ordinate Deltas) Δ northing, Δ easting and Δ elevations are all zeroes for all points. Base stations are set up with an elevation mask value of 10°. A good rule of thumb is to increase the elevation mask for the rover by one degree depending upon the distance from the base station. This is found to be approximately equal to 75 km. Setting of 5 degrees works well for the rover in general if it is within 50 km from the base station. All the fixes with their obvious errors are now within the area of one small circle–that presumably includes the true location of the antennae. The ticks again are 10 meters apart, so the amount of error reduction and the accuracy you can expect from differentially corrected data is identified. The variations and accuracies of the received signals are to be better understood as operational models that could predict errors are still not available. The differential correction patterns are significantly influenced by a variety of parameters. The errors can be minimized through proper planning of the field data collection. Proper understanding of the

* E-mail: psridhar_babu_2k@yahoo.com; Phone: 9849267760

instrument performance is required for minimizing the errors. The point error ellipses suggest clearly the amount of deviation of signal. The horizontal and vertical precisions calculated in this study are good examples of quantifiable parameters that define the signal variation.

Introduction

Space environmental effects on the satellite communication can be considered as those effects related to space segment, those effects related to ground segment and those effects on the signals propagating through the earth's lower and upper atmosphere. The atmospheric structure has significant influence on signal propagation. This has definite influence on the data processing methodologies. The specific applications with which are concerned here are essentially related to navigation and positioning. Like any conventional navigational systems the GPS is subjected to errors that can degrade the precision of the system. Troposphere and ionosphere are the two regions, which approximately correspond to broad technical categorization into lower atmosphere and upper atmosphere zones respectively, have different properties of signal propagation. The structure of atmosphere can be described, for most practical purposes, through concentric layers of atmospheric domains with different physical and chemical properties. The signal propagation results in certain errors. These errors are ionospheric errors, tropospheric errors and multipath errors apart from the clock errors. In addition, the error in the signal can be significantly increased depending upon the geometry of the satellite used to determine a position. This can be estimated through the parameter PDOP (Position Dilution of Precision). The Dilution of precision factors (DOP) explain how geometry effects to yield position accuracy and scale ranging accuracy. The optimum geometry accuracy for four GPS satellites is achieved when three satellites are equally spaced on the horizon and one directly at zenith point of observation station.

DGPS, Differential GPS is the use of differential Correction data for the satellites in view, to remove the errors in the position measured by a GPS receiver due previously to SA (Selective availability), and to other effects. The Correction data is generated by a base station, which is a high quality GPS receiver with a good antenna sited at an accurately know location. The correction data is derived from the satellites pseudo range data and the base station's position, effectively correcting the timing errors artificially introduced into the atomic clock date by SA, or atmospheric effects. Differential correction techniques are used to enhance the quality of location data gathered using global positioning system (GPS) receivers. Differential correction can be applied in real-time directly in the field or when post processing data in the office. Although both methods are based on the same underlying principles, each accesses different data sources and achieves different levels of accuracy. Combining both methods provides flexibility during data collection and improves data integrity. With SA removed, a single GPS receiver from any manufacturer can achieve accuracies of approximately 100 meters. To achieve the accuracies needed for quality GIS records–from one to two meters up to a few centimetres–requires, differential correction of the data. The majority of data collected using GPS for GIS is differentially corrected to improve accuracy.

The underlying premise of differential GPS (DGPS) as already mentioned is that any two receivers that are relatively close together will experience similar atmospheric errors. DGPS requires that a GPS receiver be set up on a precisely known location. This GPS receiver is the base or reference station. The base station receiver calculates its position based on satellite signals and compares this location to the known location. The difference is applied to the GPS data recorded by the second GPS receiver, which is known as the roving receiver. The corrected information can be applied to data from the roving receiver in real time in the field using radio signals or through post processing after data capture using special processing software.

DGPS

Differentially correcting GPS data by post processing uses a base GPS receiver that logs positions at a known location and a rover GPS receiver that collects positions in the field. The files from the base and rover are transferred to the office processing software, which computes corrected positions for the rover's file. This resulting corrected file can be viewed in or exported to a GIS.

There are many permanent GPS base stations currently operating throughout the world that provide the data necessary for differentially correcting GPS. Depending on the technology preferred by the base station owner, this data can be downloaded from the Internet or via a bulletin board system (BBS). Because base station data is consistent (with no gaps due to multipath errors) and very reliable because base stations usually run 24 hours, seven days a week, it is ideal for many GIS and mapping applications. Sources of base station data for post processing fall into four categories–public sources, commercial sources, web-based services, and base station ownership. Before purchasing a GPS receiver it is best to identify the source of base station data.

To attain accuracy levels on the order of one to 10 meters, differential correction is essential. The three main methods currently used for ensuring data accuracy are real-time differential correction, preprocessing real-time data, and post processing. Each method will achieve similar levels of accuracy, so the decision regarding which technique is appropriate will depend on factors such as project specifications, the end use of the data, and the sources available for differential correction.

In a nutshell, the differential correction process consists of setting a GPS receiver (Called a base station) at a precisely known geographic point. Since the base station knows exactly where its antenna is, it can analyze and record error in the received signals. These signal errors will be almost equivalent to the signal errors affecting other GPS receivers in the local area, so the accuracy of locations calculated by those other receivers may be improved, significantly integrating with base station data.

One measure of accuracy of GPS fixes is called circular error probable (CEP). It is the radius of a circle expressed in a linear unit, such as meters for a given situation, 50 per cent of the fixes will fall within the circle, and 50 per cent outside. Another measure of accuracy is based on two standard deviations of a normal distribution–called 2d RMS where RMS means root–mean–square. Ninety-five per cent of the fixes will lie within a circle with this radius.

Sources of GPS Error

The available data on GPS measurements in literature suggest that typical error sources and values for various types of receivers are different. There is no effort for rigorous estimate of error. However a broad estimate can be made as per data available in literature. These values could be mentioned as follows:

Satellite clocks < 1 meter

Ephemeris error < 1 meter

Receiver error < 2 meter

Ionospheric < 2 meter

Tropospheric < 2 meters

Multipath (depends on the situation and the receiver, may be large)

These values correspond to averages of many readings rather than the error that might be expected from a single reading. The field observations suggest that errors can be minimized by increasing the number of observable stations. The existence of systematic errors in a given situation make us to accept that the law of the large numbers is equally significant and relevant in the case of GPS observations also.

Table 10.1: Types of Errors

Types of Errors	*Significance*
Clock Errors	Synchronization of atomic clocks in the satellite is essential to evaluate the ability to determine the time taken for the signal to get received by the receiver from the satellite. Even a small amount of difference in the clocks can make a huge difference in the distance measurements, because the GPS signals travel at a speed 3×10^8 meters/sec.
Ephemeris Errors	The exact position of a satellite is fixed through ephemeris representation. This representation is based on mean kepler ellipse. The time dependent deviations of the predicted orbit from this reference ellipse are transmitted to the receiver at regular intervals. Every hour or so, receiver gets position information at time "t". If this ephemeris prediction is incorrect the satellite will not be where it is expected to be, even by just a meter or two–then the measurement of the range from the receiver antenna to the satellite will be incorrect.
Receiver Errors	GPS receiver errors result from a number of factors related to receiver design, cost, and quality. These are many tunes difficult to assess without, proper instrument calibration.
Atmospheric Errors	The atmosphere has significant influence on transmission of signals. For most of its trip from the satellite to the receiver antenna, the GPS signal enjoys a trip through the virtual vacuum of "Empty space". Virtually all atmospheric influence is within about 120 km of the surface. Since the calculation of the range to the satellite depends on the speed of the signal, a change in signal speed due to atmospheric effect implies an error in distance, which produces an error in position finding. Significant changes in signal speed occur throughout the atmosphere, but the primary contributions to error come from the ionosphere, which contains charged particles under the influence of the earth's magnetic field

Field Data Collection

Most of these errors can be "Calculated out" of the measurements by the process called differential correction. The results of an experiment are as follows:

Table 10.2 shows the data collected during a field investigation taken with a Trimble GPS receiver placed in DGPS mode operated during two different time periods. It should be realized that each and every fix here is attempting to approximate the same true position. They show up in different positions because of errors. The true position of the antenna was not known, but almost certainly each point misses the exact, true position by some amount. The data given in the following section is actually collected by Trimble DGPS for 4 days during day time. The specific objective of the study is to estimate the accuracies involved with DGPS analysis in post processing mode. The network adjustment and adjustment style settings are shown in the following tables (Tables 10.2 and 10.3). The data is adjusted with reference to WGS84 as reference datum.

The adjusted grid co-ordinates are shown in Table 10.4. The geodetic co-ordinates in terms of latitude, longitude and heights along with errors are shown in Table 10.5. The errors are reported using the factor 1.96σ. The deviation values (Co-ordinate Deltas) Δ northing, Δ easting and Δ elevations are all zeroes for all points.

Table 10.2: Network Adjustment Report

User name	Administrator	*Date and Time*	1 : 13 : 09 PM 7/14/2005
Coordinate System	UTM	*Zone*	44 North
Project Datum	WGS 1984		
Vertical Datum		*Geoid Model*	EGM96 (Global)
Coordinate Units	Meters		
Distance Units	Meters		
Height Units	Meters		

Table 10.3: Adjustment Style Settings–95 per cent Confidence Limits

Residual Tolerances	
To End Iterations	0.000010m
Final Convergence Cutoff	0.005000 m
Covariance Display	
Horizontal	
Propagated Linear Error (E)	U.S.
Constant Term (C)	0.00000000 m
Scale on Linear Error (S)	1.96
Three-Dimensional	
Propagated Linear Error (E)	U.S.
Constant Term (C)	0.00000000 m
Scale on Linear Error (S)	1.96
Elevation Errors were used in the calculations.	
Adjustment Controls	
Compute Correlations for Geoid	False
Horizontal and Vertical adjustment performed	
Set-up Errors	
GPS	
Error in Height of Antenna	0.000 m
Centering Error	0.000 m

Table 10.4: Adjusted Coordinates

Adjustment Performed in WGS-84

Number of Points: 32

Number of Constrained Points: 0

Adjusted Grid Coordinates

Errors are reported using 1.96σ

Point Name	Northing	N Error	Easting	E Error	Elevation	E Error
1	1928712.764 m	0.012 m	216378.474 m	0.046 m	N/A	N/A
23	1926964.613 m	0.013 m	215470.731 m	0.046 m	N/A	N/A
24	1927234.048 m	0.012 m	215360.006 m	0.046 m	N/A	N/A
25	1927132.535 m	0.018 m	215102.747 m	0.047 m	N/A	N/A
26	1927219.160 m	0.311 m	214915.783 m	1.200 m	N/A	N/A
27	1926959.781 m	0.152 m	214780.009 m	0.684 m	N/A	N/A
28	1926874.901 m	0.013 m	214592.041 m	0.046 m	N/A	N/A
29	1926913.549 m	0.012 m	214466.250 m	0.046 m	N/A	N/A
30	1927392.761 m	0.014 m	213452.124 m	0.047 m	N/A	N/A
31	1927274.493 m	0.017 m	213283.905 m	0.048 m	N/A	N/A
32	1926847.800 m	0.015 m	213364.052 m	0.047 m	N/A	N/A
10	1929389.638 m	0.015 m	214080.107 m	0.047 m	N/A	N/A
11	1929509.183 m	0.016 m	213642.365 m	0.047 m	N/A	N/A
12	1929524.877 m	0.015 m	213614.240 m	0.047 m	N/A	N/A
13	1929565.012 m	0.025 m	213333.114 m	0.053 m	N/A	N/A
14	1929618.634 m	0.014 m	212655.503 m	0.047 m	N/A	N/A
15	1929463.154 m	0.015 m	212458.098 m	0.047 m	N/A	N/A
16	1929188.037 m	0.013 m	212038.254 m	0.046 m	N/A	N/A
17	1929043.280 m	0157 m	211855.872 m	0.359 m	N/A	N/A
18	1929419.754 m	0.015 m	210122.070 m	0047 m	N/A	N/A
19	1929126.837 m	0.014 m	210025.208 m	0.047 m	N/A	N/A
20	1927796.800 m	0.019 m	210243.863 m	0.048 m	N/A	N/A
21	1927647.922 m	0.013 m	213066.021 m	0.046 m	N/A	N/A
22	1928882.689 m	0.013 m	212694.270 m	0.046 m	N/A	N/A
2	1928197.391 m	0.015 m	216310.906 m	0.047 m	N/A	N/A
3	1927444.066 m	0.013 m	216305.024 m	0.046 m	N/A	N/A
4	1926648.851 m	0.018 m	215905.484 m	0.047 m	N/A	N/A
5	1926677.355 m	0.013 m	215597.701 m	0.046 m	N/A	N/A
6	1926829.184 m	0.014 m	215601.067 m	0.047 m	N/A	N/A
7	1926949.827 m	0.014 m	215521.521 m	0.047 m	N/A	N/A
8	1929082.109 m	0.015 m	215496.074 m	0.047 m	N/A	N/A
9	1929326.357 m	0.013 m	214524.356 m	0.046 m	N/A	N/A

Table 10.5: Adjusted Geodetic Coordinates

Errors are reported using 1.96σ

Point Name	Latitude	N Error	Longitude	E Error	Height	H Error
1	17°25'35.24265"N	0.012 m	78°19'48.54665"E	0.046 m	504.537 m	0.041 m
23	17°24'38.00578"N	0.013 m	78°19'18.63817"E	0.046 m	468.091 m	0.043 m
24	17°24'46.71345"N	0.012 m	78°19'14.76144"E	0.046 m	468.969 m	0.041 m
25	17°24'43.29661"N	0.018 m	78°19'06.09916"E	0.047 m	472.828 m	0.053 m
26	17°24'46.02715"N	0.311 m	78°18'59.72773"E	1.200 m	474.771 m	0.917 m
27	17°24'37.53409"N	0.152 m	78°18'55.25384"E	0.684 m	467.323 m	0.511 m
28	17°24'34.68930"N	0.013 m	78°18'48.92999"E	0.046 m	475.878 m	0.042 m
29	17°24'35.88815"N	0.012 m	78°18'44.65259"E	0.046 m	476.278 m	0.041 m
30	17°24'51.00100"N	0.014 m	78°18'10.08782"E	0.047 m	488.229 m	0.047 m
31	17°24'47.07963"N	0.017 m	78°18'04.44869"E	0.048 m	487.228 m	0.052 m
32	17°24'33.24682"N	0.015 m	78°18'07.36603"E	0.047 m	486.265 m	0.046 m
10	17°25'56.19680"N	0.015 m	78°18'30.39864"E	0.047 m	501.005 m	0.047 m
11	17°25'59.88204"N	0.016 m	78°18'15.51880"E	0.047 m	494.126 m	0.046 m
12	17°26'00.37927"N	0.015 m	78°18'14.55893"E	0.047 m	495.088 m	0.045 m
13	17°26'01.55489"N	0.025 m	78°18'05.02025"E	0.053 m	500.483 m	0.083 m
14	17°26'02.98642"N	0.014 m	78°17'42.04936"E	0.047 m	505.917 m	0.051 m
15	17°25'57.84179"N	0.015 m	78°17'35.43941"E	0.047 m	505.532 m	0.051 m
16	17°25'48.70584"N	0.013 m	78°17'21.35501"E	0.046 m	495.874 m	0.044 m
17	17°25'43.91651"N	0.157 m	78°17'15.24894"E	0.359 m	494.122 m	0.695 m
18	17°25'55.35125"N	0.015 m	78°16'16.35963"E	0.047 m	509.429 m	0.049 m
19	17°25'45.78537"N	0.014 m	78°16'13.22146"E	0.047 m	506.568 m	0.048 m
20	17°25'02.65539"N	0.019 m	78°16'21.26762"E	0.048 m	497.440 m	0.050 m
21	17°24'59.11789"N	0.013 m	78°17'56.89302"E	0.046 m	484.845 m	0.042 m
22	17°25'39.08273"N	0.013 m	78°17'43.71471"E	0.046 m	499.984 m	0.042 m
2	17°25'18.45954"N	0.015 m	78°19'46.50242"E	0.047 m	489.117 m	0.047 m
3	17°24'53.96966"N	0.013 m	78°19'46.65940"E	0.046 m	476.443 m	0.042 m
4	17°24'27.93947"N	0.018 m	78°19'33.50757"E	0.047 m	465.714 m	0.046 m
5	17°24'28.72613"N	0.013 m	78°19'23.07321"E	0.046 m	464.565 m	0.042 m
6	17°24'33.66291"N	0.014 m	78°19'23.11526"E	0.047 m	463.678 m	0.046 m
7	17°24'37.54825"N	0.014 m	78°19'20.36481"E	0.047 m	468.869 m	0.045 m
8	17°25'46.84713"N	0.015 m	78°19'18.49219"E	0.047 m	508.527 m	0.048 m
9	17°25'54.34306"N	0.013 m	78°18'45.47191"E	0.046 m	497.410 m	0.042 m

1. The position of "T".
2. The reading "O" and
3. The "difference" between "O" and "T".

So an arrow is drawn from the measured point to the true point. This arrow which is shown in two dimensions but which would really exist in three, has both a length (called a magnitude) and a direction. An entity that has magnitude and direction is known as a vector Label the vector "E", for "error", because it represents the amount and direction by which the reading missed the true point. When GPS is used, using the reported co-ordinates, "O", as an approximation of "T'. The vector "E" was the (unknown) amount by which missing the determined position "T" equation could be written as T = O–E. Where we record "O" and we discard "E" to find an approximation of "T". That is, the true co-ordinates are the observed co-ordinates minus the error. It is better to estimate the magnitude–but not the direction of "E".

If exact value of T is known and with the measured value "O", then the above equation to find "E" is rewritten as

$$E = O - T.$$

The GPS observations and a posterior errors (calculated as 1.96σ) are given in Table 10.6. The Histogram of standardized results is shown in Figure 10.1.

Table 10.6: Adjusted Observations

Adjustment performed in WGS-84

GPS Observations

Number of Observations: 31

Number of Outliers: 23

Observation Adjustment (Critical Tau = 0.00)

Obs.	*From Pt.*	*To Pt*		*Observation*	*A-posterior Error (1.96σ)*	*Residual*	*Stand. Residual*
B17	1	11	Az.	285°25'43.7023"	0°00'00.8402"	0°00'00.0000"	0.00
			ΔHt.	–10.411 m	0.022 m	0.000 m	0.00
			Dist.	2847.941 m	0.010 m	0.000 m	0.00
B14	1	31	Az.	244°16'20.8973"	0°00'00.6591"	0°00'00.0000"	0.00
			ΔHt.	–17.309 m	0.033 m	0.000 m	0.00
			Dist.	3410.409 m	0.013 m	0.000 m	0.00
B25	1	19	Az.	272°55'43.6080"	0°00'00.2121"	0°00'00.0000"	0.00
			ΔHt.	2.031 m	0.027 m	0.000 m	0.00
			Dist.	6362.819 m	0.009 m	0.000 m	0.00
B32	1	5	Az.	200°11'12.7727"	0°00'00.4074"	0°00'00.0000"	0.00
			ΔHt.	–39.972 m	0.011 m	0.000 m	0.00
			Dist.	2178.720 m	0.004 m	0.000 m	0.00
B28	1	22	Az.	271°50'26.2464"	0°00'00.2384"	0°00'00.0000"	0.00
			ΔHt.	–4.553 m	0.011 m	0.000 m	0.00
			Dist.	3685.881 m	0.006 m	0.000 m	0.00

Contd...

Table 10.6–Contd...

Obs.	From Pt.	To Pt		Observation	A-posterior Error (1.96σ)	Residual	Stand. Residual
B33	1	6	Az.	201°37'39.4042"	0°00'00.7752"	0°00'00.0000"	0.00
			ΔHt.	–40.859 m	0.023 m	0.000 m	0.00
			Dist.	2036.486 m	0.008 m	0.000 m	0.00
B36	1	9	Az.	287°30'39.4542"	0°00'00.3806"	0°00'00.0000"	0.00
			ΔHt.	–7.127 m	0.008 m	0.000 m	0.00
			Dist.	1951.837 m	0.003 m	0.000 m	0.00
B12	1	29	Az.	225°56'39.5287"	0°00'00.1088"	0°00'00.0000"	0.00
			ΔHt.	–28.259 m	0.005 m	0.000 m	0.00
			Dist.	2624.022 m	0.002 m	0.000 m	0.00
B18	1	12	Az.	285°34'19.7592"	0°00'00.6677"	0°00'00.0000"	0.00
			ΔHt.	–9.449 m	0.019 m	0.000m	0.00
			Dist.	2879.321 m	0.008 m	0.000 m	0.00
B27	1	21	Az.	251°22'45.3649"	0°00'00.1439"	0°00;00.0000"	0.00
			ΔHt.	–19.692 m	0.009 m	0.000 m	0.00
			Dist.	3477.293 m	0.003 m	0.000 m	0.00
B13	1	30	Az.	244°55'16.3818"	0°00'00.4318"	0°00'00.0000"	0.00
			ΔHt.	–16.308 m	0.023 m	0.000 m	0.00
			Dist.	3208.344 m	0.008 m	0.000 m	0.00
B15	1	32	Az.	237°27'21.6307"	0°00'00.4378"	0°00'00.0000"	0.00
			ΔHt.	–18.272 m	0.022 m	0.000 m	0.00
			Dist.	3542.546 m	0.010 m	0.000 m	0.00
B7	1	24	Az.	213°45'26.6365"	0°00'00.2261"	0°00'00.0000"	0.00
			ΔHt.	–35.568 m	0.006 m	0.000 m	0.00
			Dist.	1794.443 m	0.002 m	0.000 m	0.00
B30	1	3	Az.	182°30'48.7669"	0°00'00.6253"	0°00'00.0000"	0.00
			ΔHt.	–28.094 m	0.009 m	0.000 m	0.00
			Dist.	1270.067 m	0.005 m	0.000 m	0.00
B16	1	10	Az.	285°36'34.4337"	0°00'00.8518"	0°00'00.0000"	0.00
			ΔHt.	–3.532 m	0.025 m	0.000 m	0.00
			Dist.	2394.522 m	0.009 m	0.000 m	0.00
B11	1	28	Az.	223°23'14"1852"	0°00'00.2790"	0°00'00.0000"	0.00
			ΔHt.	–28.659 m	0.011 m	0.000 m	0.00
			Dist.	2561"483 m	0.004 m	0.000 m	0.00
B29	1	2	Az.	186°40'09.2482"	0°00'03.0367"	0°00'00.0000"	0.00
			ΔHt.	–15.420 m	0.023 m	0.000 m	0.00
			Dist.	519.474 m	0.009 m	0.000 m	0.00
B35	1	8	Az.	291°54'44.9985"	0°00'01.9958"	0°00'00.0000"	0.00
			ΔHt.	3.990 m	0.027 m	0.000 m	0.00
			Dist.	956.009 m	0.008 m	0.000 m	0.00

Contd...

Table 10.6–Contd...

Obs.	From Pt.	To Pt		Observation	A-posterior Error (1.96σ)	Residual	Stand. Residual
B26	1	20	Az.	260°42'28.5755"	0.00'00.4971"	0°00'00.0000"	0.00
			ΔHt.	–7.097 m	0.030 m	0.000 m	0.00
			Dist.	6198.795 m	0.013 m	0.000 m	0.00
B20	1	14	Az.	282°52'30.4219"	0.00'00.4384"	0°00'00.0000"	0.00
			ΔHt.	1.380 m	0.032 m	0.000 m	0.00
			Dist.	3829.267 m	0.010 m	0.000 m	0.00
B22	1	16	Az.	275°26'56.5455"	0.00'00.2472"	0°00'00.0000"	0.00
			ΔHt.	–8.662 m	0.017 m	0.000 m	0.00
			Dist.	4363.503 m	0.005 m	0.000 m	0.00
B24	1	18	Az.	275°38'49.1020"	0.00'00.2927"	0°00'00.0000"	0.00
			ΔHt.	4.892 m	0.027 m	0.000 m	0.00
			Dist.	6292.342 m	0.010 m	0.000 m	0.00
B8	1	25	Az.	218°06'51.6818"	0°00'01.3843"	0°00'00.0000"	0.00
			ΔHt.	–31.709 m	0.035 m	0.000 m	0.00
			Dist.	2029.696 m	0.012 m	0.000 m	0.00
B34	1	7	Az.	205°07'27.7148"	0°00'00.6736"	0°00'00.0000"	0.00
			ΔHt.	–35.667 m	0.020 m	0.000 m	0.00
			Dist.	1959.012 m	0.007 m	0.000 m	0.00
B31	1	4	Az.	192°06'28.7513"	0°00'00.8669"	0°00'00.0000"	0.00
			ΔHt.	–38.823 m	0.022 m	0.000 m	0.00
			Dist.	2116.155 m	0.013 m	0.000 m	0.00
B23	1	17	Az.	273°22'46.6637"	0°00'07.1865"	0°00'00.0000"	0.00
			ΔHt.	–10.414 m	0.717 m	0.000 m	0.00
			Dist.	4531.896 m	0.369 m	0.000 m	0.00
B21	1	15	Az.	280°02'07.9366"	0°00'00.4198"	0°00'00.0000"	0.00
			ΔHt.	0.995 m	0.031 m	0.000 m	0.00
			Dist.	3989.119 m	0.010 m	0.000 m	0.00
B19	1	13	Az.	284°50'02.6757"	0°00'01.6003"	0°00'00.0000"	0.00
			ΔHt.	–4.054 m	0.074 m	0.000 m	0.00
			Dist.	3160.451 m	0.025 m	0.000 m	0.00
B10	1	27	Az.	221°33'37.8713"	0°00'47.6344"	0°00'00.0000"	0.00
			ΔHt.	–37.214 m	0.526 m	0.000 m	0.00
			Dist.	2370.926 m	0.471 m	0.000 m	0.00
B9	1	26	Az.	223°36'03.9868"	0°01'16.7901"	0°00'00.0000"	0.00
			ΔHt.	–29.766 m	0.946 m	0.000 m	0.00
			Dist.	2089.278 m	1.015 m	0.000 m	0.00
B6	1	23	Az.	206°38'28.4720"	0°00'00.4702"	0°00'00.0000"	0.00
			ΔHt.	–36.445 m	0.015 m	0.000 m	0.00
			Dist.	1968.602 m	0.003 m	0.000 m	0.00

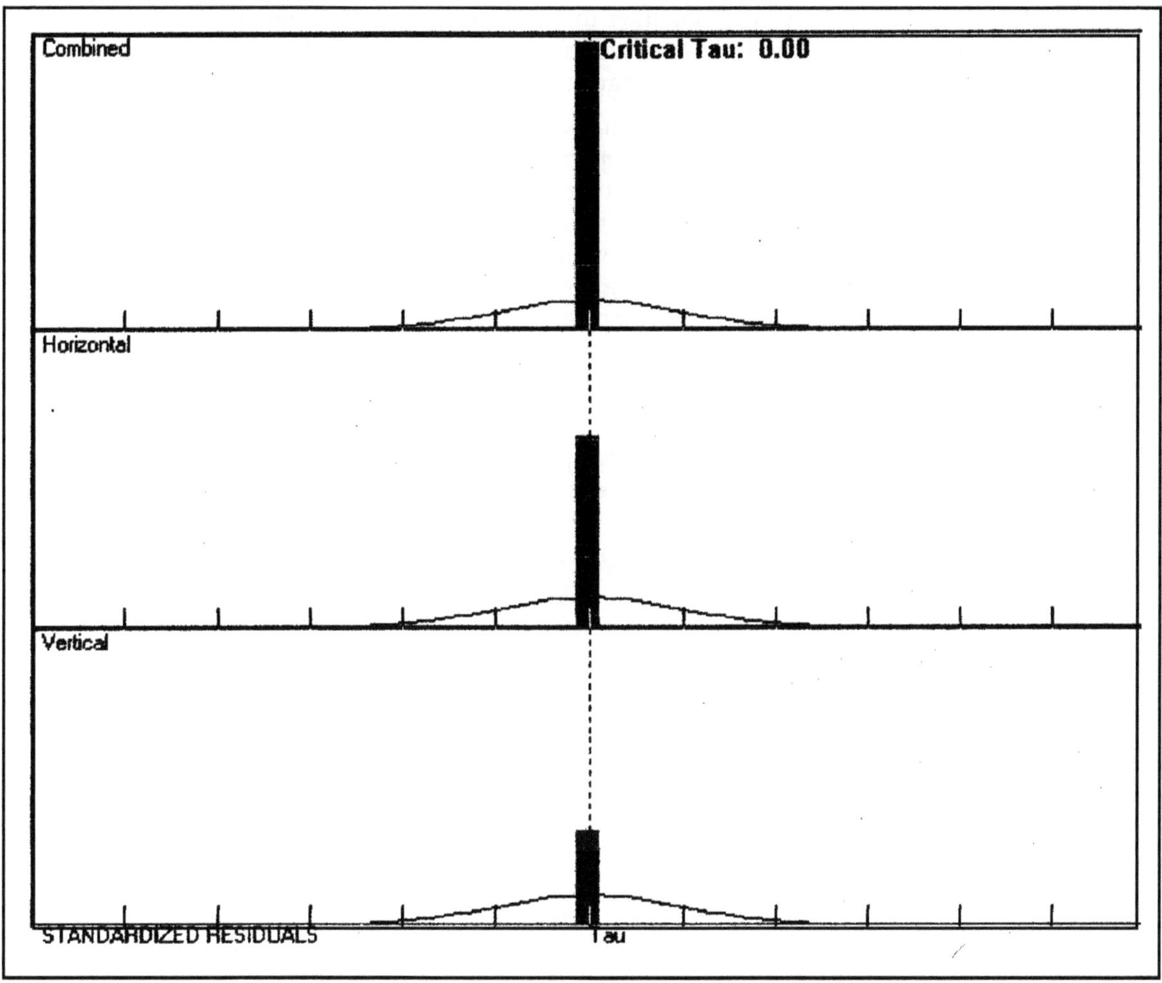

Figure 10.1: Histograms of Standardized Residuals

This technique provides an opportunity for canceling out most of the error in a GPS Position found by an antenna that is close to another antenna which is over a known point. The errors can be minimized through proper planning of the field data collection. Proper understanding of the instrument performance is required for minimizing the errors. The formula for the amount of error that might be expected with differentially corrected data is dependent on the distance between the base station antenna and the rover antenna. A rule of thumb is that the fix will be in error by one additional centimeter for each three kilometers between the two antennas. This relationship is approximately linear: 300 kilometers is found to produce error of about a meter.

Making Differential Correction Work

Practically a number of conditions have to be satisfied for the process to work. The base and rover have to be taking data at the same time, and the base has to be taking data frequently. If the base station is to serve the rover, where ever it may be (with in the 500 km limit), regardless of when data are taken, the base station must take data from all satellites the rover might see. This can cause a problem; if the base and rover are widely separated and the rover might see a satellite that the base cannot view. The point error ellipses are shown in this case in Figure 10.2.

Base stations are set up with an elevation mask value of 10°. A good rule of thumb is to increase the elevation mask for the rover by one degree depending upon the distance from the base station. This is found to be approximately equal to 75 km. Setting of 5 degrees works well for the rover in general if it is within 50 km from the base station. About fifteen degrees is good for avoiding difficulties due to terrain, that elevation angle reduces errors that tend to occur when signals come from satellites low on the horizon, since those signals must pass through more atmosphere. In any event the rover elevation mask must be set high enough so that there is no possibility for the rover to record data from a satellite that the base is not recording as shown in Figure 10.2. Correcting errors by the differential correction method implies that the base station-receiver at the known point-can communicate with the roving receiver(s).

In post processing GPS, (more correctly called post mission-processing GPS), the data from both the base and the rover are brought together later in a computer, and the appropriate correction is applied to each fix created by the rover.

Here adding all the fixes that were displayed before, but here now each fix was differentially corrected. In other words, all the fixes with their obvious errors are now within the area of one small circle–that presumably includes the true location of the antenna. The ticks again are 10 meters apart, so the amount of error reduction and the accuracy you can expect from differentially corrected data is identified. The variations and accuracies of the received signals are to be better understood as operational models that could predict errors are still not available. The differential correction patterns which are significantly influenced by a variety of parameters.

The point error ellipses suggest clearly the amount of deviation of signal. The horizontal and vertical precisions calculated in this study are good examples of quantifiable parameters that define the signal variation.

The covariant terms are shown in Table 10.7. The exact ratios for Horizontal precision and 3D precision are calculated. The exact errors in height measurement, distance and height are given in the table. The completed network of the cadastral map is shown in Figure 10.3.

Table 10.7: Covariant Terms

Adjustment performed in WGS-84

From Point	*To Point*		*Components*	*A-Posteriori Error (1.96σ)*	*Horiz. Precision (Ratio)*	*3D Precision (Ratio)*
1	23	Az.	206°38'28.4720"	0°00'00.4702"	1 : 613521	1 : 613521
		ΔHt.	–36.445 m	0.015 m		
		ΔElev.	?	?		
		Dist.	1968.602 m	0.003 m		
1	24	Az.	213°45'26.6365"	0°00'00.2261"	1 : 1165402	1 : 1165402
		ΔHt.	–35.568 m	0.006 m		
		ΔElev.	?	?		
		Dist.	1794.443 m	0.002 m		

Contd...

Table 10.7–Contd...

From Point	To Point		Components	A-Posteriori Error (1.96σ)	Horiz. Precision (Ratio)	3D Precision (Ratio)
1	25	Az.	218°06'51.6818"	0°00'01.3843"	1 : 165016	1 : 105016
		ΔHt.	–31.709 m	0.035 m		
		ΔElev.	?	?		
		Dist.	2029.696 m	0.012 m		
1	26	Az.	223°36'03.9868"	0°01'16.7901"	1 : 2058	1 : 2058
		ΔHt.	–29.766 m	0.946 m		
		ΔElev.	?	?		
		Dist.	2089.278 m	1.015 m		
1	27	Az.	221°33'37.8713"	0°00'47.6344"	1 : 5036	1 : 5036
		ΔHt.	–37.214 m	0.526 m		
		ΔElev.	?	?		
		Dist.	2370.926 m	0.471 m		
1	28	Az.	223°23'14.1852"	0°00'00.2790"	1 : 730758	1 : 730758
		ΔHt.	–28.659 m	0.011 m		
		ΔElev.	?	?		
		Dist.	2561.483 m	0.004 m		
1	29	Az.	225°56'39.5287"	0°00'00.1088"	1 : 1543071	1 : 1543071
		ΔHt.	– 28.259 m	0.005 m		
		ΔElev.	?	?		
		Dist.	2624.022 m	0.002 m		
1	30	Az.	244°55'16.3818"	0°00'00.4318"	1 : 387989	1 : 387989
		ΔHt.	–16.308 m	0.023 m		
		ΔElev.	?	?		
		Dist.	3208.344 m	0.008 m		
1	31	Az.	244°16'20.8973"	0°00'00.6591"	1 : 260172	1 : 260172
		ΔHt.	– 17.309 m	0.033 m		
		ΔElev.	?	?		
		Dist.	3410.409 m	0.013 m		
1	32	Az.	237°27'21.6307"	0°00'00.4378"	1 : 338667	1 : 338667
		ΔHt.	–18.272 m	0.022 m		
		ΔElev.	?	?		
		Dist.	3542.546 m	0.010 m		
1	10	Az.	285°36'34.4337"	0°00'00.8518"	1 : 267806	1 : 267806
		ΔHt.	–3.532 m	0.025 m		
		ΔElev.	?	?		
		Dist.	2394.522 m	0.009 m		
1	11	Az.	285°25'43.7023"	0°00'00.8402"	1 : 288317	1 : 288317
		ΔHt.	–10.411 m	0.022 m		
		ΔElev.	?	?		
		Dist.	2847.941 m	0.010 m		

Contd...

Table 10.7–Contd...

From Point	To Point		Components	A-Posteriori Error (1.96σ)	Horiz. Precision (Ratio)	3D Precision (Ratio)
1	12	Az.	285°34'19.7592"	0°00'00.6677"	1 : 351818	1 : 351818
		ΔHt.	-9.449 m	0.019 m		
		ΔElev.	?	?		
		Dist.	2879.321 m	0.008 m		
1	13	Az.	284°50'02.6757"	0°00'01.6003"	1 : 127790	1 : 127790
		ΔHt.	-4.054 m	0.074 m		
		ΔElev.	?	?		
		Dist.	3160.451 m	0.025 m		
1	14	Az.	282°52'30.4219"	0°00'00.4384"	1 : 366684	1 : 366684
		ΔHt.	1.380 m	0.032 m		
		ΔElev.	?	?		
		Dist.	3829.267 m	0.010 m		
1	15	Az.	280°02'07.9366"	0°00'00.4198"	1 : 404110	1 : 404110
		ΔHt.	0.995 m	0.031 m		
		ΔElev.	?	?		
		Dist.	3989.119 m	0.010 m		
1	16	Az.	275°26'56.5455"	0°00'00.2472"	1 : 814698	1 : 814698
		ΔHt.	-8.662 m	0.017 m		
		ΔElev.	?	?		
		Dist.	4363.503 m	0.005 m		
1	17	Az.	2.73°22'46.6637"	0°00'07.1865"	1 : 12287	1 : 12287
		ΔHt.	–10.414 m	0.717 m		
		ΔElev.	?	?		
		Dist.	4531.896 m	0.369 m		
1	18	Az.	275°38'49.1020"	0°00'00.2927"	1 : 648349	1 : 648349
		ΔHt.	4.892 m	0.027 m		
		ΔElev.	?	?		
		Dist.	6292.342 m	0.010 m		
1	19	Az.	272°55'43.6080"	0°00'00.2121"	1 : 729124	1 : 729124
		ΔHt.	2.031 m	0.027 m		
		ΔElev.	?	?		
		Dist.	6362.819 m	0.009 m		
1	20	Az.	260°42'28.5755"	0°00'00.4971"	1 : 488786	1 : 488786
		ΔHt.	–7.097 m	0.030 m		
		ΔElev.	?	?		
		Dist.	6198.795 m	0.013 m		
1	21	Az.	251°22'45.3649"	0°00'001439"	1 : 1171295	1 : 1171295
		ΔHt.	–19.692 m	0.009 m		
		ΔElev.	?	?		
		Dist.	3477.293 m	0.003 m		

Contd...

Table 10.7–Contd...

From Point	*To Point*		*Components*	*A-Posteriori Error (1.96σ)*	*Horiz. Precision (Ratio)*	*3D Precision (Ratio)*
1	22	Az.	271°50'26.2464"	0°00'00.2384"	1 : 655630	1 : 655630
		ΔHt.	–45.53 m	0.011 m		
		ΔElev.	?	?		
		Dist.	3685.881 m	0.006 m		
1	2	Az.	186°40'09.2482"	0°00'03.0367"	1 : 55113	1 : 55113
		ΔHt.	–15.420 m	0.023 m		
		ΔElev.	?	?		
		Dist.	519.474 m	0.009 m		
1	3	Az.	182°30'48.7669"	0°00'00.6253"	1 : 249404	1 : 249404
		ΔHt.	–28.094 m	0.009 m		
		ΔElev.	?	?		
		Dist.	1270.067 m	0.005 m		
1	4	Az.	192°06'28.7513"	0°00'00.8669"	1 : 163244	1 : 163244
		ΔHt.	–38.823 m	0.022 m		
		ΔElev.	?	?		
		Dist.	2116.155 m	0.013 m		
1	5	Az.	200°11'12.7727"	0°00'00.4074"	1 : 504349	1 : 504349
		ΔHt.	–39.972 m	0.011 m		
		ΔElev.	?	?		
		Dist.	2178.720 m	0.004 m		
1	6	Az.	201°37'39.4042"	0°00'00.7752"	1 : 263973	1 : 263973
		ΔHt.	–40.859 m	0.023 m		
		ΔElev.	?	?		
		Dist.	2036.486 m	0.008 m		
1	7	Az.	205°07'27.7148"	0°00'00.6736"	1 : 298200	1 : 298200
		ΔHt.	–35.667 m	0.020 m		
		ΔElev.	?	?		
		Dist.	1959.012 m	0.007 m		
1	8	Az.	291°54'44.9985"	0°00'01.9958"	1 : 113297	1 : 113297
		ΔHt.	3.990 m	0.027 m		
		ΔElev.	?	?		
		Dist.	956.009 m	0.008 m		
1	9	Az.	287°30'39.4542"	0°00'00.3806"	1 : 690946	1 : 690946
		ΔHt.	–7.127 m	0.008 m		
		ΔElev.	?	?		
		Dist.	1951.837 m	0.003 m		

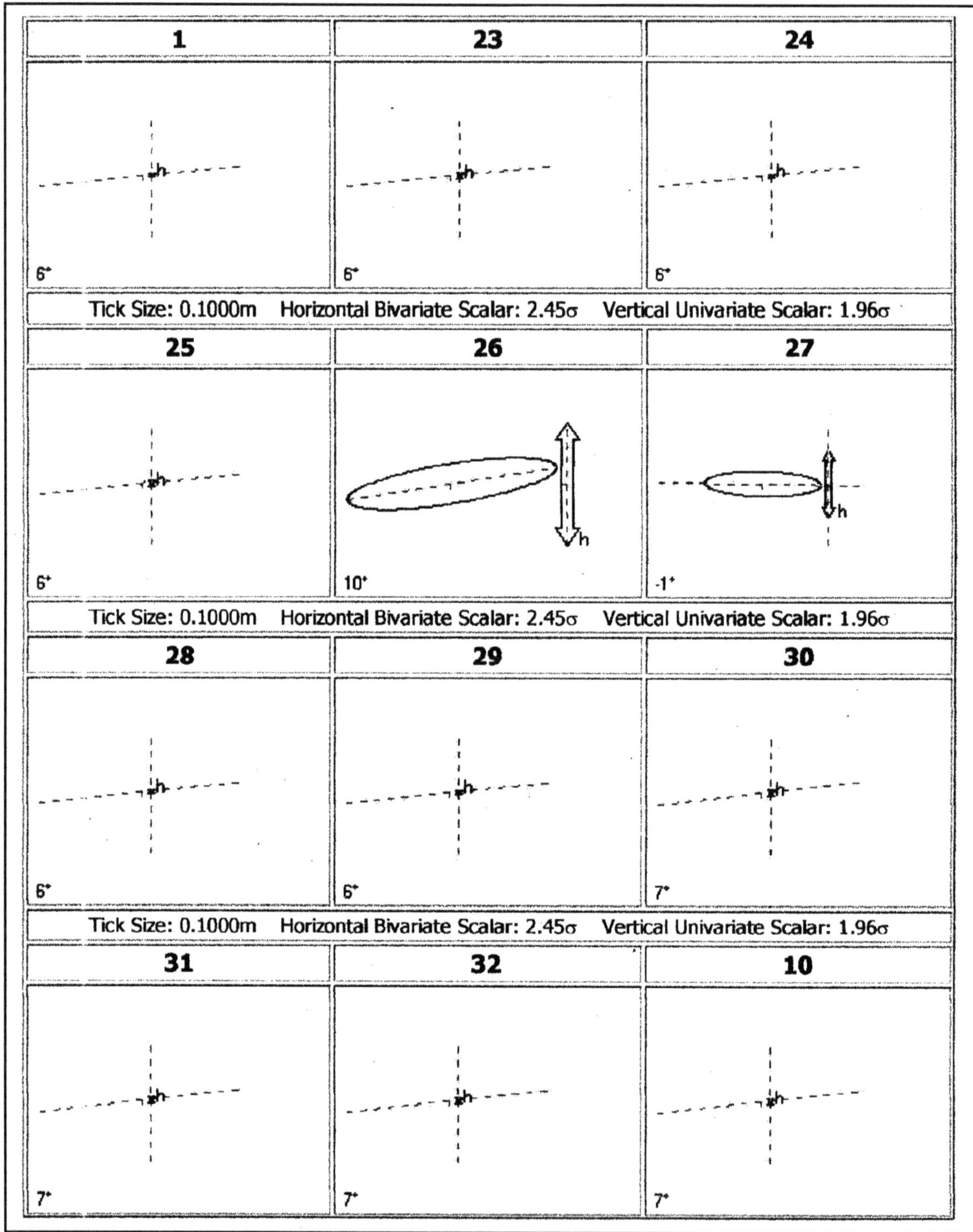

Figure 10.2: Point Error Ellipses

Figure 10.2–Contd...

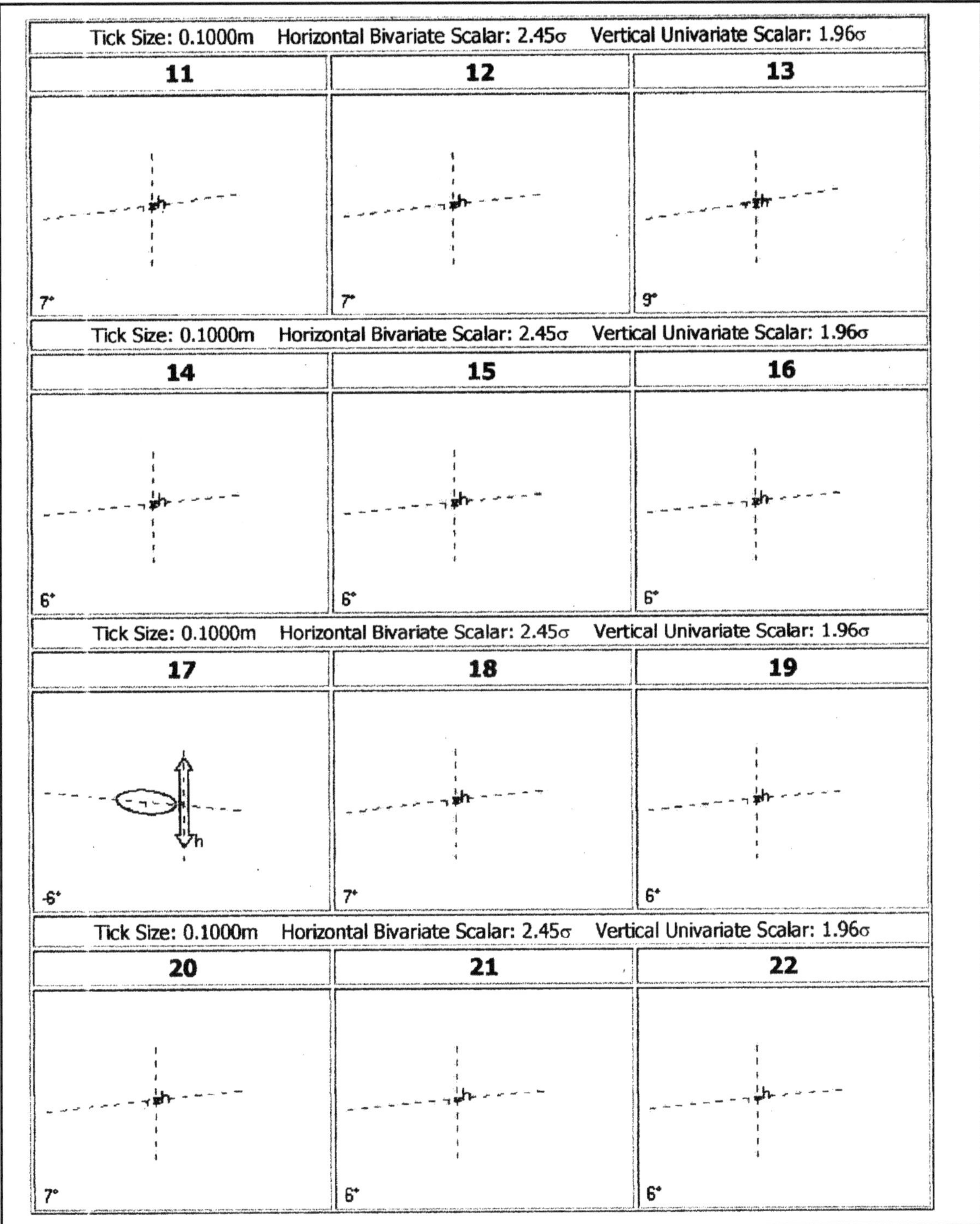

Figure 10.2–Contd...

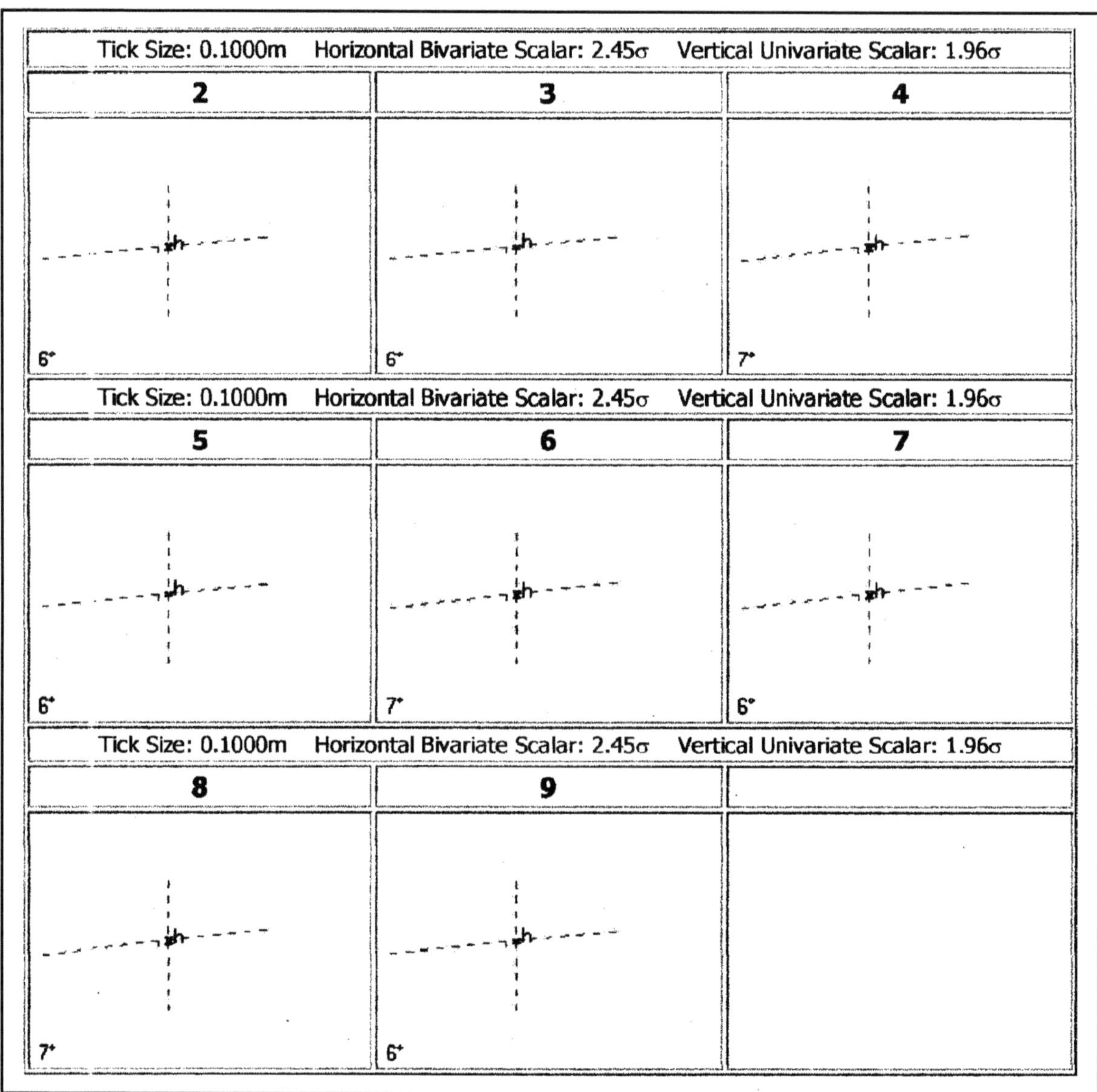

Conclusions

Extensive DGPS data has been collected and used in the present study. The variations and accuracies of the received signals are to be better understood as operational models that could predict errors are still not available. The differential correction patterns are significantly influenced by a variety of parameters. The errors can be minimized through proper planning of the field data collection. Proper understanding of the instrument performance is required for minimizing the errors. The point

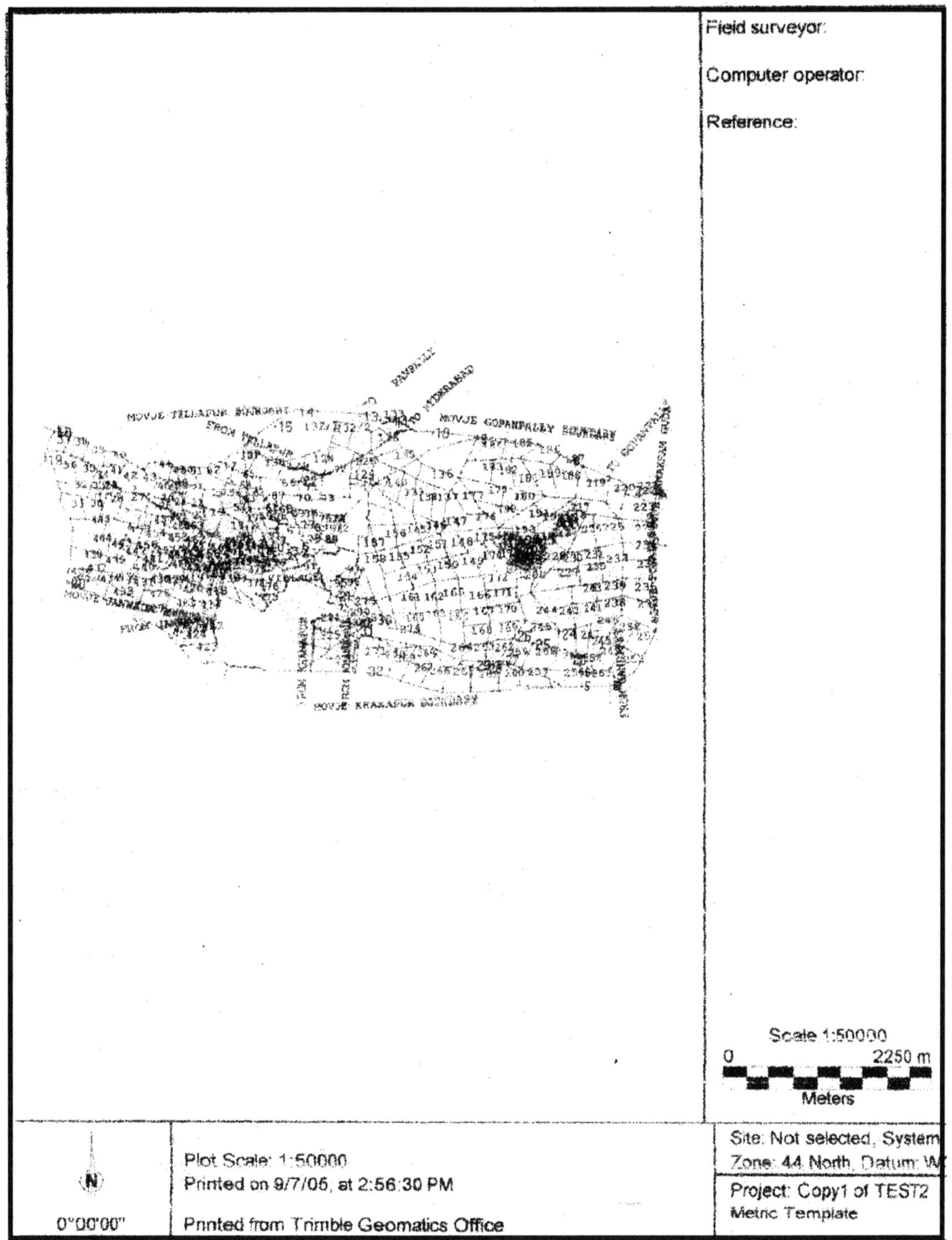

Figure 10.3: Cadastral Map

error ellipses suggest clearly the amount of deviation of signal. The horizontal and vertical precisions calculated in this study are good examples of quantifiable parameters that define the signal variation.

References

Adrados, C., Girard, I., Gendner, J. and Janeau, G., 2002. Global positioning system (GPS) location accuracy due to selective availability removal. *C.R. Biologies*, 325: 165–170.

Byun, S., G.A. Hajj and L.E. Young, 2002. Development and application of GPS signal. Multipath Simulator. *Radio Science*, 37(6): 1–23.

Misra, P. and P. Enge, 2001. *Global Positioning System Signals, Measurements, and Performance*. Ganga-Jamuna Press, Lincoln, MA.

Sardon, E., A. Rius and N. Zarraoa, 1994. Estimation of the transmitter and receiver differential biases and the ionospheric total electron content from global positioning system observations. *Radio Science*, 29(3): 577–586.

Chapter 11

Wind Energy Potential in the Aralvaimozhi Pass, Southern Tamil Nadu

J. Bensam Raj, G. Sugitha and N. Chandrasekar

Centre for Geo Technology, Manonmaniam Sundaranar University,
Abishekapatti, Tirunelveli – 627 012, Tamil Nadu

ABSTRACT

Wind power generation is estimated based on mathematical calculation. The field data obtained from the wind electric generator is compared with estimated value. The differences in the values are due to orographic induced flows in the complex topographic features. The wind energy potential is varying from day to year depending up on the conditions of the meteorological parameters. This requires exact knowledge of the special variation of the wind vector in the Aralvaimozhi pass.

Keywords: *Wind power, Wind speed, Density of air.*

Introduction

India stands today among the world largest program for renewable energy particularly wind energy. Wind as a natural phenomenon is known for change in speed and direction of motion throughout day and night which become more pronounced with seasonal and yearly cycles. Wind speed at a site is influenced by the terrain and by the height above the ground. Wind moving across the earth surface encounters friction influencing its flow, both by reducing the speed and at the same time increasing the level of turbulence. This effect becomes quite prominent while passing over or around mountains, hills, trees, buildings and other types of obstruction in its path. Wind speed generally increases with height above the ground. However, that could be exceptional cases where wind speed decreases with height due to topographical features. Aralvaimozhi pass in southern Tamil Nadu has unique topographical features such as mountains, trees etc. Further this area is more prominent to atmospheric pressure and temperature variation. Due to that the area has been identified by the wind energy producers to develop wind power generation. This chapter represents the potentiality of wind power generation in the Aralvaimozhi pass.

Facts about World Energy Consumption

In the seventies of the 20th century people worried about the shortage of energy resources and the direct consequences for the environment *e.g.*, acid rain. From the 1980's, the indirect consequences of high-energy consumption *e.g.* global warming and the ozone hole, became a greater concern (Elliott, 1997).

The outlook for future demand is that it is predicted to rise quite dramatically. It is supposed that from 1998 to 2010 the world wide yearly electricity demand will rise but by amount 30 per cent to 20,852 Twh and by nearly 50 per cent to 27,326 Twh by 2020, with an average annual growth rate of 2 per cent. Until 2025, the consumption of the developing countries is estimated to rise on the basis of population growth and increasing industrialization by more than 100 per cent, followed by the former soviet union and eastern Europe with an increase of about 80 per cent although the share of fossil fuels used in the production sector will shrink remarkably, the CO_2 emissions will increase by amount 50 per cent due to overall rising consumption (Daniels *et al.*, 1997).

The aforementioned risk to the biosphere and the limit on fossil fuel energy resources will force new solutions. One of the best solutions on the basis of today's technologies consists of a combination of saving energy by different measures and the exploitation of additional sustainable resources of energy such as solar energy and wind energy.

The Global Situation in the Use of Wind Energy

Wind energy is the kinetic energy content of a moving air mass. The kinetic energy content of the global atmosphere equals on average a turn over time of about 7 days of kinetic energy production or dissipation, also assuming average rates (Sorensen, 2000.) The annual streaming energy of the atmosphere lies between 8.2 and 13.6×10^{22} Joule. The entire electricity demand of the earth in 1990 would have been covered by the use of only 0.04 per cent of this energy. The world's total onshore wind resources are estimated to be about 53,000 TWh with the following distribution: Western Europe 4,800 TWh, Eastern Europe and former Soviet Union 10,600 TWh, Rest of Asia 4,600 TWh, Latin America 5,400 TWh, North America 14,000 TWh, Australia 3,000 TWh and Africa 10,500 TWh.

Materials and Methods

Wind power details were collected for 12 months between January 2005 and January 2006 from the various windmills installed in Aralvaimozhi pass. Similarly the types of wind electric generators with their capacity and tower heights were collected to assess the wind power generation in the region. The readings were taken at a height of 52 meters above the ground level and the capacity of machine is 750 Kilowatts. The wind power was calculated by using Gamma distribution (Jayakumar *et al.*, 2006). The mathematical analysis of wind power calculations and fitting procedure of Gamma distribution to observe data are as follows:

The probability density function of the two parameter Gamma distribution is defined by:

$$f(x) \quad \frac{\alpha(\alpha x)^{\beta \ 1}}{(\beta)} e^{\ \alpha x}, x, \alpha, \beta \quad 0 \tag{1}$$

where, $\Gamma(\beta)$ is the gamma function defined by

$$\Gamma(\beta) = \int_0^\infty e^{-x} x^{\beta-1} dx \tag{2}$$

The gamma distribution has a single peak at $x = (\beta - 1)/\alpha$, $(\beta < 1)$ and it takes the variety of shapes, depending on the values of b, ranging from reverse J shaped for $b < 1$ to single peaked for $b > 1$. The constant b is called shape parameter and a is called scale parameter. The cumulative gamma distributive function is defined by

$$F(x) = \int_0^\infty \frac{\alpha(\alpha t)^{\beta-1} e^{-\alpha t}}{\Gamma(\beta)} dt \tag{3}$$

The value of the above function cannot be given in a closed form and it can be evaluated numerically using tables of incomplete gamma function. The mean wind speed says *wbar*, in any locality is defined by the mean value of the gamma distribution and can be shown as

$$wbar = \frac{\alpha}{\beta} \tag{4}$$

The wind power at any location is given by $u = \frac{\rho w^3}{2}$ where ρ is the air density. So the available mean wind power at any location is given by

$$E(U) = \rho \, E(w^3)/2 \tag{5}$$

To estimate this value, we have to find the value of $E(w^3)$. This is done as follows. If $G(V)$ is the cumulative distribution function of w^3, then

$$G(V) = P\{w^3 \le v\} = P\{w \le v^{1/3}\} = \int_0^{v^{\frac{1}{3}}} \frac{\alpha(\alpha t)^{\beta-1} e^{-\alpha t}}{\Gamma(\beta)} dt \tag{6}$$

Hence if $g(v)$ is the density function of v then,

$$g(v) = \frac{dG(v)}{dv} = \frac{v^{\frac{-2}{3}} \alpha(\alpha v^{\frac{1}{3}})^{\beta-1} e^{-\alpha v^{\frac{1}{3}}}}{\Gamma(\beta)} = \frac{\alpha^\beta v^{\frac{\beta}{3}-1} e^{-\alpha v^{\frac{1}{3}}}}{3\Gamma(\beta)} \tag{7}$$

So,

$$E(w^3) = \int_0^\infty v g(v) dv = \frac{\alpha^\beta \int_0^\infty v^{\frac{\beta}{3}} e^{-\alpha v^{\frac{1}{3}}} dv}{3\Gamma(\beta)} \tag{8}$$

It can be shown that

$$\int_0^\infty v^{\frac{\beta}{3}} e^{-\alpha v^{\frac{1}{3}}} dv = \frac{3\Gamma(\beta+3)}{\alpha^{\beta+3}} \tag{9}$$

Hence, $E(w^3) = \dfrac{(\alpha+2)(\alpha+1)\alpha}{\beta^3}$ and so the mean wind power is given by

$$E(U) = \frac{\rho(\alpha+2)(\alpha+1)\alpha}{2\beta^3} \tag{10}$$

The above equation gives an expression for available wind power at any location. As an example, for Aralvaimozhi pass, the estimates of values of α and β are 15.8 and 3.9 respectively. Hence the average wind speed is given by

E (w) = 15.8/3.9 = 4.05 and the mean power, assuming the value of air density to be 1.2 is

$$E(U) = \frac{1.2 \times 15.8 \times 16.8 \times 17.8}{2(3.9)^3} = 47.7$$

In a similar manner, the mean wind speed and wind power were estimated for the remaining stations. Wind speed and electrical power are shown in the Table 11.1.

Similarly the observed wind powers from the installed windmills are shown in the Table 11.2.

Table 11.1: Estimated Wind Speed and Electrical Power

Wind Speed in m/s	*Electrical Power in KW*
4.0	47.7
5.0	87.7
6.0	152.0
7.0	229.0
8.0	346.6
9.0	435.9
10.0	548.2
11.0	625.1
12.0	685.7
13.0	728.2
14.0	742.1
15.0	770.0

Table 11.2: Observed Wind Power Energy

Wind Speed in m/s	*Electrical Power in KW*
4.0	42.4
5.0	76.9
6.0	138.4
7.0	226.1
8.0	324.4
9.0	427.5
10.0	536.1
11.0	609.1
12.0	675.2
13.0	721.0
14.0	732.0
15.0	750.4

Comparison of estimated and observed electrical power from the installed wind mills at various places in Aralvaimozhi pass are shown in the Table 11.3.

Table 11.3: Comparison Between Estimated and Observed Electrical Power

Wind Speed in m/s	*Estimated Electrical Power in KW*	*Observed Electrical Power at Various Places in KW*		
		Muppandal	*Vadakangulam*	*Radhapuram*
4.0	47.7	40.2	42.2	45.1
5.0	87.7	75.0	78.1	80.2
6.0	152.0	139.7	140.2	140.0
7.0	229.0	220.5	226.0	225.3
8.0	346.6	330.2	330.0	335.2
9.0	435.9	428.2	430.0	430.0
10.0	548.2	530.2	540.0	535.5
11.0	625.1	620.1	615.0	615.0
12.0	685.7	655.0	665.0	672.0
13.0	728.2	720.2	722.2	722.7
14.0	742.1	735.2	730.5	730.0
15.0	770.0	750.2	745.8	748.2

Results and Discussion

Wind energy is one of the important free renewable source, which vigorously taped in our country. Perhaps it becomes an alternative energy source in the cost effective manner. This system has associated with the positive and negative impacts on the environments. Though wind energy itself produces no gaseous emissions, during its construction and installation, impacts are generated. But these impacts are local and relatively predictable. The main impacts on environment are mostly sound emission and the electro magnetic interference. However the monitory values of the energy produced is higher than the environmental impacts expected. The power of energy estimated based on mathematically and fields observations are compared for validation. The data clearly indicates (Tables 11.1, 11.2 and 11.3) that the estimated and observed electrical power at any wind power locations installed in Aralvaimozhi pass is almost similar. The minor differences in the estimated and observed values are due to more complex topographic changes. The wind resource varies with time of day and season of year. It has inherent variants but the estimated wind power potential using Gamma distribution has attested the sustainability of wind energy to be exploited continuously. Further the efficiency of wind energy is mainly depending up on wind speed, air density, hub height, rotor diameter and topography (Burton *et al.*, 2001). Here the possibility for high and continuous wind speed in Aralvaimozhi pass is due to the main wind direction lies parallel to the axis of Aralvaimozhi valley. Hillside winds blow parallel to the inclination of the hillside and consist of an up-slope flow along the hillside blowing during the day, and down hillside flow blowing during the night. Valley winds blow parallel to the longitudinal axis of the valley, during the day in a downward direction and in the night in the up-valley direction. During heavy rain the yield of wind energy has been reduced due to the disturbance in the flow around the profile. However, an impressive representation by mathematically has been formulated for assessing the status of wind energy in the study area. The wind power generation is high during the periods from April to October. The remaining periods are low due to shift in wind direction. This chapter concludes that the mathematical evaluation and field observation reveals the significant parameters such as air density, wind speed, wind direction, topographic features and the type of wind generator required for the sustainable growth of wind energy to be produced from Aralvaimozhi pass.

Acknowledgement

We acknowledge the RS Wind Tech Engineers and S. Ponnaiyan Associates for providing the data base and extending necessity facilities to carryout the field work in their boundary.

References

Burton, T. and Sharpe, D., 2001, *Hand book of Wind Energy*. John Wiley and Sons Ltd., New York, pp. 10–26, 55–70.

Daniels, A. and Schroeder, T., 1997. *Meteorological Aspects and Recommendations for Assessing and Using the Wind as an Energy Source in the Tropics*, WCASP–43, WMO/TD–No.826, Geneva.

Elliott, D., 1997. Energy, society and environment: Technology for a sustainable future. *Routledge Introductions to Environment Series*, London.

Jayakumar, D., Prashanthi Devi, M., Suriyanarayanan, S., Balasubramanian, S., Ranganathan, C.R., 2006. Wind Energy Potential in Tamil Nadu India: Predication and mapping using GIS. www.gisdevelopment.net/application/nrm/overview/nrm0l0.htm

Sorensen, B., 2000. *Renewable Energy: Its Physics, Engineering, Use, Environmental Impact, Economy and Planning Aspects*. Academic Press.

Chapter 12

Acoustic Parameters of Ortho-Nitrophenol in Solutions

R. Palanivel, M.E. Rajasaravanan and K. Balamurugan

Department of Physics, Annamalai University, Annamalainagar – 608 002, Tamil Nadu

ABSTRACT

Application of ultrasonic interferometer technique to study the molecular interactions in the system containing ortho-nitrophenol (ONP) as solute in three different solvents namely 1,4-Dioxane, n-propanol and n-Butanol.

Keywords: *Ultrasonic velocity, Adiabatic compressibility, Intermolecular free length, Specific acoustic impedance.*

Introduction

The ultrasonic velocity measurements provide qualitative information of the elastic forces existing in the solutions in combination with the other physically measured data. The acoustic parameters and their variations with temperature over the selected range are used with good precision to understand the structural interactions of 2-nitrophenol (ONP) in different solute concentrations in three different solvents, (*i*) $C_4H_8O_2$ (DXE); (*ii*) C_3H_8O (NPL) and (*iii*) $C_4H_{10}O$ (NBL). The variations of acoustic parameters with concentrations are considered as an index of molecular interactions (Arul and Palaniyappan, 2005; Kannappan and Jayasanthi, 2005).

Application of the interferometer to specimens in which temperature is varied yields the temperature dependence of the adiabatic elastic constants (Espinola and Watermann, 1958). Ultrasonic velocity changes with concentration and the temperature influence on the acoustic parameters explains many of the properties of liquids and solutions including the microscopic changes occurring in the medium like molecular orientations (Dhanalaksmi and Sekar, 1996).

Atoms and molecules are the fundamental building blocks of all states of matter. Their arrangement and the way in which they interact with one another in the same phase and in different phases define their properties and characteristics.

Materials and Methods

Mononitrophenol exists in three isomeric forms. 2-nitrophenol (*ortho* or O-nitrophenol) is one of the isomers known for its chemical identity and its trade name is Atonik. It is light yellow colour crystalline solid, peculiar aromatic odour and volatile in steam. It is soluble in water and organic solvents. Molecular weight, melting point and density of the ortho-nitrophenol are 139.11 gm, 44–45°C and 1495 kg/m^3 at 14°C and its purity is checked by melting point method respectively. The OH group in the compound is susceptible to substitution reaction with the formation of ethers and esters. The nitro group can be reduced to the amino group under strong reducing conditions. The nitrophenols may also undergo ring substitution reactions.

An account of the solvents chosen and their physical properties are given in Table 12.1.

Table 12.1: Chemical Identity and Physical Properties of the Solvents

Solvent	*Ultrasonic Velocity ms^{-1}*	*Density kgm^{-3}*	*Viscosity Nsm^{-2} 10^{-3}*	*Molecular Weight*	*Boiling Point °C*	*Dipole Moment Debye*
1,4-Dioxane	1308.3	1032.9	1.200	88.11	101.2	3.9
n-propanol	1121.1	803.7	2.237	60.09	97.2	1.66
n-Butanol	1233.4	809.7	3.379	74.12	117.7	1.66

The ultrasonic velocity of the solution in different concentration from 0.001 to 0.005m were measured using single frequency continuous wave ultrasonic interferometer. (Model, Mittel Enterprise, New Delhi) at 3 MHz, using standard procedures. The density and viscosity were measured using specific gravity bottle and Ostwald's viscometer for each concentration at each temperature.

The accuracy of the velocity, density and viscosity measurements are ±0.1 m/s, 2 in 10^3 and 0.1 per cent respectively. All the measurements were carried out at three different temperatures 303K, 313K, and 323K using a thermostatically controlled water path with an accuracy of ±0.1°C.

Results and Discussion

The measured values of ultrasonic velocity, density and viscosity for solute concentrations from 0.001 to 0.005 molality at temperatures 303K, 313K and 323K and the calculated acoustic parameters namely Adiabatic Compressibility (β_s), Intermolecular Free-length (L_f) and Specific Acoustic Impedance (z) and their variations with temperatures are given in Tables 12.2 and 12.3 for the three solvents respectively.

The adiabatic compressibility is calculated using the formula $\beta_s = (C^2\rho)^{-1}$ where ρ is the density of the solution and C is the ultrasonic velocity in the solution. The intermolecular free length is calculated using the formula $L_f = K(\beta_s)^{1/2}$ where K is a constant, the values of K for different temperature are used from the values available in the literature (Jacobson, 1952). The specific acoustic impedance of the solution was calculated using the relation. $Z_{sol} = \rho_{sol} C_{sol}$ where ρ_{sol} is the density of the solution and C_{sol} is the ultrasonic velocity in solution.

Table 12.2: Measured Values of C, e [rho] and η

Temperature K	Molality (m)	C ms^{-1}	1,4-Dioxane ρ Kgm^{-3}	$\eta \times 10^3$	C ms^{-1} Nsm^2	n-Propanol ρ Kgm^{-3}	$\eta \times 10^3$ Nsm^2	C ms^{-1}	n-Butanol ρ Kgm^{-3}	$\eta \times 10^3$ Nsm^2
	0.001	1326.6	1054.87	1.2945	1140.3	819.75	1.5697	1254.9	850.90	2.0434
	0.002	1337.4	1073.34	1.3740	1160.4	850.45	1.5925	1262.1	879.34	2.0466
303	0.003	1350.3	1092.36	1.4330	1179.6	882.24	1.6240	1271.3	907.15	2.0537
	0.004	1360.5	1111.47	1.5404	1194.1	912.58	1.6412	1277.4	935.22	2.0579
	0.005	1369.2	1127.87	1.6227	1210.8	943.74	1.6773	1283.4	962.58	2.0672
	0.001	1298.4	1047.08	1.1410	1119.0	814.04	1.3348	1225.2	846.83	1.7111
	0.002	1308.6	1062.20	1.2136	1138.1	844.93	1.3497	1233.6	876.08	1.7332
313	0.003	1320.2	1085.75	1.3094	1156.6	876.35	1.3721	1244.6	903.16	1.7676
	0.004	1332.6	1104.41	1.4021	1172.4	906.33	1.3903	1253.4	931.06	1.7828
	0.005	1342.5	1122.25	1.4841	1190.1	937.04	1.4176	1263.9	958.68	1.7850
	0.001	1263.9	1038.38	0.9667	1086.3	810.60	1.1062	1192.8	841.21	1.4150
	0.002	1275.3	1057.86	1.0520	1106.1	840.58	1.1205	1201.2	869.56	1.4535
323	0.003	1285.2	1078.32	1.1522	1122.0	871.55	1.1525	1213.2	897.64	1.4909
	0.004	1296.3	1096.98	1.2418	1138.5	901.17	1.1726	1221.8	925.53	1.5275
	0.005	1305.0	1115.82	1.3339	1155.6	931.60	1.1926	1231.5	952.79	1.5624

Table 12.3: Calculated Values of β, L_f and Z

Temperature K	Molality (m)	1,4 Dioxane			n-Propanol			n-Butanol		
		$\beta \times 10^{10}$ $N^{-1}s^2$	$L_f \times 10^{11}$ m	$Z \times 10^6$ $Kgm^{-2}s^{-2}$	$\beta \times 10^{10}$ $N^{-1}s^2$	$L_f \times 10^{11}$ m	$Z \times 10^6$ $Kgm^{-2}s^{-1}$	$\beta \times 10^{10}$ $N^{-1}s^2$	$L_f \times 10^{11}$ m	$Z \times 10^6$ $Kgm^{-2}s^{-2}$
303	0.001	5.3867	4.6326	1.3994	9.3812	6.1136	0.9348	7.4628	5.4527	1.0678
	0.002	5.2040	4.5553	1.4355	8.7324	5.8983	0.9869	7.1393	5.3332	1.1 098
	0.003	5.0208	4.4725	1.4750	8.1460	5.6968	1.0407	6.8206	5.2128	1.1533
	0.004	4.8608	4.4006	1.5121	7.6851	5.5333	1.0897	6.5529	5.1095	1.1946
	0.005	4.7294	4.3407	1.5443	7.2277	5.3661	1.1427	6.3072	5.0128	1.2354
313	0.001	5.6650	4.8321	1.3595	9.8106	6.3589	0.9109	7.8666	5.6941	1.0375
	0.002	5.4977	4.7602	1.3900	9.1373	6.1368	0.9616	7.5008	5.5602	1.0807
	0.003	5.2843	4.6669	1.4334	8.5301	5.9294	1.0136	7.1467	5.4273	1.1242
	0.004	5.0988	4.5842	1.4717	8.0272	5.7520	1.0626	6.8366	5.3083	1.1670
	0.005	4.9440	4.5141	1.5066	7.5349	5.5728	1.1152	6.5298	5.1878	1.2117
323	0.001	6.0286	5.0800	1.3124	10.4543	6.6606	0.8805	8.3553	5.9545	1.0034
	0.002	5.8123	4.9664	1.3491	9.7237	6.4237	0.9298	7.9702	5.8157	1.0445
	0.003	5.6145	4.8812	1.3858	9.1143	6.2191	0.9779	7.5689	5.6674	1.0890
	0.004	5.4249	4.7980	1.4220	8.5611	6.0274	1.0260	7.2378	5.5421	1.1308
	0.005	5.2624	4.7256	1.4561	8.0381	5.8404	1.0765	6.9204	5.4192	1.1734

The variation of the calculated acoustic parameters with respect to various concentrations of orthonitrophenol (ONP) at different temperatures are represented graphically in Figure 12.1 respectively.

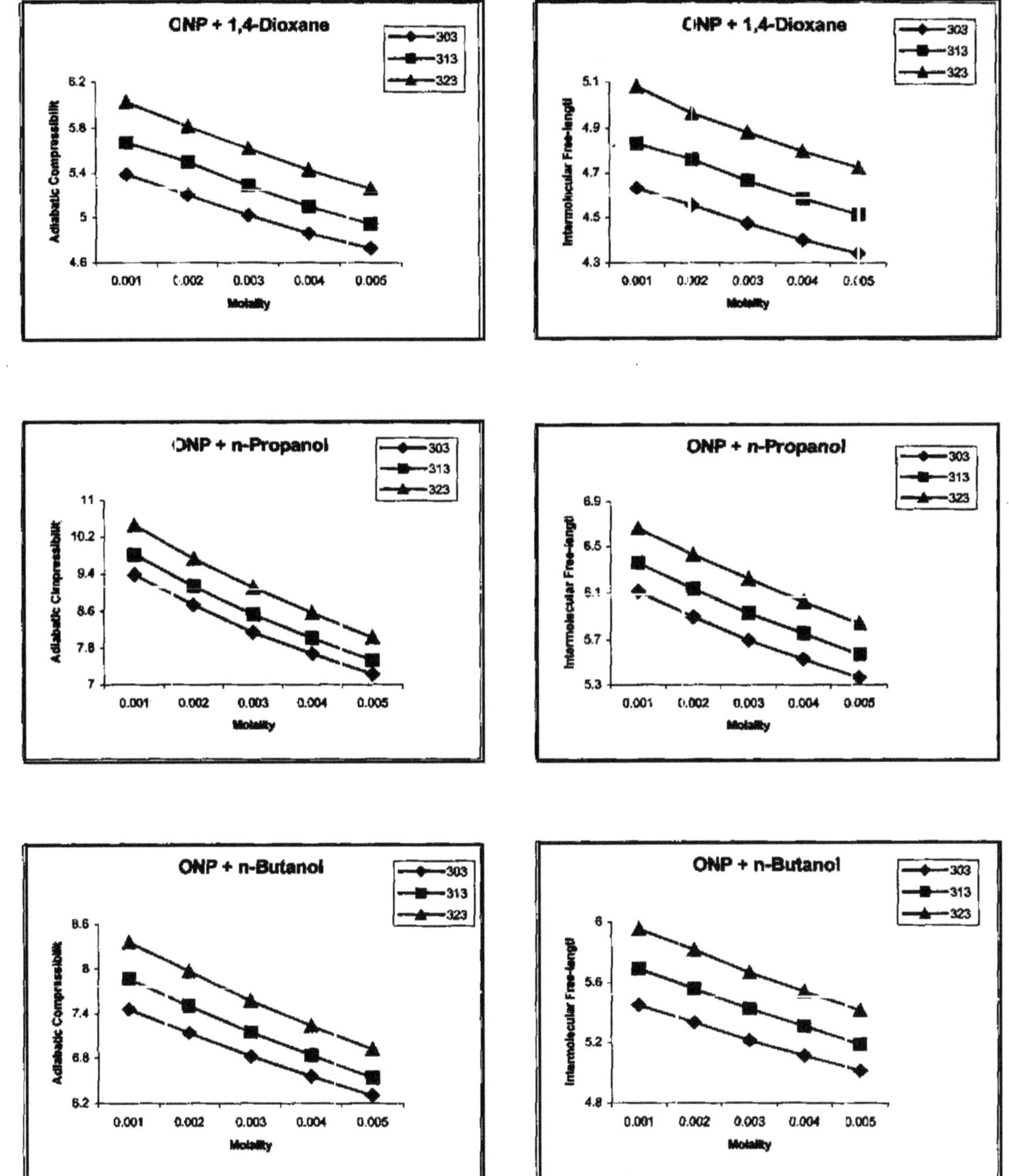

Figure 12.1: Variation of Adiabatic Compressibility and Intermolecular Free length with Concentration of ONP in 1,4-dioxane, n-proponal and n-butanol

In the present study the molecular interactions of ortho-nitrophenol (ONP) is investigated in solutions using three bidistilled organic solvents 1,4-Dioxane, n-propanol and n-Butanol from the measurement of ultrasonic velocity. The ultrasonic velocity of all the three solvents is found to increase considerably in solutions with orthonitrophenol. The pure solvents respectively have the velocity values 1308 ms^{-1}, 1121.1 ms^{-1} and 1233.4 ms^{-1} at room temperatures 303 K. In solutions of each fixed concentration the velocity value increased in its value.

On increasing the temperature to 313 K the solution on each concentration from 0.001 to 0.005m ultrasonic velocity decreases consistently in all the three solvents. On raising the temperature further to 323 K, the trend in the velocity change is similar as in the earlier rise in temperature. But the velocity change is considerably higher that the earlier temperature change. On increasing the concentration from of the solute 0.001 to 0.005 m in all three different temperatures, the velocity shows increase as the concentration is increased. It may be due to the intermolecular forces. The increase in velocity with increase in concentration suggests that there are solute-solvent interaction in the form of attraction in all the three cases (Kannappan and Hemalatha, 2005). Eswari *et al.* (2004) have reported that the acoustical studies of molecular interactions in binary liquid mixtures of butoxy ethanol with some amines at 308 K the variation of the adiabatic compressibility β with concentrations and temperature showed the similar behaviour. The adiabatic compressibility (β) decreases uniformly with increase of concentration in all the three systems in each temperatures of study. But for the same concentration at each temperature the increase in β value is almost the same in all the three systems studied.

Changes in the ultrasonic velocity and the compressibility observed in the systems of the present study are agreeing well with the result of the earlier workers (Srivastava *et al.*, 1983). Jacobson (1952) stated the decrease of compressibility with increase in concentration of the first component is the indication of the association existing between solute and solvent molecules. The tendency of the adiabatic compressibility value decrease with increase of solute concentration can be interpreted in terms of the effect of solvent molecules on the surrounding solute molecules, which results into relatively incompressible solution. The intermolecular free length also changes in the same manner as that of the adiabatic compressibility of the systems of study on varying the concentration and temperature.

The decrease of L_f on increasing the concentration of solute resulted the increase of ultrasonic velocity in the same manner. The change in L_f is in agreement with the result reported by Eyring and Kincaid (1938). This is the indication of the significant interactions existing between the solute and solvent molecules due to which the structural arrangement is affected considerably.

With the increase in temperature the freelength tends to increase due to thermal agitation, it may results in the decrease in the adiabatic compressibility due to the increase of solute concentration and hence the velocity is increased.

By increase the temperature for each concentration, it is supposed that the intramolecular hydrogen bonding may get broken due to thermal agitation. That is resulted in increasing the intermolecular free length. As expected due to the thermal agitation the increase in velocity value clearly reflects the change in the intermolecular free length L_f (Sundaresan and Srinivasa Rao, 1994).

The specific acoustic impedance is a highly reasonable tool with good precision in understanding the structural interactions in solutions. These constant are used as index of molecular interactions (Dhanalaksmi and Sekar, 1996). The acoustic impedance is a significant factor stands to affect the reflection and transmission of sound waves. Its behaviour with increase of concentration and temperature will be the indication of the interaction between solute and solvent molecules.

References

Arul, G. and Palaniyappan, L., 2005. Ultrasonic study of I-butanol in pyridine with benzene. *Indian J. of Pure and Appl. Phys.*, 43: 755–758.

Dhanalakshmi, A. and Sekar, S., 1996. *Proceedings of Solvents Methanol Symposium on Ultrasonics.* Ultrasonic Society of India. Applied Publishers Ltd., p. 331.

Espinola, R.P. and Watermann, P.C., 1958. Ultrasonic interferometer for the measurement of the temperature dependence of elastic constants. *J. Appl. Phy.*, 29(4): 718–721.

Eswari Bai, M., Subha, M.C.S., Narayana Swamy, G. and Chowdoji Rao, K., 2004. Acoustical studies of molecular interactions in binary liquid mixtures of butoxy ethanol with some amines at 308.15K. *J. Pure Appl. Ultrason.*, 26: 79–83.

Eyring, H. and Kincaid, J.F., 1938. Free volumes and free angle ratios of molecules in liquid. *J. Chem. Phys.*, 6: 620.

Jacobson, B., 1952. Intermolecular free length in liquid state, I. Adiabatic and isothermal compressibilities. *Acta Chem. Scand.*, 6: 1485.

Kannappan, G. and Hemalatha, G., 2005. Ultrasonic studies on the molecular interaction of 1-chlorobenzotriazole with aromatic compounds in solution. *Indian J. of Pure and Appl. Phys.*, 43: 849–853.

Kannappan, V. and Jaya Santhi, R., 2005. Ultrasonic study of induced dipole-dipole interactions in binary liquid mixtures. *Indian J. of Pure and Appl. Phys.*, 43: 750–754.

Srivastava, T.N., Singh, R.P. and Bhagwan Swaroop, 1983. Ultrasonic velocity and adiabatic compressibility of triphenyltin chloride in acetone at various temperatures. *Indian J. of Pure and Appl. Phys.*, 21: 67–68.

Sundaresan, B. and Srinivasa Rao, A., 1994. Intermolecular free length study of PVC in DMF and in chlorobenzene. *Polym. Int.*, 33: 4253.

Chapter 13

Comparative Study of Alkaloid Extraction from Different Part of *Rauvolfia tetraphylla*

O.P. Verma[1], Onkar Chaudhary[2], Nirmala Singh[2], Ajay Singh[1], A.K. Gupta[1], and B.N. Mishra[3]

[1]*Allahabad Agricultural Institute-Deemed University, Allahabad – 211 007*
[2]*J.D.M.D.C., Kanpur*
[3]*IET, Lucknow*

ABSTRACT

Rauvolfia tetraphylla is a rich source of unique natural products which are known as secondary metabolites. Alkaloids are a class of secondary metabolites. Six alkaloids were isolated from different parts of *Rauvolfia tetraphylla* which were ajmalicine, reserpine, rescinnamine, yohimbine, ajmaline and serpentine respectively. It was found that all these alkaloids were present in good amount only in root and flower stem samples whereas samples as mature leaf and flower stem had only reserpine as detectable alkaloid compounds whereas flower stem and roots had only rescinnamine while ajmaline and serpentine are present in all samples. The highest yield 9.0 per cent, shown by flowers whereas lowest yield 0.22 per cent shown by roots. The objectives of present investigation were to isolate different alkaloids from different parts of *Rauvolfia tetraphylla.*

Keywords: *Ajmalicine, Rescinnamine, Yohimbine, Ajmaline, Serpentine and yohimbine.*

Introduction

Rauvolfia tetraphylla a medicinal plant of apocyanaceae family is indigenous to India and neighboring countries. It has deep root system, which allows it to resist long period of drought. Fruits of this plant are a drupe, ovoid in shape and have hard seed (Chakarbarty, 2003). Alkaloids are an important class of basic organic compounds that occurs in higher plants consists of one or more nitrogen heterocyclic rings as an integral part of their structure. These are known to produce striking physiological effects when administered to human (Bahl and Bahl, 2000). Alkaloid is of much importance for its remarkable use in medicine. Many alkaloids such as ajmaline, serpentine, yohimbine and heteroyohimbine type are known to be isolated from *rauvolfia* (Stockigt *et al.*, 1981). These alkaloids work by controlling nerve impulse along certain pathways that effects heart and blood vessels, thus lowering blood pressure. They also deplete serotonin from nerves of the central nervous system and have potent antirrhythmic effect. The plant is also medicinally important in the treatment of cardio vascular diseases, hypertensions and various psychiatric diseases (Sahu, 1983).

Materials and Methods

Plant Material

Rauvolfia tetraphylla were used for alkaloid extraction. *Rauvolfia* are grown and maintained in the field at Allahabad Agricultural Institute-Deemed University, Allahabad. The plants were grown following standard agronomic practices. The plant material was freshly harvested for use.

Chemicals and Reagents

All the chemicals and reagents were supplied by Himedia. Dragendroff reagent was used for alkaloid extraction.

Alkaloid Extraction

For alkaloid extraction, each plant part of the *Rauvolfia tetraphylla* was used. Samples were taken and dried completely. Powder was made by grinding and dried. Thereafter, methanol (10 ml) was mixed to the sample. Filtration was done after 12 hours. This process was repeated three times. Methanol was evaporated and thereafter 1 M dilute HCI was added. The contents were warmed in a water bath for 5 minutes at 50°C. Thereafter, basification was done by mixing 5 M cone. NaOH. This solution was mixed with di chloromethane (20ml); the separation was done by using separating funnel. This process was repeated three times. Afterwards, its evaporation was done and the dried residue is weighed and expressed as crude alkaloids.

TLC Analysis

TLC plates were prepared by layering silica over the plate. Thereafter, it was dried completely in oven. The alkaloids were then spotted on plate with the help of a capillary. After it, plate was run in the development jar with solvent for running. After completely running, plate was removed from the jar and it was dried again. Thus the alkaloids obtained were seen in UV light. Thereafter, spraying was done using dragendroff reagent. Fluoroalkanes, cyc1ohexane, n-hexane, I-chlorobutane, carbon tetra chloride, I-propyl ether, toluene, diethyl ether, tetrahydrofuran, chloroform, ethanol, ethyl acetate, dioxane, methanol, acetonitrile, nitro methane, ethylene glycol and water were used as solvents for better separation of molecule.

Results and Discussion

Total alkaloids were extracted from each part of the. plant of *Rauvolfia tetraphylla* andgravimetric estimation of indole alkaloids are presented in Table 13.1. The percentage yield of crude alkaloids were calculated and shown in Table 13.1.

Table 13.1: Gravimetric Estimation of Indole Alkaloids from Different Parts of *Rauvolfia tetraphylla*

Sl.No.	Parts	Wt. of Tissue (mg)	Wt. of Alkaloid (mg)	Yield (per cent)
1.	Fruit	1000	2.2	0.22
2.	Flower	20	1.8	9.0
3.	Flower stem	685	7.4	1.08
4.	Very young leaf	182	13.9	7.64
5.	Young leaf	352	15.1	4.28
6.	Mature leaf	1000	17.6	1.76
7.	Old leaf	185	10.8	5.83
8.	Stem	1000	5.8	0.58
9.	Root	1000	21.6	2.16

In *Rauvolfia tetraphylla*, total alkaloids from different parts of plant ranged from 0.22 to 9.0 per cent. Flower and very young leaf have the highest yield of alkaloids 9.0 per cent and 7.64 per cent respectively. While fruit has the lowest yield of alkaloids, 0.22 per cent. Among all the four leaves that are very young, young, mature as well as old, very young leaf showed the highest yield of alkaloids as shown in Table 13.1.

Yield (Per cent) of Alkaloids

The yield (per cent) of alkaloids shown in graphical form in Figure 13.1. Flower and very young leaf have the highest yield of alkaloids 9.0 per cent and 7.64 per cent respectively in *Rauvolfia tetraphylla*.

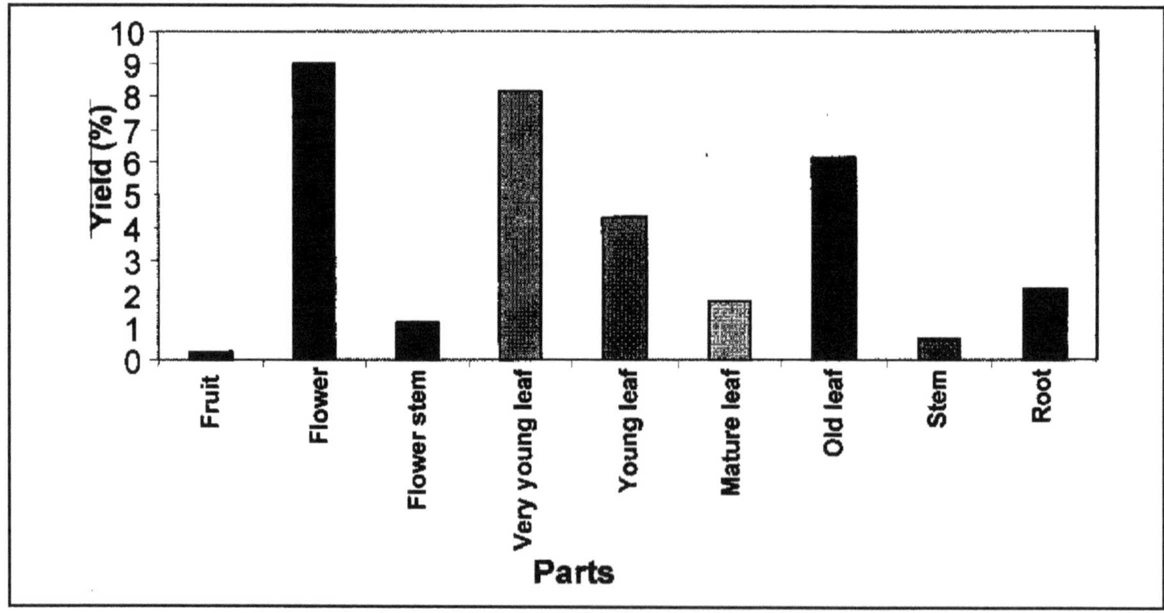

Figure 13.1: Yield of Alkaloids in *Rauvolfia tetraphylla*

Alkaloid Composition by TLC

The composition of alkaloids extracted by TLC analysis has shown in Table 13.2. In *Rauvolfia tetraphylla* when samples were analyzed under TLC, it was detected that root have alkaloids Ajmalicine, Reserpine, Yohimbine, Ajmaline, Serpentine while mature leaf, and flower stem have respine, flower stem and fruit have only recinnamine. It is important that all the parts of *Rauvolfia tetraphylla* have alkaloids ajmaline and serpentine. RTR, RTS, RTM, RTFS, and RTF represented as *Rauvolfia tetraphylla* root, stem, mature leaf, fruit stem and flower respectively. It is apparent from the Table 13.2 that the two *Rauvolfia* alkaloids were present in good amounts in the all samples (Table 13.2 and Figure 13.1). Whereas samples as RTM, and RTFS had only reserpine as detectable alkaloid compounds while RTFS and RTF had only rescinnamine. Ajmalicine and serpentine were present in all samples shown in (Table 13.2).

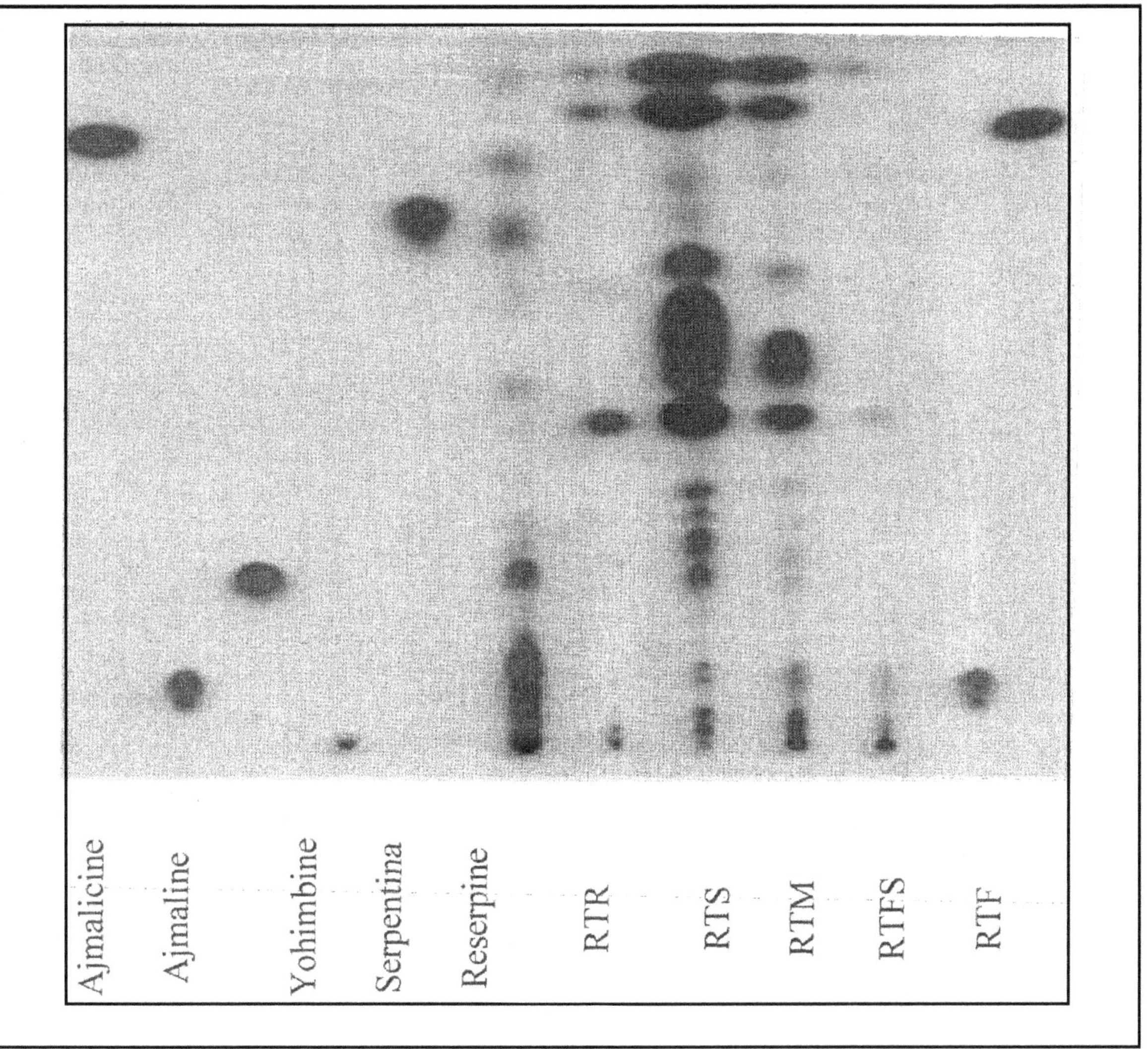

Figure 13.2: TLC Analysis of Alkaloids for *Rauvolfia tetraphylla*

Table 13.2: Qualitative Detection of *Rauvolfia* Alkaloid Constituents versus Plant Parts in *Rauvolfia tetraphylla*

Alkaloids	*Parts*				
	RTR	*RTS*	*RTM*	*RTFS*	*RTF*
Ajmalicine	+	–	–	–	–
Reserpine	+	–	+	+	–
Rescinnamine	–	–	–	+	+
Yohimbine	+	–	+	+	–
Ajmaline	+	+	+	+	+
Serpentine	+	+	+	+	+

References

Bahl, A. and Bahl, B.S., 1948. Nature of alkaloids. In: *Alkaloids*, 15th edn. S. Chand and Comp, pp. 741.

Chakarvarty, A., 2003. Rauvolfia. In: *Handbook Horticulture*, 3rd edn. ICAR Publications, p. 601–602.

Sahu, B.N., 1983. *Rauvolfia serpentina (Sarpagandha): Chemistry and Pharmacology*. Today and Tomorrow's Printers, 2: 595.

Stockigt, J., 1995. Biosynthesis in *Rauvolfia serpentine:* Modern aspects of an old medicinal plant. In: *The Alkaloids: Chemistry and Pharmacology,* (Ed.) G.A. Cordell, p. 115–172.

Stockigt, J., 1995. Natural products and enzymes from plant cell cultures. *Plant Cell Tissue and Organ Culture*, 43(2).

Stockigt, J., Pfitzner, A. and Firl, J., 1981. Indole alkaloids from cell suspension cultures of *Rauvolfia serpentina* Benth. *Plant Cell Reports*, 1(1): 36–39.

Chapter 14

Effect of *Parthenium hysterophorus* Linn. Seedling Vigour and Yield of Some Legumes

K. Aruna Lakshmi and M.P. Kusuma

Department of Biotechnology, College of Engineering, GITAM, Rushikonda, Andhra Pradesh

ABSTRACT

The allelopathic effect of *Parthenium hysterophorus* on the growth and yield of three different legumes *viz.*, Red gram, Green gram and Black gram is evaluated. Aqueous crude extracts of *Parthenium* shoots and roots diluted in different ratios *viz.*, 1 : 1 to 1 : 20 (crude extract : water v/v) are sprayed on 15day old seedlings of the different legumes. The response of the three legumes to foliar spraying varied considerably. Aqueous shoot extract in the concentration of 1 : 10 increased growth by 38 per cent over controls in *Cajanus* and 81 per cent in Green gram. But high concentration of root extract (1 : 1) and shoot extract (1 : 3) stimulated vegetative growth of Black gram. The positive allelopathic effects of *Parthenium* on different legumes can be exploited to prepare cheap, ecofriendly growth promoting substances.

Keywords: *Parthenium hysterophorus, Red gram, Green gram and Black gram.*

Introduction

Parthenium hysterophorus Linn. (Asteraceae) commonly called as Congress weed has been described as world's worst weed for agriculture, environment and human health (Bhan *et al.*). The leaves of the plant cause severe dermatitis and the tiny light pollen cause naso-bronchial allergies due to the presence of sesquiterpene-lactone called Parthenin (Subbarao and Towers, 1992). The crop loss due to this plant is immense. In India, the weed causes yield losses of up to 40 per cent in several crops and is reported to reduce production by 90 per cent (Adkins *et al.*, 1997). The fight against *Parthenium* using various weapons like herbicides, mechanical uprooting and biological control agents has been lost as the weed continues to increase and spread into remote areas of our country. Manual uprooting is effective but is expensive. Attempts are now being made to utilize the weed profitably. It has been reported that different parts of the plant show different allelopathic effects in crop plants (Oudhia 2000 a,b,c,d). *Parthenium* has tremendous fecundity and a single plant can produce up to 50,000 flowers thus a very rich source of phytohormones (Ramaswamy, 1997). It is therefore hypothesized that a crude aqueous extract of *Parthenium* would increase the flower production and subsequently yield. The stimulatory effect of foliar extracts of *Parthenium* on growth and yield have been reported in crop plants like chickpea, linseed, kasturi bhendi, rice (Oudhia 2000 a,b,c,d). An attempt has therefore been made to study the effect of *Parthenium* extract on some legumes like *Cajanus cajan, Phaseolus mungo and Phaseolus radiatus.*

Materials and Methods

Seeds of Red gram (*Cajanus cajan* var. ICPL 85063) obtained from ICRISAT (International Crops Research Institute for the Semi Arid Tropics, Hyderabad), Black *gram* (*Vigna mungo* var. IPC-8863) and Green gram (*Vigna radiatus* var. T-9) purchased from Andhra Pradesh Seed Corporation were used for different experiments in the present study. Healthy vegetative (pre-flowering) and flowering plants of *Parthenium* were collected at random from different localities of Visakhapatnam. An aqueous crude extract was prepared by grinding various parts of *Parthenium* plant like shoot+leaf and roots at a rate of 1gm/ml. The crude extract was filtered through muslin cloth. Aqueous dilute solutions of the crude filtrate were made in the following ratio 1 : 1, 1 : 2, 1 : 10 and 1 : 20 (crude filtrate: water, v/v).

For pre-testing the effect of aqueous *Parthenium* extracts on the germination and seedling vigor of legumes, experiments were carried under controlled conditions in the laboratory in Petridishes. About 20ml of diluted extract was dispensed into glass petridishes lined with filter paper. In each Petri dish 10 seeds were placed equidistantly and 20ml of distilled water was dispensed into the control petridish. The petridishes were incubated at 20±2°C and the germination of seeds was recorded five days after incubation (DAI) and the root and shoot length was noted at the end of seven days. All experiments were replicated thrice.

All field experiments were conducted at Gandhi Institute of Technology and Management (GITAM) College campus. The experimental field is a sandy loam having good drainage. The field was thoroughly tilled and beds of 1m × 0.5m were prepared and 20kgs of farmyard manure + vermicompost was added to each bed. Healthy, uniform sized seeds were sown in 3 rows in each bed. Seed beds were irrigated every day in the morning and on every second day after the germination of seeds. Healthy seeds were washed thoroughly with tap water to remove debris on the surface and were planted equidistantly in a randomized plot design.

Results and Discussion

The results of the effect of extracts of *Parthenium* on the growth and yield of Black gram, Green gram and Red gram were tabulated in Tables 14.1–14.3. It is evident from the tables that three legumes differed considerably in their response to spraying.

Table 14.1: Effect of Different Concentration of *Parthenium hysterophorus* on Black Gram

		Root Extract				Shoot Extract				
		R_1	R_2	R_{10}	R_{20}	S_3	S_4	S_5	S_{10}	S_{20}
Percentage	Treated	70	–	–	–	50	50	70	–	–
germination	control	100	–	–	–	100	100	90	–	–
+ % increase in growth		39.2	–7.14	6.0	0	20	28.5	–7.14	–80.9	–53.5
+ % increase in yield		9.2	51.3	–	–	191.8	90.1	35.7	–	155

+: Over controls.

Table 14.2: Effect of Different Concentration of *Parthenium hysterophorus* on Green Gram

		Root Extract				Shoot Extract				
		R_1	R_2	R_{10}	R_{20}	S_3	S_4	S_5	S_{10}	S_{20}
Percentage	Treated	80	80	–	–	60	60	70	–	–
germination	control	100	100	–	–	90	100	90	–	–
+ % increase in growth		22.3	–4.8	–	–	23.0	22.7	4.5	81	66.6
+ % increase in yield		44.4	–	–	–	16.6	–	87.5	–	–

+: Over controls.

Table 14.3: Effect of Different Concentration of *Parthenium hysterophorus* on Red Gram

		Root Extract					Shoot Extract				
		R	R_1	R_2	R_{10}	R_{20}	S	S_1	S_2	S_{10}	S_{20}
Percentage	Treated	100	100	100	–	–	100	100	–	–	–
germination	control	100	100	100	–	–	100	100	–	–	–
+ % increase in growth		–1.08	–22.8	4.83	–80	45	0.2	–33.8	7.09	38.6	15.6

+: Over controls.

S	1 : 0	Crude shoot+leaf extract of Parthenium
S_1	1 : 1	Dilution of shoot+leaf extract of Parthenium
S_2	1 : 2	Dilution of shoot+leaf extract of Parthenium
S_{10}	1 : 10	Dilution of shoot+leaf extract of Parthenium
S_{20}	1 : 20	Dilution of shoot+leaf extract of Parthenium
R		Crude Root extract of Parthenium
R_1	1 : 1	Dilution of root extract of vegetative plants
R_2	1 : 2	Dilution of root extract of vegetative plants
R_{10}	1 : 10	Dilution of Root extract of Parthenium
R_{20}	1 : 20	Dilution of Root extract of Parthenium

Extracts of *Parthenium* did not effect seed germination in *Cajanus*, but did inhibit (20 per cent) germination in Black and Green gram in the laboratory experiments.

Root extract enhanced the vegetative growth in Blackgram and *Cajanus* while shoot extract performed better in improving the growth of Green gram.

In Black gram root extract has found to be more stimulatory (40 per cent) than shoot extract at 1 : 1 concentration and shoot extract at 1 : 3 and 1 : 4 concentrations increased shoot length by 30 per cent and yield by 191 per cent over the controls (Arunalakshmi, APCOST, 2002).

Highly diluted shoot extract (1 : 10) stimulated vegetative growth of Green gram by 81 per cent over controls. However further dilutions were found to inhibit shoot growth. Root extract of *Parthenium* showed less stimulatory effect. High concentrations of shoot extract (S_1–S_5) inhibited vegetative growth but had a positive effect on the yield of the Green gram. Among the totaled, S_5 is having maximum yield (87 per cent) over controls.

Hundred per cent of the seeds germinated and there was no difference in germination between treatments and control. Crude extract of the shoot had the greatest inhibitory effect reducing the growth by 75–78 per cent compared to the controls. Dilution 1 : 1 and 1 : 2 also reduced growth than in controls and showed an inhibitory or phytotoxic effect. However further dilution of the extracts to 1 : 10; 1 : 20 and 1 : 30 had a positive effect on the growth. Shoot extract diluted to 1 : 10 had a stimulatory effect on shoot growth (38.6 per cent) of *Cajanus*. The root extract at a dilution of 1 : 20 showed maximum growth (45 per cent) than controls. Although not significant, *P. hysterophorus* extracts were found to increase plant height and leaf size of *Cajanus* over their controls. The leaves of the treated plants were dark green in color over their controls. This suggests that the allelochemicals present in the extract had not effected or interfered with the germination and vegetative growth of *Cajanus*.

The factor promoting or inhibiting growth in the crude extract has not been quantified or characterized. Such studies would probably throw more light on the precise mechanism involved in allelopathy. The positive allelopathic effects of Parthenium on black gram can be exploited to prepare a cheap, ecofriendly growth promoting substance. The weed can profitably utilized as a potential source of growth promoters, which is a very useful product from the standpoint of a farmer.

Acknowledgement

We thank Andhra Pradesh Council of Science and Technology, Hyderabad for financial assistance. We are thankful to the Management, GITAM for providing facilities. The technical help provided by Mrs. M. Kusuma and Mr. K. Bhaskara Rao is greatly acknowledged.

References

Adkins, S.W., Navie, S.C., Graham, G.C. and McFadyen, R.E., 1997. *Parthenium* weed in Australia: Research underway at the Co-operative Research Centre for Tropical Pest Management. In: *Proc. 1st International Conference on Parthenium Particulate Management*, (Eds.) M. Mahadevappa and V.C. Patil. University of Agricultural Sciences, Dharwad, India, 1: 13–17.

Aruna Lakshmi, K., 2002. Isolation of growth substances from *Parthenium hysterophorus* L. making virtue out of waste weed. *APCOST Project Report.*

Bhan, V.M., Kumar, Sushil and Raghuvanshi, M.S., 1997. Future strategies for effective *Parthenium* Management. In: *Proc. 1st International Conference on Parthenium Management*, (Eds.) Mahadevappa, M. and V.C. Patil. University of Agricultural Sciences, Dharwad, India, 1: 90–95.

Manikandan, R. and Abdul Hakim, S., 1999. The effect of shoot and root extracts of *Parthenium* on the growth and yield of ground nut and black gram. *Plant Sci.*, 12: 345–348.

Oudhia, P., 2000a. Allelopathy between Rice and *Parthenium*. *Int. Rice Res. Notes*, 25: 34.

Oudhia, P., 2000b. Allelopathic effect of *Parthenium* on Chickpea. *Crop Res.*, 19: 221–224.

Oudhia, P., 2000c. Allelopathic effect of *Parthenium* on Linseed. *Crop Res.*, 20: 563–566.

Ramaswamy, P.P., 1997. Potential uses of *Parthenium*. In: *Proc. 1st International Conference on Parthenium*, (Eds.) Mahadevappa, M. and V.C. Patil. University of Agricultural Sciences, Dharwad, India, 1: 77–80.

Rao, M., Prakash, O. and Subba Rao, P.V., 1985. Reaginic allergic to *Parthenium* pollen, Evalution by skin test and rast. *Clinical Allergy*, 15(5): 449–454.

Chapter 15

Parametric Studies of Pollutants from Copper Coated Spark Ignition Engine with Catalytic Converter with Gasoline Blended with Methanol

M.V.S. Murali Krishna, K. Kishor, P.R.K. Prasad, A. Sahithy and G.V.V. Swathy

Department of Mechanical Engineering, Chaitanya Bharathi Institute of Technology, Gandipet, Hyderabad – 500 075, Andhra Pradesh

ABSTRACT

The major pollutants emitted from spark ignition engine run with gasoline fuel are carbon monoxide (CO) and unburnt hydrocarbons (UBHC), which are hazardous to human beings and environment. However, if the engine is run with alcohol, aldehydes are also to be checked which are carcinogenic in nature and cause eye irritation, severe headache and vomiting sensation and hence control of these pollutants call for immediate attention. Experiments are carried out for controlling the pollutants from a variable-compression ratio, copper-coated spark ignition engine run with methanol-blended gasoline (20 per cent V/V) fitted with catalytic converter containing sponge iron catalyst. The influence of the engine parameters such as speed, compression ratio and configuration of the engine with and without catalytic converter on the pollutants is studied. The speed of the engine has marginal effect, while compression ratio has strong influence on reduction of pollutants. Air injection into the catalytic converter has further reduced the pollutants. Copper-coated spark ignition engine with methanol-blended gasoline found to decrease the pollutants considerably when compared to conventional engine with pure gasoline operation.

Keywords: *Spark ignition engine, Pollutants, Catalytic converter, Air injection, Copper coating.*

Introduction

CO is major pollutant from spark ignition engine, breathing of which causes many health disorders, like reduction of haemoglobin content in the blood, increases dizziness, breathing and respiratory problems, eye irritation, loss of appetite etc. (Fulekar, 1999). It also cause detrimental effects on other animal and plant life besides, environmental disorders (Sharma, 1996). This pollutant is considerably high during idling and peak load operation of the engine. Hence globally, stringent regulations are made for permissible CO levels in the exhaust of 2 and 4-stroke spark ignition engines. The formation of unburnt hydrocarbons is due to incomplete combustion. The two important reasons for incomplete combustion of the fuel are cool metal surfaces of the combustion chamber and imperfect mixture ratio. Aldehydes are intermediate compounds of partial combustion. They are carcinogenic in nature. They too cause health problems. Of many methods available for reduction of these emissions, the one employing a catalytic converter is more effective (Vara Prasad *et al.*, 1997). The use of platinum group metals as catalysis is quite expensive and hence efforts are on for search of cheaper catalysts (Luo and Zheng, 1999; Murali Krishna *et al.*, 2000). Further modification of engine design (Nedunchezhian and Dhandapani, 2000) and fuel composition (Pundir and Abraham, 1985; Sharma *et al.*, 1985) is also found to be advantageous in controlling the pollutants in the exhaust of the engine. The use of catalysts to promote combustion is an old concept. More recently copper is coated over piston crown and inside of cylinder head wall (Dhandapani, 1991) and it is reported that the copper coating improves the fuel economy and increased combustion stabilization. In the context of depletion of fossil fuels due to increase of fuel consumption, the search for alternate and renewable fuels has also become pertinent. The properties of methanol are very close to those of gasoline. In addition, no major modification in the engine is required if low quantities of methanol are blended with gasoline ill spark ignition engine. In the present study, the effect of various engine parameters on the control of these emissions is reported with different versions of the engine such as conventional engine and copper coated engine with catalytic converter with sponge iron as catalyst and methanol blended gasoline (methanol 20 per cent V/V) as fuel.

Experimental Programme

The experimental set-up employed in the present study is shown in Figure 15.1. A four-stroke, single-cylinder, water-cooled, spark ignition engine of brake power 2.2 kW at rated speed of 3000 rpm is used. The engine is coupled to an eddy current dynamometer for measuring its brake power. The compression ratio of the engine is varied from 3 to 9 with the change of the clearance volume by adjustment of cylinder head, threaded to the cylinder of the engine. The engine speeds are varied from 2200 to 3000 rpm. In the present investigations, in order to improve the effective combustion of fuel and to reduce the emission of pollutants, the piston crown and inside surface of the cylinder head is coated (Nedunchezhian and Dhandapani, 2000) with copper by plasma spraying. A bond coating of NiCoCr alloy is applied for a thickness of about 100 microns using an 80 kW METCO plasma spray gun. Over the bond coating, copper (89.5 per cent), aluminium (9.5 per cent) and iron (1.0 per cent) is coated for 300 microns thickness. The coating has very high bond strength and does not wear off even after 50 hrs of operation. A catalytic converter shown in Figure 15.2 is fitted to the exhaust pipe of the engine.

Provision is made to inject a definite quantity of air into the catalytic converter. The converter is filled with sponge iron catalyst with void ratio of 0.7 : 1 (void ratio is the ratio between the volume occupied by the catalyst to the volume of the catalytic chamber), where the pollutants are found (Murali Krishna *et al.*, 2000) to be minimum. The CO and UBHC emissions in the exhaust of the engine

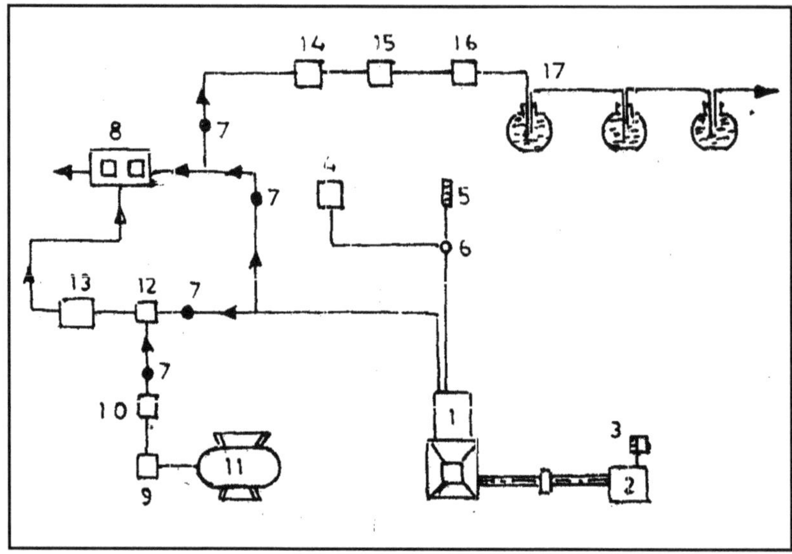

Figure 15.1: Experimental Set-up

(1) Engine, (2) Eddy current dynamometer, (3) Loading arrangement, (4) Fuel tank, (5) Burette, (6) Three-way valve, (7) Directional valve, (8) CO and UBHC analyzer, (9) Rotometer, (10) Heater, (11) Air-compressor, (12) Air chamber, (13) Catalyst chamber, (14) Rotometer, (15) Heater, (16) Round bottom flask containing 2-DNPH solution

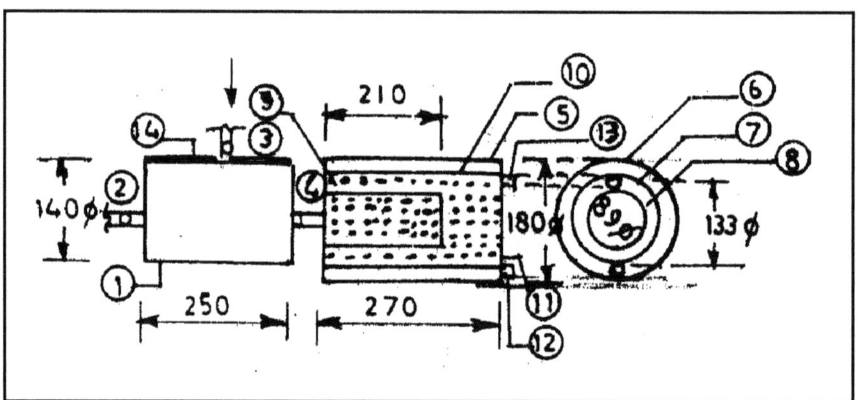

All dimensions are in mm

Figure 15.2: Details of Catalytic Converter

(1) Air chamber, (2) Inlet for air chamber from the engine, (3) Inlet for air chamber from the compressor, (4) Outlet for air chamber, (5) Catalytic chamber, (6) Outer cylinder, (7) Intermediate-cylinder, (8) Inner-cylinder, (9) Inner sheet, (10) Intermediate sheet, (11) Outer sheet, (12) Outlet for exhaust gases, (13) Provision to deposit the catalyst and (14) Insulation

are measured with Netel Chromatograph analyzer. For measuring aldehydes from the exhaust of the engine, DNPH method (Inoue *et al.*, 1980) is employed in the experimentation. The exhaust of the engine is bubbled through 2,4-dinitrophenyl hydrazine (2,4 DNPH) solution. The hydrazones formed are extracted into chloroform and are analyzed by employing high performance liquid chromatography (HPLC) to find the percentage concentration of formaldehyde and acetaldehyde in the exhaust of the engine. Various sets of the exhaust gases are drawn at three different locations, (1) immediately after the exhaust valve of the engine, (2) after the catalytic converter, and (3) at the outlet after air injection into the catalytic converter. The quantity of air drawn from the compressor and injected into the converter is kept constant so that the backpressure does not increase and reverse flow is not created in the converter. Experiments are carried out on different configurations of the engine *i.e.*, conventional engine and copper coated engine with different test fuels like pure gasoline and methanol blended gasoline under different operating conditions of the catalytic converter like set-A-without catalytic converter and without air injection, set-B-with catalytic converter and without air injection and set-C with catalytic converter and with air injection.

Results and Discussion

Table 15.1 shows the data of carbon monoxide (CO) emissions at peak load at different compression ratios and speeds with different versions of the engine at different operating conditions of the catalytic converter. A reduction in CO emissions is found to increase with increase of the speed of the engine for all configurations with different test fuels. Improved combustion with the increase of turbulence decreased CO emissions. It is observed that at each speed, the CO content in the exhaust decreased considerably with the use of catalyst, which is more pronounced with the air injection into the converter. Methanol blended gasoline decreased CO emissions considerably when compared to pure gasoline. The combustion of alcohol produces more water vapor than free carbon atoms as methanol has lower C/H ratio of 0.25 (where C and H represent the number of carbon and hydrogen atoms respectively in the composition of the fuel) against 0.44 of gasoline. Methanol has oxygen in its structure and hence its blends have lower stoichiometric air requirements compared to gasoline. Therefore more oxygen is available for combustion with the methanol-blended gasoline, which leads to reduction of CO emissions. Methanol dissociates in the combustion chamber of the engine forming hydrogen, which helps the fuel-air mixture to burn quickly and thus increases combustion velocity, which brings about complete combustion of carbon present in the fuel to carbon dioxide and also carbon monoxide to carbon dioxide and thus makes leaner mixture more combustible, causing reduction of CO emissions. Copper-coated engine reduced CO emissions; when compared to conventional engine. Catalytic activity increases with temperature as combustion temperature increases with the increase of the speed of the engine. Hence there is reduction of CO emissions with copper-coated engine when compared to conventional engine. As the compression ratio decreased from 9 : 1 to 8 : 1, CO emissions decreased with different test fuels with both versions of the engine. This is due to increase of exhaust gas temperature with the decrease of compression ratio leading to oxidation of CO in the exhaust manifold with different versions of the engine.

The data of un-burnt hydro carbon (UBHC) emissions are shown in Table 15.2 while Table 15.3 and Table 15.4 show the data of formaldehyde and acetaldehyde emissions at peak load for different test fuels at different compression ratios, speeds with different versions of the engine respectively. The trend exhibited by UBHC emissions is the same as that of CO emissions. As speed decreased from 3000 rpm to 2600 rpm, the turbulence of combustion decreased and hence the speed of the flame decreases. The gas layer is entrapped between the piston and combustion chamber walls leading to increase in quench area. The presence of quench area inhibits the spreading of the flame, thereby increasing the hydrocarbon emissions. Formaldehyde concentration is observed to be more in comparison with acetaldehyde concentration in both versions of the engine when they are run with methanol-blended gasoline, as it is evident from the Table 15.3 and Table 15.4. Copper coated engine decreased aldehydes emissions considerably when compared with conventional engine. Catalytic converter with air injection drastically decreased aldehyde emissions in both versions of the engine.

Table 15.1: Data of Carbon Monoxide (CO) Emissions (%) at Peak Load for Different Test Fuels at Different Compression Ratios, Speeds with Different Versions of the Engine

Fuels →		*Gasoline*						*Methanol Blended Gasoline (20 per cent V/V)*					
Engine Version →		*CE*			*CCE*			*CE*			*CCE*		
Speed (rpm) →		*2600*	*2800*	*3000*	*2600*	*2800*	*3000*	*2600*	*2800*	*3000*	*2600*	*2800*	*3000*
Set A	8 : 1	3.55	3.35	3.25	2.72	2.62	2.5	2.32	2.22	2.11	1.53	1.43	1.33
	9 : 1	3.95	3.85	3.75	3.23	3.13	3.0	2.81	2.71	2.61	2.04	1.94	1.83
Set B	8 : 1	2.05	1.95	1.85	1.66	1.56	1.4	1.33	1.23	1.12	0.72	0.61	0.51
	9 : 1	2.55	2.35	2.25	2.02	1.92	1.8	1.74	1.64	1.52	1.12	1.01	0.91
Set C	8 : 1	1.52	1.42	1.3	1.24	1.14	1.0	0.83	0.72	0.61	0.52	0.45	0.34
	9 : 1	1.71	1.61	1.5	1.43	1.33	1.2	1.02	0.92	0.81	0.71	0.65	0.54

Set A: Without catalytic converter and without air injection; Set B: With catalytic converter and without air injection; Set C: With catalytic converter and with air injection; CE: Conventional engine; CCE: Copper coated engine.

Table 15.2: Data of Unburnt Hydro Carbon (UBHC) Emissions (ppm) at Peak Load for Different Test Fuels at Different Compression Ratios, Speeds with Different Versions of the Engine

Fuels →		*Gasoline*						*Methanol Blended Gasoline (20 per cent V/V)*					
Engine Version →		*CE*			*CCE*			*CE*			*CCE*		
Speed (rpm) →		*2600*	*2800*	*3000*	*2600*	*2800*	*3000*	*2600*	*2800*	*3000*	*2600*	*2800*	*3000*
Set A	8 : 1	550	510	450	425	375	325	320	295	270	195	175	155
	9 : 1	610	560	500	445	420	375	370	345	320	240	225	205
Set B	8 : 1	340	300	250	240	215	155	165	140	95	102	80	65
	9 : 1	390	350	300	310	265	206	205	180	135	142	120	105
Set C	8 : 1	240	200	150	310	165	80	125	100	70	81	60	45
	9 : 1	290	250	200	195	152	105	145	120	90	103	80	65

Set A: Without catalytic converter and without air injection; Set B: With catalytic converter and without air injection; Set C: With catalytic converter and with air injection; CE: Conventional engine; CCE: Copper coated engine.

Conclusions

The pollutants of CO, UBHC and aldehyde emissions at peak load decreased by 20–45 per cent relatively with the change of the engine configuration from conventional version to catalytic coated engine with different test fuels. The pollutants decreased by 25–45 per cent relatively with the change of fuel from gasoline to methanol-blended gasoline in both versions of the engine under different sets. They increased by 10–13 per cent relatively with the change of compression ratio from 8 : 1 to 9 : 1 while they decreased by 3–12 per cent relatively with the change of speed from 2600 rpm to 3000 rpm with different test fuels in both configurations of the engine under different operating conditions of the catalytic converter. Air injection decreased the emissions by 60 per cent with different test fuels with different configurations of the engine.

Table 15.3: Data of Formaldehyde Emissions (%) at Peak Load for Different Test Fuels at Different Compression Ratios, Speeds with Different Versions of the Engine

Fuels →		*Gasoline*						*Methanol Blended Gasoline (20 per cent V/V)*					
Engine Version →		*CE*			*CCE*			*CE*			*CCE*		
Speed (rpm) →		*2600*	*2800*	*3000*	*2600*	*2800*	*3000*	*2600*	*2800*	*3000*	*2600*	*2800*	*3000*
Set A	8 : 1	8.5	7.5	6.5	6.5	5.5	4.5	16.2	17.5	16.5	11.4	10.5	9.5
	9 : 1	9.2	8.2	7.5	7.1	6.2	5.5	19.1	18.2	17.4	12.2	11.2	10.4
Set B	8 : 1	6.6	5.5	4.5	4.4	3.5	2.5	9.3	8.5	7.5	7.5	6.5	5.5
	9 : 1	7.5	6.5	5.5	5.4	4.5	3.5	10.4	9.5	8.5	8.4	7.5	6.5
Set C	8 : 1	4.5	3.5	2.5	2.4	1.5	0.75	7.3	6.5	5.5	5.4	4.5	3.5
	9 : 1	5.0	4.0	3.0	3.1	2.0	1.25	8.1	7.0	6.0	6.1	5.0	4.0

Set A: Without catalytic converter and without air injection; Set B: With catalytic converter and without air injection; Set C: With catalytic converter and with air injection; CE: Conventional engine; CCE: Copper coated engine.

Table 15.4: Data of Acetaldehyde Emissions (%) at Peak Load for Different Test Fuels at Different Compression Ratios, Speeds with Different Versions of the Engine

Fuels →		*Gasoline*						*Methanol Blended Gasoline (20 per cent V/V)*					
Engine Version →		*CE*			*CCE*			*CE*			*CCE*		
Speed (rpm) →		*2600*	*2800*	*3000*	*2600*	*2800*	*3000*	*2600*	*2800*	*3000*	*2600*	*2800*	*3000*
Set A	8 : 1	7.5	6.5	5.5	5.5	4.5	3.5	10.5	9.5	8.5	8.4	7.5	6.5
	9 : 1	7.9	7.2	6.5	5.9	5.2	4.5	10.9	10.2	9.5	8.9	8.2	7.5
Set B	8 : 1	5.2	4.5	3.5	3.5	2.5	1.5	6.5	5.5	4.5	4.5	3.5	2.5
	9 : 1	6.2	5.5	4.5	6.5	4.5	2.5	7.0	6.0	5.0	5.0	4.0	3.0
Set C	8 : 1	3.2	2.5	1.5	0.75	0.5	0.25	4.4	3.5	2.5	1.75	1.5	1.25
	9 : 1	3.7	3.0	2.0	1.5	1.0	0.5	5.1	4.0	3.0	2.25	2.0	1.75

Set A: Without catalytic converter and without air injection; Set B: With catalytic converter and without air injection; Set C: With catalytic converter and with air injection; CE: Conventional engine; CCE: Copper coated engine.

Acknowledgements

The authors are thankful to the authorities of Chaitanya Bharathi Institute of Technology, Hyderabad, for the facilities provided. The financial assistance provided by Andhra Pradesh Council of Science and Technology (APCOST) for this project is greatly acknowledged.

References

Dhandapani, S., 1991. Theoretical and experimental investigation of catalytically activated lean burn combustion. *Ph.D. Thesis*, Indian Institute of Technology, Chennai.

Fulekar, M.H., 1999. Chemical pollution: A threat to human life. *Indian J of Env. Prot.*, 1(3): 353–359.

Inoue, T., Oishi, K. and Tanaka, T., 1980. Determination or aldehydes in the automobile exhaust by HPLC. *Toyota Gijistu*, 21(4): 500–506.

Luo, M.F. and Zheng, X.M., 1999. CO oxidation activity and TPR characteristics of CeO_2: Supported manganese oxide catalyst. *Indian J. Chem.*, 38(1): 703–707.

Murali Krishna, M.V.S., Vara Prasad, C.M. and Ramana Reddy, Ch.V., 2000. Studies on control of carbon monoxide emissions in spark ignition engine using catalytic converter. *Ecol. Env. and Conser.*, 6(4): 377–380.

Nedunchezhian, N. and Dhandapani, S., 2000. Experimental investigation of cyclic variation or combustion parameters in a catalytically activated two-stroke SI engine combustion chamber. *Eng. Today*, 2(1): 11–18.

Pundir, B.P. and Abraham, M., 1985. Performance of methanol gasoline blends in Indian passenger cars. In: *Proc. 9th National Conference on I.C. Engine and Combustion*. Indian Institute of Petroleum. Dehradun.

Sharma, B.K., 1996. *Engineering Chemistry*. Pragathi Prakashan (P) Ltd., Meerut.

Sharma, J. *et al.*, 1985. Fleet study of small two stroke engine vehicles fuelled with methanol gasoline blends. In: *Proc. 9th National Conference on I.C. engines and Combustion.*

Vara Prasad, C.M., Murali Krishna, M.V.S. and Prabhakar Reddy, C., 1997. Reduction of CO in petrol engine exhaust using catalytic converter. In: *Proc. 15th National Conference on IC Engines and Combustion*, College of Engineering, Gundi, Anna University, Chennai, pp. 372–377.

Chapter 16

Use of RAPD and ITS-PCR Typing for Studying Genetic Diversity of *Bacillus cereus*

T. Anju[1], *P. Dhevagi*[2]* *and T.C. Joseph*[3]

[1]*Vivekanada College of Engineering,*
[2]*Assistant Professor, Department of Environmental Sciences,*
Tamil Nadu Agricultural University, Coimbatore – 3
[3]*Scientist, CIFT, Cochin*

ABSTRACT

DNA based methods are used to trace genetic relationships and to identify strains rapidly. A comparative study was planned to study the genetic diversity of *B. cereus* using RAPD and ITS–PCR. About eight RAPD primers (S1442, S1443, S1444, S1445, S1457, S1441, S1448 and S1451) were tested for their efficiency and discrimination. The primer S1448 gave a high degree of discrimination between the isolates than other primers. All the 18 isolates of *B.cereus* were subjected to PCR amplification using the primers: SD–Bact–1494-a-s-20 and LD-Bact-0035-a-A-15. The results of ITS-PCR fingerprinting of *B.cereus* showed identical banding patterns in all the isolates tested. The above study reveals that RAPD typing discriminates the DNA polymorphism among isolates of *B. cereus* better than ITS–PCR typing.

Introduction

The genus *Bacillus* is very important because of their resistant spores and the capability of vegetative cells to secrete wide variety of enzymes, commercially important biomolecules and as a source of

* Corresponding Author; E-mail: devagihfr@yahoo.com.

spoilage or pathogenic organisms (Shewan, 1962; Kramer and Gillbert, 1989 and Drobniewski, 1993). *Bacillus cereus* food poisoning is due to diarrhoeagenic in proteinaceous foods and emetic entertoxinsin farinaceous foods (Kramer and Gillbert,1989; Granum *et al.*, 1993 and Granum, 1994). Besides it also causes non-gastrointestinal disease like endocartitis and endophthalmitis (Drobniewski, 1993). It shows a great diversity among the species. Comparison of genomic structure and organization between different strain of *B. cereus* and *B. thuringiensis* indicates great diversity within these species is as large as the diversity between them (Carlson *et al.*, 1994). Identification at strain level is important not only for epidemiological and phylogenetic studies but also for ecological and industrial purposes.

Genotyping that involves direct DNA based analysis of chromosome or extrachromosome has many advantages over other traditional typing methods. Both genotyping and phenotyping methods have been utilized for sub typing of *B. cereus* but these methods are complex, interpretation is difficult. Hence there is a need for a suitable discriminatory typing method for this bacterium to facilitate phylogenetic and epidemiological investigations.

Internal Transcribed Spacer Typing (ITS)

ITS sequencing has been used commonly in introgeneric phylogenetic studies of alga (Lakshmanan and Kumar, 2003). Many scientists(Marks and Cummings, 1996) used the ITS region to demonstrate low genetic diversification among fresh water isolates of the common and widespread alga *Cladophora* obtained from a wide range of habitats and geographical locations. Peter (1997) used ITS sequences to compare species within the widespread algal family Desmarestiaceae.

The rRNA genetic locus rrn is found in both prokaryotic and eukaryotic organisms (Mark *et al.*, 1993; Amman *et al.*, 1995 and Watson *et al.*, 1987) and are highly conserved in bacteria and other kingdoms. ITS typing method takes advantage of some of the heterogeneity found in the spacer regions, which exist between genes coding for rRNA in microorganisms. In this method, oligomer primes to the conserved regions of the genes coding for 16S and 23S rRNA is used, so that the intergeneric spacer region between 16S and 23S rRNA could be amplified. To obtain increased discrimination, the amplicons from the initial PCR ribotyping can be cut using restriction endonucleases (Kotsman *et al.*, 1992 and Kotsman *et al.*, 1995).

Random Amplified Polymorphic DNA (RAPD)

RAPD is a modification of polymerase chain reaction with a single primer able to anneal primer at multiple locations throughout the genome, can produce a spectrum of amplified products that are characteristics of the template DNA (Welsh and McClelland, 1990; Williams *et al.*, 1990; Welsh and McClelland, 1993 and Welsh *et al.*, 1992).

RAPD method has proved useful for obtaining fingerprints of organisms for epidemiological investigations in medicine or food science (Nilsson *et al.*, 1998). RAPD has been used for the typing of food borne microorganisms such as *E.coli* and to compare intra and interspecific differences in bacteria. The RAPD fingerprinting technique has also been proposed as a tool for generating taxon specific markers with different specificities (Welsh and McClelland 1990, Fani *et al.*, 1993, Bazzicalupo and Fani, 1995 and Day, 1997). RAPD in conjunction with PCR has been employed to identify many organisms to the strain level of classification. This technique is sensitive and specific because the entire genome of an organism is used as the basis for generating a DNA profile. The combination of PCR and random oligonucleotide primers has provided a method for the rapid and sensitive delineation of animal, plant, fungal, algal and bacterial strains (Scott *et al.*, 1992 and Akopyanz *et al.*, 1992). Hence a comparative study was planned to study the efficiency of RAPD and ITS-PCR typing for studying genetic diversity in *B. cereus* isolates.

Materials and Methods

Soil and vegetable samples collected from different market environments around Salem, Tamil Nadu, were used for the isolation of *B. cereus*. *B. cereus* has been isolated in Polymixin-Pyruvate-Egg-Yolk-Mannitol-Bromothymol blue Agar (PEMBA). The biochemical tests *viz.*, Starch hydrolysis, Fermentation of carbohydrates, MRVP Test, Nitrate reduction test, Gelatin liquefaction, Catalase test, Temperature tolerance, Motility test were done to confirm the cultures as *B. cereus*.

RPLA Detection of *B. cereus* Enterotoxin

Enterotoxin producing ability of *B. cereus* isolates was tested by using *B. cereus* enterotoxin (diarrhoel type) test kit (Oxoid kit). The Reversed Passive latex agglutination test (RPLA, where the antibody which is attached to particles reacts with the soluble antigen) was performed as per the manufacturer's instructions. Enterotoxin production was indicated by the formation of latex agglutination due to crosslinking of the latex particles by the specific antigen.

DNA Extraction and PCR Typing

DNA was extracted from *B. cereus* stains (Jacek Majewskl *et al.*, 2000) and used for PCR amplification and restriction analysis. RAPD-PCR was performed using a method (Bazzicalupo and Fani, 1995) for large-scale typing of *B. cereus*. ITS-PCR was performed by using method described by Daffonchio *et al.* (1998).

Amplification

The PCR products were run on 2 per cent agarose gel with a molecular weight marker (M) as standard and visualized under UV light. Every site that is amplified will lead to the appearance of a band on the stained agarose gel.

Restriction Digestion

To obtain increased discrimination the samples were restricted with restriction enzymes like Alu I, Hind III and Saul. The restricted samples were run in 2 per cent agarose gel and results were observed under UV light (Stephan,1996 and Sambrook and Russell, 1991).

Results and Discussion

Recently molecular sequencing techniques have been viewed as highly superior to structure based phylogenies (Swofford *et al.*, 1996). The employment of molecular sequence methods has not resulted in widespread overturn of previous evolutionary hypothesis based; on structural data (Moritz and Hillis, 1996) are often results in greater phylogenic resolution than the use of morphological data alone.

Bacillus cereus Isolates

Polymixin-Pyruvate-Egg-Yolk-Mannitol-Bromothymol blue Agar (PEMBA) is a selective media, used to detect small numbers of *B. cereus* cells and spores in the presence of large numbers of other organisms was used in this study to isolate *B. cereus* isolates. Flat, crenate to slightly rhizoid colonies 2-5mm in diameter, of ground-glass surface appearance, Turquoise peacock blue in colour, surrounded by a zone of egg-yolk precipitation caused by the lecithinase activity was isolated and maintained for further work.

PEMBA was preferred because incubation period was from 24h to 48h and improved discrimination of presumptive isolates was observed. The capacity to produce an egg-yolk precipitin reaction in these media is one of the diagnostic features of *B. cereus*. Here low levels of peptone were added to promote sporulation. The pure colonies were further confirmed with specific biochemical tests (Table 16.1). Of the 30 cultures isolated, 18 cultures were positive for *B. cereus* and were used for further molecular typing analysis.

Table 16.1: Biochemical Reactions of *Bacillus cereus*

1.	Gram reaction	Gram Positive, rods
2	Spores	Spore forming rods arranged in chains with sporangia not swollen
2.	Motility	Motile
3.	Catalase production	Positive
4.	Methyl Red Reaction	Negative
5.	VP test	Positive
6.	Growth at 50°C	Negative
7.	Growth 65°C	Negative
8.	Growth in 7 per cent NaCl	Positive
9.	Starch hydrolysis	Positive
10.	Gelatin liquefaction	Positive
11.	Nitrate reduction	Positive
12.	Acid from glucose	Positive
13.	Gas from glucose	Negative
14.	Mannitol fermentation	Negative

RPLA Enterotoxin Detection

The biochemically screened 18 isolates of *B. cereus* were tested for diarrhoeal enterotoxin production. Polystyrene latex particles are sensitized with purified antiserum taken from rabbits immunized with purified *B. cereus* diarrhoeal enterotoxin. These latex particles will agglutinate in the presence of *B. cereus* enterotoxin. A control agent was provided which consists of latex particles sensitized with non-immune rabbit globulins. The test was performed in V-well microtitre plates. Dilutions of the food extract or culture filtrate was made in two rows of wells, a volume of the appropriate latex suspension was added to each well and the contents were mixed. If the wells containing *B. cereus* enterotoxin, agglutination occurs due to the formation of a lattice structure which confirms the toxin production. Upon settling, this forms a diffuse layer on the base of the well. Out of the eighteen isolates evaluated only five isolates gave positive results for enterotoxin production.

Random Amplified Polymorphic (RAPD) Analysis

Eighteen mesophilic *B. cereus* isolates were analysed by RAPD-PCR to study the genetic diversity. About eight RAPD primers (S1442, S1443, S1444, S1445, S1457, S1441, S1448 and S1451) were tested for their efficiency and discrimination. The primer S1448 gave a high degree of discrimination between the isolates than other primers. So this primer was taken up for the RAPD analysis of all the 18 strains.

RAPD patterns using primer S1448 are reported in Figures 16.1a, b, and c. Few strains showed some similarity but in general a high level of variability was observed. There were ten different banding patterns observed showing good diversity among isolates.

Studies report that PCR fingerprinting by using RAPD of *B. cereus* show a wised genotypic heterogeneity among isolates (Daffonchio *et al.*, 1998; Stephan 1996 and Sabine Lectner, 1998). In a study to differentiate *B. anthracis* from the members of the *B. cereus* group using PCR fingerprinting, Henderson *et al.* (1994) showed a high degree of polymorphism among the strains of *B. cereus* studied.

In another study to evaluate the strain diversity in *B. cereus* by RAPD analysis using five different primers, 25 strains were grouped into 22 different pattern types (Stephan 1996). RAPD was further proved to be discriminatory for *B. cereus* (Bazzicalupo and Fani, 1995; Christiansson *et al.*, 1997 and Christiansson *et al.*, 1998). This technique has been used to type organisms such as *Staphylococcus aureus, Enterococcus faecium, E. coli and Entrobacter spp.*, as well as *L. monocytogenes* (Kotsman *et al.*, 1995

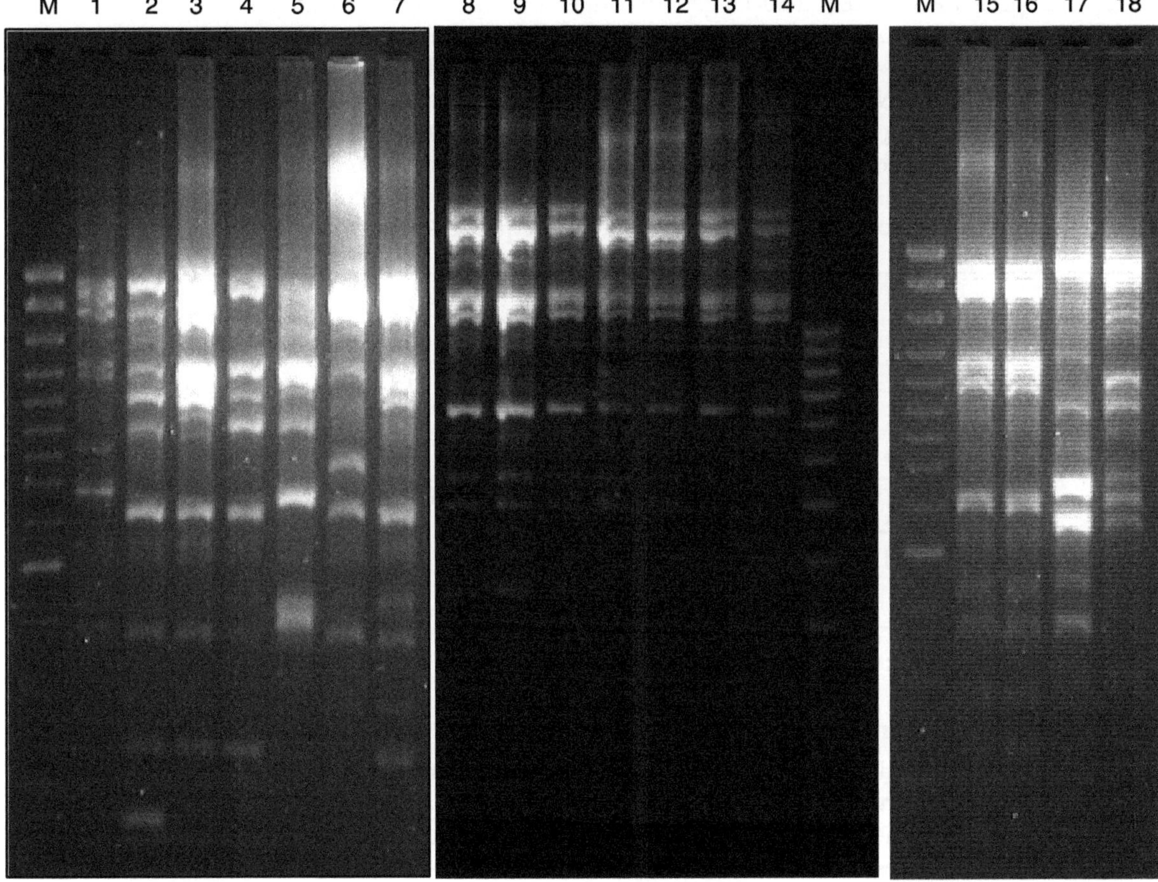

Figure 16.1a: RAPD Analysis of *B. cereus* Isolates
Lanes: M, 100bp ladder; 1–7: *B. cereus* isolates

Figure 16.1b: RAPD Analysis of *B. cereus* Isolates
Lanes: M, 100bp ladder; 8–14: *B. cereus* isolates

Figure 16.1c: RAPD Analysis of *B. cereus* Isolates
Lanes: M, 100bp ladder; 15–18 *B. cereus* isolates

and Sontakke and Fabe, 1995). All these investigators have clearly indicated, RAPD as good discriminatory method for studying heterogeneity of genome within the species. This was also confirmed by the RAPD profile of the data here. However this technique must be well optimized and standardized to obtain reproducible results (Bazzicalupo and Fani, 1995 and Welsh and McClelland 1990).

ITS-PCR Typing

To investigate the discriminatory ability of Internal Transcribed Spacer region for studying diversity, all 18 isolates of *B. cereus* were subjected to PCR amplification using the primers: SD–Bact–1494-a-s-20 and LD-Bact-0035-a-A-15. The results of ITS-PCR fingerprinting of *B. cereus* showed identical banding patterns in all the isolates tested. ITS results of *B. cereus* showed four banding patterns. Two major bands of 250bp and 450bp were seen in all strains studied here. Fifth strain had extra bands at 550bp and 750bp. The 550bp length band was observed in the eighteenth strain also.

In prokaryotes, spacer regions show a large degree of sequence and length variation at both the genus and species level. Most bacterial genera contain multiple copies of the operon for rRNA (Gurtler and Stanisich, 1996; Jenson *et al.*, 1993 and Leblond-Bourget *et al.*, 1996), so that the spacer regions within a single strain may differ in the length and or sequence. Thus, multiple bands obtained from a particular strain after DNA amplification would represent spacer regions of different length in different ribosomal-RNA-coding operons. So the ribosomal operon is used as a classical molecular marker to trace genetic relationships and to identify strains rapidly (Amman *et al.*, 1995). The rRNA operon is transcribed into one pre-rRNA transcript that contains the following components in the order (5′-3′): 16S, Spacer, tRNA, Spacer, 23S, Spacer and 5S rRNA sequences (Watson *et al.*, 1987) (Figure 16.2).

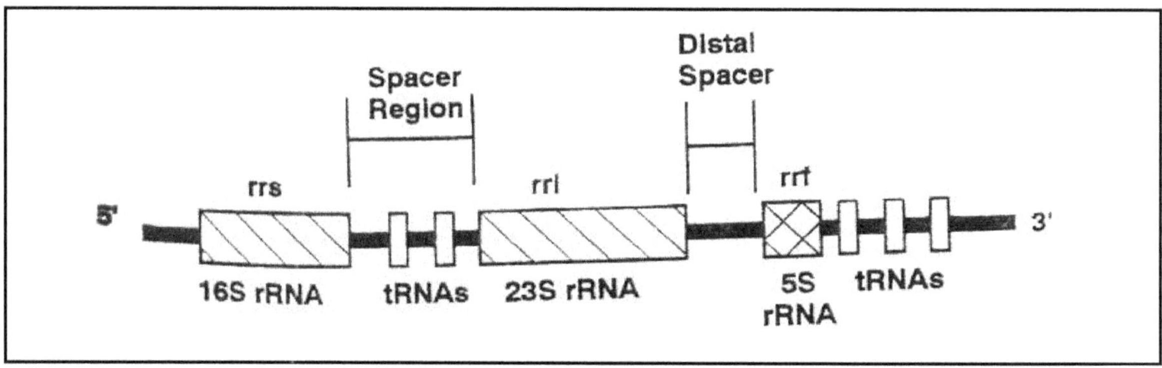

Figure 16.2: Schematic Representation of a Typical Ribosomal rRNA Operon

Restriction Digestion of ITS-PCR Products

For further discrimination, the ITS-PCR products were cut with restriction enzymes, to analyse the diversity based on the different restriction sites in the ITS spacer region of *B. cereus* rDNA operon. Three restriction enzymes, two-tetra cutter (Sau3Al and AluI) and one hexa cutter (Hind III) enzyme were involved for the restriction. There was no digestion with AluI enzyme in any of the strains. The restriction digestion bands with Hind III gave identical banding patterns in all the strains. Bands for digestion with the Sau3AI also had identical banding patterns showing no variations among isolates. The restricted banding patterns are given the Figure 16.3a, b; Figure 16.4a, b and Figure 16.5a, b.

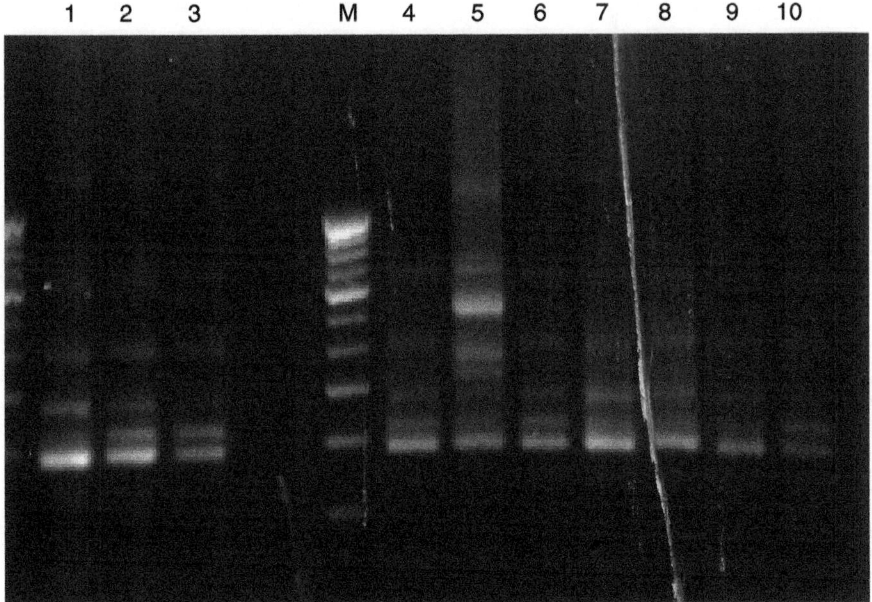

Figure 16.3a: Restriction Digestion of ITS–PCR Products with Sau3AI
Lanes: M, 100bp ladder; 1–10 *B. cereus* isolates

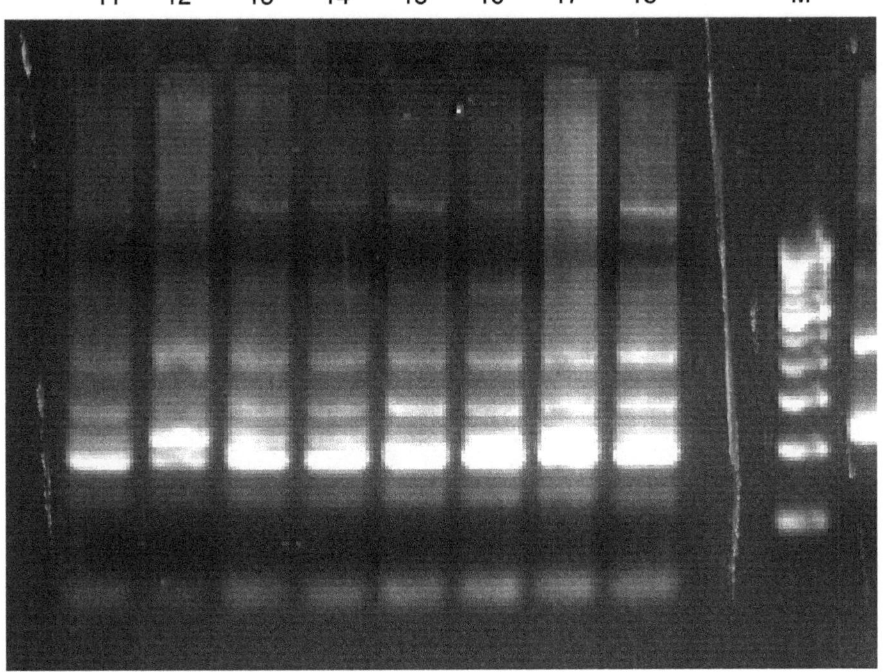

Figure 16.3b: Restriction Digestion of ITS–PCR Products with Sau3AI
Lanes: M, 100bp ladder; 11–18 *B. cereus* isolates

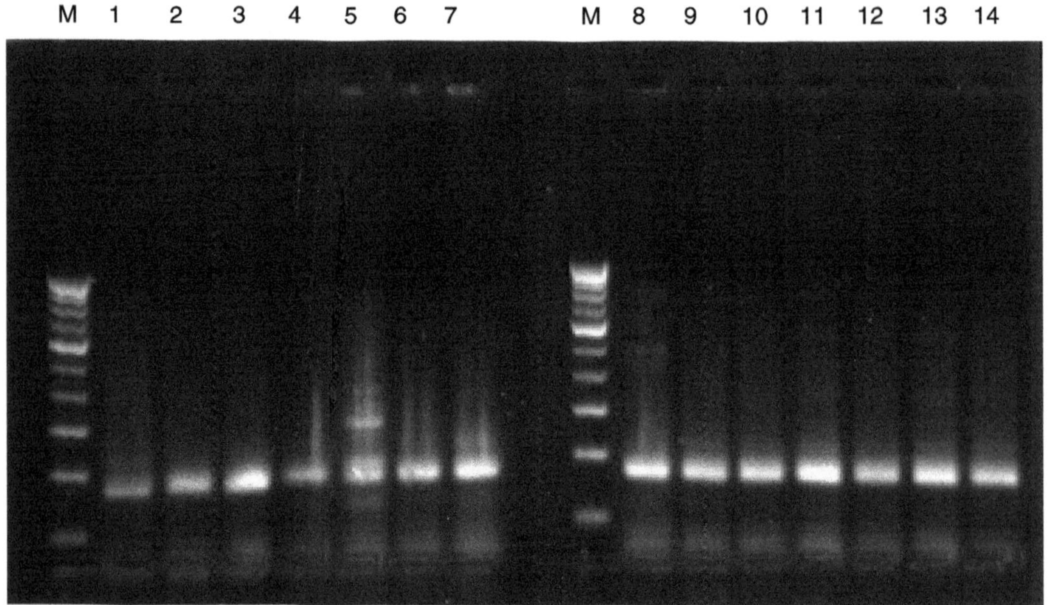

Figure 16.4a: Restriction Digestion of ITS–PCR Product with Alu I

Lanes: M, 100bp ladder; 1–14 *B. cereus* isolates

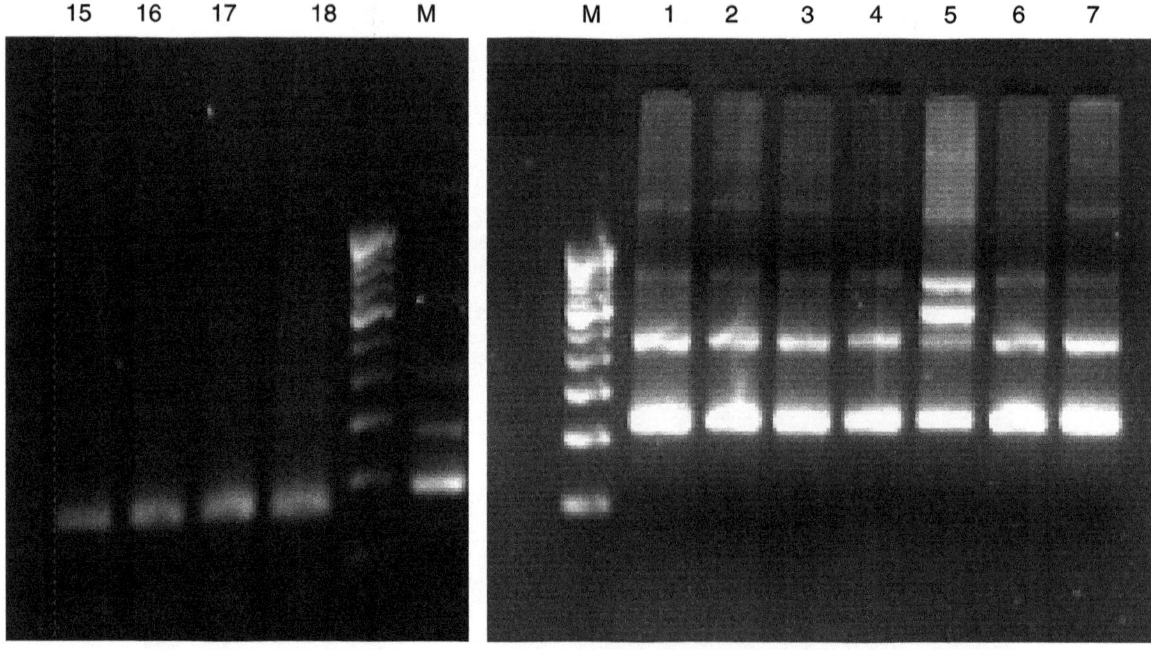

Figure 16.4b: Restriction Digestion of ITS–PCR product with Alu I

Lanes: M, 100bp ladder; 15–18 *B. cereus* isolates

Figure 16.5a: Restriction Digestion of ITS–PCR Products with Hind III

Lanes: M, 100bp ladder; 1–7 *B. cereus* isolates

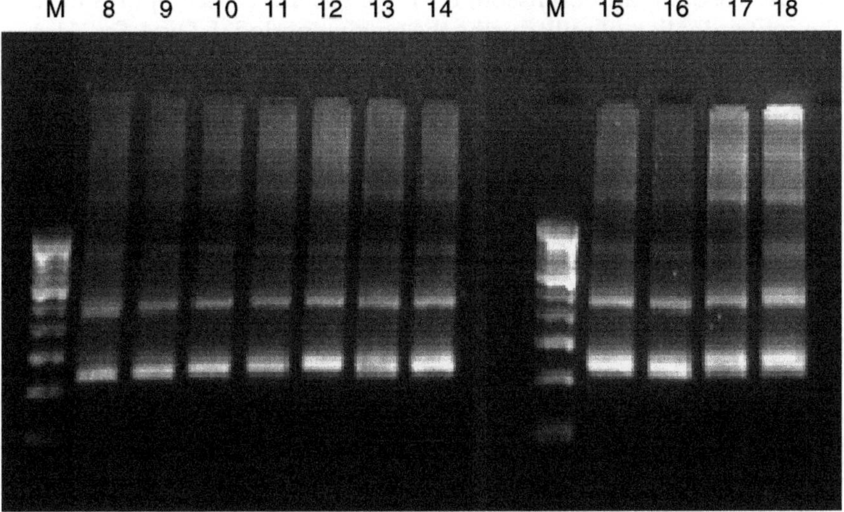

Figure 16.5b: Restriction Digestion of ITS–PCR Products with Hind III
Lanes: M, 100bp ladder; 8–18 *B. cereus* isolates

The 16S-23S rDNA intergeneric transcribed spacers (ITS) are considered most variable region of the ribosomal operon, which is used to type bacteria at species level (Gurtler and Stanisich, 1996). The small differences observed in amplicon length are due to the different primer positions on the 16S and 23S rRNA gene sequences. The result of present study suggests that in *B. cereus* the 16S-23S ITS are well conserved in terms of length (Daffonchio *et al.*, 1998). The ITS-PCR fingerprinting suggests that the evolution of rRNA operons in *B. cereus* doses not reflect the whole DNA. The above study reveals that RAPD typing has better discrimination in the DNA polymorphism among isolates of *B.cereus* than ITS-PCR typing.

References

Akopyanz, N., Bukanov, N.O., Westblom, T.U., Kresovich, S. and Berg, D.E., 1992. DNA diversity among clinical isolates of *Helicobacter pylori* detected by PCR-base RAPD fingerprinting. *Nucleic Acids Res.*, 20: 5137.

Amman, R.I., Ludwig, W. and Schleifer, K.H., 1995. Phylogenetic identification and *in situ* detection of individual microbial cells without cultivation. *Microbiol. Rev.*, 59: 143–169.

Bazzicalupo, M. and Fani, R., 1995. The use of RAPD for generating specific DNA probes for microorganisms. In: *Methods in Molecular Biology, Vol. 50: S-species Diagnostic Protocols: PCR and other Nucleic Acid Methods*, (Ed.) J.P. Clapp. Humana Press Inc., Totowa, New Jersy.

Carlson,C .R., Cougant, D.A. and Kolstf, A.B., 1994. Genotypic diversity among *B. cereus* and *B. thuringiensis* strains. *Appl. Environ. Microbiol.*, 60: 1719–1725.

Christiansson, A., Svensson, B., Envioth, A. and Brendehang, J., 1997. Tracing contamination routes for *B. Cereus* in the dairy process with RAPD–DCR. In: *Sustainable Food Production and Competitive Industries.* (Nord Food Copenhagen) 23–25 November Denmark, p. 108.

Christiasson, A., Bertilsson, J. and Srensson, B., 1998. *Bacillus cereus*: Spores in raw milk. Factors affecting the contamination of milk during the grazing period. *J. Diary Sci.*, 11.

Daffonchio, D., Borin, A.S., Consolandi, D., Mora, P.L., Manachini and Sorlini, C., 1998. 16S-23S rRNA internal transcribed spacers as molecular markers for the species of the 16S rRNA group I of the genes *Bacillus*. *FEMS Microbiol. Lett.*, 163: 229–236.

Day, W.A., Pepper, I.L. and Joens, L A.1997. Use of arbitrarily primed PCR product in the development of a *Campylobacter jejuni*-specific PCR. *Appl. Environ. Microbiol.*, 63: 1019–1023.

Drobniewski, F.A., 1993. *B. cereus* and related species. *Clinical Microbiology Reviews*, 6: 324–338.

Fani, R., Damiani, G., Di Serio, C., Gallori, E., Grifoni, A. and Bazzicalupo, M., 1993. Use of random amplified polymorphic DNA (RAPD) for generating specific DNA probes for microorganisms. *Mol. Ecol.*, 2: 243–250.

Granum, P.E., Brynested, S. and Kramer, J., 1993. Analysis of entrotoxin production by *B. cereus* from dairy products, food poisoning incidents and non-gastrointestinal infections. *Int. J. Food Microbiol.*, 17: 269–279.

Granum, P.E., 1994. *Bacillus cereus* and its toxins. *J. Appl. Bacteriol.*, 76: 61S–66S.

Gurtler, V. and Stanisich, V.A., 1996. New approaches to typing and identification of bacteria using the 16S-23S-rDNA-spacer region. *Microbiol.*, 142: 3–16.

Henderson, I., Duggleby, C.J. and Turnbull, P.C. B., 1994. Differentiation of *Bacillus anthracis* from other *Bacillus cereus* group bacteria with the PCR. *Int. J. Syst. Bacteriol.*, 44: 99–105.

Jacek, Majewskl, Piotr Zawdzkr, Paul Pickerill, Fredcrick, M.C. and Christopher, G.D., 2000. Barriers to genetic exchange between bacterial species : *Streptococcus pneumoniae* Transformation. *J. Bacteriol.*, 182: 1016.

Jenson, M.A., Webster, J.A. and Straus, N., 1993. Rapid identification of bacteria on the basis polymerase chain reaction amplified ribosomal DNA spacer polymorphism. *Appl. Environ. Microbiol.*, 59: 945–952.

Kotsman, J.R., , Alden, M.B., Mair Edluid, T.D., Lipuma, J.J. and Stull, T.L, 1995. An universal approach to bacterial molecular epidemology by polymerase chain reaction ribotyping. *J. Infect. Dis.*, 171 204–208.

Kotsman, J.R., Edlind, T.D.J., Lipuma, J. and Stull, T.L., 1992. Molecular epidemiology of *Pseudomonas cepacia* determined by polymerase chain reaction ribotyping. *J. Clin. Microbiol.*, 30: 2084–2087.

Kramer, J.M. and Gillbert, R.J., 1989. *B. cereus* and other *Bacillus* species. In: *Foodborne Pathogens*, (Ed.) M.P. Doyle. Marcel Dekker Inc., New York, p. 21.

Lakshmanan, K. and Kumar, K., 2003. Nucleic acid studies in the phylogenies and strain identification of cyanobacterial taxa. *Madras Agric. J.*, 90(10–12): 579–588.

Leblond-Bourget, N., Philippe, H., Mangin, I. and Decaris, B., 1996. 16S r DNA and 16 S to 23S internal transcribed spacer sequence analysis reveal inter and introspecific bifidobacterium phylogeny. *Int. J. Sys. Bacteriol.*, 4: 102–111.

Mark, A.J., Webster, John and Straus, Neil, 1993. Rapid identification of bacteria on the basis of polymerase chain reaction: Amplified Ribosomal DNA spacer polymorphisms. *Appl. Environ. Microbiol.*, 59(4): 945–952.

Marks, J.C. and Cummings, M.P., 1996. DNA sequence variation in the ribosomal internal transcribed spacer region of freshwater, *Cladosphora* species. *J. Phyco.*, 32: 1035–1042.

Moritz, C. and Hillis, D.M., 1996. Molecular systematics: Context and controversies. In: *Molecular Systematics*, 2nd edn., (Eds.) D.M. Hillis, C. Moritz and B.K. Mable. Sinauer Associates, Sunderland MA, pp. 407.

Nilsson, J., Svensson, B., Ekelund, K. and Christiansson, A., 1998. A RAPD-PCR method for large-scale typing of *B. Cereus. Lett. Appl. Microbiol.*, 27: 168.

Peters, E.C., 1997. Diseases of coral-reef organisms. In: *Life and Death of Coral Reefs*, (Ed.) C. Bireland. Chapman and Hall, NewYork, p. 114–139.

Sabine, Lectner, Ralf Mayr, Francis, K.P., Bright, M., Thomas, Kalpan, Elke, Wiebner-Gunkel, Gordon, S.A.B., Stewart and Scherer, Siegfried, 1998. *Bacillis weihenstephannsis* sp. is a new psychrotolerant species of the *Bacillus cereus* group. *Int. J. Syst. Bacteriol.*, 48: 1373–1382.

Sambrook, J. and Russell, R.W.D., 1991. *Molecular Cloning: A Laboratory Manual*. Cold Spring Harbor Laboratory Press, New York, p. 2.25–2.33.

Scott, M.P., Haynes, K.M. and Williams, S.M., 1992. Parentage analysis using RAPD-PCR. *Nucleic Acids Res.*, 20: 5493.

Shewan, J.M., 1962. The bacteriology of fish and spoiling fish and some related chemical changes. *Recent Adv. Food Sci.*, 1: 167–193.

Sontakke, S. and Fabe, J.M., 1995. The use of PCR ribotyping for typing strains of *Listeria* spp. *Eur. J. Epidemiol.*, 11: 1–9.

Stephan, R., 1996. Randomly amplified polymorphic DNA (RAPD) assay for genomic fingerprinting of *B. cereus* isolated. *Int. J. Food Microbiol.*, 31: 311–316.

Swofford, D.L., Olsen, G.L., Waddell, P.J. and Hillis, D.M., 1996. Phylogenetic inference. In : *Molecular Systematics*, 2nd edn., (Eds.) D.M. Hillis, C. Moritz and B.K. Mable. Sinauer Associates, Sunderland MA, pp. 407.

Watson, J.D., Hopkins, N.H., Roberts, J.W., Steitz, J.A. and Weiner, A.M., 1987. *Molecular Biology of Gene*. Ed Menlo Park, CA. The Benjamin/Cummings Publishing Company.

Welsh, J. and McClelland, M., 1990. Fingerprinting genomes using PCR with arbitrary premiers. *Nucleic Acids Res.*, 18: 7213–7218.

Welsh, J. and Mc Clelland, M., 1992. PCR amplified length polymorphism in tRNA intergenic spacers for categorizing *Staphylococci. Mol. Microbiol.*, 6 : 1673.

Welsh, J., Pretzman, C., Postic, D., Saint Girons, I., Barantons, G. and Mc Clelland, M., 1992. Genomic fingerprinting by arbitrary primed polymerase chain reaction resolves *Borrelia burgalorferi* in to three distinct phyletic groups. *Int. J. Syst. Bacteriol.*, 42: 370.

Williams, J.G.K., Kubelik, A.R., Livak, K.J., Rafaski, J.A. and Tingey, S.V., 1990. DNA polymorphisms amplified by arbitrary primes are useful as genetic markers. *Nucleic Acids Res.*, 18: 6531–6535.

Chapter 17

Comparison of Rate of Copper Ion Induced Oxidation of Lipoprotein in End Stage Renal Diseased and Renal Transplant Patients: An *in vitro* Study

C.S. Parameswari[1]*, B. Vijaya Geetha[2], R. Vijaya Kumar[3]

[1]Reader, [2]Research Scholar
Post Graduate Department of Biochemistry, Bharathi Women's College (Autonomous), North Chennai – 600 108, Tamil Nadu
[3]Professor and Head, Department of Nephrology, Stanley Medical College and Hospital, Chennai – 600 001, Tamil Nadu

ABSTRACT

The Cardiovascular Disease (CVD) accounts for 15–30 per cent of deaths in patients suffering from End Stage Renal Disease (ESRD) and about 47 per cent of deaths in patients having renal transplantation. The increased lipoprotein oxidation is found to be the main risk factor for CVD and thereby the progression of atherosclerosis. The purpose of the present study is to assess the rate of lipoprotein oxidation among the ESRD and renal transplant patients. The level of Malondialdehyde (MDA) and Conjugated Dienes (CD) were estimated in the lipoprotein fraction before and after an *in vitro* copper ion induced oxidation and the rate of formation of CD was also estimated. The levels of lipid peroxides and CD were significantly increased in oxidized LDL + VLDL fractions of renal transplant patients. When compared to normal subjects and ESRD patients, the lag time for CD formation and the time taken to form the maximum level of CD were significantly reduced for renal

* Corresponding Author: Phone: 044-28487064, 044-28487060; E-mail: vijhmphil_2002@yahoo.com

transplant patients. The shorter lag time of LDL + VLDL lipoprotein oxidation in renal transplant patients may be due to high levels of cholesterol and decreased levels of serum antioxidants among these patients. Further it proves that the increased susceptibility of the LDL + VLDL fraction of renal transplant patients to oxidation make them more prone to the progression of atherosclerosis and arteriosclerosis leading to Chronic rejection among these patients.

Keywords: *Cardiovascular disease, Lag time, Lipoprotein oxidation, Antioxidants, Atherosclerosis and Chronic rejection.*

Introduction

Cardiovascular disease is a significant cause of morbidity and mortality in ESRD patients and renal transplant patients (Leskinen *et al.*, 2003). The impaired renal function promotes atherogenesis by increasing the oxidation of lipoproteins in both the ESRD and renal transplant patients (Luc and Fruchart 1992; Parfery *et al.*, 1999). Cyclosporine A contributes significantly to the cardiovascular risk in transplant patients because of its hypertensive and hyperlipidemic effects (Kahan 1989). The oxidative modification of LDL which is most prevalent among ESRD and renal transplant patients is an important factor in the initiation and development of atherosclerosis and arteriosclerosis. The arteriosclerosis leads to chronic rejection in renal transplanted patients (Linton MF and Fazio 2001).

The aim of the present study involves the measurement of lipoprotein oxidation in *in vitro* condition in ESRD and renal transplant patients. The adequate markers of oxidatively modified LDL are studied. The kinetics of formation of CD during copper ion induced lipoprotein oxidation is studied and compared with the normal subjects.

Subjects and Methods

Patients

The blood samples for the present study were collected from Nephrology Department, Stanley Medical College Hospital, Chennai, Tamil Nadu. The study was carried out in 20 ESRD patients with a creatinine level > 5mg/dl of blood who were undergoing hemodialysis using polysulphane membrane and acetate buffered dialysate for two or more sessions of each for at least 4 hours of dialysis per week and in twenty patients who underwent live related renal transplantation since 2 years, at the Nephrology Department, Stanley Medical College Hospital, Chennai. They were under treatment with Cyclosporine A at a dosage of 6 mg/kg body weight initially and adjusted depending on the Cyclosporine A level in blood and tapered the utilization of Cyclosporine A and steroid 1 mg/kg body weight, the level reduced by 6 months with Azathioprine of 1–1.5 mg/kg body weight increased to 1.5–2.4 mg/kg body weight as life long treatment. The patients had no clinical or laboratory evidence of fever, diabetes mellitus, elevated liver enzyme levels and infective disorders. They were non-alcoholic, non-smokers and without antilipidemic drug treatment. The normal healthy volunteers served as control subjects. The patients and the normal subjects were of male in the age group between 20 and 45 years. Informed consent was obtained from each patient before the sample collection. The study was approved by the committee of Ethics at Stanley Medical College Hospital, Chennai.

Blood Sampling

The ESRD patients were inpatients and the renal transplanted patients were treated as the outpatients in the Nephrology Department. From each patient, 10ml of blood was collected after a 12 hours of overnight fasting in sterile tubes and the serum obtained was used for the following studies.

Isolation of LDL + VLDL from Serum

The LDL + VLDL lipoprotein fraction was isolated from the serum according to the method of Burstein and Scholnick (Burstein and Scholnick, 1973). The LDL + VLDL was isolated using sucrose, sodium salt of heparin and Magnesium chloride under the high speed centrifugation of about 6000g for 30 minutes. The isolated LDL + VLDL were subjected to dialysis using Tris-HCl buffer and Barium chloride at 4°C to remove impurities like protein and heparin. The resultant was a clear yellow solution of LDL + VLDL. The protein in LDL + VLDL fraction was estimated according to the method of Markwell *et al.*, 1978. 20 µl of LDL + VLDL fraction was made up to 1 ml using distilled water. Then 3 ml of Markwell reagent was added and the tube was incubated at room temperature for 10 minutes. Then 0.3 ml of Folin Phenol reagent was added and the blue colour developed was read at 660 nm after 45 minutes. The protein content of LDL + VLDL fraction was expressed as mg/ml of LDL + VLDL fraction.

Assay of Lipid Peroxidation in LDL + VLDL Fraction

LDL + VLDL fraction was oxidized in *in vitro* condition by the method of Wallin *et al.* with some modifications (Wallin *et al.*, 1993). The extent of lipid peroxidation in the LDL + VLDL fraction before and after an *in vitro* induction of oxidation was studied by measuring Thiobarbituric Acid Reacting Substances (TBARS) and Conjugated Dienes (CD). The lipid peroxidation in terms of TBARS was quantitated as follows; about 200 µg (in terms of LDL + VLDL protein) of LDL + VLDL isolated from the sample was suspended in 50 µl of 5 µM copper sulphate solution. Then the volume of the tube was made up to 1 ml using Phosphate Buffered Saline (PBS) (0.001 M pH 7.2). At the end of one hour of incubation at room temperature, 100 µl of 50 per cent Trichlroacetic acid was added to arrest the reaction. The TBARS formed was estimated by adding 100 µl of 1.3 per cent of TBA in sodium hydroxide and heated at 60°C for 30 minutes. The tube was cooled and the pink colour developed was extracted using 4ml of n-butanol and pyridine mixture (15 : 1) and the absorbance of butanol layer was read at 535 nm. The amount of MDA formed was calibrated from a standard curve prepared using 1, 1′, 3, 3′-tetramethoxy propane. The CD in the LDL + VLDL fraction was assayed by the method of Esterbauer *et al.*, 1989. 200 µg of LDL + VLDL fraction (in terms of protein) was suspended in 50 µl of 5 µM copper sulphate for *in vitro* oxidation. After 10 minutes, the oxidation was stopped by the addition of 100 µl of 0.1 per cent Butylated hydroxytoluene in alcohol. To this 1.5 ml of cyclohexane was added and the absorbance was read at 233 mm against the cyclohexane blank. The rate of copper ion induced *in vitro* oxidation of LDL + VLDL lipoprotein fraction was determined by estimating CD in oxidized LDL + VLDL fraction as mentioned above at a time interval of 5 minutes for a period of 2 hours. A graph depicting time in minutes against absorbance at 233 nm was plotted and the indices or oxidation in terms or lag time and the time taken for maximum oxidation were calculated.

Statistical Analysis

The numerical data were presented as mean value ± standard deviation. The values were subjected to analysis of normal tests of significance. The statistical significance was evaluated by One-way ANOVA (Analysis of Variance).

Results and Discussion

The death rate due to complications of vascular disease ranges from 5/1000 deaths per patient year in younger transplanted patients to 18.5/1000 deaths in the older age group (Raine and Ledingham, 1984). Renal transplant patients have infact the accelerated atherosclerosis because of arterial hypertension, dyslipidemia and other vascular risk factors (Japichino *et al.*, 2001). In 44 per cent of renal transplant patients the hypercholesterolemia is due to increased hepatic synthesis of apoB containing lipoproteins and is an independent risk factor for kidney graft loss by chronic rejection in male patients with previous acute rejection (Isabel Beneyto, Castello, 2002). The *in vitro* oxidation of lipoprotein using the copper ion exhibits the same scenario as it is present in *in vivo* condition.

The levels of TBARS and conjugated dienes in native and oxidized forms of LDL + VLDL fraction of serum from the normal subjects, ESRD and renal transplant patients were presented in Table 17.1. When compared to normal and ESRD patients, the renal transplant patients showed a significant increase in TBARS and conjugated diene levels.

Table 17.1

The levels of TBARS and Conjugated dienes in the native and Cu^{2+} ion induced *in vitro* oxidized forms of LDL + VLDL fractions from the serum of normal subjects, ESRD and renal transplant patients. Each Group consists of 20 subjects.

Subjects	*TBARS*		*Conjugated Dienes*	
	Native	*Oxidized*	*Native*	*Oxidized*
Normal Subjects	23.6±0.63	43.5±0.76	$2.44\times10^3\pm187.06$	6.24×103±273.25
ESRD	***A	***A	***A	***A
Patients	29.19±0.65	72.75±1.44	$3.56.10^3\pm256.03$	$10.85\times10^3\pm125.46$
Renal	***B	***B	***B	***B
Transplants	***C	***C	***C	***C
	81.4±1.71	116.15±2.89	$7.83\times10^3\pm2.89$	$19.38\times10^3\pm155.16$

***F < 0.001–highly significant by ANOVA.

The levels of TBARS are expressed as nanomoles of MDA formed/mg of LDL+ VLDL protein.

The levels of Conjugated diene formed are expressed as $\Delta233\times10^3$/mg of LDL + VLDL protein.

The Values are mean ± SD

A: Comparison of normal subjects and ESRD patients

B: Comparison of normal subjects and Renal transplant patients

C: Comparison of ESRD and Renal transplant patients.

Figure 17.1 represented the rate of Copper ion induced *in vitro* oxidation of LDL + VLDL. The graph showed the level of CD formed, when LDL + VLDL fraction isolated from the serum of normal subjects, ESRD patients and renal transplant patients were subjected to 5 μM copper ion induced oxidation at a time level of 5 minutes for 2 hours. The duration of lag time and the time of maximum lipid peroxidation are obtained for normal subjects and both the patients by kinetic modeling analysis. The lag time of oxidation for normal subjects, ESRD and renal transplant patients were found to be 20 min, 15 min and 10 min respectively. The renal transplant patients had significantly shorter lag time when compared to normal and ESRD patients. Similar changes have also been reported already (McEneny *et al.*, 1997). The time taken to form maximum CD formation was found to be 70 min, 50 min and 25 min for healthy subjects, ESRD patients and renal transplant patients respectively. The shorter duration was utilized for the maximum amount of CD formation in renal transplant patients when compared to ESRD patients and normal subjects.

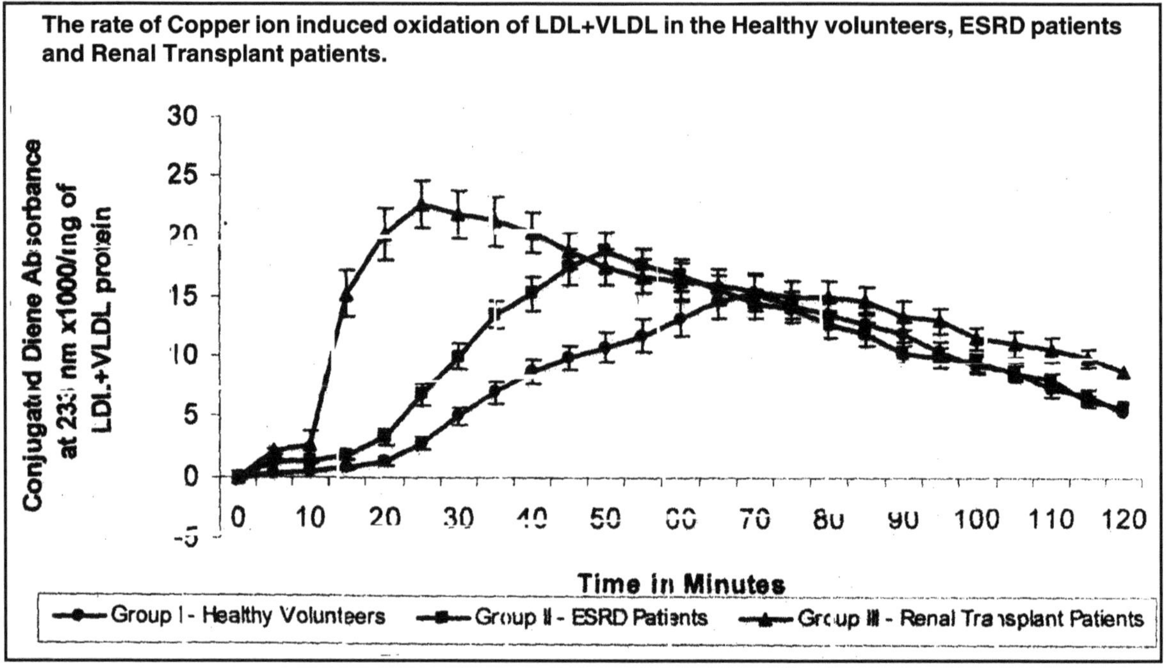

Figure 17.1: Estimation of Rate of Copper Ion Induced Oxidation of LDL + VLDL

The levels of Conjugated diene formed are expressed as absorbance at 233 nm × 1000/mg of LDL + VLDL.

The values are expressed as Mean ± Standard Deviation.

Legend:

Healthy volunteers: The normal healthy male subjects in the age group of 20–45 years.

ESRD Patients: The ESRD patients with blood creatinine level greater than 5 mg/dl of blood

Renal transplant patients: Live related renal Transplant Patients since two years.

The lag phase in transplant patients is shorter due to the fact that there is low level of antioxidants in LDL which have been fully used up, not remained to prevent LDL oxidation and hence the lag phase is shorter in renal transplant patients when compared to ESRD and healthy controls which is depicted in the Figure 17.1. The low level of serum antioxidants in renal transplant patients have been reported in our previous studies (Parameswari *et al.*, 2004). The prolonged dialytic treatment in ESRD patients causes a decrease in vitamin E concentration of LDL and increases their susceptibility to oxidation (Maggi *et al.*, 1994).

Cyclosporine A, the drug used in the immunosuppressive therapy of renal transplants is the glycogenic immunosuppressor that can induce post transplant hyperlipidemia which may contribute to less favorable plasma lipid profile and contribute to high incidence of CVD by facilitating lipid peroxidation of LDL thereby increasing the uptake of LDL by LDL receptors in arterial walls. The renal transplant patients with higher concentrations of Cyclosporine A have significantly higher oxidisability of LDL (Venkiteswaran *et al.*, 2001).

The decreased serum antioxidants level of the ESRD and renal transplant patients mainly contributes to the shorter lag time of lipoprotein oxidation. Hence it may be concluded that the antioxidant therapy for renal transplant patients would extend the lag phase. It may further reduce the mortality and morbidity of renal diseased patients due to CVD and the graft loss caused by chronic rejection among the renal transplants.

References

Burstein, M. and Scholnick, H.R., 1973. Lipoprotein–Polyanion metal interactions. *Adv. Lipid Res.*, 11: 68–108.

Esterbauer, H., Streigl, G., Pulh, H. and M. Rothender, 1989. Continuous monitoring of *in vitro* oxidation of human low density lipoprotein. *Free Radic. Res. Commun.*, 6: 67–75.

Isabel Beneyto, Castello, 2002. Hyperlipidemia: A risk factor for chronic allograft dysfunction. *Kidney Int.*, 61 Suppl. (80): 873–877.

Japichino, G.G., Bonati, L., Rubini, P. and E. Capocasale, 2001. Prevalence of atherosclerosis in renal transplant patients. *Minerva Cardio Angiol.*, 49(4): 229–238.

Kahan, B.D., 1989. Cyclosporine. *N. Engl. J. Med.*, 321: 1725–1738.

Leskinen, Y., Groundstroem, K., Viratenen, V., Letimaki, T., Huhtala, H. and H. Saga, 2003. Risk factors for aortic atherosclerosis determined by transesophageal echocardiography in patients with chronic renal failure. *Am. J. Kidney Dis.*, 42(2): 277–285.

Linton, M.F. and S. Fazio, 2001. Class A: Scavenger receptors, macrophages and atherosclerosis. *Curr. Opin. Lipidol.*, 12(5): 489–495.

Luc, G., and J.C. Fruchart, 1992. Oxidation of lipoproteins and atherosclerosis. *Am. J. Clin. Nutrition*, 53: 2065– 2095.

Maggi, E., Bellazzi, R., Falaschi, F., Frattoni, A., Perani, G., Finardo, G., Gazo, A., Nai, M., Romanini, D. and G. Bellomo, 1994. Enhanced LDL oxidation in uremic patients: An additional mechanism for accelerated atherosclerosis? *Kidney Int.*, 45(3): 876–883.

Markwell, M.A.S., Haas, S.M., Bieber, Z.L. and N.E. Tolbert, 1978. A modification of Lowry procedure to simplify protein determination in membranes and lipoprotein sample. *Anal. Biochem.*, 87: 206–210.

McEneny, J., Loughrey, C.M., Trimble, E.R., McName, P.T., and I.S. Young. Susceptibility of VLDL to oxidation in patients on regular hemodialysis. *Atherosclerosis*, 29(2): 215–220.

Parameswari, C.S., Vijayachamundeeswari, D. and R. Vijaya Kumar, 2004. Antioxidant status in the renal transplant patients with chronic rejection. *SFRR–India Bulletin*, 3(2): 17–22.

Parfery, P.S., Foley, R.N. and C. Rigatto, 1999. Risk issue in renal transplantation: Cardiac aspects. *Transplant Proc.*, 31: 91.

Raine, A.E.G. and J.G.G. Ledingham. Cardiovascular complications after renal transplanation. In: *Morris Kidney Transplanation*, II edition.

Venkiteswaran, K., Sgoutas, C.S., Santana, N. and J.F. Neylan, 2001. Tacrolimus, Cyclosporine and plasma lipoproteins in renal transplant recipients. *Transplant Int.*, 14(6): 405–410.

Wallin, B., Rosengren, B., Shetzer, H.G. and G. Comejo, 1993. Lipoprotein oxidation and measurement of thiobarbituric acid reactive substance formation in a single microtitre plate: Its use for evaluation of antioxidants. *Anal. Biochem.*, 208: 10–15.

Chapter 18

Comparative Study of Different Lab Diagnostic Methods for the Diagnosis of Pulmonary Tuberculosis

Sweta Bharadwaj[1] *and Nighat Hussain*[2]

[1]*Department of Biotechnology/Zoology, Government V.Y.T.P.G. College, Durg – 491 001, C.G.*
[2]*Department of Microbiology, Apollo Hospitals, Bilaspur, C.G.*

ABSTRACT

The accuracy and utility of various lab diagnostic methods for Tuberculosis detection was determined in 200 clinical specimens referred for the diagnosis of possible active TB. Patients underwent sputum induction, CXR, Mantoux testing and had blood drawn for serological testing. After final diagnosis of the participants the sensitivity and Specificity of culture, PCR, Serology, Mantoux, and CXR respectively were found to be 73–100 per cent, 42–100 per cent, 33–87 per cent, 94–20 per cent, 67–77 per cent respectively. It was concluded that culture still remains the gold standard for tuberculosis identification and utilization of a combination of tests together could improve diagnostic accuracy.

Introduction

Tuberculosis is a problem of global concern. Early diagnosis of active cases of pulmonary tuberculosis is essential for initiating prompt treatment and containment of the disease. Commonly used lab methods for the detection and identification of *Mycobacterium tuberculosis,* the causal organism of tuberculosis are: Microscopy, Culture, Serological test and Rapid Amplification test method. Smear Microbiology by Zeihl Neelson method is rapid, economical and convenient but is not a very sensitive test as 40–50 per cent of patients with pulmonary tuberculosis have smear negative specimens (Kim *et al.*, 1984) Culture is considered as the reference method for the detection of tubercle bacilli, but mycobacterial culture is laborious and time consuming. Traditional culture on L-J media takes 2–6 weeks. 40–60 per cent patients with pulmonary tuberculosis are smear negative and in this situation, culture method takes several weeks to become positive. Serological tests using enzyme linked immunosorbent assay (ELISA) techniques to detect antibodies to MTB are relatively simple and economical but to date have had poor specificity and sensitivity and cannot substitute traditional diagnostic tests (Pottumarthy *et al.*, 2000). Nucleic acid amplification tests in smear-negative, culture positive samples have shown low sensitivity and optimal specificity (Clarridge *et al.*, 1993; Doern, 1996). Mantoux has very poor specificity and will be false negative in 20–30 per cent patients newly diagnosed cases with active TB (Starke, 1996).Thus, there is need to evaluate the sensitivity, specificity and Clinical importance of different methods used for the diagnosis of tuberculosis.

Objective

The principal aim of assessment is the early recognition of disease by the use of better diagnostic methods available. The study was conducted in the Department of Microbiology, Apollo Hospitals, Bilaspur (C.G.) from October 2003 to September 2005.

Experimental Method

The study material comprised of 200 cases of Pulmonary tuberculosis coming to the hospital which included 24 specimens from patients with active pulmonary tuberculosis, 110 cases of inactive tuberculosis and the rest comprising of NTM (Piersimoni *et al.*, 1997) and other pulmonary diseases (36 cases) (Table 18.1).

Table 18.1: Final Diagnoses of Participants

Sl.No.	*Disease*	*N*	*Per cent*
1.	Active pulmonary TB	24	12
	A. 1.Culture +ve TB	18	75
	B. 2.Culture –ve TB	6	25
2.	Inactive TB	110	55
3.	NTM	6	3
4.	Other pulmonary D	36	30
	Total	200	100

Specimens

With all standard aseptic precautions, the Sputum and blood samples were taken. Around 10ml of early morning sputum (brought by deep coughing) was collected from each person in sterile disposable 50ml Screw capped tubes. These specimens were digested and decontaminated by N-acetyl L-cysteine NaOH method and concentrated by centrifugation.

Methods

1. *CXR*: Chest X-Ray was taken and studied if any lesion was present or any other abnormality seen.

Figure 18.1: Mycobacterium Tuberculosis Colonies in L-J Medium

2. *Microscopy*: AFB smear was prepared and stained by Zeihl Neelson Method.
3. *Mantoux Test*: This test was performed by intradermal injection of 0.1 ml of intermediate strength (5 TU in childrens below 10 yrs age and 10 TU and all age group persons above 10 yrs of age) PPD-S.
4. *Culture Method*: The suspended sediment was inoculated in solid Lowenstein Jensen medium and liquid Kirchner medium and incubated at 37°C (Figure 18.1).
5. *Serological Procedure*: IgM, and IgA determination was done by andelisa kits and IgG determination was done by Pathozyme-myco IgG Kits.
6. *Nucleic Acid Amplification Reaction*: done in reference lab.

Results

Majority of male patients suffering from the disease belonged to the age group 45–60 and above. While in female patients the majority belonged to a relatively younger age group of 30–45 yrs (Table 18.2).

Maximum patients with Active TB were quite younger, symptomatic and showed past TB record and 80 per cent of all cases of NTM were of relatively old aged females (Tables 18.3 and 18.4).

The positive predictive value for Elisa is more in case of age <40 of symptomatic person but in this case the negative predictive value of CXR is more *i.e.* in persons of age < 40 the findings of Elisa are more reliable. Same is true is case of persons above 40 years of age. The positive predictive value for PCR is more as compared to the negative predictive value. The CXR predictive value is more in symptomatic young persons as compared to asymptomatic and in old aged persons (Table 18.5).

Table 18.2: Distribution of Cases According to Age and Sex in Various Subgroups

Age Group	*Total*	*M*	*Per cent*	*F*	*Per cent*
0–15	18	12	10	6	7
15–30	28	18	15	10	12
30–45	60	23	19	37	46
45–60	72	42	35	20	25
>60	32	25	21	7	1
Total	200	120	100	80	100

Table 18.3: Sex and Age Group Distribution of Patients in Different Disease Groups

Age	*ATB*		*NTM*		*Other Pul. D.*	
	M	*F*	*M*	*F*	*M*	*F*
0–15	0	0	0	0	0	0
15–30	1	2	0	0	6	2
30–35	7	8	0	1	8	4
45–60	4	1	0	2	6	4
> 60	1	Nil	2	1	3	3

Table 18.4: Characteristics of Subjects by Final Diagnosis

	Active TB	*ITB*	*NTM*	*Other Pul. D.*	*P Value*
n	24	110	6	60	
Mean Age	38	42	53	47	0.001
Sex					
Female n (per cent)	8 (33)	34 (30)	4 (66)	26 (43)	0.05
Male n (per cent)	16 (66)	6 (69)	2 (33)	34 (56)	
Past Diagnosis of TB					
Yes n (per cent)	21 (87)	18 (16)	3 (8)	4 (15)	0.05
No (n, per cent)	3 (87)	92 (83)	3 (50)	56 (93)	
Past contact with TB					
Yes (n, per cent)	18(75 per cent)	27 (27)	1 (16)	15 (25)	
No (n, per cent)	6 per cent (6 per cent)	80 (72)	5 (8)	45 (75)	NS
Symptoms					
None (n, per cent)	5 (20)	44 (40)	2 (66)	43(71)	
Any (n, per cent)	19 (79)	11 (10)	4 (33)	17 (28)	0.001
CXR Infiltrates (n, per cent)	18 (75 per cent)	19 (17)	3 (50 per cent)	20 (33)	0.001

Table 18.5: Comparative Predictive Value of Elisa and PCR

All Patients	Predictive Value of +ve Test			Predictive Value of -ve Test		
	Elisa Elisa +	PCR PCR +	Elisa and PCR PCR (–) Elisa (+)	Elisa Elisa–	PCR PCR–	Elisa and PCR Elisa and PCR (–) (+)
Symptomatic						
Age <40 suspicious	0.66	0.99	0.42	0.55	0.76	0.79
Other CXR	0.21	0.99	0.11	0.90	0.95	0.95
Age >=40 suspicious	0.45	0.99	0.31	0.74	0.84	0.85
Other CXR	0.10	0.99	0.07	0.96	0.97	0.97
Asymptomatic						
Age <40 suspicious	0.52	0.99	0.38	0.69	0.80	0.81
Other CXR	0.13	0.99	0.10	0.94	0.96	0.96
Age >=40 suspicious	0.31	0.99	0.28	0.84	0.87	0.87
Other CXR	0.06	0.99	0.06	0.98	0.97	0.97

Culture showed 73 per cent sensitivity and 100 per cent specificity. Mantoux test showed good sensitivity 94 per cent, but specificity was very low 20 per cent, infact poor in most cases. The sensitivity of PCR was 57 per cent, when compared with culture and was much more reduced to 42 per cent, when clinical diagnosis was also included. The sensitivity of Elisa for serological testing was much lower at 38 per cent and 33 per cent which may be due to the use of a single antigen (A-60) in our case. The sensitivity of Elisa can be increased when multiantigen system is used. In our findings, the sensitivity and specificity of both PCR and Serological test when either of them were positive came to be much better as 75 per cent and 90 per cent *i.e.* if the test are used in combination the sensitivity and specificity is increased (Table 18.6).

Table 18.6: Sensitivity and Specificity of Different Diagnostic Tests

Diagnosis Test	Culture		Culture and Diagnosis	
	Sensitivity (Per cent)	Specificity (Per cent)	Sensitivity (Per cent)	Specificity (Per cent)
Culture	100	100	73	100
PCR	57	100	42	100
ELISA Ratio of reading to cut off 0.8	38	87	33	87
PCR and ELISA (ratio/cut off 0.8)				
Either +ve	75	90	60	90
Both +ve	14	100	10	100
Mantoux	91	20	94	20
Probable active TB	57	96	48	96

Discussion

Smear microscopy is not very sensitive as 40–60 per cent cases of pulmonary tuberculosis have smear negative specimens (Gordin and Slutkin, 1990). Specificity of culture is high but sensitivity is reduced to 73 per cent. This may be attributed to false results which may be obtained due to non-viable organisms. Elisa has relatively low sensitivity and specificity which may be due to the use of single antigen in our case. The sensitivity of Elisa can be increased when multi antigen system is used. PCR has excellent specificity but false results lead to low sensitivity. Negative results obtained by amplification assays for culture positive specimens may be explained by the presence of inhibitors of enzymatic amplification and/or by small number of mycobacteria, unequally distributed in the test suspension (Devallois *et al.*, 1996 and Piersimoni *et al.*, 1997). Thus, it is concluded that no currently available tests has sensitivity and specificity high enough for the accurate diagnosis of pulmonary tuberculosis and culture still remains the gold standard. Results of Elisa may provide additional clinical information to guide management in patients with Pulmonary tuberculosis. Amplification assess could be used for testing in smear negative specimens because rapid diagnosis will benefit those patients. Thus, use of a combination of test is superior to any single test.

References

Clarridge, J.E. III, Shawar, R.M., Shinnick, T.M. and Plikaytis, B.B., 1993. Large scale use of polymerase chain reaction for detection of *Mycobacterium tuberculosis* in a routine mycobacteriology laboratory. *J. Clin. Microbiol.*, 31: 2049, 2056.

Devallois, A., Legrand, E. and Rastogi, N., 1996. Evaluation of Amplicor MTB test as adjunct to smear and culture for direct detection of *Mycobacterium tuberculosis* in the French Caribean. *J. Clin. Microbiol.*, 34: 1065–1068.

Doern, G.V., 1996. Diagnostic mycobacteriology: Where are we today? *J. Clin. Microbiol.*, 34: 1873–1876.

Gordin, F. and Slutkin, G., 1990. The validity of acid fast smears in diagnosis of Pulmonary tuberculosis. *Arch. Patho. Lab. Med.*, 114: 1025–1027.

Kim, T.C., Blackman, R.S. and Heatwole, K.M., *et al.*, 1984. Acid fast bacilli in sputum smears of patients with tuberculosis. *Am. Rev. Respir. Dis.*, 129: 264.

Piersimoni, C., Callegaro, A., Nista, D., Bornigia, S., Conti, F., Santini, G. and Sio, G., 1997. Comparative evaluation of two commercial amplification assays for direct detection of *Mycobacterium tuberculosis* complex in respiratory specimens. *J. Clin. Microbiol.*, 35: 193.

Pottumarthy, S., Wells, V.C., and Morris, A.J., 2000. A comparison of seven tests for serological diagnosis of tuberculosis. *J. Clin. Microbiol.*, 38: 2227–2231.

Starke, J.R., 1996. Tuberculosis skin testing: New schools of thought. *Pediatrics*, 98: 123–125.

Chapter 19

Standardization of the Ration Size of Artemia Feed for Efficient Growth and Energy Budget in an Ornamental Fish *Etroplus Suratensis*

***A. Amutha Jaisheeba*[1], *R. Sornaraj*[1] *and A.J.A. Ranjit Singh*[2]**

[1]*Research Department of Zoology, Kamaraj College, Thoothukudi*
[2]*Reader and Head, Department of Biology, S.P.K. College, Alwarkurichi.*

ABSTRACT

A growth trial was conducted using the fry of an ornamental fish pearl spot (*Etroplus suratensis*) to study the effect of different ration levels of *Artemia* as feed. Four rations of *Artemia* 100 per cent, 75 per cent, 50 per cent and 25 per cent and *ad libitum* were given as feed. After 30 days growth Food Conversion Ratio (FCR), Specific Growth Rate (SGR) and Average Daily Growth (ADG) were calculated. A good growth gain, SGR and ADG (0.53±0.05, 2.79±0.73 and 0.09±0.01) were observed in the fish fed with 100 per cent rations. A significant high food conversion ratio value was observed in 75 per cent ration fed group fish (72.08±0.49). In 25 per cent ration fed group, a decrease was noted in the growth (27.7 per cent). The present study suggests that *E. suratensis* were able to utilize the feed effectively at 75 per cent of ration level.

Keywords: *Feed ration, Ornamental fish, Artemia, Etroplus suratensis.*

Introduction

Nutritionally balanced diet is an important factor for better growth and survival of fish. The feed is one of most important input in aquaculture. It accounts for more than 30–60 per cent of total cost of fish production. Moreover, feeding the fish at an optimum rate provides maximum growth per unit feed cost (Hung *et al.*, 1995). According to Bromley (1987), the over riding factor governing growth is the rate of food intake. Studies on the relationship between growth rate and the rate of food intake are essential for understanding the bioenergetics of fish in natural systems (Hansen *et al.*, 1993) and for designing the feeding regimes in intensive aquaculture system.

Most of the workers fixed the ration size based on percentage of the live body weight per day (Hung *et al.*, 1995; Fontaine *et al.*, 1997) or on dry weight of fish (Wurtsbaugh and Davis, 1977). The present work was undertaken to standardize the ration level of feed required to induce maximum growth in a culturable green chromide, *E. suratensis*, a cichlid widely distributed all over India.

Methodology

The growth experiments were carried out for a period of 30 days in plastic troughs of 20 litre capacity. The troughs were drained, cleaned and dried before filling it with water. Juveniles of *E. suratensis* of uniform size (0.400 mg) were collected and reared for a period of two weeks in the laboratory. The fish were randomly stocked in the experimental troughs, provided with aeration. Three replicates for each treatment were maintained.

Control group was fed *ad libitum*. Four ration levels *viz.*, 100 per cent, 75 per cent, 50 per cent and 25 per cent of artemia were given to experimental fish. Quantity of feed was adjusted after every sampling. The unconsumed feed and faecal matter were removed separately everyday, and the pooled samples were dried and weighed. Fishes were sampled on 15th day to assess their growth. Thirty days after the experiment, fish were weighed using electronic digital balance. Water quality parameters are within the range, suitable for fish culture. Weight gain, feed conversion ratio, Specific Growth Rate and Average Daily Growth rate were calculated at the end of the experiment.

Result and Discussion

The Table 19.1 shows the growth rate of the *E. suratensis* reared in brackish water medium fed with different rations of feed the *Artemia*. The growth rate of the fish showed a declining trend in relation to the quantity of the feed ration when compared to control. The average growth performance in the fish fed with different rations showed a significant difference in their growth performance, when compared to the control. As for as 100 per cent ration size the growth was very high both on the 15th and after the 30th day of feeding.

On the 15th day of feeding, the 75 per cent feed size given group showed a good growth rate than the control group. In case of 100 per cent fed group, the growth rate had decreased to about 8.1 per cent than the control and in case of 50 per cent fed group there was 16.6 per cent reduction in growth rate than control. In 25 per cent fed group a decrease in the growth was about 27.7 per cent than the control.

After the 30th day of feeding about 0.53 per cent of decrease was observed in the growth of the fish in 100 per cent feed size group. In case of 75 per cent feed size group, the growth rate had decreased about 9.15 per cent than the control. In case of 50 per cent feed size group about 18.41 per cent reduction was seen in growth than in the control and in case of 25 per cent fed group there was about 22.46 per cent growth decrease was observed than the control. Statistical analysis was carried out using students 't' test. Correlations between the growths of fishes were found using regression analysis (Zar, 1984). All the values obtained in this experiment showed a significant difference from the control group.

Table 19.1: Shows the Growth Performance of *Etroplus suratensis* Fed with Different Ration Levels of *Artemia* for a Period of 30 Days. Each value represents the wet weight of the average of 3 groups of individuals.

Sl.No.	Feed Ration	Initial Weight	Weight After		Growth Per cent		't' Value		'p' Value
			15th Day	30th Day	15th Day	30th Day	15th Day	30th Day	
1.	*Ad libitum* (Control)	2.021	3.367	4.731	66.60	134.092	6.897	13.897	$P < 0.05$ S
2.	100 per cent	1.995	3.484	4.613	74.636	131.228	6.082	10.694	$P < 0.05$ S
3.	75 per cent	2.062	2.792	3.935	35.40	90.834	2.474	6.356	$P < 0.05$ S
4.	50 per cent	2.082	2.749	3.406	32.036	63.593	7.882	15.588	$P < 0.05$ S
5.	25 per cent	1.956	2.661	3.001	36.042	53.425	25.178	37.321	$P < 0.05$ S

The average daily growth (ADG) of the experimental fish varied according to food rations. The average daily growth was high in 100 per cent fed group and it gradually showed a declining trend in case of 75 per cent, 50 per cent and 25 per cent ration fed groups (Table 19.2). The average daily growth was 0.09±0.0039 mg/gm/day in case of the control fish but it declined in experimental fish

Table 19.2: Shows the FCR, ADG, SGR, Ae (per cent) and Fe (per cent) of *Etroplus suratensis* Fed with Different Ration Levels of *Arrtemia* for a Period of 30 Days Each value (X±SD) represents an average of 3 groups of individuals.

Sl.No.	Feed Ration	Food Conversion Ratio (FCR)	ADG (Per cent)	SGR (Per cent)
1.	*Ad libitum* (Control)	83.025±0.0241	0.090±0.0039	2.83±0.7321
2.	100 per cent	68.755±0.179 t = 1295.545 $P < 0.05$ S	0.087±0.015 t = 9.412 $P < 0.05$ S	2.7977±0.723 t = 1.5068 $P > 0.05$ Ns
3.	75 per cent	72.076±0.493 t = 1020.545 $P < 0.05$ S	0.0062±0.0078 t= 14.118 $P < 0.05$ S	2.3977±0.0077 t = 0.9467 $P > 0.05$ Ns
4.	50 per cent	67.975±0.444 t = 1372.727 $P < 0.05$ S	0.044±0.0043 t = 26.470 $P < 0.05$ S.	1.993±0.038 t = 0.6940 $P > 0.05$ Ns
5.	25 per cent	43.062±0.071 t = 3630.909 $P < 0.05$ S	0.034±0.00.37 t = 7.465 $P < 0.05$ S	1.784±0.0135 t = 0.6295 $P > 0.05$ Ns

The rate of food conversion ratio (FCR) in the experimental fish also varied according to feed size. The rate of food conversion ratio was very high in 100 per cent fed group and it gradually showed a declining trend in case of 75 per cent, 50 per cent, 25 per cent ration fed groups. The food conversion ratio was about 83.025±0.024 mg/gm/day in the control fish. The food conversion ratio value observed in 100 per cent, 75 per cent, 50 per cent and 25 per cent ration fed groups were about 68.755±0.179, 72.076±0.493, 67.975±0.444 and 43.062±0.071 mg/g/day respectively (Table 19.2).

The specific growth rate (SGR) increased as the rations given to various groups of fishes were increased. The highest SGR value obtained in 100 per cent fed group fish was 2.79±0.73 mg/g/day followed by 2.39±0.007, 1.99±0.038, 1.78±0.013 mg/g/day in 75 per cent, 50 per cent and 25 per cent respectively (Table 19.2).

Feeding is a potent factor governing the growth of an animal. Ration is an important factor affecting the growth of a fish. The growth and ration relationship had been established for several species by several authors (Brett and Groves, 1979; Jobling, 1994; Nimmi and Beamish, 1974). In the present investigation the *Etroplus suratensis* showed the maximum growth performance in the higher level of ration than the lower levels. It was observed that the weight of fish increases with length, which is an indicator of normal growth pattern. Arunachalam and Ravichandra Reddy (1981) observed enhanced consumption of *Tubifex* with an increase in ration in *Mystus vittatus*. In the present investigation the *Etroplus* showed the maximum growth performance in the higher level of ration than the lower levels. When the feeding ration was lower than that of optimal growth, higher weight gain of fish was achieved at a higher feeding ratio (Arzel *et al.*, 1998; Ballestrazzi *et al.*, 1998; Van Ham *et al.*, 2003 and Eroldogen *et al.*, 2004).

The low FCR obtained in the present investigation was due to low availability of feed and low feeding rate in the low ration fed groups, such as 50 per cent and 25 per cent. Sethuramalingam *et al.* (1996) showed an increase of FCR in *Labeo rohita* by the increase of ration. When the feeding ratio was lower than that needed for optimal growth, higher weight gain of fish was achieved at the higher feeding ratio. When high FCR value is obtained the general growth performance of the fishes will decline (Arzel *et al.*, 1998; Ballestrazzi *et al.*, 1998; Van Ham *et al.*, 2003 and Eroldogan *et al.*, 2004). According to Pandian (1967) and Kinne (1960), the FCR is affected by a number *of* factors such as body weight; ration size, temperature and salinity.

In the present investigation, increased SGR was observed in higher ration fed group (Table 19.2). This may be due to the sufficient availability of the feed in the higher ration to the fishes and that could have met the protein requirement of the fish. Jauncey (1982a) has the opinion that the SGR increased linearly with increase in dietary protein up to 40 per cent. Eroldogan *et al.* (2004) obtained the highest weight gain and SGR in European sea bass with an initial weight of 2.6g was obtained at 100 per cent of satiation in both freshwater and seawater when fish were fed a commercial diet at 36 per cent, 45 per cent, 54 per cent, 63 per cent, 71 per cent and 100 per cent of satiation daily.

References

Arunachalam, S. and Reddy, S. Ravichandra, 1981. Interactions of feeding rates on growth, food conversion and body composition of the fresh water cat fish, *Mystus vittatus* (Bloch). *Hydrobiologia*, 78: 25–32.

Arzel, J., Metailker, R., Le Gall, P. and Guillaume, J., 1998. Relationship between ration size and dietary protein level varying at the expense of carbohydrate and lipid in triploid brown trout fry, *Salmon trutta*. *Aquaculture*, 162: 259–268.

Ballestrazzi, R., Lanari, D. and D'Agaro, E., 1998. Performance, nutrient retention efficiency, total ammonia and reactive phosphorous excretion of growing European sea bass (*Dicentrarchus labrax*, L) as affected by diet processing and feeding level. *Aquaculture*, 161: 55–65.

Brett, J.R. and Groves, T.D.D., 1979. Physiological energetics. In: *Fish Physiology Vol. 8: Bioenergetics and Growth Fish Physiology*, (Eds.) W.S. Hoar, D.J. Randall and J.R. Brett. Academic Press, New York, pp. 599–675.

Bromley, P.J., 1980. Effect of dietary protein, lipid and energy content on the growth of turbot (*Scophthalmus maximus* L.). *Aquaculture*, 19: 359–369.

Bromley, P.J., 1987. The effects of food type, meal size and body weight on digestion and gastric evacuation in turbot, (*Scophthalmus maximus* L.). *J. Fish. Biol.*, 30: 501–512.

Eroldgan, O.T., Kumlu, M. and Aktas, M., 2004. Optimum feeding rate for European sea bass, *Dicentrarchus labrax* L reared in seawater and freshwater. *Aquaculture*, 231: 501–515.

Fontaine, P., Jean Noel Gardeur, Kestemont, Patrick and Georges, Andrew, 1997. Influence of feeding level on growth, intraspecific, weight variability and sexual growth dimorphism of Eurasian Perch, *Perca fluviatilis* L. reared in a recirculation system. *Aquaculture*, 157: 1–98.

Hansen, M.J., Boisclair, D., Brandt, S.B., Hewett, S.W., Kitchell, I.F., Lucas, M.C. and Ney, J.J., 1993. Applications of bioenergetics model to fish ecology and management: Where do we go from here? *Trans. Am. Fish. Soc.*, 122: 1019–1030.

Hung, S.S.O., Fred, S. Contre and Lutes, Paul B., 1995. Optimum feeding rate of white sturgeon, *Acipensor transmontanus* yearlings under commercial production conditions. *J. App. Aquaculture*, (5): 1.

Jauncey, K., 1982a. Carp (*Cyprinus carpio* L.) Nutrition: A review. In: *Recent Advances in Aquaculture*, (Eds.) J.F. Muir and R.J. Roberts. Croom Helm, London, pp. 215 –263.

Jobling, M., 1994. Ingestion, absorption and digestion. In: *Fish Bioenergetics*, (Ed.) M. Jobling. Chapman and Hall. London, pp. 99–119.

Kinne, O., 1960. Growth, food intake and. food conversion in a euryplastic fish exposed to different temperatures and salinities. *Physiol. Zool.*, 33: 288–317.

Nimmi, A.J. and Beamish, F.W.H., 1974. Bioenergetics and growth of large mouth bass (*Micropterus salmonoides*) in relation to body weight and temperature. *Can. J. Zool.*, 521: 447–456.

Pandian T.J., 1967. Intake, digestion, absorption and conversion of food in the fishes, *Megalops cyprinoids and Ophiocephalus striatu*. *Mar. Biol.*, 1: 16 – 32.

Ravichandra Reddy, S. and Shakuntala, Katre, 1979. Growth rate and conversion efficiency of the air breathing catfish, *H. fossilis* in relation to ration size. *Aquaculture*, 18: 35–40.

Ricker, 1973. *W.E. Fish. Res. Bd.*, Canada, 30: 409.

Van Ham, E.H., Bemtssen, M.H.G., Imsland, A.K., Porpoura, A.C., Wendelaar Bonga, S.E. and Stetansson, S.O., 2003. The influence of temperature and ration on growth, feed conversion, body composition and nutrient retention of juvenile turbot, *Scophthalmous maximus*. *Aquaculture*, 217: 547–558.

Wurtsbaugh, W.A. and Davis, G.E., 1977. Effects of temperature and ration level on the growth and food conversion of *Salmo gairdneri* Richardson. *J. Fish. Biol.*, 11: 87–98.

Zar, J.H., 1984. *Biostatistical Analysis*, 2nd edn., (Ed.) B. Kurtz. Prentice Hall, Int. Inc., New Jersey.

Chapter 20

Effect of Food Ration on Energetics of Marine Ornamental Fish, *Etroplus suratensis* Fed with *Artemia*

A. Amutha Jaisheeba[1], *R. Sornaraj*[1] *and A.J.A. Ranjit Singh*[2]

[1]*Research Department of Zoology, Kamaraj College, Thoothukudi*
[2]*Reader and Head, Department of Biology, S.P.K. College, Alwarkurichi*

ABSTRACT

Juveniles of *Etroplus suratensis* were fed with the live feed *Artemia*. Four different ration size of *Artemia viz.,* 100 per cent, 75 per cent, 50 per cent and 25 per cent were given for 30 days. The rate of food consumption and absorption were changed according to feed size Feed consumption rates were 170.678±1.4, 125.217±1.08, 79.151±0.50 and 35.521±0.8 mg/g/day respectively in 100 per cent, 75 per cent, 50 per cent and 25 per cent ration of diets. The absorption rate also increased, when the ration size was increased from 25 per cent to 100 per cent *viz.,* 0.605±0.012, 1.294±0.018, 2.029±0.418 and 2.759±0.068 mg/g/day respectively. The absorption efficiency was reduced to 87.9 per cent and 78.9 per cent in 50 per cent and 25 per cent ration size fed group. Gross Conversion Efficiency (GCE) in different ration fed groups was 2.32, 1.4, 1.3 and 1.4 mg/g/day in case 25 per cent, 50 per cent, 75 per cent and 100 per cent fed groups respectively. The Net Conversion Efficiency (NCE) values observed were 2.9, 1.6, 1.4 and 1.5 mg/g/day in case 25 per cent, 50 per cent, 75 per cent and 100 per cent fed groups respectively. The present investigation observed that maximum consumption of food in lower ration fed group and the result suggested that it may be due to puccity of food availability and increased appetite in the lower ration fed groups.

Keywords: *Etroplus suratensis, Rations of feed, Absorption rate, GCE, NCE.*

Introduction

Due to the high cost of fish feed, fish farmers embark on restricted feeding of fish stocks to reduce the cost of production. A balanced feed with proper nutrient proportions and appropriate protein to energy ratio matching the nutritional requirements of the species reared, can substantially improve its growth. Studies on the relationship between growth rate and the rate of food intake are essential for understanding the bioenergetics of fish in natural systems (Hansen *et al.*, 1993) and for designing feeding regimes in intensive aquaculture system. The past four decades have seen a continued increase in research devoted to the study of fish growth and bioenergetics. Energetics studies in fish had been reviewed by many workers (Cowey and Sargent, 1979; Jauncey, 1982a and Jobling, 1983b).

Feeding frequency had been reported to affect growth and food conversion efficiency (Boujard and Leatherland, 1992a and Jarboe and Grant, 1997). The effect of varying feeding rates on growth, feed consumption and its efficiency and body composition in a number of species had been studied (Pandian and Raghuraman, 1972; Choe *et al.*, 1985 and De Silva *et al.*, 1986). However there is no work on the marine ornamental fish *E. suratensis,* which has a good market demand of proper diet dose is fixed for the culture of *E. suratensis,* this will promote the rearing practices of this fish to a great extent.

Methodology

In the present study the juveniles of *E. suratensis* (0.485±0.08) were collected from the shore line of Thoothukudi, East coast of India, and they were carefully transported to the laboratory in polythene bags containing oxygenated water. The period of acclimation lasted for about fifteen days. During this period the fish were fed *ad libitum* with live *Artemia,* thrice a day, which were collected from nearby saltpans. After acclimation the experimental fishes were sorted out into groups including 5 fishes in each. Triplicates were maintained. Before starting of the experiment the fishes were not given feed for one day. Care was taken in choosing the fish with same size and weight range. The fishes were fed thrice a day (8.00 am, 12 noon and 4.00 pm). The unfed were collected after one hour.

The feed given to the experimental fish was determined on the basis of 20 per cent of the body weight (Tacon and Coway, 1985), from this 100 per cent, 75 per cent, 50 per cent and 25 per cent rations of food were calculated for further studies. The water quality parameters were maintained properly. The bioenergetics parameters of the fish were calculated following the IBP (Petrusewicz and Macfayden, 1970) formula.

The experiments were run to a period of 30 days. Initially before introducing the fish into the rearing trough their wet weight of each of the individuals were determined and on the 15th and the 30th day also weight were recorded.

Results and Discussion

The total food absorption by the experimental fish changed according to the ration size given to them. The rate of food absorption was about 214.565±2.912 mg/g/day in control group of fishes that were fed with *Artemia ad libitum*. The rate of food absorption gradually showed a declining trend in case of 100 per cent, 75 per cent, 50 per cent and 25 per cent rations fed groups. It was about 170.678±1.4, 125.217±1.08, 79.151±0.50 and 35.521±0.8 mg/g/day respectively in 100 per cent, 75 per cent, 50 per cent and 25 per cent ration of diets (Table 20.1).

The metabolic rate increased as the quantity of the ration increased. The metabolic value obtained during the experimental regime was 34.476±0.296, 77.827±0.303, 123.344±0.238 and 168.06±1.674 mg/g/day for the different rations fed groups such as 25 per cent, 50 per cent, 75 per cent and 100 per cent respectively. In case of the test fishes, the metabolic rate was increased from 0.59, 1.27, 1.99 and 2.72 mg/g/day, when the ration was increased from 25 per cent, 50 per cent, 75 per cent and 100 per cent respectively (Table 20.2).

Table 20.1: Rates of Food Consumption and Absorption of *Etroplus suratensis* Fed with Different Ration Levels of Live Feed *Artemia* for a Period of 30 Days. Each value (X±SD) represents an average of 3 groups of individuals.

Sl.No.	*Feed Ration*	*Initial Weight (g)*	*Dry Weight of Total Food Intake (mg/fish)*	*Food Consumption (mg/g/day)*	*Dry Weight of Total Food Absorbed (mg/fish)*	*Food Absorption Rate (mg/g/day)*
1.	*Ad libitum* (Control)	2.02±0.09	33.03±0.377	3.71±0.06	22.366±0.225	3.43±0.05
2.	100 per cent	1.99±0.011	26.46±0.391	3.02±0.05	17.138±0.309	2.76±0.07
3.	75 per cent	2.06±0.133	19.80±0.249	2.35±0.05	10.017±0.050	2.03±0.42
4.	50 per cent	2.08±0.38	13.23±0.246	1.60±0.016	2.381±0.052	1.29±0.02
5.	25 per cent	1.96±0.013	6.57±0.288	0.79±0.033	2.909±0.040	0.60±0.012

Table 20.2: Shows the Rate of Food Conversion and Metabolic Rate of *Etroplus suratensis* Fed with Different Ration Levels of *Artemia* for a Period of 30 Days. Each value (X±SD) represents an average of 3 groups of individuals.

Sl.No.	*Feed Ration*	*Food Conversion (mg/fish)*	*Conversion Rate (mg/g/day)*	*Metabolism*	*Metabolic Rate (mg/g/day)*
1.	*Ad libitum* (Control)	0.28±0.006	3.490±0.059	211.855±1.357	3.38±1.407
2.	100 per cent	0.285±0.013* (8.571)	2.81±0.081^{Ns} (0.008)	168.06±1.2978* (57.642)	2.72±0.09* (1.133)
3.	75 per cent	0.283±0.0003* (4.286)	1.993±0.093^{Ns} (0.062)	123.344±0.1844* (111.905)	1.99±0.11* (2.315)
4.	50 per cent	0.296±0.017* (3.928)	1.246±0.1688^{Ns} (0.125)	77.827±0.2345* (169.280)	1.27±0.05* (3.383)
5.	25 per cent	0.288±0.0033* (4.643)	0.588±0.0212^{Ns} (0.212)	34.476±0.2298* (225.341)	0.59±0.0045* (4.437)

*: Significant at 0.05; Ns: Non significant at 0.05 per cent.

The rate of food consumption was very high in 100 per cent fed group and it gradually showed a declining trend in case of 75 per cent, 50 per cent and 25 per cent ration fed groups (Table 20.1). The food consumption rate was about 3.7±0.59 mg/g/day in case of the control fish. The rate of consumption gradually showed a declining trend in relation to the size of the ration given to the different groups. It was about 3.01±0.051, 2.3±0.052, 1.6±0.016 and 0.8±0.033 mg/g/day respectively in 100 per cent, 75 per cent, 50 per cent and 25 per cent ration fed group of fishes.

The rate of absorption clearly followed the trend of food consumption. As the ration size increased the absorption rate also increased. The absorption rate in case of control was 3.4±0.049 mg/g/day.

The absorption rate in the test fishes were increased when the ration size increased from 25 per cent to 100 per cent as 0.605±0.012, 1.294±0.018, 2.029±0.418 and 2.759±0.068 respectively (Table 20.1).

The rate of food conversion was very high in low ration fed groups such as 25 per cent and 50 per cent (conversion 0.288 and 0.296). As the ration size increased to 7S per cent, 100 per cent and *ad libitum* the conversion level decreased in an descending order as 0.283, 0.285, and 0.285 respectively (Table 20.2). The food conversion rate showed a decreasing trend as the ration size increased. The conversion rate obtained was 0.588±0.0212, 1.246±0.1688, 1.993±0.093 and 2.81±0.081 mg/g/day in different ration fed group such as 25 per cent, 50 per cent, 75 per cent and 100 per cent respectively (Table 20.2).

There was no much difference in the efficiency of food absorption as a function of the ration size. The highest absorption efficiency was observed in the 100 per cent ration fed group and it was about 94.82 per cent. No significant difference was observed in absorption efficiency in the feed groups 75 per cent, 100 per cent and control group (Table 20.3). As the ration size decreased to 50 per cent and 25 per cent the absorption efficiency also reduced as 87.9 per cent and 78.9 per cent respectively.

Table 20.3: Shows the Ae (per cent) Fe (per cent) GCE and NCE of *Etroplus suratensis* Fed with Different Ration Levels of *Artemia* for a Period of 30 Days. Each value (X±SD) represents an average of 3 groups of individuals.

Sl.No.	*Feed Ration*	*Absorption Efficiency (Ae%)*	*Feeding Efficiency (Fe%)*	*Gross Conversion Efficiency (K_1%)*	*Non Conversion Efficiency (K_2%)*
1.	*Ad libitum* (Control)	95.36±0.014	8.205±0.0089	1.204±0.087	1.266±0.105
2.	100 per cent	94.82±0.288* (4823.134)	9.894±0.161* (390.00)	1.454±0.090* (17.5)	1.534±0.0437* (10061.90)
3.	75 per cent	92.75±0.431* (1758.805)	9.459±0.0086* (198.25)	1.360±0.204* (42.0)	1.496±0.333* (1023.81)
4.	50 per cent	87.95±0.233* (5859.552)	10.007±0.067* (465.25)	1.471±0.063* (92.5)	1.672±0.373* (670.0)
5.	25 per cent	78.93±0.499* (7248.060)	15.908±0.152* (1877.00)	2.322±0.099* (236.0)	2.942±0.0157* (895.24)

The gross and net conversion efficiencies showed a declining trend when the ration level increased from 25 per cent to 100 per cent. The GCE and NCE are very low in case of the control group and it is about 1.204 and 1.266 mg/g/day respectively. The GCE observed in other ration fed groups were 2.32, 1.4, 1.3 and 1.4 mg/g/day in case of 25 per cent, 50 per cent, 75 per cent and 100 per cent fed groups respectively. The NCE values observed were 2.9, 1.6, 1.4 and 1.5 mg/g/day in case of 25 per cent, 50 per cent, 75 per cent and 100 per cent fed groups respectively (Table 20.3).

In the present investigation an increase in the ration size increased the feeding rate, absorption rate and growth performance of the fish, *Etroplus suratensis.* Arunachalam and Ravichandra Reddy (1981) observed an enhanced consumption of *Tubifex* with increase of ration in *Mystus vittatus.* Feeding frequency is also known to affect growth and food conversion efficiency (Boujard and Leatherland, 1992a and Jarboe and Grant, 1997). For example, food conversion efficiency of channel catfish, *Ictalurus punctatus,* increased when the number of meals was increased from two to four at a fixed ration 2 per cent body weight per day (bw/day) (Greenland and Gill, 1979). Several authors have studied the effects of varying ration size on growth, food consumption and conversion efficiency and body composition of many species. *Ictalurus punctatus* (Andrews and Stickney, 1972), *Tilapia mossampica* (Pandian and Raghuraman, 1972); *Heteropneustes fossilis* (Ravichandra Reddy and Shakuntala, 1979); *Mystus vittatus* (Arunachalam and Reddy, 1981); *Cirrhinus mrigala* (Das and Ray, 1989); *Ctenophoryngodon idella* (Cui *et al.*, 1994) and *Perca fluviatilis* (Fontaine *et al.*, 1997).

In the present study, the absorption rate and absorption efficiency did not vary significantly between the groups receiving different rations which confirmed the view of number of earlier workers (Kelso, 1972; Vivekanandan, 1976; Ponniah and Pandian, 1977 and Russel *et al.*, 1996). In some species, absorption efficiency decreased with increasing rations (Elliot, 1976). In contrast, absorption efficiency increased with increasing ration in *Minnows* (Cui and Wootton, 1988). Feeding rate determines the conversion rate (Raghuraman, 1973).

Conversion rate of the selected weight class of *C. mrigala* increased with the increase in ration level which confirmed the results obtained in *Polycanthus cupanus* (Ponniah and Pandian, 1977); *Ophiocephalus striatus* (Vivekanandan, 1976); *Heteropneustes fossilis* (Ravichandra Reddy and Shakuntala, 1979). Though the conversion rate increased with the ration size, statistically significant change in conversion rate was observed when the ration was increased from 50 to 75 per cent level in all weight classes. Hence a ration of 75 per cent was fixed as the optimum dose. The group, which received low rations (25 per cent), exhibited less increase in body weight, which may be due to the oxidation of the low level of food consumed by the candidate species for energy production. Savitz (1971) supported this view and he stated that the fish *Lepomis macrochirus* fed with zero ration-exhibited loss of body weight, which may be due to the oxidation of fat and protein stored in the body.

There are contradicting views regarding the effect of ration on conversion efficiencies of fish. While Kelso (1972) reported that the ration size has no influence on conversion efficiency. Ravichandra Reddy and Shakuntala (1979) stated that the food consumption and conversion efficiency depend on the ration of feed provided to the fish.

The rate of absorption very clearly followed the trend of food consumption. As the ration size increased, the absorption rate also increased. In the present investigation there was no much difference in the food consumption rate among the 100 per cent fed group and *ad libitum* fed (control) group. The result suggested that maximum consumption of food was observed in lower ration fed group than the higher ration fed group and this may be due to the puccity of feed availability and increased appetite in the lower ration fed group of fishes. Several previous workers also had the opinion of the present investigation. Shimeno *et al.* (1981) reported that feeding ratio could be lowered to the 80 per cent satiation without reduction in weight gain of common carp when the fish were fed a commercial diet 5–6 times a day at various feeding rations (100 per cent to 30 per cent) for 30 days.

As far as the absorption efficiency was concerned there was no much change in various group of fishes, which were given the different rations. Very high value was observed in the control fish and 100 per cent fed fish, where as in other groups, the observed absorption efficiency value was very close to the control and 100 per cent fed group (Table 20.3). The results indicated that, the absorption efficiency was not influenced by any of the bioenergetics parameter in the candidate species, *Etroplus suratensis,* Sampath (1984) had the opinion that the absorption rate and absorption efficiency were reduced when the rations exceed the optimal level. However, there are information, which stated that the absorption efficiency was least influenced by the various ration size (Gerking, 1971; Vivekanandan, 1976; Ponnaiah and Pandian, 1977 and Narayanan, 1980). Maria Charles *et al.* (1984) reported that the fish *Cyprinus carpio* fed with different rations showed a very high absorption efficiency (97 per cent) and further stated that the absorption efficiency was not influenced by any bioenergetics parameters. The present investigation confirmed the view of Maria Charles *et al.* (1984) by showing constant level of feeding limits.

The conversion rates are directly influenced by the food rations given to the fishes. Maximum conversion rate was observed in 100 per cent fed group (Table 20.2). The findings were supported by Elliott (1975 and 1976) and he proposed that conversion rates are directly influenced by the food rations given to the fishes, higher the ration, the greater was the growth rate. In *Cyprinus carpio* the maximum conversion rate was observed when it was given the maximum ration of food. Maria Charles *et al.* (1984).

References

Andrews, J.W. and Stickney, R.R., 1972. Interactions of feeding rates and environmental temperature on growth. food conversion and body composition of channel cat fish. *Trans. Am. Fish. Soc.*, 101: 94–99.

Arunachalam, S. and Reddy, S. Ravichandra, 1981. Interactions of feeding rates on growth, food conversion and body composition of the fresh water cat fish, *Mystus vittatus* (Bloch). *Hydrobiologia*, 78: 25–32.

Boujard, T. and Lcatherland, J.F., 1992a. Circadian rhythm and feeding time in fishes. *Environmental Biology of Fishes*, 35: 109–131.

Cho, C.Y., Cowey, C.B. and Watanabe, T., 1985. Influence of level and type of dietary protein and level of feeding on feed utilization of rainbow trout. *J Nutri.*, 106: 1547–1556.

Cowey, C.B. and Sargent, J.R., 1979. Nutrition. In: *Fish Physiology*, Vol. 8, (Eds.) W.S. Hoar, D.J. Randall and J.R. Brett. Academic Press, New York, pp. 279–352.

Cui, J. and Wootton, R.J., 1988, Effects of ration, temperature and body size on the body composition of the minnow (*Phoxinus phoxinus* L.). *J. Fish. Biol.*, 32: 749–764.

Cui, Y., Chen, S. and Wang, S., 1994. Effect of ration size on the growth and energy budget of the grass carp, *Ctenopharyngodon idella* (Val.). *Aquaculture*, 123: 95–100.

Das, J. and Ray, A.K., 1989. Effect of feeding levels on the growth performance of (C. *mrigala*) (Ham) fingerlings. In: Paper Presented in the *National Seminar on Forty Years of Freshwater Aquaculture in India*, organized by Association of Aquaculturists and Central Institute of Freshwater Aquaculture 7–9, November, Bhubaneswar, India.

De Silva, S.S., Gunasekara, R.M. and Keembiyahetty, C., 1986. Optimum ration and feeding frequency in *Oreochromis niloticus*. In: *The first Asian Fisheries Forum*. Asian Fisheries Society, Manila, pp. 559–564.

Elliot, J.M., 1975. Weight of food and time required to satiate brown trout (*Salmon trutta* L). *Freshwater Biol.*, 5: 51–64.

Elliott, J.M., 1976. Body composition of brown trout, *Salmo trutta* L. in relation to temperature and ration size. *J. Anim. Ecol.*, 45: 273–289.

Fontaine, P., Gardeur, Jean Noelm Kestemont, Patrick and Georges, Andrew, 1997. Influence of feeding level on growth, intraspecific weight variability and sexual growth dimorphism of Eurasian Perch, *Perca fluviatilis* L. reared in a recirculation system. *Aquaculture*, 157: 1–98.

Gerking, S.D., 1971, Influence of rate of feeding and body weight on protein metabolism of bluegill sun fish. *Physiol. Zool.*, 44: 9–19.

Greenland, D.C. and Gill, R.E., 1979. Multiple daily feed with automatic feeders improve growth and feed conversion rates of channel catfish. *The Progressive Fish Culturist,* 41: 151–153.

Hansen, M.J., Boisclair, D., Brandt, S.B., Hewett, S.W., Kitchell, J.F., Lucas, M.C. and Ney, J.J., 1993. Applications of bioenergetics model to fish ecology and management. Where do we go from here? *Trans. Am. Fish. Soc.,* 122: 1019–1030.

Jarboe, Time and Grant, W.Y., 1997. The influence of feeding time and frequency on the growth survival, feed conversion and body composition of channel catfish, *Ictalurus punctatus,* cultured in three tier closed recirculating raceway system. *J. App. Aquaculture,* 7: 43–52.

Jauncey, K., 1982a. Carp (*Cyprinus carpio* L.) Nutrition: A review. In: *Recent Advances in Aquaculture,* (Eds.) J.F. Muir and R.J. Roberts. Croom Helm, London, pp. 215–263.

Jobling, M., 1983b. A short review and critique of methodology used in fish growth and nutrition studies. *J. Fish. Biol.,* 23: 683.

Kelso, J.R.M., 1972. Conversion, maintenance and assimilation for walleye, stizostedion vitreum as affected by size, diet and temperature. *J. Fish. Res. Bd. Can.,* 29: 1181–1192.

Maria Charles, P., Sebastian, S. Maria, Raj, M.C.V. and Marian, M.P., 1984. Effect of feeding frequency on growth and feed conversion of *Cyprinus carpio* fry. *Aquaculture,* 40: 293–300.

Narayanan, M., 1980. Study of food utilization by multi-species laboratory population of fish. *Ph.D., Thesis,* M.K. University, India, pp. 182.

Pandian, T.J. and Raghuraman, 1972. Effect of feeding rate on conversion efficiency and chemical composition of the fish *T. mossambica. Mar. Biol.,* 12: 129–136.

Ponniah, A.G. and Pandian, T.J., 1977. Surfacing activity and food utilization in the air-breathing fish, *Polycanthus cupanus* exposed to constant PO_2. *Hydrobiologia,* 53: 221–227.

Raghuraman, R., 1973. Effect of body weight on intake and conversion of food in the fish, *Rosbora daniconius. Curr. Sci.,* 42: 24–25.

Ravichandra Reddy, S. and Shakuntala, Katre, 1979. Growth rate and conversion efficiency of the air breathing catfish, *H. fossilis* in relation to ration size. *Aquaculture,* 18: 35–40.

Russel, N.R., Fish, J.D. and Wootton, R.J., 1996. Feeding and growth of juvenile seabass: The effect of ration and temperature on growth rate and efficiency. *J. Fish. Biol.,* 49: 206–220.

Sampath, K. and Premila, C. Lily, 1995. Optimum food requirement of the cat fish *Mystus keletius* as a function of body weight. *J. Aqua. Trop.,* 10: 139–149.

Sampath, K., 1984, Preliminary report on the effects of feeding frequency in *Channa striatus. Aquaculture,* 40: 301–306.

Savitz, J., 1971. Effects of starvation on body protein utilization of blue gill sun fish, *Lepomis macrochirus* with a calculation of caloric requirements. *Trans. Am. Fish. Soc.,* 100: 18.

Shimeno, S., Takada, M., Takayama, S., Fukur, A., Saraki, H. and Kajiyama, M., 1981. Adoption of hepatopancreatic enzymes to dietary carbohydrate in carp. *Bull. Jpn. Soc. Sci. Fish.,* 47: 71–77.

Vivekanandan, E., 1976 Effects of feeding on the swimming activity and growth of *Ophiocephalus striatus. J. Fish. Biol.,* 8: 321–330.

Chapter 21

Waste Management by Nitrogen-Fixing Cyanobacteria

A.K. Bastia

P.G. Department of Botany, North Orissa University, Baripada – 757 003, Orissa

ABSTRACT

Colour is one of the major problems in the treatment of distillery waste. The present study aims to develop an environmental management model that utilizes nitrogen-fixing cyanobacteria for waste management and colour reduction. Three species of nitrogen fixing cyanobacteria *viz. Nostoc linckia, Calothrix parietina* and *Westiellopsis* sp. of local rice fields capable of growing mixotrophically were studied in glass jars by supplementing low concentrations (1.5 per cent, v/v) of waste of a Molasses distillery. The waste was rich with various algal nutrients including reducing sugars (4.5–6.0 per cent, w/v). The diluted waste (1.5 per cent, v/v) supplemented medium was used for their mixotrophic growth for biomass production, in addition, reduced the colour of the industrial waste to a significant extent.

Keywords: *Waste management, Cyanobacteria, Mixotrophic, Biomass, Colour reduction.*

Introduction

Cyanobacteria a fascinating groups of prokaryotic organisms, play an important role in environment managemental (Singh, 1950 and 1961; Venkatraman, 1981; Rai *et al.*, 2000; Whitton and Potts, 2000). They have also a significant role in the ecology of aquatic ecosystems for environment clean-up and assessment of water quality, as test systems for evaluating nutrient status of effluents and also as indicators of habitat condition (Palmer, 1969; Rai and Kumar, 1976; Adhikary *et al.*, 1992; Rana *et al.*, 1971; Palmer, 1980). The oxygenic photosynthetic property, high level of tolerance to various pollutants and large surface area are some of the striking features which make them highly suitable for waste management (Bastia *et al.*, 2007) and to reduce their colour to certain extent (Thangeswaran, 1991; Subramanian, 1991).

Previous studies showed that a number of heterocystous cyanobacteria were found to grow and fix nitrogen appreciably mixotrophically and heterotrophically in dark utilizing low concentrations of neutralised distillery wastes (Adhikary, 1978a and b; 1990). In the present work an attempt has been made for mass cultivation of three selected nitrogen-fixing cyanobacteria of local rice fields, in grass jars by supplementing low concentration (1.5 per cent, v/v) of distillery waste to maximize their biomass production and to find out their capability to reduce the colour.

Materials and Methods

The Molasses distillery, Aska (19°16'N and 84°55'E) discharged large quantities of untreated waste to the surrounding rice-fields since 1970. Physico-chemical analysis of the waste conducted previously (APHA, 1971) is given in Table 21.1. The waste was rich with various algal nutrients like N, P, Ca, Fe, K salts and reducing sugars (4.5 to 6.0 per cent), mostly glucose and fructose (Adhikary, 1987a). Previous study has emphasized that waste organic matters serve as the main nutrient influencing heterotrophic growth of certain eukaryotic algae and cyanobacteria in natural environments (Zajic and Chiu, 1970). Experiments were conducted in 5 litre capacity grass jars containing 1.5 per cent (v/v) distillery waste in nitrogen free BG 11 medium and inoculating 30 mg sun dried inoculum of each organism (Plate 21.1) *viz. Nostoc linckia, Calothrix parietina* and *Westiellopsis* sp. The glass jars were incubated up to 30 days at room temperature of 32±2°C under continuous light intensity of 7.5 W/m^2. Dark experiments were discontinued due to bacterial growth. The initial pH of the culture medium in the glass jars were adjusted to 7.8 and the final pH was recorded after 30 days of incubation. The growth of the test organisms in the autotrophic and mixotrophic cultures in presence of distillery waste was estimated on dry weight basis. The cyanobacterial materials of the glass jars were filtered through a silk cloth, transferred to pre-weighed Whatman No 1 filter paper, dried in hot air oven at 100°C for 24 hours and weighed again. Experiments were repeated thrice and the mean value±S.D. has been given in Table 21.2. Reduction in the intensity of colour of the distillery waste was measured in a Hitachi Model 200-20 Spectrophotometer.

Results and Discussion

Effect of Diluted Distillery Waste on Growth and Biomass Production of Nitrogen-fixing Cyanobacteria

All the test cyanobacteria grew well in the glass jars containing nitrogen free BG 11 medium up to 30 days of incubation. Presence of 1.5 per cent (v/v) distillery waste in the basal medium enhanced the dry weight yield of *Westiellopsis* sp. and *Calothrix parietina* up to 80 per cent and 53 per cent respectively over their control value. However, distillery waste did not stimulate the growth of *Nostoc linckia* in comparison to its autotrophic culture (Plate 21.2). Chemical analysis of distillery waste, Aska (Adhikary, 1987a) showed that it was composed of 4.5 to 6.0 per cent sugars being mostly glucose or fructose. The earlier experiments (Bastia, 1994) showed that mixotrophic growth of *Calothrix parietina* and *Westiellopsis* sp. in presence of exogenous sugars like glucose, fructose and sucrose was significantly higher than respective autotrophic growth where as in *Nostoc linckia* neither of these substrates supported higher growth than the control value. Since the waste contained glucose or fructose, they might be stimulating the mixotrophic growth of *Calothrix parietina* and *Westiellopsis* sp. (Table 21.2) and hence increased their biomass.

Table 21.1: Physico-chemical Characteristics of the Molasses Distillery Waste, Aska (Orissa state, India)

Characteristics	*Mean±S.D. Values*	*Tolerance Limit for Industrial Effluents Discharged on Land for Irrigation Purpose I.S: 3307 (1965)*
Colour	Reddish brown	
Odour	Smell of burnt sugar	
Total solids	5225.0±3.08	2100 max
pH	4.8±0.3	5.5–9.0
Dissolved oxygen	0.5±0.1	
Biochemical oxygen demand (B.O.D) for 5 days at 20°C	3660.0±3.28	500 max
Chemical oxygen demand (C.O.D.)	2321.5±5.75	
Total nitrogen	86.7±2.23	
Sulphate	208.6±1.4	1000 max
Chloride	745.2±2.55	600 max
Phosphate	9.3±0.28	
Sodium	25.0±0.72	60 max
Calcium	68.2±0.66	
Iron	5.6±0.13	
Magnesium	16.4±0.18	
Potassium	89.3±3.15	
Reducing sugar*	4.5–6.0 per cent	

*: The effluent was first hydrolyzed to breakdown the polysaccharides and disaccharides and analyzed. The result given is the amount of total reducing sugars present in the waste. All values are mean±S.D. Values except pH and reducing sugar content are in mg/litre.

Table 21.2: Mass Cultivation of Three Rice-Field Cyanobacteria in Presence or Absence of a Molasses Distillery Waste. Final dryweight yield and pH of the culture medium after growing the organisms at room temperature (32±2°C) under 7.5 W/m² up to 30 days is given.

Growth Condition	*Starter Culture (Dry weight of each organism = 30 mg)*					
	Nostoc linckia		*Calothrix parietina*		*Westiellopsis sp.*	
	Growth (mg dry weight)	*pH of the medium*	*Growth (mg dry weight)*	*pH of the medium*	*Growth (mg dry weight)*	*pH of the Medium*
Autotrophic (BG 11 medium)	690±5	9.5	500±7	9.8	450±10	9.1
Mixotrophic (BG 11 medium + distillery waste; 1.5 per cent, v/v)	700±6(1.45)	8.5	765±4(53)	10.5	810±12(80)	9.8

pH of the medium in glass jars for mass cultivation was adjusted to 7.8 before sterilization; Growth values represent mean±S.D. and pH value represent mean of three determinations; Figures in parenthesis show percent increase in growth over the value obtained in autotrophic culture.

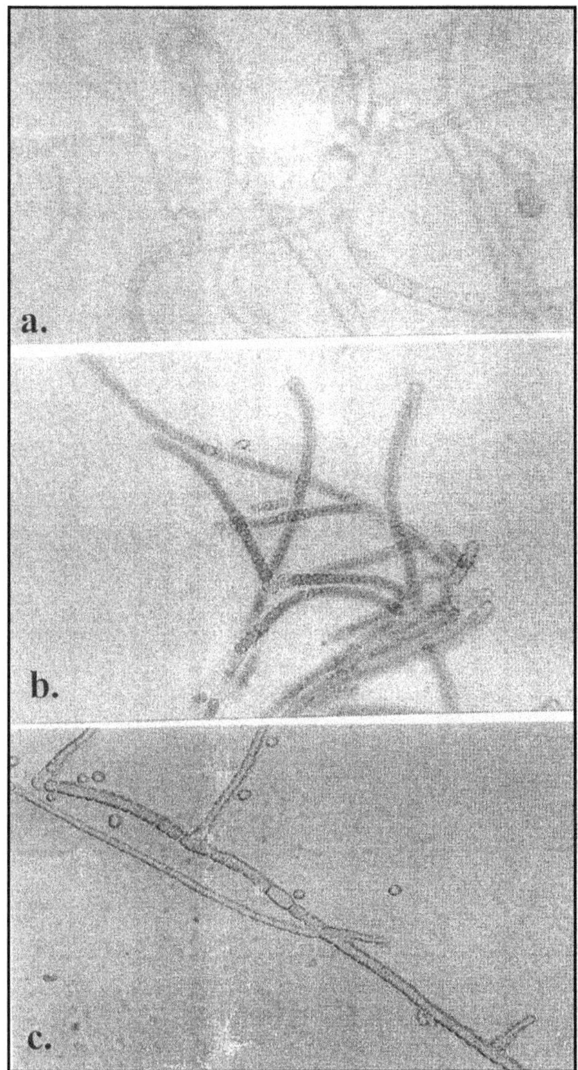

Plate 21.1: Light Microscopic Photographs of Three Rice-field Cyanobacteria

(*a*) *Nostoc linckia* (x 2348); (*b*) *Calothrix parietina* (x 1565); (*c*) *Westiellopsis* sp. (x 2348)

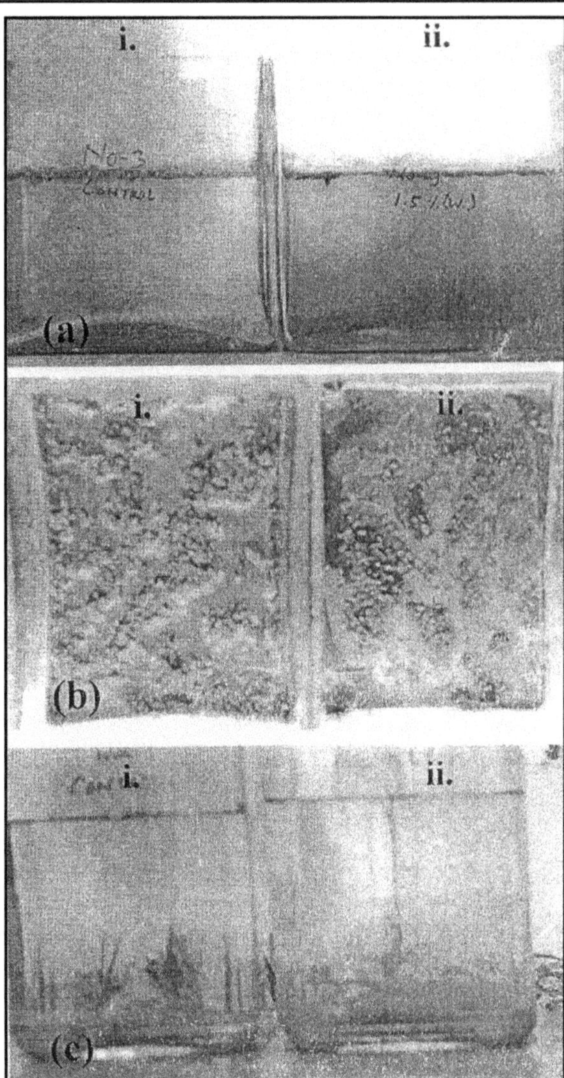

Plate 21.2: Growth of (*a*) *Nostoc linckia* (*b*) *Calothrix parietina* and (*c*) *Westiellopsis* sp. in 5 litre capacity glass jars at 32±2°C under 7.5W/m^2 light intensity (*i*) Culture vessels containing nitrogen free BG 11 medium (control) (*ii*) Culture vessels containing 1.5 per cent (v/v) distillery waste

Reduction of Waste Colour Utilizing Nitrogen-fixing Cyanobacteria

The pure distillery waste was reddish in colour. Experiments were initially designed to use 3–4 per cent (v/v) distillery waste prepared with BG 11 medium which can be equivalent to the concentration of the exogenous sugars, glucose or fructose (0.2 per cent approx.) used in the experiments. But the colour of the medium in presence of 34 per cent (v/v) distillery waste became too reddish to penetrate light up to bottom of the culture vessels. Thus 1.5 per cent (v/v) waste was used for the experiments. Though the concentration of reducing sugars in the waste supplemented culture was approximately half than that of the concentration used for heterotrophic growth studies, even lower concentration of distillery waste significantly increased the growth of *Calothrix parietina* and *Westiellopsis* sp. Further, increase in the pH of the waste supplemented culture over that of the basal medium was observed only in cultures of those species which showed appreciable mixotrophic growth (Table 21.2).

Reduction of waste colour with increasing period of incubation up to 30 days was also observed in the cultures of *Calothrix parietina* and *Westiellopsis* sp. which showed appreciable growth in presence of diluted distillery waste. No such reduction was observed in mixotrophic culture of *Nostoc linckia*. Absorption spectra of the 1.5 per cent (v/v) distillery waste prepared in basal medium and of the same medium after growing *Calothrix parietina* and *Westiellopsis* sp. up to 30 days was measured in the visible range of spectrum (Figure 21.1). This gave an indirect evidence about the preliminary trial exhibited ability to reduce the colour by two such tested cyanobacterial strains.

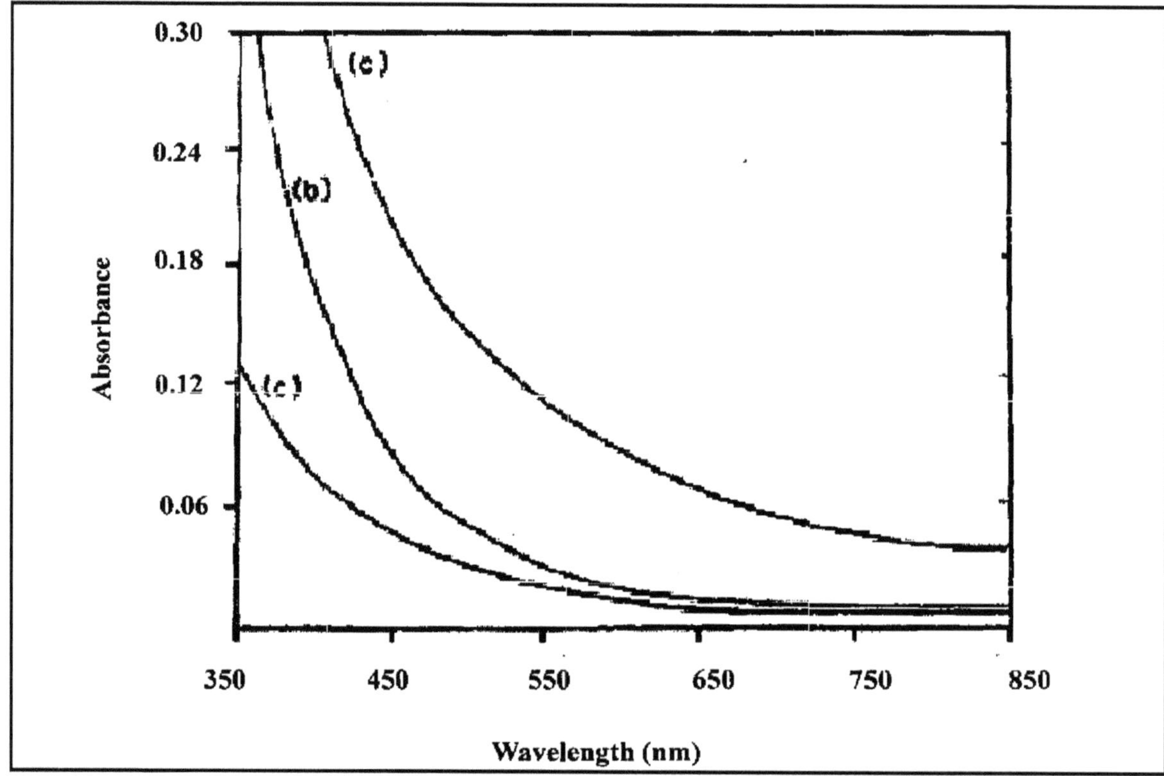

Figure 21.1: Absorption Spectra of the Waste Supplemented Culture Medium (Without Cyanobacterial Cell) Containing 1.5 per cent (v/v) Distillery Waste Before and After Growing *Calothrix parietina* and *Westiellopsis* sp. up to 30 Days at 32±2ºC Under 7.5 W/m² Light Intensity.

(*a*) Distillery waste (1.5 per cent, v/v) supplemented culture medium before growing test organisms; (*b*) After growing *Calothrix parietina* in culture containing 1.5 per cent (v/v) distillery waste up to 30 days; (*c*) After growing *Westiellopsis* sp. in culture containing 1.5 per cent (v/v) distillery waste up to 30 days.

Pollution control through bioremediation is the most economical and eco-friendly approach. In developing countries like India, most industries belong to medium and small scale sectors and cost is the key factor in setting up waste treatment systems. Biological treatment although least expensive, has not been effective with many effluents as there is no willful introduction of choosen microbes to treat specific types of pollutants. The available indications emphasize the need to take up extensive investigations both under laboratory and subsequently field conditions to gainfully employ mixotrophically growing nitrogen fixing cyanobacteria in bulk not only for their use as biofertilizer but also to treat and recycle wastes thus controlling pollution at an affordable cost.

Acknowledgement

I would like to thank Prof. and Head P.G. Dept. of Botany, North Orissa University, Baripada for comments and suggestion and Prof. S.P. Adhikary, P.G. Department of Botany, Utkal University for providing facilities and encouragement.

References

Adhikary, S.P., 1987a. Growth response of *Calothrix marchica* Lemm. var. intermedia Rao to exogenous organic substrate and distillery effluent in the light and dark. *J. Basic Microbiol.*, 27: 475–481.

Adhikary, S.P., 1987b. Open air cultivation of nitrogen-fixing blue-green algae utilizing distillery effluent. *Environ. Ecol.*, 5: 192–194.

Adhikary, S.P., 1990. Effect of distillery effluent on the growth of the blue-green alga *Calothrix marchica* Lemm. In: *Perspectives in Phycology*, (Ed.) V.N. Rajarao. Today and Tomorrow's Printers and Publishers, New Delhi, pp. 453–459.

Adhikary, S.P., Bastia, A.K. and Tripathy, P.K., 1992. Growth response of the nitrogen fixing cyanobacterium *Westiellopsis prolifica* Janet to fertilizer factory effluents. *Bull. Environ. Contam. Toxicol.*, 49: 137–144.

APHA (American Public Health Association), 1971. *Standard Methods for Examination of Water and Wastewater*, 13th Edition. APHA, NY, USA.

Bastia, A.K., 1994. Heterotrophic potentialities of nitrogen-fixing blue-green algae (cyanobacteria) of local rice-fields. *Ph.D Thesis*, Utkal University, Orissa, India.

Bastia, A.K., Mohapatra, P.M. and Adhikary, S.P., 2007. Management of flyash using cyanobacteria. In: *Pollution Management*, (Eds.) P.C. Trivedi and S. Nehra. Pointer Publishers, Jaipur, Rajsthan, India, pp.159–169.

Palmer, C.M., 1969. A composite rating of algae tolerating organic pollution. *J. Phycol.*, 5: 78–82.

Palmer, C.M., 1980. *Algae and Water Pollution*. Castle House Publications, England, pp. 31–78.

Rai, L.C. and Kumar, H.D., 1976. Algal growth as a means of evaluation of nutrient status of the effluents of a fertilizer factory near Sahaupuri, Varanasi. *Tropical Ecol.*, 17: 50–56.

Rai, L.C., Kumar, H.D., Mohn, F.H. and Soeder, C.J., 2000. Services of algae to the environment. *J. Microbiol. Biotech.*, 10: 119–136.

Rana, B.C., Gopal, T. and Kumar, H.D., 1971. Studies on the biological effect of industrial wastes on the growth of algae. *Indian J. Environ. Health*, 13: 138–143.

Singh, R.N., 1950. Reclamation of usar lands in India through blue-green algae. *Nature*, 165: 325–326.

Singh, R.N., 1961. *Role of Blue-green Algae in the Nitrogen Economy of Indian Agriculture*, ICAR, New Delhi, India.

Subramanian, G., 1991. Cyanobacterial immobilization for distillery effluent treatment. *M.Phil. Thesis*, Bharathidasan University, Tiruchirapalli, India.

Thangeswaran, A., 1991. Bioflocculant production by marine cyanobacteria, *M.Phil. Thesis*, Bharatidasan University, Tiruchirapalli, India.

Venkataraman, G.S., 1981. Blue-green algae for rice production: A manual for its promotion. *FAO Soil Bulletin*, 46: 1–102.

Whitton, B.A. and Potts, M., 2000. *The Ecology of Cyanobacteria*. Kluwer Academic Publisher, Dordrecht, The Netherlands.

Zajic, J.E. and Chiu, Y.S., 1970. Heterotrophic culture of algae. In: *Properties and Products of Algae*, (Ed.) J.E. Zajic. Plenum Press, New York and London, pp. 1–47.

Chapter 22

Health Hazards of Foundries and Forges

Anuj Kumar Shrivastava

Department of Applied Sciences and Humanities,
National Institute of Foundry and Forge Technology (NIFFT),
Hatia, Ranchi – 834 003, Jharkhand

ABSTRACT

A huge amount of dust, SPM, RSPM and gaseous pollutants is released into the environment both workplace and ambient environment from various foundries forges and Iron and Steel industries and other manufacturing industries. These pollutants cause health hazards. Many health hazards are reported in the foundry and forge shops workers of the industries.

Keywords: *Foundries, Forges, Health hazards.*

Introduction

In Iron and Steel and other manufacturing industries, foundries and forges produce a lots of pollutants in the environment–both working and ambient environment. In there processes, metals are extracted and produced form ores by various metallurgical processes and processes for moulding, melting and castings etc. are accompanied by evolution of heat, noise, dust fines, fly-ash, oxides of Nitrogen, Sulphur and metals. Particulate matters are generated in large quantities when preparing mould core sands and moulds melting metals, pouring metal, knocking out poured moulds and loading and unloading raw materials. Here, metals are given a specific shape by metals castings for various engineering purposes.

Gaseous matters like gases, vapours, fumes and smoke are produced during melting and pouring operations. The major pollutants emitted from various work areas in Foundry *i.e.* Pattern shop, Sand preparation, moulding and core making mould drying and ladle heating. Cupola, electric arc furnace, pouring and mould cooling, knock out, Fettling, Heat treatment etc. In addition, various air pollutants and noise pollutants (Davis, 2002) are produced from Forge shops and other manufacturing industrial units. Pollutants are also emitted in sintering, pelletization, rolling mills, coke-over plants, refractories etc. in steel making and by products manufacturing. These pollutants cause various health hazards in living beings.

General Health Hazards of Foundries and Forges

Many people are exposed to common air pollutants in their occupations *e.g.* smoke, dust, SPM, RSPM, carbon mono-oxide, sulphur dioxide, oxides of nitrogen (No_x), hydrocarbons, and heavy metals like Pb, Cd, Cr, As, Ni etc. Their prolonged exposure causes various health hazards. Heavy metals cause acute and chronic poisoning. Some disastrous episodes have focused attention upon air pollution as a health hazard.

Some fugitive gaseous and dust emissions may cause primary occupational health problems to the workers engaged in the foundries, gaseous and dust emissions from the foundries pose potential health risk to the populations residing in the surrounding areas. Black lung, metal fume fever, silicosis, pneumoconiosis etc. are all occupational maladies which are attributed to inhalation of one or other type of fine dust particles (Khoshoo, 1984).

Health Hazards (Effects)

The effects of air pollution on human health generally occur as a result of contact between the pollutants and the body. These hazards are:

1. Eye irritation
2. Headache
3. Nose and throat irritation
4. Irritability of respiratory tract
5. Gases like hydrogen sulphide, ammonic and mercaptans cause odour nuisance even at low concentrations.
6. Increase in mortality rate and morbidity rate.
7. A variety of particulates particularly pollens initiate aesthetic attacks.
8. Chronic pulmonary diseases like Bronchitis and asthma, are aggravated by a high concentration of SO_2, NO_2, particulate matter and photochemical smog.
9. Carbon monoxide combines with the hemoglobin in the blood and consequently increases stress on those suffering from cardio-vascular and pulmonary diseases.
10. Dust particles cause respiratory disease. Diseases like silicosis, asbestosis etc. result form specific dust.
11. Carcinogenic agents like PAH's, Cr(VI), Cd etc. cause cancer.
12. Hydrogen fluoride causes diseases of bone (fluorosis) and mottling of teeth.
13. Certain heavy metals like lead, cadmium, mercury, chromium etc. enter into body by inhalation, skin absorption and through food chain. They cause acute and chronic poisoning.

Specific Health Hazards of Foundries and Forges

Apart from the general health hazards described above, some specific health hazards are prevalent amongst workers in foundries and forges of Iron and Steel industries and other manufacturing industries.

Dust, SPM, noise and gaseous pollutants pose a potential threat to health of workers in industries and populations residing in the surrounding areas. Dust also absorb gases and in such a combination prove to be a more serious health hazards due to synergism. It has recently been demonstrated that SO_2 absorbed on submicroscopic particles penetrate deep into the lung and this is a greater danger to health (Khoshoo, 1984).

Silicosis is found in Foundry workers. Most of the cases of silicosis, however, has arisen in the manufacture of silica containing materials. Pottery industry got such a bad reputation for silicosis. Grinders in the cutlery industry using sandstone wheels died in large numbers from silicosis. Pneumoconiosis and black lung diseases are caused due to coal dust. Asbestosis is common in asbestos workers in industries.

Metals

Heavy metals pollution either directly or indirectly entering the food chain are becoming an increasing threat to health. Mercury metal is toxic for living beings. It enters into the body through either food chain or by skin or inhalation, which causes mercury poisoning. The symptoms of mercury poisoning are tremor, gingivitis and loss of teeth, together with a condition called "erethism", an individual shyness (Murrary, 1988).

Organic mercurials are used as bactericides and fungicides and are used in the paint, paper and alkali industries. Methyl mercury is more poisonous than mercury itself, which is used as a catalyst in a factory in Minamata. Tetra ethyl lead (TEL) used as antiknock agent in gasoline and TML cause potential health hazard. The harmful effects include blocking of spindle fibre mechanism in cell division, uncoupling oxidative phosphorylation and altered rapid eye movement phase of sleep leading to insomia. Other metals with toxic effects, known to occur as pollutants in some localized areas are cadmium, Be, Pb, Cu, Zn, Cd, Fe, Ni etc.

Vanadium levels in the environment are rising as a consequence of the buring of vanadium containing fossil fuels and of mining and processing in order to meet growing needs for the metal in industries. Both acute and chronic effects of occupational exposure to V compounds are manifested in the respiratory tract by irritation, including bronchitis and pneumonia.

Some studies on health hazards of dust, fume, gaseous pollutants, noise heavy metals etc. from foundry, forge of Iron and Steel and other manufacturing industries have been reported *e.g.* Risk of bladder cancer in foundry workers (Gaertner *et al.*, 2002).

In this epidemiological studies, a meta analysis of foundry workers was done. Summary risk estimates (SRE) were calculated form 40 systematically extracted results. Weakly increased risk were overall with an SRE of 1.11. Exposure response finding showed significantly increase risk of about 1.6 to 1.7 after 20 or more years of employment. Occupation specifies SREs showed a 40–50 per cent increased risk among moulders, caster and unskilled foundry labourers. There was limited evidence that bladder cancer risk correlated with lung cancer risk, which is a more established risk among foundry workers.

Mortality from lung cancer disease in steel industries workers had been reported due to chromium and PAH (NIFFT, Internet website).

Silicosis and cardio-vascular diseases have been reported. Public health assessment in forges and foundries was done. High burn hazards were seen in longitudinal wear study of four work shirts in ferrous metals in forge and foundry workers. Some health hazards are seen in forges and foundries of industries from emissions of their propane burner. Sometime explosion hazards are seen in propane burner. Noise from forges and foundries cause loss of hearing and disorder in central nervous system.

Conclusion

Various health hazards are observed in persons who are regularly exposed to different kinds of pollutants *e.g.* dust, fume, gaseous pollutions, noise in forges and foundries of Iron and Steel and other manufacturing industries. Industrial safety officers/ environ managers tackle environmental problems of the industries. In this way, environmental, health and safety management playa strategic role in organizational governance, environmental risk, and compliance.

Acknowledgement

The author is thankful to the Director, NIFFT and HOD, DASH, NIFFT, Hatia, Ranchi for providing necessary facilities.

References

Davies, B., 2002. Reducing noise and from forges and foundries: The handbook of black country forging and foundry project, p. 1–2.

Gaertner, R.R.W. and Theriault, G.P., 2002. Risk of bladder cancer in foundry workers: A meta analysis. *J. Occupational and Environmental Medicine*, 59: 655–663.

Khoshoo, T.N., 1984. *Integrated Approach to Foundry Pollution Abatement*. Annual Convention of the Institute of Indian Foundryman, New Delhi, 19th March.

Murray, R., 1988, Health hazards of mining. *J. Environmental Geochemistry and Health*, 10(3/4): 71–73.

Website, www.niffi.com.

Chapter 23

Phenology and Changes in Population Size of Some Poorly Perpetuated Plants in the Sholas of Nilgiris, the Western Ghats

S. Padmavathy[1], S. Paulsamy[1]* and P. Senthilkumar[2]

[1]Department of Botany, Nirmala College, Coimbatore – 641 018
[2]Department of Botany, Kongunadu Arts and Science College, Coimbatore – 641 029

ABSTRACT

Observations on phenophases and population size for six poorly perpetuated economically important plant species such as *Acmella calva, Cayratia pedata, Clinopodium umbrosum, Cynoglossum zeylanicum, Fragaria* vesca and *Lycianthes bigeminita* were made in four sholas of Nilgiris namely, Thiashola, Korakundah, Kammand and Kolacombai for a period of one year from April 2003 to March 2004 at monthly intervals. All the six species exhibited peak of bud formation between April and June and bud break in July months. Most of the leaves were produced in single flush. The leaf expansion continued upto August in *Clinopodium umbrosum* and Cynoglossum *zeylanicum.* The flowering phenophase was observed from July to September but in *Clinopodium umbrosum*, it was extended upto October. The active period of fruit formation occurred during September to November; seed maturation in November and December and seed dispersal during the months between February and March. The study of population dynamics showed that there was a net decrease in population of the species, *Cayratia pedata* and *Fragaria vesca* over a period of one year at Thiashola where they were present. The other species, *Clinopodium umbrosum, Cynoglossum zeylanicum*

---

* Corresponding Author; E-mail: paulsami@yahoo.com.

and *Lycianthes bigeminata* maintained their population at the same level during the study period of one year. On the other hand, the species, *Acmella calva* has slightly lost its population size in Kammand shola where as, it maintained the character in Kolacombai shola without any major changes.

Keywords: *Phenology, Population dynamics, Nilgiris, Sholas.*

Introduction

Plants respond biologically to various parameters of the holocoenotic environment (Sundriyal *et al.*, 1987). Apart from the many extrinsic factors, the survival and establishment of a species in a community are also depending upon the time of phenophases particularly, the seed dispersal time. The plant-animal interactions in the community is based on the knowledge of seasonal production of plant parts. Considering these facts, it is known that the phenological studies are important for the conservation of genetic resources and forest management as well as for a better understanding of ecological adaptations of plant species and community level interactions. Further, the other important attribute, which decides the establishment of a species. The population size is regulated by an array of environmental factors. Hence, the present study was aimed at to observe the time of phenophases and to determine the influence of environmental variables on the population density of six weakly perpetuated economic plants of some shola understories in Nilgiris, the Western Ghats.

Materials and Methods

Study Area

Four shola forests namely Thiashola, Korakundah, Kammand and Kolacombai located between the altitudes 1600 and 2300m above msl in Nilgiris were selected for the present study. The climatic conditions and the values of soil moisture for the four study sholas are given elsewhere (Paulsamy, 2006). The annual rainfall is ranging between 1500–2800 mm/year across the sholas and the temperature is varied from 4 to 29°C. The relative humidity is around 90 per cent in all the four sholas. The soil is sandy loam and acidic with high contents of nitrogen, phosphorus and potassium.

Methods

Detailed phenological records of the six poorly established economically important plants such as *Acmella calva, Cayratia pedata, Clinopodium umbrosum, Cynoglossum zeylanicum, Fragaria vesca* and *Lycianthes bigeminata* were prepared from April, 2003 to March, 2004 at monthly intervals. However, during high activity periods, observations were made more frequently. Results were recorded on each sampling date by marking 20 randomly selected individuals in the study sholas. Since all the 4 study sholas are located at more or less similar altitude and geographical position with uniform macroclimatic conditions, observations for phonological behaviours were made at one shola only where the respective species were present. When a phenophase was noticed in about 10 per cent individuals, then it was considered to be initiated, and to be in peak when it occurred in more than 80 per cent individuals. The phenograms were drawn according to phenophases occurred in more than 80 per cent individuals as followed by Lodhiyal *et al.* (1998).

The study on population density was carried out for a period of 13 months from April, 2003 to April, 2004 and the results were expressed in 100m^2 scale. In the respective sholas of the species presence, five permanent quadrats of 1m^2 were delineated. In case of adults, each plant with a height of 1feet was considered as an individual. For the study of seedling population, seedling cohorts were marked in each of the five permanent quadrats in the April month of the years 2003 and 2004. Individuals arising from the rhizome and seeds were marked with point of different colours. The survival of adults and seedlings were recorded at monthly intervals.

Results and Discussion

Phenological observations such as bud formation, vegetative growth, flowering, fruiting, seed maturity and seed dispersal for the six poorly established economically important species present in the 4 study sholas are presented in Figure 23.1. Generally, for all the six species studied, the bud formation took place during the months between April and June. Oberbauer andBillings (1981) pointed out that more abundant and shallow roots in the upper soil layers contributed by the herbs favours the sudden appearance of buds immediately after adequate rain in thick forest. The bud break was started during July, and most of the leaves were produced in this single flush. The leaf expansion continued upto August for certain species like *Clinopodium umbrosum* and *Cynoglossum zeylanicum* in the study sholas. Similar pattern of observations on the bud formations and vegetative growth have been observed already for certain forest understorey species (Auld and Martin, 1975; Senthilkumar, 2000) The flowering period, after the vegetative growth was noted generally during the months of July through September for all the species. However, for *Clinopodium umbrosum*, the flowering phenophase was extended upto the month of October. The flowering response of plants are generally related with the temperature factors and it is species specific also (Holway and Ward, 1965). Generally, the period between September and November was noted to be the time of active period of fruit formation for all the six species and the other months November and December were being the period of seed maturation. The seed dispersal was found to be happened mostly during the months, February and March. The reproductive phases like seed maturation and seed dispersal are appeared with the onset of cold and dry season when the cessation of growth in plants are normally happened by this time (Parihar *et al.*, 1999)

The variation in population density of six poorly established species for the study period of 13 months from April 2003 to April 2004 is presented in Table 23.1. The density of *Acmella calva*, the species present in Kammand and Kolacombai sholas only varied between 0.02 (in Kolacombai during April, 2004) and 0.51/m^2 (in Kammand during Sep., 2003). The species, *Cayratia pedata* present only in Thiashola showed its density variation between 0.03 (April, 2004) and 0.21/m^2 (Sep. 2003). The two common species available in all the four sholas, *Clinopodium umbrosum* and *Cynoglossum zeylanicum* registered their densities between 0.02/m^2 at Thiashola (June, 2003) and 0.26/m^2 at Kammand (Sep. 2003), and 0.03/m^2 at Korakundah and Kolacombai (April, 2003) and 0.21/m^2 at Kammand (Nov. 2003) respectively. The species, *Fragaria vesca and Lycianthes bigeminata* present only in Thiashola recorded their densities from 0.02/m^2(April, 2004) to 0.19/m^2 (Oct. 2003) and 0.01 (June, 2003) and 0.04/m^2 (Oct. 2003) respectively. The overall lower population of all these six species in the shola forest is due to the collective influence of several factors like poor adaptation, narrow range of ecological amplitude and failure in interspecific competition (Tripathi and Dwivedi, 1978). A considerable period of dry condition after seed dispersal in the months of February through April perhaps also be a reason for the low population density for these species.

The positive correlations existed between the population size of the studied species and the environmental variables such as temperature, rainfall, relative humidity and soil wetness showed that all these factors have involved collectively and largely in the determination of population size (Table 23.2). Similar type of relationship has already been observed for certain weed species in the protected forests of Anaimalais, the Western Ghats by Senthilkumar (2000). In the similar macroclimatic conditions at Grass Hills, Anaimalais, Paulsamy *et al.* (2000) reported that the factor, relative humidity

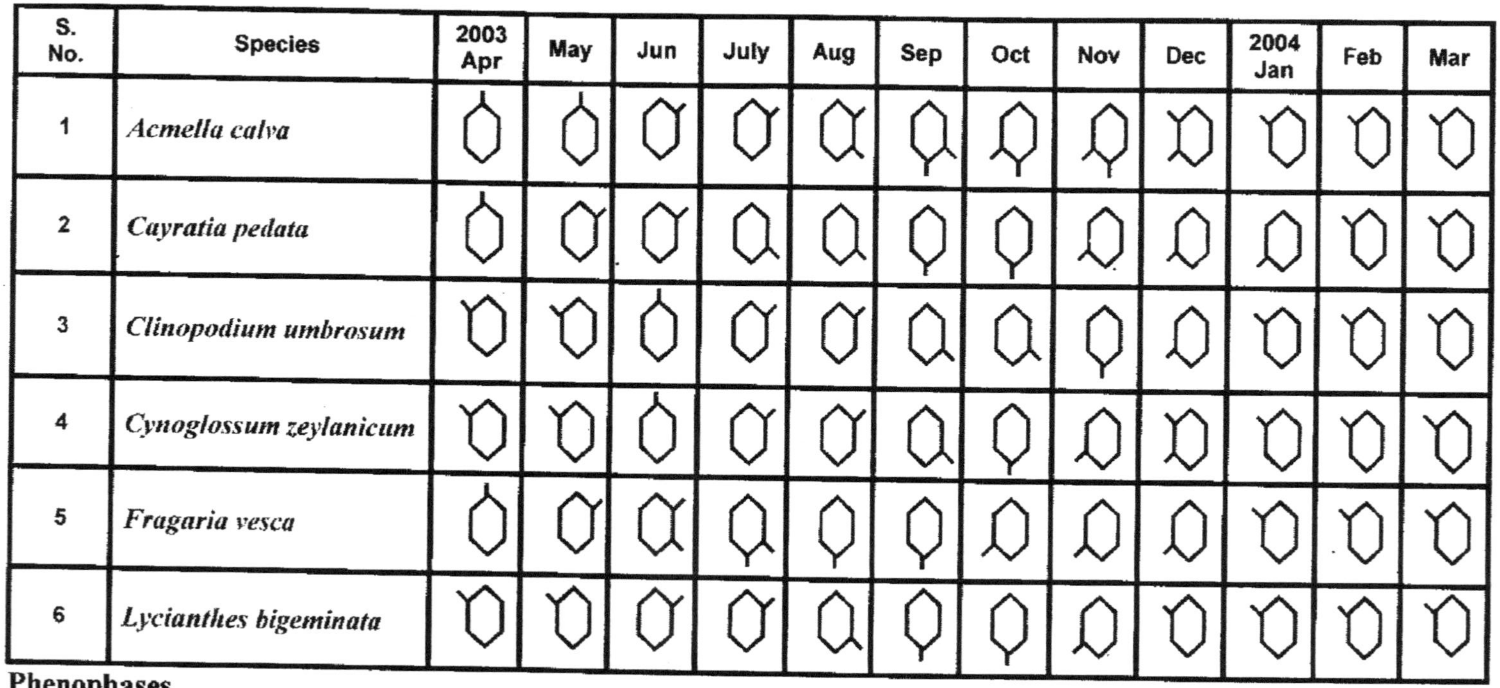

S. No.	Species	2003 Apr	May	Jun	July	Aug	Sep	Oct	Nov	Dec	2004 Jan	Feb	Mar
1	*Acmella calva*												
2	*Cayratia pedata*												
3	*Clinopodium umbrosum*												
4	*Cynoglossum zeylanicum*												
5	*Fragaria vesca*												
6	*Lycianthes bigeminata*												

Phenophases

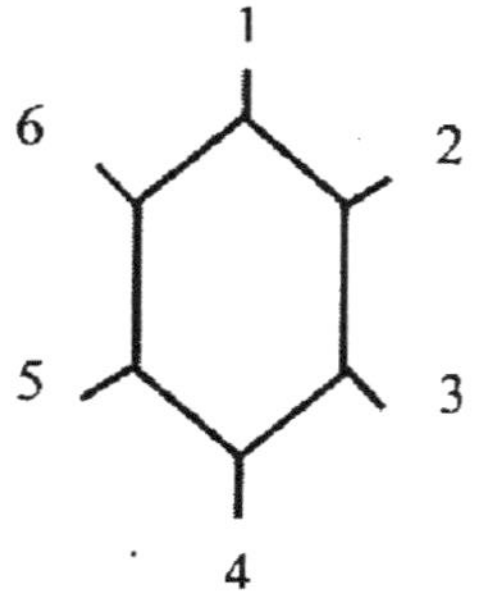

1. **Budding**
2. **Vegetative growth**
3. **Flowering**
4. **Fruiting**
5. **Seed maturity**
6. **Seed dispersal**

Figure 23.1: Phenograms of the Ecologically Weaker Plants with Economic Importance in the Lower Stratum of the Studied Sholas

Table 23.1: Population Density (individuals/m^2) of the Species, *Acmella calva*, *Cayratia pedata*, *Clinopodium umbrosum*, *Cynoglossum zeylanicum*, *Fragaria vesca* and *Lycianthes bigeminata* (± SD) in the Respective Sholas during April, 2003 to March, 2004

Year of Month	Shola												
	Thiashola					Korakundah		Kammand				Kolacombai	
	Ca.pe	*Cl.um*	*Cy.ze*	*Fr.ve*	*Ly.bi*	*Cl.um*	*Cy.ze*	*Ac.ca*	*Cl.um*	*Cy.ze*	*Ac.ca*	*Cl.um*	*Cy.ze*
2003													
Apr	0.05±0.00	–	–	0.03±0.00	–	–	–	0.04±0.00	–	–	0.03±0.00	–	–
May	0.10±0.01	–	–	0.06±0.00	–	–	–	0.06±0.00	–	–	0.06±0.00	–	–
Jun	0.12±0.01	0.02±0.00	0.06±0.00	0.08±0.00	0.01±0.00	0.06±0.00	0.03±0.00	0.08±0.00	0.07±0.00	0.04±0.00	0.06±0.00	0.04±0.00	0.03±0.00
Jul	0.15±0.01	0.07±0.00	0.08±0.01	0.11±0.01	0.02±0.00	0.08±0.00	0.08±0.01	0.26±0.02	0.15±0.01	0.07±0.00	0.08±0.01	0.08±0.00	0.05±0.00
Aug	0.17±0.01	0.13±0.01	0.12±0.01	0.15±0.01	0.03±0.00	0.11±0.01	0.14±0.01	0.38±0.03	0.16±0.01	0.09±0.01	0.15±0.01	0.13±0.01	0.08±0.01
Sep	0.21±0.02	0.16±0.01	0.15±0.01	0.14±0.01	0.03±0.00	0.20±0.02	0.17±0.02	0.51±0.04	0.26±0.02	0.18±0.02	0.22±0.02	0.16±0.01	0.11±0.01
Oct	0.18±0.01	0.19±0.02	0.12±0.01	0.19±0.02	0.04±0.00	0.18±0.02	0.16±0.01	0.36±0.04	0.23±0.02	0.15±0.01	0.19±0.02	0.14±0.01	0.13±0.01
Nov	0.19±0.02	0.12±0.01	0.15±0.01	0.11±0.01	0.03±0.00	0.15±0.01	0.12±0.01	0.26±0.02	0.15±0.01	0.21±0.02	0.14±0.01	0.11±0.01	0.11±0.01
Dec	0.12±0.01	0.09±0.01	0.08±0.00	0.09±0.01	0.02±0.00	0.09±0.01	0.08±0.00	0.16±0.01	0.11±0.01	0.12±0.01	0.12±0.01	0.10±0.01	0.10±0.01
2004													
Jan	0.08±0.00	0.07±0.00	0.06±0.00	0.08±0.00	0.02±0.00	0.08±0.00	0.06±0.00	0.13±0.01	0.09±0.01	0.07±0.00	0.08±0.00	0.07±0.00	0.06±0.00
Feb	0.06±0.00	0.05±0.00	–	0.04±0.00	–	–	–	0.08±0.00	–	–	0.06±0.00	–	–
Mar	0.04±0.00	–	–	0.03±0.00	–	–	–	0.05±0.00	–	–	0.03±0.00	–	–
Apr	0.03±0.00	–	–	0.02±0.00	–	–	–	0.04±0.00	–	–	0.02±0.00	–	–

Ca.pe: *Cayratia pedata*; Cl.um: *Clinopodium umbrosum*; Cy.ze: *Cynoglossum zeylanicum*; Ly.bi: *Lycianthes bigeminata*.

Table 23.2: Correlation Coefficient (r) Between Certain Environmental Variables and Population Size of Six Species of Poor Ecological Status in the Studied Sholas

Variables	*Shola*												
	Thiashola					*Korakundah*		*Kammand*				*Kolacombai*	
	Ca.pe	*Cl.um*	*Cy.ze*	*Fr.ve*	*Ly.bi*	*Cl.um*	*Cy.ze*	*Ac.ca*	*Cl.um*	*Cy.ze*	*Ac.ca*	*Cl.um*	*Cy.ze*
Temperature (°C)	0.037	0.237	0.186	0.014	0.118	0.216	0.263	–0.587	0.44	–0.117	–0.117	0.189	0.209
Rainfall (mm)	0.831	0.473	0.537	0.818	0.402	0.416	0.486	0.743	0.569	0.08	0.664	0.374	0.088
Relative humidity (%)	–0.134	0.193	0.217	–0.093	0.216	0.331	0.193	–0.112	0.037	0.353	0.155	0.608	0.445
Soil moisture (%)	0.78	0.276	0.293	0.733	0.093	–0.893	0.234	0.721	0.444	–0.011	0.747	0.586	0.191

Ca.pe: *Cayratia pedata*; Cl.um: *Clinopodium umbrosum*; Cy.ze: *Cynoglossum zeylanicum*; Ly.bi: *Lycianthes bigeminata*.

largely controlled the structure and function of *Cymbopogon flexuosus* dominated grassland community. From the study it is suggested that artificial seeding at the time of monsoon setting (June) may aid more appearance of individuals and hence the higher population size for all the six species, since the existing low population density of all these species is climate controlled.

References

Auld, B.A. and Martin, P.M., 1975. The autecology of *Eupatorium adenophorum* Spreng. in Australia. *Weed Res.*, 15: 27–31.

Holway, J.G. and Ward, R.T., 1965. Phenology and alpine plants in Northern colorado. *Ecology*, 46: 73–83.

Lodhiyal, L.S., Singh, S.P. and Lodhiyal, N., 1998. Phenology, population structure and dynamics of Ringal Bamboo (*Arundinaria falcata*) in Nainital Hill of Central Himalaya. *Trop. Ecol.*, 39: 109–115.

Oberbauer, S.F. and Billings, W.D., 1981. Drought tolerance and water use by plants along an alpe topographic gradient. *Oecologia*, (*Berl.*), 50: 325–331.

Parihar, S.S., Agarwal, P. and Vinodshankar, 1999. Seed production and seed germination in blue panic grass (*Panicum antidotale* Retz.). *Trop. Ecol.*, 40(1): 75–78.

Paulsamy, S., Rangarajan, T.N., Arumugasamy, K., Manian, S., Udaiyan, K., Sivakumar, R. and Senthilkumar, D., 2000. Effect of habitat variation on the structure, herbage production and oil yield of *Cymbopogon flexuosus* Stapf. domated grassland in Anaimalais, the Western Ghats. *J. Environ. Biol.*, 21: 85–94.

Paulsamy, S., 2006. *Annual Progress Report of the Project, 'Evaluation of Conservation Strategies for the Sustainable Utilization of Herbaceous Bioresources in the Sholas of Nilgiris, the Western Ghats'*, sponsored by Ministry of Environment and Forests, Govt. of India, New Delhi.

Senthilkumar, D., 2000. Ecological status of weeds in the understorey of three different habitats in Topslip of Anaimalais, the Western Ghats, India. *Ph.D. Thesis*, Bharathiar University, Coimbatore, India.

Sundriyal, R.C., Joshi, A.P. and Dhasimana, R., 1987. Phenology of high altitude plants at Tungnath in the Garhwal Himalaya. *Trop. Ecol.*, 28: 289–299.

Tripathi, R.S. and Dwivedi, R.P., 1978. On the dynamics and regulation of population with special reference to plants. In: *Glimpses of Ecology*, (Eds.) J.S. Singh and B. Gopal. International Scientific Publications, Jaipur, pp. 425–437.

Chapter 24

Effect of K_2O Levels, Sources and Time of Application on Curcumin Content of Turmeric cv. Salem

V.T. Yamgar, M.S. Shirke and S.M. Salunkhe

Agricultural Research Station, K. Digraj, Dist. Sangli – 416 305, M.S.

ABSTRACT

An experiment was conducted during the *kharif* season of three consequent years 1997–2000. The soil under experiment was medium black. The data revealed that K_2O both @ 150 and 175 kg/ha was found significantly superior over other doses. With same maximum curcumin content (4.50 per cent). The source *viz.*, muriate of potash and sulphate of potash has affected the curcumin content as it is significant higher (4.40 per cent) with sulphate or potash than muriate of potash (4.00 per cent). Curcumin content was found to be differed according to time of application of K_2O also as it was 4.30 per cent when applied in two splits (50 per cent at planting while 50 per cent at reathingup) and only 4.10 per cent if entire does as single basal dose. The curcumin content was also found varied due to interaction effect of different main and subtreatments.

Keywords: *Curcumin, K_2O, Turmeric, Salem, Potassium.*

Introduction

The curcumin content in turmeric is higher in the range of 2.5 to 5.5 per cent. There are varieties with curcumin as high as 10.9 per cent (George 1990). Curcumin has been recently used by the pharmaceutical industry as a stabilizer for certain formulations of drugs like nifedipine and Bangladesh has Started using it as dye for wool (Dhawan, 1993). The average curcumin content is 3.1 per cent (Manjunath *et al.*, 1989).

Potassium (K_2O) was though to be involved in transportation of carbohydrates to rhizomes (Murlidharan and Balakrishanan, 1972). Hence, an experiment was carried out to asses the effect of K_2O levels, its sources and time of application on curcumin content in produce under Sangli conditions.

Materials and Methods

The studies were carried out on the farm of Agricultural Research Station K. Digraj, Dist. Sangli with Variety 'Salem' for three years 1997–2000. The experiment was laid out in split plot design in vertisols with pH–8.2, EC–0.38, organic carbon–0.52 and available NPK as 218, 8.42 and 5.24 kg/ha respectively. There were 100, 125, 150 and 17:5 kg/ha, respectively. Four subtreatments constituting sources of potassium *i.e.* muriate of potash and sulphate of potash and the time of application *i.e.* a full dose once at the time of planting (T_1) while 50 per cent dose at planting and remaining 50 per cent at the time of earthingup.

The gross plot size was 6 × 3.75 m² while net plot size was 4.80 × 3 m². The healthy rhizomes were planted at the spacing of 37.5 × 30 cm. The crop was provided with organic manure FYM @ 25 tonnes/ha and inorganic fertilizers NPK @ 200 : 100 : 100 kg/ha as RDF respectively. The curcumin at harvest was extracted by 'Reflx' method of extraction in the laboratory. The pooled data after statistical analysis is presented in Table 24.1.

Table 24.1: Effect of Potash Levels, Sources and Time of Application on Curcumin Content (per cent) of Turmeric

Sl.No.	*Potash Levels*	*Curcumin Content (Per cent)*			*Pooled Mean*
		1997–98	*1998–99*	*1999–2000*	
I.	**Effect of Potash Levels**				
1.	K_1 (100 kg/ha)	4.85 (11.48)	3.40 (11.38)	3.37 (11.09)	3.39 (11.36)
2.	K_2 (125 kg/ha)	4.10 (11.57)	4.30 (11.97)	3.26 (10.94)	4.00 (11.58)
3.	K_3 (150 kg/ha)	4.10 (11.60)	5.0 (12.84)	4.20 (11.73)	4.50 (12.06)
4.	K_4 (175 kg/ha)	3.90 (11.40)	5.30 (13.26)	4.40 (12.03)	4.50 (12.23)
	SE±	0.049	0.159	0.235	0.214
	CD at 5%	0.159	0.481	0.713	0.650
	CV	7.404			
II.	**Effect of Sources**				
1.	MOP	3.78 (11.08)	4.11 (12.15)	3.80 (11.26)	4.00 (11.52)
2.	SOP	4.30 (11.99)	4.70 (12.57)	4.10 (11.64)	4.40 (12.09)
	SE±	0.034	0.112	0.166	0.152
	CD at 5%	0.106	0.340	N.S.	0.460
	CV%	7.404			

Contd...

Table 24.1–Contd...

Sl.No.	Potash Levels	Curcumin Content (Per cent)			Pooled Mean
		1997–98	1998–99	1999–2000	
III.	**Effect of Time of Application**				
1.	T_1 (at planting)	3.80 (11.22)	4.60 (12.36)	3.90 (11.34)	4.10 (11.64)
2.	T3 (½ at planting and ½ at earthingup)	4.20 (11.80)	4.60 (12.36)	4.0 (11.56)	4.30 (11.97)
	SE±	0.037	0.132	0.133	0.097
	CD at 5%	0.122	N.S.	N.S.	0.29
	CV%	7.404			

Results and Discussion

Curcumin content of turmeric applied with graded levels of additional K_2O was found varied. The maximum curcumin content (4.50 per cent) was extracted from turmeric produced due to K_2O @ 175 kg/ha which was higher amongst all other treatments (Table 24.1). These results are on the line of those reported by Sudha *et al.* (2001) who have tested higher levels of potassium @ 120 kg/ha than state recommendation of 60 kg/ha K_2O and obtained better yield of turmeric with 6.5 per cent curcumin. Jayraj (1990) has also reported the increased curcumin content with increased dose of K_2O and supported the results of present studies. However, in contrast with these results Rao *et al.* (1975) has reported the higher levels of NPK to decrease the curcumin content, however, later studies by Mohanbabu and Muthuswamy (1984) and Ahmedshah *et al.* (1988) showed that graded levels of K_2O significantly improved the curcumin content at harvest of turmeric. Maryunathgoud *et al.* (2002) have also observed the increase in curcumin content upto 3.65 per cent with increased levels of NPK.

The data from Table 24.1 also revealed that the sources of K_2O applied has the significant effect on curcumin content. Among the sources tried, sulphate of potash (SOP) recorded significantly higher curcumin content (4.40 per cent) over the muriate of potash (4.0 per cent). The curcumin content was also found deviated due to differential time of application as it was significantly better (4.30 per cent) due to application of K_2O in split doses over basal dose (4.10 per cent).

Interaction Effects

The interaction effects were found Non-significant in all the treatments.

References

Ahmed Shah, H., Vedamuthu, P.G.B., Abdul Khader Md and Prakasan, V., 1988. Influence of different levels of potassium on yield and curcumin content of turmeric. In: *Proc. Nat. Seminar on Chillies, Ginger and Turmeric,* (Eds.) G. Satyanarayana, M. Sugunakar Reddy, M. Rama Rao, X.AI. Azam and R. Naidu, 11–12 January. Spices Board and Andhra Pradesh Agri. Univ., Hyderabad, pp. 109–113.

Dhawan, B.N., 1993. Turmeric: A gold mine. *Indian Spices,* 30(2 and 3): 19–20.

George, C.K., 1990. India the best source of spices. *Indian Spices,* 27(3): 7–11.

Jayraj, P., 1990. Effect of potassium in mitigating the effect of shade in intercrops. *M.Sc. (Agri.) Thesis*, Kerala Agric. Univ., Vellanikkara, Trichur.

Manjunath, M.N., Sattigeri, V.D. and Nagraja, K.V., 1989.Curcumin in turmeric. *Indian Spices*, 25(4) and 26(1): 9–10.

Mohanbabu, N. and Muthuswamy, S., 1984. Influence of potassium on the quality of turmeric. *South Indian Hort.*, 32: 343–346.

Murlidharan, A. and Balakrishanan, S., 1972. Studies on the performance of some varieties of turmeric (*Curcuma* spp.) and its fertilizer requirements. *Agric. Res. J. Kerala*, 10: 112–115.

Rao Rama, M. Reddy Ramakrishna, K. and Subharayadu, M., 1975. Promising turmeric types of Andhra Pradesh. *Indian Spices*, 12: 2–5.

Sudha, B., Meerabai, M. and Raj, K.A., 2001. More curcumin from turmeric. *Indian Spices*, 38(2): 16–17.

Chapter 25

Studies on Effect of Levels, Sources and Time of Applications of K_2O on Yield of Turmeric (*Curcuma longa*) cv. Salem

V.T. Yamgar, M.S. Shirke and S.M. Salunkhe

Agricultural Research Station, K. Digraj, Dist. Sangli – 416 305, M.S.

ABSTRACT

The field experiment was undertaken at Agricultural Research Station K. Digraj for assessment of effects of graded levels of potassium (K_2O) on the yield of, turmeric. Among the different doses of K_2O when applied @ 175 kg /ha registered significantly higher yields (84.93 qt/ha). It is followed by the dose of 150 kg/ha which remained at par with 175 kg/ha. The different sources of K_2O has also found to affect the yield as there was increased yield (83.95 qt/ha) with K_2O in the form of sulphate of potash (SOP) than in the form of muriate of potash (MOP) 79.99 qt/ha.

The yields were also found affected and varied according to time application of K_2O. Significantly higher yield (83.12 qt/ha) was obtained when K_2O is applied in split [50 per cent at planting and 50 per cent at the time of earthingup then that of only is basal dose (80.83 qt/ha)] for better production of turmeric under Sangli District.

Keywords: *Turmeric, Potassium, SOP, MOP, Curcuma longa.*

Introduction

Turmeric is a coloured, versatile natural product combining the properties of a spice and brilliant yellow dyestuff. India is the largest producer (3,790,00 MT) and exporter of cured turmeric (28,199 MT) earning a foreign exchange of Rs. 44.59 crores (Anonymous, 1995). The average productivity of the crop is however very low, 2, 108 kg/ha as against a potential yield of 14500 kg/ha (Anonymous, 1991). Lack of suitable agro-techniques and high yielding cultivars for particular agro-climatic conditions are reported to be important reasons, for low productivity. Among different agronomic practices, optimum level of NPK largely determine the productivity in a particular agro-climatic region.

Potassium alongwith 'N' and 'P' plays a major role in growth and yield as it is involved in assimilation, transport and storage tissue development (Tisdale *et al.*, 1985). The response of applied nutrients in turmeric is influenced by a number of factors such as soil type, variety, spacing cropping system, time of application of nutrients and their sources. Hence, the present investigations were taken up to find out optimum levels of potassium, its sources and time of application for better yields of turmeric.

Materials and Methods

The experiment was laid out in split-plot design in medium black soils at Agricultural Research Station K. Digraj, Dist. Sangli (M.S.). The initial status of soil was as, pH 8.2, EC–0.38. Organic carbon–0.52 and available average NPK to the extent of 218, 8.42 and 524 kg/ha respectively. The experiment was conducted during *kharif* seasons of three consequent years (1997–2000). There were four main treatments of potassium levels *viz.*, 100, 125, 150 and 175 kg/ha. The sub-treatments were muriate of potash and sulphate of potash as source and applied in split (50 per cent at planting and 50 per cent at the time of earthing up) and full dose only at planting.

The healthy rhizomes of variety 'Salem' were planted in two replications of each main and sub treatments. The plot size 6×3.75 m^2 for main treatments while 4.80×3 m^3 for sources of potassium. While 37.5×30 m^2 for time of application treatment was maintained. Each plot was provided with FYM @ 25 tonnes/ha. The required quantity of K_2O were calculated and applied as per the treatments. Other cultural practices were carried out as per the recommendations. The observations were recorded on yields at harvest and statistically analysed pooled data are presented in Table 25.1.

Results and Discussion

The data in Table 25.1 revealed that the yield of turmeric was found significantly influenced by K_2O levels, their sources and also the time of application.

Effect of K_2O Levels

Potash when applied @ 175 kg/ha had registered significantly higher yields (84.93 qt/ha) as compare to rest of the levels except @ 150 kg/ha (83.74 qt/ha) which was remained at par with dose of 175 kg/ha. The results are in confirmity with Rethinam *et al.* (1994) as they have reported that of all nutrients the uptake of potassium was highest and could be responded in increased yields. However Rathinavel (1983) has also described that application of 'K' significantly increases plants height, tiller number, number of leaves and rhizomes. In number, number of leaves and rhizomes. In support of results of present study they have also reported that highest levels of 'K' (240 kg/ha) recorded highest number of above parameters at all growth stages but increase were not significant beyond 180 kg/ha for all characters including yield. Behura (2001) supported the results of present study regarding increase in yield with increased 'K' level in mango of increased yield at higher level of 'K' are not on the line of Ahmed Shah *et al.* (1988) who reported that there was little response to incremental dose of 'K' up to 180 kg/ha in respect of rhizome characters.

Table 25.1: Effect of Potash Levels, Sources and Time of Application on Yield of Turmeric Cv. Salem

Sl.No.	Treatment	Turmeric Yield (q/ha)			Pooled Mean q/ha
		1997–98	1998–99	1999–2000	
I.	**Potash levels**				
1.	K_1 (100 kg/ha)	79.72	79.02	77.01	78.88
2.	K_2 (125 kg/ha)	80.34	83.12	75.97	80.41
3.	K_3 (150 kg/ha)	80.55	89.16	81.45	83.74
4.	K_4 (175 kglha)	79.16	92.08	83.54	84.93
	SE±	0.340	1.104	1.63	1.48
	CD at 5 per cent	1.100	3.34	4.95	4.51
II.	**Effect of sources**				
1.	MOP	76.59	84.37	78.19	79.99
2.	SOP	83.26	87.29	80.83	83.95
	SE±	0.24	0.77	1.15	1.05
	CD at 5 per cent	0.73	2.36	N.S.	3.19
	C.V. per cent	0.0195			
III.	**Effect of Time of Application**				
1.	T_1 (at planting)	77.91	85.83	78.74	80.83
2.	T_2 (90 days after planting)	81.94	85.83	88.27	83.12
	SE±	0.256	0.91	0.78	0.67
	CD at 5 per cent	0.777	N.S.	N.S.	2.01
	C.V. per cent	4.0195			

Manjunathgoud *et al.* (2002) reported the linear increase in fresh and cured rhizome yield with increased levels of NPK upto 150 : 125 : 250 kg/ha which accounted one and half time more than that of control. However, Raghava Rao and Swamy (1984) recorded better growth, fresh rhizome yield and cured yield of turmeric provided wth high NPK level of 312.5 + 112.5 + 200 kg/ha respectively under Andhra Pradesh condition as compare to control as found in present studies. Turmeric being an exhaustive crop responded well to higher levels of NPK which is reflected by the increased uptake of NPK and consequently increased yield Venkatesha *et al.* (1998).

Effect of Sources of K_2O

The data revealed that when K_2O was applied in the form of sulphate of potash (SOP) the yields were increased significantly (83.95 qt/ha) as compare to muriate of potash (MOP) (79.99 qt/ha). Though the yield obtained was more in case of sulphate potash it will compensate the prices of these two forms as the prices of sulphate of potash were much more than muriate of potash. Hence, the source of potash to be supplied should not consider strictly.

Effect of Time of Application of K_2O

Yield of turmeric on weigh basis was found to increased significantly when it is supplied in split doses as 50 per cent at the time of planting and remaining 50 per cent while earthing up. The yield obtained by split application was 83.12 qt/ha while was only 80.83 qt/ha as it is applied as a basal dose only at planting. The results are not in agreement with Jayraj (1990) as he reported that turmeric performed will under 25 per cent shade with only 90 kg/ha K_2O of which entire quantity was applied as single basal dose at planting.

Interaction Effect

Interaction effect of levels of K_2O, Sources of K_2O and time of application of K_2O was found non-significant.

References

Ahmed Shah, H., Vedamuthu, P.G.B., Abdul Khader Md. and Prakasam, U., 1988. Influence of different levels of potassium on yield and cuxumin content of turmeric. In: *Proc. Nat. Seminar on Chillies,* Ginger, (Eds.) Reddy, M. Rama Rao, K.M. Azam and R. Naidu, 11–12 January. Spices Board and Andhra Pradesh, Agril. Univ. Hyderabad, pp. 109–133.

Anonymous, 1991. Survey of Indian Agriculture. *The Hindu,* Madras, pp. 35–57.

Anonymous, 1995. All time record in export of spice products. *Spice India,* (8): 19–20.

Behura, S., 2001. Effect of nitrogen and potassium of growth parameters and rhizomatic characters of mango ginger (*Curcuma amada*). *Indian J. Agronomy,* 46(4): 747–751.

Jayraj, P., 1990. Effect of potassium in mitigating the effect of shade in intercrop. *M.Sc. (Agri.) Thesis,* Kerala Agril. Univ., Vellanikkara, Trichur.

Manjunath Goud, B., Venkatesh, J. and Bhagavantagouda, K.H., 2002. Studies on plant density and levels on growth, yield and quality of turmeric Cv. Bangalore local. *Mysore J. Agric. Sci.,* (36): 31–35.

Raghava Rao, D.V. and Swamy, G.S., 1984. Studies on the effect of N, P and K on growth yield and quality of turmeric. *South Indian Horti.,* 32: 288–291.

Rathinavel, M., 1983. Effect of potash application and time of harvest on growth, yield and quality of turmeric. *M.Sc. (Agri.) Thesis,* TNAU, Coimbtore.

Rethinam, P., Sivaraman, K. and Sushama, P.K., 1994. Nutrition of turmeric. In: *Advances in Hort., Vol. 9: Plantation and Spice Crops, Part–I,* (Eds.) K.L. Chadha and P. Rethinam. Malhotra Publ. House, New Delhi, India, pp. 477–488.

Tisdale, S.L., Nelson, W.L. and Beaton, J.D., 1985. *Soil Fertility and Fertilizers,* 4th edn. McMillan Pub. Co., New York.

Venkatesha, J., Khan, M.M. and Chandrappa, H., 1998. Studies on uptake of NPK nutrients by turmeric cultivars. *J. Maharashtra Agric. Univ.,* 23(1): 12–14.

Chapter 26

Integrated Management of Major Diseases of Betelvine

S.G. Magdum, M.S. Shirke and S.M. Salunkhe

All India Networking Project on Betelvine, Agricultural Research Station, K. Digraj, Dist. Sangli – 416 305

ABSTRACT

Field experiments were conducted for three years at Agricultural Research Station, K. Digraj, Dist. Sangli for management of two major, foot rot and leaf rot diseases of betelvine using sanitation of field, Bordeaux mixture and *Trichoderma harzianum*. The results showed that the minimum per cent disease incidence (PDI) of foot rot and leaf rot was found (6.84 and 1.85 respectively) due to treatment with sanitation + one soil application of 1 per cent Bordeaux mixture at pre-monsoon + one soil application of *Trichoderma harzianum* one month after. application of Bordeaux mixture and again one soil application of 1 per cent Bordeaux mixture two month after its first application (T_5). This treatment also resulted highest (58.93 lac leaves/ha) yield and C:B ratio (1:2.60). However, the treatment remained statistically at par with T_4, *i.e.* sanitation + one soil application of 1 per cent Bordeaux mixture + one soil application of *Trichoderma harzianum* one month after Bordeaux mixture application with PDI of 7.32 and 2.10 for foot rot and leaf rot respectively and also in respect of yield (55.10 lac leaves /ha) and C:B ratio of 1:2.40.

Keywords: *PDI, Foot rot, Leaf rot, Piper betel L. Bordeaux mixture, Phytophothora.*

Introduction

A number of diseases have been reported from betelvine growing areas of India. Amongst those foot rot and leaf rot caused by *Phytophothora* spp. reported to cause losses to the extent of 30-100 per cent (Saksena, 1977; Dasgupta and Sen, 1999). Lacking appropriate management practices, diseases continue to pose a serious threat to betelvine cultivation. Some approaches for cultural and chemical management have been investigated in pest (Maiti *et al.*, 1978; Dasgupta *et al.*, 1988; Dasgupta and Sen, 1999 and Mahanty *et al.*, 2000). A complicated situation is that the betel leaves being chewed directly without any preperative steps, precludes use of chemical pesticides against pests and diseases. Due to residual toxicity, chemical pesticides are restricted as those causes health hazards to human being (Nag *et al.*, 1993).

The usefulness of *Trichoderma* spp. As bio-agent in disease management of betelvine was reported (Tiwari and Mehrotra 1968; Vyas *et al.*, 1981). Antagonists against major betelvine diseases have been screened (D'souza *et al.*, 2001). The relative efficacy of fungicides and *Trichoderma* spp. in management of foot rot has been reported (Mahanty *et al.*, 2000). Hence an attempts were made to integrate the bioagent, *Trichoderma harzianum* and Bordeaux mixture along with field sanitation for management of the disease under open field condition.

Materials and Methods

The experiment was conducted under open field condition in complete randomized block design. The treatments were replicated four times for the variety 'Kapoori.' Field was laid in plot size of 3×1.5 m^2 with 40 cm spacing and started before onset of the monsoon. Before starting the experiment all the infested vines were removed and destroyed by burning. All the recommended agronomic practices were carried in time. The treatments were:

Tr_1: Sanitation

Tr_2: Sanitation + one soil application of bordeaux mixture 1 per cent at monthly interval from pre-monsoon

Tr_3: Sanitation + one soil application of bordeaux mixture 1 per cent after onset of monsoon.

Tr_4: Sanitation + one soil application of bordeaux mixture 1 per cent + *Trichoderma harzianum* @ 5gm/lit water for drenching one month after application of bordeaux mixture.

Tr_5: Sanitation + one soil application of bordeaux mixture 1 per cent at pre-monsoon + soil application of *T. varidae* one month after bordeaux mixture application + one soil application at bordeaux mixture 1 per cent two month after first application of bordeaux mixture.

Tr_6: : Untreated (control).

The mortality of vines was recorded 30 days after application of each treatment. The per cent disease incidence (PDI) of foot rot and leaf rot was calculated. The leaf yield on number basis was recorded at each picking and by comparing market price of healthy and infested leaves the C:B ratio was worked out. The statistically analysed pooled data are presented in Table 26.1.

Results and Discussion

The pooled results indicated that the minimum PDI of foot rot and leaf rot (6.84 and 1.85 respectively) was found from plots treated with T_5 (Table 26.1) which remained at par with that of T_4, (7.32 and 2.10 respectively). However, in general all treatments were found to control the diseases effectively over untreated control. The treatment with *Trichoderma* (T_5) resulted in highest yield (58.92 lac leaves/ha) with maximum C:B ratio of 1:2.60 and also remained at par with (T_4) (55.10 lac leaves/ha and 1:2.40 respectively). The similar results were obtained by Dasgupta *et al.* (2000) who reported the *Trichoderma* as an effective antagonist against foot rot pathogen in betelvine and supported these results. The results due to T_5 may be attributed to reduction of disease incidence as experienced by

Mahanty *et al.* (2000) who reported that drenching of bordeaux mixture at monthly interval is comparatively more efficacious in terms of PDI than bioagent. However they have reported the *Trichoderma harzianum* as much more safer and productive of equivalent yield on number and weight basis.

Table 26.1: Integrated Disease Management in Betelvine (Pooled)

Treatment	*Mean PDI of Foot Rot*	*Mean PDI of Leaf Rot*	*Leaf Yield/h(lac)*	*Wt. of 100 Leaves (gm)*	*C:B Ratio*
T_1	10.46 (18.88)	3.01 (10.06)*	46.46	222.40	1:2.12
T_2	11.12 (19.46)	3.67 (10.90)	46.66	215.00	1:1.75
T_3	9.02 (17.15)	3.83 (1 1.12)	46.30	208.56	1:2.11
T_4	7.32 (15.73)	2.10 (8.32)	55.10	217.96	1:2.40
T_5	6.84 (14.88)	1.85 (7.70)	58.93	212.63	1:2.60
T_6	15.56 (23.12)	5.52 (13.50)	37.82	184.4	1: 1.44
SE±	1.42	0.91	3.04	10.25	–
CD at 5 per cent	4.31	2.75	9.19	N.S	–

*: Figures in parenthesis are arcsin values.

The importance of sanitation in betelvine realized by us is also an experience of Dastur (1935) and McRae (1934). The results of effectiveness of *Pseudomonas fluorescens* can not be discussed in case of betelvine due to unavailability of results by previous workers. The results thus concluded that the integration of bioagent and fungicide with specific time interval for soil application is the way for management of foot and leaf rot in betelvine.

References

Dastur, J.F., 1935. Diseases of pan (*Piper betel* L.) in the central provinces. In: *Proc. Indian Acad. Sci.*, B(1):778–815.

Dasgupta, B., Sengupta, K. and Karmarkar, S., 1988. *Indian Agric.*, 32: 99–105.

D'souza, A., Roy, J.K., Mahanty, B. and Dasgupta, B., 2001. Screening of *Ichoderma hterzianum* against major fungal pathogens of betelvine. *Indian Phytopathol.*, 54: 340–345.

Dasgupta, B. and Sen, C., 1999. Assessment of *Pltytopltthora* root rot of betelvine using chemicals. *J. Mycol. Plant Pathol.*, 29: 91–95.

Dasgupta, B., Roy, J.K. and Sen, C., 2000. Two major fungal diseases of betel vine. In: *Diseases of Plantation Crops, Spices, Betelvine and Mulberry*, (Ed.) M.K. Dasgupta, pp. 133–137.

McRae, W., 1934. Foot rot disease of (*Piper betel* L.) in Bengal. *Indian J. Agric. Sci.*, 4: 585–617.

Maiti, S., Khatua, D.C. and Sen, C., 1978a. Chemical control of two major diseases of betelvine. *Pesticides*, 12: 45–47.

Mahanty, B., Roy, J.K., Dasgupta, B. and Sen, C., 2000. Relative efficacy of promising fungicides and biocontrol agents *Trichoderma* in the management of foot rot of betelvine. *Journal of Plantation Crops*, 28(3): 179–184.

Nag, S.K., Chawdhary, A., Das, A.K. and Dasgupta, B., 1993. Potentials of chlorothalonil in management of major diseases of betelvine: Its persistence and residue levels on foliage. *Pestology*, 17: 37–43.

Saksena, S.B., 1977. *Phytophthora parasitica* the scourage of Pan (*Piper betel* L.). *Indian Phytopath.*, 30: 1–16.

Tiwari, D.P. and Mehrotra, R.S., 1968. Rhizosphere and rhizoplane studies of (*Piper betel* L.) with special reference to biological control of root rot disease. *Bull Indian Phytopath. Soc.*, 4: 79–89.

Vyas, K.M., Chile, S.K. and Vyas, R.S., 1981. Effect of culture filtrate of various fungi on sporulation of betelvine *Phytophtltora*. *Hindustan Antihio. Bull.*, 23: 25–26.

Chapter 27

A Comparative Study of the Exhaust Emissions of a Low Heat Rejection Engine with Two Different Levels of Insulation with Jatropha Oil

M.V.S. Murali Krishna[1]*, *S. Naga Sarada*[2],
G. Sudha Rani[3], *K. Kalyani Radha*[2] *and P.V.K. Murthy*[4]

[1]*Mechanical Engineering Department, Chaitanya Bharathi Institute of Technology, Gandipet, Hyderabad – 500 075*

[2]*Mechanical Engineering Department, J.N.T.U College of Engineering, Kukatpally, Hyderabad – 500 085*

[3]*Mechanical Engineering Department, Jaya Prakash Narayan College of Engineering, Dharmapur, Mahabubnagar – 509 001*

[4]*Gnayana Saraswathi College of Engineering and Technology, Dharmaran (B), Dichpally, Nizamabad – 503 230*

ABSTRACT

Crude jatropha oil, a non-edible vegetable oil shows a greater potential for replacing conventional diesel fuel quite effectively, as its properties are compatible to that of diesel fuel. But low volatility and high viscosity of jatropha oil call for low heat rejection diesel engine. Investigations have been carried out on low heat rejection engine with air gap insulated piston with superni (an

* Corresponding Author; E-mail: mvsmk_in@yahoo.com.

alloy of nickel) crown and air gap insulated liner with superni insert with varied air gap thickness at different injection pressures. Pollution levels decreased with insulated engine with higher air gap thickness with vegetable oil in comparison with conventional engine with pure diesel operation.

Introduction

The major pollutants emitted from diesel engine are smoke and oxides of nitrogen (NO_x). Excessive breathing of smoke causes (Sharma, 1996; Fulekar, 1999) tuberculosis and may also lead to death. It also causes detrimental effects on animal and plant life besides environmental (Khopkar, 1993) disorders. Inhaling of oxides of nitrogen causes dizziness, vomiting sensation, severe headache etc. Hence globally, stringent regulations are made for permissible pollutants in the exhaust of the engines.

It is well known fact that about 30 per cent of the energy supplied is lost through the coolant and the 30 per cent is wasted through friction and other losses, thus leaving only 30 per cent of energy utilization for useful purposes. In view of the above, the major thrust in engine research during the last two decades has been on development of low heat rejection engines. Several methods adopted for achieving low heat rejection to the coolant were using ceramic coatings (Krishnan *et al.*, 1980) on piston, liner and cylinder head and creating air gap in the piston (Rama Mohan, 1995) and other components with low-thermal conductivity materials like superni, mild steel etc. However, this method involved the complications of joining two different metals. A few researchers (Jabez Dhinagar, 1993) used different crown materials with different thickness of air gap in between the crown and the body of the piston. As the cetane number of the vegetable oils is closer to that of diesel fuel, attempts were made to induct vegetable oils in the forms of crude vegetable oil (Rehaman, 1995) emulsified form (Kiannejad *et al.*, 1993) and bio-diesel (Annapurna Devi, *et al.*, 2006) in the conventional engine. However, due to high viscosity of the vegetable oil, these vegetable oils were used in low heat rejection (Bhaskar *et al.*, 1993; Murali Krishna, 2005) diesel engines. However, no systematic investigations are reported so far about the use of crude vegetable oil in low heat rejection diesel engine with varied air gap thickness. Hence attempt is made in this direction and the pollution levels are reported and compared with pure diesel operation.

Experimental Program

The low heat rejection (LHR-1) diesel engine contains a two-part piston-the top crown made of low thermal conductivity material, superni-90 (an alloy of nickel) screwed to aluminum body of the piston, providing a 3 mm-air gap in between the crown and the body of the piston. A superni-90 insert is screwed to the top portion of the liner in such a manner that an air gap of 3-mm is maintained between the insert and the liner body. Investigations are also conducted on LHR-2 engine which consists of similar arrangement of piston and liner with 3.2-mm air gap thickness. At 500°C the thermal conductivity of superni-90 and air are 20.92 and 0.057 W/m-K respectively.

Figure 27.1 shows the experimental set up. The conventional engine is four-stroke, water cooled diesel engine with 3.68 k W at a rate speed of 1500 rpm having compression ratio of 16:1. The engine is connected to electric dynamometer for measuring brake power of the engine. Burette method is used for finding fuel consumption of the engine. Air-consumption of the engine is measured by air-box method. The naturally aspirated engine is provided with water-cooling system in which inlet temperature of water is maintained at 60°C by adjusting the water flow rate. The injection pressures are varied from 190 bars to 270 bars (in steps of 40 bars) using nozzle-testing device. The exhaust gas temperature (EGT) is measured with thermocouple made of iron and iron-constantan. Pollution levels

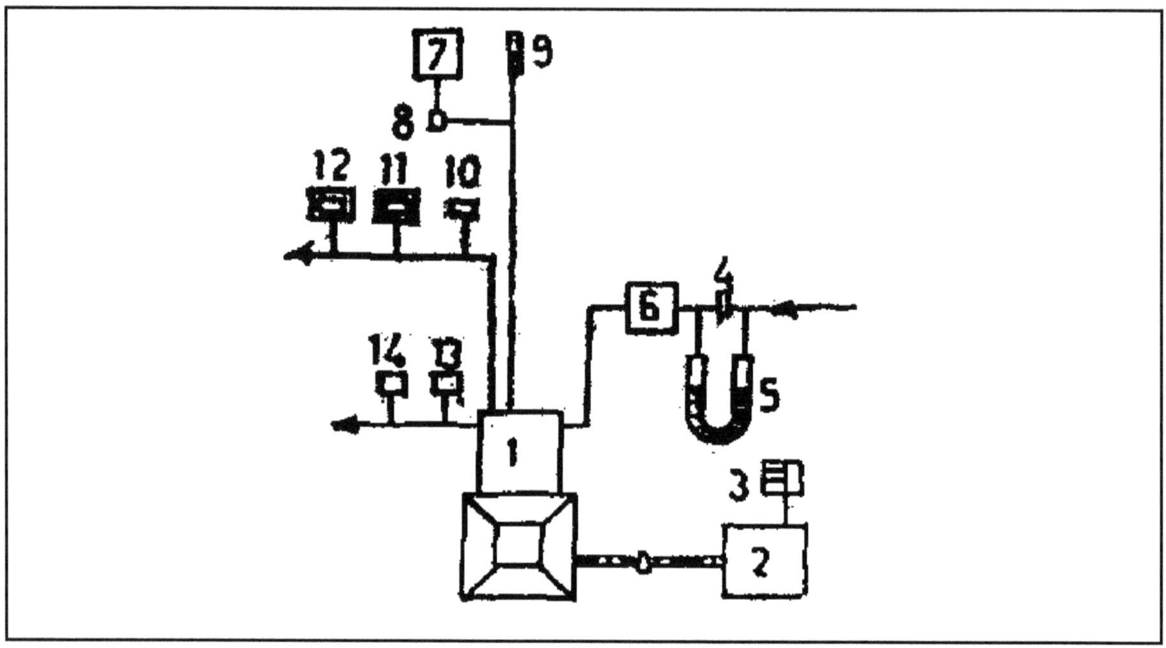

Figure 27.1: Experimental Set-up

1: Engine; 2: Electrical dynamometer; 3: Load box; 4: Orifice meter; 5: U-tube water manometer; 6: Air box; 7: Vegetable oil tank; 8: Three-way valve; 9: Burette. 10: Exhaust gas temperature indicator; 11: A VL Smoke meter; 12: Netel Chromatograph NO_x analyzer; 13: Outlet jacket water temperature indicator; and 14: Water flow meter;

of smoke and oxides of nitrogen (NO_x) are recorded by AVL smoke meter and Netel Chromatograph NOx analyzer respectively at the peak load operation of the engine. Jatropha oil substituted 100 per cent for diesel fuel is used as fuel in the investigations. The experiments are conducted on conventional engine (CE)–engine with air gap insulated piston and air gap insulated liner with 3 mm air gap thickness (LHR-I)–engine with air gap insulated piston, air gap insulated liner with 3.2 mm air gap thickness (LHR-2). Results are compared with pure diesel operation on conventional engine. The properties of crude vegetable oil are given in reference (Murali Krishna, 2005).

Results and Discussions

Smoke Levels

The data of variation of smoke levels at peak load with different versions of the engine at normal temperature of the vegetable oil and at different injection pressures is shown in Table 27.1. Drastic increase of smoke levels is observed at the peak load operation in both versions of the engine at different operating conditions of the vegetable oil, compared to pure diesel operation on conventional engine. This is due to the higher magnitude of the ratio of C/H, (C-Number of carbon atoms, H-Number of hydrogen atoms in the composition of the fuel) when compared with pure diesel. The increase of smoke levels is also due to decrease of air-fuel ratios and volumetric efficiency with vegetable oil operation compared with pure diesel operation.

Table 27.1: Data of Smoke Levels with Different Versions of the Engine at Peak Load Operation with Different Test Fuels

Engine Version	Smoke Levels (Hartridge Smoke Units, HSU) — Pure Diesel Operation, Injection Pressure (Bar)			Crude Oil Operation, Injection Pressure (Bar)		
	190	230	270	190	230	270
CE	43	38	34	68	63	58
LHR-1	55	50	45	63	58	53
LHR-2	68	63	58	58	53	48

CE: Conventional engine; LHR-I: Air gap insulated piston and air gap insulated line with 3 mm air gap; LHR-2: Air gap insulated piston, air gap insulated liner with 3.2 mm air gap.

Smoke levels are proportional to the density of the fuel. Since vegetable oils have higher density compared to diesel fuels, smoke levels are higher with vegetable oils. Due to higher molecular weight, crude vegetable oils have low volatility and because of their unsaturation, crude vegetable oils are inherently more reactive than diesel fuels, which results that they are more susceptible to oxidation and thermal polymerization reactions. However, LHR engines marginally decreased smoke levels due to efficient combustion and less amount of fuel accumulation on the hot combustion chamber walls of the LHR engines compared to the conventional engine. Increase of air gap thickness in LHR engines decreased smoke levels marginally due to decrease of ignition delay and combustion period. Smoke levels decreased with increase of injection pressure, in both versions of the engine, at normal temperature of the vegetable oil. This is due to improvement in the fuel spray characteristics at higher injection pressures and increase of air entrainment causing lower smoke levels.

Table 27.2: Data of NO_x Levels with Different Versions of the Engine at Peak Load Operation with Different Test Fuels

Engine Version	NOx Levels (ppm) — Pure Diesel Operation, Injection Pressure (Bar)			Crude Oil, Injection Pressure (Bar)		
	190	230	270	190	230	270
CE	850	890	930	700	720	730
LHR-1	1300	1280	1260	1245	1230	1180
LHR-2	1400	1350	1300	1300	1250	1200

CF: Conventional engine; LHR-1: Air gap insulated piston and air gap insulated liner; LHR-2: Air gap insulated piston, air gap insulated liner and ceramic coated cylinder head.

The data of variation of oxides of nitrogen (NO_x) levels at peak load with different versions of the engine at normal temperature of the vegetable oil and at different injection pressures is shown in Table 27.2. NO_x levels are lower in the conventional engine" while they are higher in the LHR engines with crude vegetable oil at the peak load when in comparison with pure diesel operation. This is due to lower heat release rate because of high duration of combustion causing lower gas temperatures with the vegetable oil operation on conventional engine, which reduced NO_x levels. Increase of combustion temperatures with the faster combustion and improved heat release rates in the LHR engines cause higher NO_x levels. LHR-2 engine increased NO_x levels in comparison with LHR-I engine due to provision of higher degree of insulation.

Increased injection pressures decreased NO_x levels in LHR engines while they increased in conventional engine due to decrease of gas temperatures in LHR engines and increase of the same in the conventional engine.

Conclusions

Increase of air gap thickness decreased smoke levels and marginally increased NOx levels in LHR engine with vegetable oil operation. Increase of injection pressure decreased pollution levels in LHR engine. However, increase of air gap thickness increased pollution levels with pure diesel operation in LHR engine.

Acknowledgements

The authors are thankful to the authorities of Chaitanya Bharathi Institute of Technology, Hyderabad for providing facilities to carryout this work. The financial assistance provided by the All India Council for Technical Education, New Delhi is also gratefully acknowledged.

References

Annapurna Devi, N. *et al.*, 2006. Production and evaluation of bio-diesel from sunflower and nigerseed oil with performance studies on a diesel engine. *International Congress on Renewable Energy*, pp. 478–484.

Bhaskar, T. *et al.*, 1993. The effect of two ignition improving additives on the performance of oil in low heat rejection diesel engine. In: 4th *International Conference on Small Engines and their Fuels*, Thailand, pp. 14–19.

Fulekar, M.H., 1999. Chemical pollution: A threat to human life. *Indian J. Environ. Prot.*, 1(3): 353–359

Jabez Dhinagar, S. *et al.*, 1993. A comparative study of the performance of a low heat rejection engine with four different levels of insulation. In: *International Conference on Small Engines and Fuels*, Chang Mai, Thailand, pp. 21–126.

Khopkar, S.M., 1993. *Environmental Pollution Analysis*. New Age International (P) Ltd., Publishers, New Delhi.

Kiannejad, F. *et al.*, 1993. Performance and emissions of a 1.5 litre single cylinder diesel engine with low cetane number vegetable oil fuel and emulsification with water. In: 4th *International Conference on Small Engines and Fuels*. Chang Mai, Thailand, pp. 32–39.

Krishnan, D.B. *et al.*, 1980. Performance of an Al-Si graphite particle composite piston in a diesel engine. *Transactions of Wear*, 60(2): 205–215.

Murali Krishna, M.V.S., 2005. Investigations on low heat rejection diesel engine with alternate fuels. *Ph.D Thesis*, J.N.T. University, Hyderabad.

Rama Mohan, K., 1995. Performance evaluation of an air gap insulated piston engine. *Ph.D. Thesis,* Kakatiya University.

Rehman, A. and Singhai, K.C., 1995. Vegetable oils as alternate fuels for diesel engine. In: 5th *Asian-Pacific International Symposium on Combustion and Energy Utilization,* Hong Kon, pp. 924–928.

Sharma, B.K., 1996. *Engineering Chemistry.* Pragathi Prakashan (P) Ltd., Meerut.

Chapter 28

Larvicidal and Ovicidal Effect of a Neem Product on Aedine and Culex Mosquitoes

Kiran Joseph, R.S. Mohanraj*, B. Dhanakkodi and K. Sathya

Department of Zoology, Kongunadu Arts and Science College (Autonomous), Coimbatore – 641 029, Tamil Nadu, India

ABSTRACT

An *in vitro* study was carried out to understand the effect of a neem based insecticide on the larvae and egg hatchability of Aedine and Culex mosquitoes. The LC_{50}/24 hour values against the different instar stages indicated larvicidal effect of the formulation. Hatchability of mosquito eggs was considerably reduced when treated with the botanical insecticide.

Keywords: *Aedes aegypti, Culex quinquefasciatus, Neem product, Toxicity, Hatchability.*

Introduction

For 260 million years, the two-winged insect has been on the face of this earth. It has outlived the dinosaurs and faced all the rough weather from volcanoes to human onslaught and emerged as a true nemesis to the human race. No other insect may have caused us so much trouble as this tiny perpetrator whom we encounter everyday. This common day foe is none other than the Mosquito.

---

* Corresponding Author.

Aedes aegypti is the only known potential vector of dengue and urban yellow fever (Mekuria, 1991; Mazzarri and Georghiou, 1995). This species of mosquito was shown to be a competent laboratory vector of Chikungunya (CHIK) virus (Jupp and McIntosh, 1990). *Ae. aegypti* has also been noted to transmit filariasis and encephalitis (Dorland's illustrated Medical Dictionary, 1982). The *Culex quinquefasciatus,* a known vector of Bancroftian filariasis, is an important pest in all the tropical towns and cities in Asia (Bowers *et al.,* 1995).

Among the various control measures for mosquitoes, use of synthetic insecticides is still the most convenient and indispensable method all over the world. The extensive application of synthetic pesticides has however caused irrepairable hazards to the environment and humans. Moreover, insecticide resistance has also become the greatest problem in the control of many medically important mosquito species. *Ae. aegypti* in particular has developed resistance to a variety of organochloride, organophosphorus, carbamate and pyrethroid insecticides (Kumar *et al.,* 1991; Mekuria *et al.,* 1991; Mazzarri and Georghiou, 1995; Rawlins and Wan, 1995). Neem products are also relatively advantageous over synthetic insecticides in being eco-friendly and non-toxic to non-target organisms including humans. Purohit *et al.* (1989) demonstrated that plant products are easily biodegradable. It was also emphasized that they do not leave poisonous residues and insects do not develop resistance. Oroumchi and Lorra (1993) stated that neem products may be a source of inexpensive and environmentally sound pesticides.

Neem derivatives obtained from different parts of the plant were documented to produce diverse toxic and biological effects, on a variety of insect species (Kaur *et al.,* 2001; Ramarethinam *et al.,* 2002; Mohanraj and Dhanakkodi, 2004, 2005 a, 2005b, 2006, 2007 a, 2007b; Neloilya, 2007). However, investigations on the influence of neem derivatives on development aspects of mosquitoes are only a few that too fragmentary.

Reports on the toxicity of neem as mosquitocide (adulticide and larvicide) are available. Attri and Prasad (1980) observed the toxic effects of neem oil extractive on larvae of *Cx. fatigans.* Chavan and Nikam (1988) demonstrated that petroleum ether extract and alkanes from neem leaves were toxic to *Cx. pipiens fatigans.* Neem leaf extracts of polar and non-polar solvents (Deshmuhk and Renapurkar, 1992; Obomanu *et al.,* 2006) and neem oil (Das *et al.,* 1995) have been tested against larvae of *Culex* mosquito. The susceptibility of *Cx. quinquefasciatus* to acetone extract of neem seed coat (Sagar and Sehgal, 1997), NSK and neem gum extracts (Babu and Murugan, 1998) and Nimbex, Nimbecidine, Neem gold, Salotrap (Thamil Solai *et al.,* 1999) and that of *Ae. aegypti* to acetone extract fraction of the fruit of *Melia volkensii* and *Azadirachta indica* (Awala *et al.,* 1998; Umar *et al.,* 2007), resinous exudate from the tender leaves of neem (Babu *et al.,* 2000) and Econeem, a neem based insecticide (Mohanraj *et al.,* 2000) has been found. Larvicidal effect of neem products on the malarial vector, *Anopheles* have also been reported earlier (Aliero, 2003; Vatandoost and Vaziri, 2004; Okumu *et al.,* 2007). However, conducting more investigations on toxicity of different forms of neem to mosquitoes is therefore the need of the hour.

Hatchability of eggs is one of the biological features that ascertain the renewal of a species. The ovicidal effect produced by any means on egg hatching is likely to reduce the population of harmful insects. Neem derivatives were found to lessen the hatching of eggs of many species of insects (Ekesi, 2000; Sundar *et al.,* 2000, Kaur *et al.,* 2001). Although, ovicidal activities of various chemicals on mosquito eggs including *Ae. aegypti* were reported earlier (Judson, 1967; Miura *et al.,* 1976; Fournet *et al.,* 1993; Sahgal and Pillai, 1993), reports on the ovicidal effects of neem substances on mosquito eggs are meagre. It has been stated that no information is available regarding ovicidal activity of AZ and related products in mosquitoes Su and Mulla (1998a). This raises an ample chance to carry out investigations on this aspect.

Materials and Methods

Animals

Aedine and *Culex* eggs were collected, hatched, reared and maintained for many generations in the laboratory. The eggs and larvae obtained from this stock were used for different experiments.

Test Compound

A neem based insecticide, EC, containing 0.15 per cent w/w azadirachtin (1500 ppm).

Different concentrations of the neem formulation was prepared in parts per million (ppm) basis, based on the concentration of azadirachtin through serial dilution with unchlorinated filtered water.

Bioassay Test

Bioassay tests were carried out for testing the efficacy of the neem formulation *Ae. aegypti* and *Cx. quinquefasciatus* at different stages of development *viz.*, I, II, III and IV instars and pupae. Instructions of WHO (1960) as detailed by Pampana (1963) for conducting bioassay experiment with mosquito larvae were carefully followed.

Different concentrations of the test compounds were prepared using unchlorinated water. Clean plastic tubs of 500 ml capacity were used as test containers. Batches of 20 larvae were exposed to 200 ml of a particular concentration of test solution. The larvae were collected with an eye dropper placed onto filter paper strips and immediately transferred to test cup containing test solution according to Cilek *et al.* (1991).

Mortality rates of larvae were recorded after 24 hours. Five or more concentrations of a test compound giving between 0 and 100 per cent mortality for larvae at different instar stages were tested. Parallel controls were maintained. Two replicates were done at each concentration. In recording the percent mortality for each concentration, the moribund and dead larvae in both replicates were combined.

For computing LC_{50}, the data were subjected to Finney's method of probit analysis as detailed by Regupathy and Dhamu (1990).

Determination of Effect of Neem Formulation on Hatching of *Ae. aegypti*, *Cx. quinquefasciatus* Eggs

Twenty eggs of aedine were exposed to a particular concentration of the neem product tested. The hatchability was recorded until 96 hours from the initial time of the experiment. The time was fixed so, because it was demonstrated that the completion of embryogeny occurs within 4 days (Judson and Gojrati, 1967). In case of *Culex*, the egg rafts with known number of eggs were carefully exposed to different concentrations.

Hatching rate was calculated on the basis of non-hatchability of eggs according to Sahgal and Pillai (1993). To ensure non-hatchability, the eggs from any test container were collected after 96 hours. Unhatched and decapped eggs were separated and counted using dissection microscope. Five replications were conducted at each concentration of neem product. The data were statistically examined using student 't' test.

Results and Discussion

LC_{50}/24 hour values of the neem formulation to I instar larvae was 63.095 ppm (*Aedes*) 25.118 ppm (*Culex*) and this was found to gradually increase with the age of larvae. The LC_{50} values of the neem formulation to the II, III and IV instar stages were 199.526, 316.277, 501.187 ppm of (*Aedes*); 39.810, 79.432, 158.489 ppm (*Culex*) respectively. Pupae showed the highest resistance to the neem formulation as is evident from the relatively higher LC_{50} values 794.328 ppm (*Aedes*) 501.187 ppm (*Culex*) (Tables 28.1 and 28.2).

Table 28.1: LC_{50}/24 Hour Values of a Neem Formulation to the Pre-adult Stages (I, II, III, IV Instar and Pupae) of *Aedes aegypti*

Stages of Development (Instars)	*Number of Larva/ Trial*	*LC_{50}/ 24 Hour Log Dose*	*Confidence Interval*		*Variance*	*Chi-Square*	*Degree of Freedom*
			LL	*UL*			
I	20	1.8 (63.095 ppm)	1.664 (46.131 ppm)	1.936 (86.297 ppm)	0.00486	4.964	3(16.266)
II	20	2.3 (199.526ppm)	2.2362 (172.266 ppm)	2.3638 (231.100 ppm)	0.00106	1.915	3(16.266)
III	20	2.5 (316.277ppm)	2.4283 (268.101 ppm)	2.5717 (372.992 ppm)	0.00134	0.2741	3(16.266)
IV	20	2.7 (501.187 ppm)	2.6692 (466.874 ppm)	2.7308 (538.021 ppm)	0.000247	4.531	3(16.266)
Pupa	20	2.9 (794.328 ppm)	2.8562 (718.124 ppm)	2.9438 (878.617 ppm)	0.000501	7.8021	3(16.266)

Table 28.2: LC_{50}/24 Hour Values of a Neem Formulation to the Pre-adult Stages (I, II, III, IV Instar and Pupae) of *Culex quinquefasciatus*

Stages of Development (Instars)	*Number of Larva/ Trial*	*LC_{50}/ 24 Hour Log Dose*	*Confidence Interval*		*Variance*	*Chi-Square*	*Degree of Freedom*
			LL	*UL*			
I	20	1.4 (25.118 ppm)	1.3996 (25.095 ppm)	1.4003 (25.136 ppm)	0.0003749	4.032	3 (16.266)
II	20	1.6 (39.810 ppm)	1.5048 (31.974 ppm)	1.6952 (49.567 ppm)	0.002370	6.171	3 (16.266)
III	20	1.9 (79.432 ppm)	1.812 (64.868 ppm)	1.988 (97.274 ppm)	0.002022	3.871	3 (16.266)
IV	20	2.2 (158.489 ppm)	2.115 (129.270 ppm)	2.2885 (194.312 ppm)	0.002051	2.264	3 (16.266)
Pupa	20	2.7 (501.187 ppm)	2.6716 (469.461 ppm)	2.7284 (535.056 ppm)	0.0002122	0.091	3 (16.266)

Average number of eggs hatched in control medium was 18.4 out of 20 (92 per cent). This was reduced to the lowest level of 0 among eggs of *Aedes* and *Culex* placed in medium of 400 ppm, neem formulation (Tables 28.3 and 28.4).

Table 28.3: Alteration in the Hatchability of *Aedes aegypti* Eggs Exposed to Different Concentrations of a Neem Formulation and Control

Parameters	Control	Concentrations (ppm)				
		100	150	200	300	400
Number of eggs introduced	20	20	20	20	20	20
Mean number of eggs hatched*	18.4	14.0	9.6	5.2	3.0	0
S.D.	±0.894	±0.836	±0.836	±0.836	±0.707	0
Pe c ent hatchability	92	70#	48#	26#	15#	0
Percent reduction over control		23.91	47.82	71.73	83.69	100

*: Mean (±SD) of 5 replicates.

#: Significantly different from control; P < 0.001 per cent.

Table 28.4: Alteration in the Hatchability of *Culex quinquefasciatus* Eggs Exposed to Different Concentration of a Neem Formulation and Control

Parameters	Control	Concentrations (ppm)				
		100	150	200	300	400
Number of eggs introduced/trial	20	20	20	20	20	20
Mean number of eggs hatched*	18.4	15.8	11.0	7.7	3.4	0
SD	±0.894	±0.386	±0.942	±0.833	±0.547	
Percent hatchability	92	79#	55#	38#	17#	0
Percent reduction over control		14.13	40.21	58.15	81.52	100

*: Mean (±SD) of 5 replicates.

#: Significantly different from control; P < 0.001 per cent.

Attri and Prasad (1980) demonstrated that neem oil extractive acted as strong toxicant to larval mosquito. Zebitz (1984) recorded 78.2 ppm as LC_{50} value of aqueous neem seed kernel extracts to fourth instar larvae of *Ae. aegypti*. Neem oil was found to kill mosquito larvae in respect to the concentration (Rawat, 1994). Mortality of 50 per cent and more occurred in larval *Ae. aegypti* (Sinniah *et al.*, 1994) and *Cx. quinquefasciatus* (Amorose, 1995) exposed to relatively low concentrations (0.02 per cent as LC_{100}/24 hour as well as 0.99 and 1.2 ppm as LC_{50}/24 hour to 3rd and 4th stage larvae) of neem extract and neem oil. Acetone extract of neem seed coat at 40 ppm caused 100 per cent mortality of the first instar larvae of *Ae. aegypti* (Sagar and Sehgal, 1997). In a previous study on the toxicity of a neem formulation azadirachtin against larval stages of *Ae. aegypti* (Mohanraj *et al.*, 2000) the LC_{50}/24 hour values recorded against I, II, III, IV instar and pupal stages were 7.94, 12.365, 14.565, 14.642 and 17.38 ppm, respectively.

The difference in the values between the present and previous studies may be due to the difference in the composition of pesticides. The toxic effect may mainly be due to the active component azadirachtin present in neem. This compound was demonstrated to be the major active constituent of neem and was well known for its toxic, antifeedant, growth regulating effect on insects (Schmutterer, 1990). Xie *et al.* (1995) confirmed that azadirachtin was largely responsible for toxic actions of neem on insects. The species and formulation–dependent differences in the larvicidal activity of neem products was noticed (Mulla *et al.*, 1997 cit. Su and Mulla, 1999a). Death of mosquito larvae was probably due to the poisoning effect of neem ingredients (Friend, 1998) and deficiency of dissolved oxygen in the neem treated water (Aliero, 2003).

In the current study, a considerable increase in the LC_{50} values with the age of larvae was observed, *i.e.*, early instar larvae were more sensitive to the neem product than late instars. It was recognized that fourth stage larvae and pupae of mosquitoes were more tolerant to toxicants, than early instars (Jebanesan *et al.*, 1999). Low sensitivity with increase in age among different pre adult stages of *Ae.aegypti* exposed to various neem extracts as well as formulations (Mohanraj *et al.*, 2000). The decreased susceptibility may be attributed to increased size and weight of older larvae (Helson and Surgenoner, 1983).

Mohanraj *et al.* (2000) observed a considerable dose-based reduction in the hatchability of eggs treated with a neem formulation. One ppm of AZ found to produce hundred percent mortality in eggs (exposed for 4 hour) of *Cx. tarsalis* and *Cx. quinquefasciatus* (Su and Mulla, 1998a). Ovicidal activity of growth regulators/inhibitors (Miura *et al.*, 1976) and insecticides (Saghal and Pillai, 1993) in mosquitoes of different species was reported. Severe reduction in hatching of eggs from Lepidopteran and Hymenopteran insects on treatment with aqueous extracts of black pepper, neem seed and garlic bulb at concentrations 5, 10 and 15 per cent and also with their volatiles was recorded (Ekesi, 2000).

Boradent and Free (1984) based on a study on the ovicidal activity of insect growth regulators, Diflubenzuron and Bay SIR 8514 opined that the chemicals diffuse into eggs and affect the vital physiological and biochemical processes associated with embryonic development, thereby inhibiting eclosion of eggs. Penetration of ovicides through chorion was demonstrated earlier (Smith and Salkerd, 1966). The mode of ovicidal action of an IGR (Dimilin) was explained as the inhibition of chitin synthesis, thus interfering with the formation of endocuticular deposition (Miura *et al.*, 1976). A similar mode of action may be suggested for neem products as they were observed to affect chitin synthesis (Sharma and Chander sheikher, 1999a).

It is therefore suggested that the neem derivatives can also be considered for mosquito control operation, besides their use against agricultural pests.

References

Aliero, B.L., 2003. Larvicidal effects of aqueous extracts of *Azadirachta indica* (neem) on the larvae of *Anopheles* mosquito. *African Journal of Biotechnology*, 2: 325–327.

Amorose, T., 1995. Larvicidal efficacy of neem (*Azadirachta indica*) oil and defatted cake on *Culex quinquefasciatus* Say. *Geobios*, 22: 169–173.

Attri, B.S. and Prasad, R., 1980. Neem oil extractive: An effective mosquito larvicide. *Indian J. Ent.*, 42: 371–374.

Awala, P., Mwangi, R.M. and Irungu, L.W., 1998. Larvicidal activity of a granular formation of a *Melia volkensii* (Gurke) acetone extract against *Aedes aegypti* L. *Insect Sci. Applic.*, 18: 225–228.

Babu, R. and Murugan, K., 1998. Interactive effect of neem seed kernel and neem gum extracts on the control of *Culex quinquefasciatus* Say. *Neem Newsletter*, 15: 9–10.

Babu, R., Murugan, K. and Mohan, P.S., 2000. Larvicidal effect of resinous exudate from the tender leaves of *Azadirachta indica*. *Neem Newsletter*, 17: 1–3.

Boradbent, A.B. and Free, D.J., 1984. Effects of dyflobenzuron and Bay SIR 8514 on the oriental fruit moth (Lepidotera : Olethreutidae) and the oblique banded leaf roller (Lepidotera : Tortricidae). *J. Econ. Entomol.*, 77: 194–197.

Bowers, W.S., Sener, H., Evans, F., Bingal, A. and Erdogon, J., 1995. Activity of Turkish medical plant against mosquitoes *Aedes aegypti* and *Anopheles gambiae*. *Insect Sci. Applic.*, 16: 339–342.

Chavan, S.R. and Nikam, S.T., 1988. Investigation of alkanes from neem leaves and their mosquito larvicidal activity. *Pesticides*, 22: 32–33.

Cilek, J.W., Webb, J.D. and Knapp, F.W., 1991. Residual concentration and efficacy of three temphos formulations for control of larval *Aedes aegypti*. *J. Am. Mosq. Control Assoc.*, 7: 310–312.

Das, T.K., Somchoudhury, A.K., Saha, K. and Sarkar, P.K., 1995. Persistant toxicity of modern insecticides, germicides and plant extracts against mosquito larvae. *Pestology*, 19: 5–12.

Deshmukh, P.B. and Renapurkar, D.M., 1992. Insect repellent activity of *Azadirachta indica* Juss against mosquito *Culex pipiens fatigans*. *Pestology*, 16: 33–36.

Dorland's Illustrated Medical Dictionary, 1982, 26th Edition. W.B. Saunder Company, Igaku-Shoin/ Saunders, London, p. 35.

Ekesi, S., 2000. Effect of volatiles and crude extracts of different plant materials on egg viability of *Maruca vitrata* and *Clavigralla tomentosicollis*. *Phytoparasitica*, 28: 305–310.

Fournet, F., Sannier, C. and Monteny, N., 1993. Effects of the insect growth regulators OMS 2017 and diflubenzuron on the reproductive potential of *Aedes aegypti*. *J. Am. Mosq. Control Assoc.*, 9: 426–430.

Friend, J.A., 1998. Twenty years of horticulture research with neem against diamond backmoth, *Plutella xylostella* L. *Pestology*, 22: 33–40.

Helson, B.V. and Surgeoner, G.A., 1983. Effect of temperature and stage of development on susceptibility of *Aedes euedes* and *Aedes stimulans* (Diptera : Culicidae) larvae to temephos. *Can. Ent.*, 115: 623–628.

Jebanesan, A., Venkatachalam, M.R. and Jagadeesan, G., 1999. Toxicity evaluation of certain neem based bio–insecticides against the larval mosquitoes of *Culex quinquefasciatus* Say. *Pestology*, 23: 72–75.

Judson, C.L., 1967. Alternation of feeding behaviour and fertility in *Aedes aegypti* by the chemosterilant apholate. *Entomol. Exp. Appl.*, 10: 387–394.

Judson, C.L. and Gojrati, H.A.N., 1967. The effects of various oxygen tensions on embryogeny and larval responses of *Aedes aegypti*. *Entomol. Exp. Appl.*, 10: 181–188.

Jupp, P.G. and McIntosh, B.M., 1990. *Aedes furcifer* and other mosquitoes as vectors of Chikungunya virus at mica, northeastern transvaal, South Africa. *J. Am. Mosq. Control Assoc.*, 12: 415–419.

Kaur, J.J., Rao, D.K., Sehgal, S.S. and Seth, R.K., 2001. Effect of hexane extract of neem seed kernel on development and reproductive behaviour of *Spodoptera litura* Fabr. *Ann. Pl. Protec. Sci.*, 9: 171–178.

Kumar, S., Thomas, S. and Pillai, M.K.K., 1991. Involvement of mono-oxygenases as a major mechanism of deltamethrin resistance in larvae of three species of mosquitoes. *Indian J. Exp. Biol.*, 29: 379–384.

Mazzarri, M.B. and Georghiou, P., 1995. Characterization of resistance to organophosphate, carbomate and pyrethroid insecticides in field populations of *Aedes aegypti* from Venezuela. *J. Am. Mosq. Control Assoc.*, 11: 315–322.

Mekuria, Y., Gwinn, T.A., Williams, D.C. and Tidwell M.A., 1991. Insecticide susceptibility of *Aedes aegypti* from Santo Domingo, Dominican Republic. *J. Am. Mosq. Control Assoc.*, 7: 69–72.

Miura, T., Schaefer, H., Takahashi, R.M. and Mulligan, F.S., 1976. Effects of the insects growth inhibitor, dimilin on hatching of mosquito egg. *J. Econ. Entomol.*, 69: 655–658.

Mohanraj, R.S. and Dhanakkodi, B., 2004. Effect of neem extracts on ovarian length and protein of *Aedes aegypti. J. Exp. Zool., India*, 7: 125 – 128.

Mohanraj, R.S. and Dhanakkodi, B., 2005a. Ovicidal effects of neem products on mosquito eggs. *J. Ecotoxicol. Environ. Monit.*, 15: 73–77.

Mohanraj, R.S. and Dhanakkodi, B., 2005b. Growth inhibiting effect of crude neem seed kernel extract of *Aedes aegypti. Indian J. Ent.*, 67(3): 241–243.

Mohanraj, R.S. and Dhanakkodi, B., 2006. Influence of neem extracts of ovipositional activity of *Aedes aegypti. Pestology*, 30: 38–41.

Mohanraj, R.S. and Dhanakkodi, B., 2007a. Delayed effect of neem extracts on fitness parameters of *Aedes aegypti. Journal of Current Sciences*, 10: 161–165.

Mohanraj, R.S. and Dhanakkodi, B., 2007b. Interference of botanical extracts on oviposition of *Aedes aegypti. Adv. Bio. Sci.*, 6: 11–17.

Mohanraj, R.S., Saraswathy, S., Thangavel, K., Logaswamy, S. and Dhanakkodi, B., 2000. Effect of neem formulation on pre-adult stages and egg hatchability of *Aedes aegypti. Neem Newsletter*, 17: 4–6.

Neoliya, N.K., Singh, D. and Sangwan, R.S., 2007. Azadirachtin-based insecticides induce alteratioin in *Helicoverpa armigera* Hub. Head polypeptides. *Current Science*, 92: 94–99.

Obomanu, F.G., Ogbalu, O.K., Gabriel, U.U., Fekarurhobo, G.K. and Adediran, B.I., 2006. Larvicidal properties of *Lepiadagathis alopecuroides* and *Azadirachta indica* on *Anopheles gambiae* and *Culex quinquefasciatus. African Journal of Biotechnology*, 5: 761–765.

Okumu, F.O., Knols, B.G.J. and Fillinger, U., 2007. Larvicidal effects of a neem (*Azadirachta indica*) oil formulation on the malaria vector *Anopheles gambiae. Malaria Journal*, 6: 63–73.

Oroumchi, S. and Lorra, C., 1993. Investigation on the effects of aqueous extracts of neem and Chinaberry on development and mortality of the alfalfa weevil *Hypera postica* Gyllenh (Col., Curculionidae). *J. Appl. Ent.*, 116: 345–351.

Pampana, E., 1963. *A Textbook of Malaria Eradication.* Oxford University Press, London, pp. 459–462.

Purohit, P., Jyotsna, D. and Srimannarayana, G., 1989. Anti-feedant activity of indigenous plant extracts against larvae of castor semi-looper. *Pesticides*, 25: 23–26.

Ramarethinam, S., Loganathan, S., Marimuthu, S. and Murugesan, N.V., 2002. Studies on the evaluation of neem oil based EC formulation (0.03 per cent Azadirachtin) on the semilooper caterpillar,

Achaeajanata L. (Lepidoptera : Noctuidae) infesting castor plant, *Ricinus communis* L. *Pestology*, 26: 9–14.

Rawat, S., 1994. Neem products and their pesticidal characteristics. *Everyman's Sciences*. December–January, 1993–94.

Rawlins, S.C. and Wan, J.O.H., 1995. Resistance in some caribbean populations of *Aedes aegypti* to several insecticides. *J. Am. Mosq. Control Assoc.*, 11: 59–65.

Regupathy, A. and Dhamu, K.P., 1990. *Statistic Workbook for Insecticide Toxicology*. Surya Desktop Publishers, India, pp. 4–7, 16–22.

Sagar, S.K. and Sehgal, S.S., 1997. Toxicity of neem seed coat extract against mosquitoes. *Indian J. Ent.*, 59: 215–223.

Sahgal, A. and Pillai, M.K.K., 1993. Ovicidal activity of permethrin and deltamethrin on mosquitoes. *Entomon.*, 17: 149–154.

Schmutterer, H., 1990. Properties and potential of natural pesticides from the neem tree, *Azadirachta indica*. *Ann. Rev. Ent.*, 35: 271–297.

Sharma, P.L. and Chander Sheikher, 1999a. Toxic and morphogenic effects of neem leaf extracts on *Helicoverpa armigera* Hubner. *Pest Management and Economic Zoology*, 5: 95–100.

Sinniah, B., Sinniah, D. and Ibrahim, J., 1994. Effect of neem oil on mosquito larvae. *Mosq. Borne. Dis. Bull.*, 11: 90–94.

Smith, E.H. and Salkerd, E.H., 1966. The use and action of ovicides. *A. Rev. Ent.*, 11: 331–368.

Su, T.Y. and Mulla, M.S., 1998a. Ovicidal activity of neem products (Azadirachtin) against *Culex tarsalis* and *Culex quinquefasciatus* (Diptera : Culicidae). *J. Am. Mosq. Control Assoc.*, 14: 204–209.

Su, T.Y. and Mulla, M.S., 1999a. Oviposition bioassay responses of *Culex tarsalis* and *Culex quinquefaciatus* to neem products containing azadirachtin. *Entomol. Exp. Appl.*, 91: 337 – 345.

Sundar, M.S., Krishnayya, P.V. and Rao, P.A., 2000. Evaluation of lethal concentrations of azadirachtin based proprietary formulation for larvicidal and ovicidal effect against *Spodoptera litura* Fab. *J. Ent. Res.*, 24: 147–150.

Thamil Solai, P.K., Selvaraj Pandian, R. and Charles Manoharan, A., 1999. Toxicity evaluation of the pesticides Salotrap, Neemgold, Nimbecidine and Nimbex against the larvae and pupae of *Culex quinquefasciatus* Say. In: *Biopesticides in Insect Pest Management*, (Eds.) S. Ignacimuthu and Alok Sen. Phoneix Publishing House Pvt. Ltd., New Delhi, pp. 46–49.

Umar, A., Kela, S.L. and Ogidi, J.A., 2007. Effects of extraction solvents on the toxicity of *Azadirachta indica* A. Juss (Neem) seed kernel extracts to *Aedes aegypti* (Diptera : Culicidae) larvae. *Int. Jor. P. App. Scs.*, 1: 32–38.

Vatandoost, H. and Haziri, V.M., 2004. Larvicidal activity of a neem tree extract (Neemarin) against mosquito larvae in the Islamic Republic of Iran. *Eastern Mediterranean Health Journal*, 10: 573–581.

Xie, Y.S., Fields, P.G. and Isman, M.B., 1995. Repellency and toxicity of azadirachtin and neem concentrates to three stored-product beetles. *J. Econ. Entomol.*, 80: 1024–1031.

Zebitz, C.P.W., 1984. Effect of some crude and azadirachtin enriched neem (*Azadirachta indica*) seed kernel extracts on larvae of *Aedes aegypti*. *Entomol. Exp. Appl.*, 35: 11–16.

Chapter 29

Isolation of Pathogenic Bacterial Flora in Dairy Products

S.A. Karmakar, P.M. Tumane and B.J. Wadher

Post Graduate Department of Microbiology,
R.T.M. Nagpur University, Nagpur – 440 033, M.S.

ABSTRACT

Dairy products samples of Curd, Shrikhand, Icecream and Cheese were randomly collected from the different local shops in Koradi, Khaperkheda, Gittikhadan and Laxminagar area in Nagpur and the samples were thoroughly analyzed for presence of bacterial flora-both pathogenic and nonpathogenic. Of these 12 samples, 3 were found to be positive for *E. coli,* 6 were found to be positive for *B.cereus,* 4 were found to be positive for *Pseudomonas* spp. 2 were found to be positive for *Salmonella* spp. 2 were found to be positive for *Enterococcus* spp and 1 sample was found to be positive for *Staphylococcus* spp. Results were discussed based on both the positive health and industrial aspects of the beneficial organisms and pathogenic ill-effects of the non-beneficial organisms.

Keywords: *Bacterial flora in dairy products.*

Introduction

Milk being a nutritious food for humans and other animals also provide an ideal environment for microbial growth thus the microorganisms can gain entry into milk and its products and can multiply and bring about spoilage of these products or render them unsafe due to potential health hazards. Apart from the harmful microorganisms there is another group of dairy microorganisms called starter cultures (*Lactobacilli* spp, *Streptococcus lactis*) which bring about desirable changes in milk and thus can be used for the preparation of a variety of fermented dairy products. The scope of dairy microbiology stretches right from production of milk, to preparation of products and till it is utilized by consumers (Yadav, Grover, Batish, 1992)

The milk when secreted in healthy udder is almost sterile. While some organisms *e.g.* the starter cultures it is only during subsequent milking and post milking operations that raw milk gets various microorganisms due to contamination from surroundings these microorganisms can range from few hundreds to few thousands per ml of milk products. Contamination of milk with spoilage and disease producing microorganisms can occur at any stage during production, processing, marketing and utilization. The extent of contamination and subsequent microbial multiplication determine the microbial quality of product (Batish, 1982).

Predominating flora include *Streptococci, Bacilli, Coliforms, Pseudomonas, Micrococcus, Enterococci.* Others include yeasts molds viruses etc. among these bacteria are most important group following yeasts molds and viruses.

While bacteria and yeasts are prokaryotic, yeasts and molds are eukaryotic.

Bacteria have the ability to utilize various milk constituents to grow and multiply their numbers by binary fission in which one cell divides into two which in turn follows the same pattern leading to exponential growth. The time required for each division is called as generation time. This generation time varies greatly with organism to organism. *e.g.* 20 mins for *E. coli* means they can grow quite rapidly in and reach high numbers in a short period.

The starter microorganisms used in manufacture of different fermented dairy products for *e.g.* Lactic Acid Bacteria (LAB) *Propionibacteria, lactose fermenting Yeasts, Penicillum* spp, constitute the desirable microflora of these products LAB (*Streptococci, Lactobacilli, Pediococci, Leuconostocs* etc.) constitute the starter culture–(the one that starts the process) for preparation of curd, yoghurt, lassi, cultured buttermilk and a numbers of cheese varieties.

Besides milk and milk products, also undergo a numbers of changes–Souring, Proteolysis, Lipolysis, etc. Hence it is very necessary to prevent microbial spoilage of dairy products as it can cause heavy commercial loss for producers and a feeling of distaste for consumers (Anmika Mayra Mackinen, 1996).

But the notorious ones are the pathogenic organisms which enter at the manufacturing processing and packaging level. Among the pathogenic organisms *Salmonella typhi* causes typhoid, *Salmonella paratyphi* cause paratyphoid, while *E. coli* causes food poisoning, *Bacillus cereus* also causes acute poisoning while *Streptococci* causes Scarlet fever, Tonsillitis etc. (Bergey, 1986).

Milk and milk products are popular because of their unique flavour and texture so much so that if these attributed are not attractive, the product can be rejected by the customers out rightly. Besides the potential diseases and toxicosis cause by the pathogens can be disastrous to human health targeting a large number of individuals simultaneously who use such products. Hence prevention of contamination is of utmost importance in the dairy industry (Levine Anderson, 1932).

Materials and Methods

For these various dairy samples were collected from local shops in 4 places–Koradi, Khaperkheda, Gittikhadan and Laxminagar area.

Sampling

Samples of the local products were collected under aseptic conditions from local shops of Koradi, Khaperkheda, Gittikhadan and Laxminagar areas in previously autoclaved glass bottles, while the branded samples were brought in previously packed containers and used for further procedures.

Detection

Each of the samples were first gram stained and then were streaked on various selective media and nutrient agar media (Hi-Media Laboratories) and the organisms present were detected in each of the samples. The positive results were isolated on nutrient agar slants and these isolates were further confirmed of their identity on the basis of biochemical characteristics which included IMViC, Sugar fermentations, Triple sugar iron agar, Catalase test, Urease test (Hi-Media Laboratories media).

Thus a bacterial profile of the samples collected was made and the each of these organisms was classified upto the genetic level.

Results and Discussion

In the present investigation, samples of Curd, Shrikhand, Cheese, Icecream were obtained from various places in and around Nagpur such as Koradi, Khaperkheda, Gittikhadan area, Laxminagar area and obtained the following bacterial flora *Lactobacillus* spp, *Streptococcus lactis, Staphylococcus aureus, Escherichia coli, Bacillus cereus, Pseudomonas* spp, *Salmonella* spp, *Enterococci* spp. etc.

Table 29.1

Isolate No.	*Lacto-bacillus*	*Strep-tolactis*	*S. aureus*	*E. coli*	*B. cereus*	*Salmonella*	*Pseudo-monas*	*Entero-cocci*	
C1	+	+	+	−	−	+	−	+	−
C2	+	+	−	+	+	−	+	+	−
C3	+	+	−	−	+	−	−	−	−
C4	+	+	−	−	+	−	+	−	−
Cb1	−	−	−	−	−	−	−	−	−
Cb2	−	−	−	−	−	−	+	−	−
SB1	−	−	−	−	−	−	−	−	−
S2	+	+	−	+	+	−	−	−	−
Pb1	+	+	−	−	−	−	−	−	−
P2	+	+	−	+	+	+	+	−	−
Ib1	−	−	−	−	+	−	−	−	−
I2	+	−	−	−	−	−	−	−	−

+: Indicates presence of organism; −: Indicates absence of organism.

C1: Koradi; C2: Khaperkheda; Cb1, Cb2: Branded samples (Amul, Haldiram) (Curd); Sb1: Branded (Amul); S2: Koradi (Shrikhand); Pb1: Branded (Amul); P2: Koradi Local (*Paneer*); Ib1: Branded (Dinshaws); I2: Koradi Home made Icecream; C3: Gittikhadan Area; C4: Laxminagar Area.

While *Lactobacillus* and *Streptococcus lactis* were found in almost all the 4 curd samples, they were found to be present mainly only in the local samples of Paneer, Shrikhand and Icecream, this could be due to processing conditions of Shrikhand and Icecream and the acidic conditions of *Paneer* which destroys most of the microbes (Yadav, Grover, Batish, 1992).

Pathogenic organisms were found almost intermittently in all the local samples obtained from the four places. Pathogenic bacteria were not observed in any of the branded samples. This could be attributed to the fact to the stringent processing and packaging of the branded goods besides the local samples are prepared and handled by local people who are illiterate and do not feel the facilities for proper hygiene maintenance.

A similar work of detection of bacterial flora was done by Dr. Sunita Grover *et al.* in 1993 in Delhi by collecting samples of dairy products from local shops of Karnal,Himachal Pradesh and found *Lactobacillus* spp, *Streptococcus lactis, Staphylococcus aureus, Escherichia coli, Bacillus cereus, Pseudomonas* spp, *Salmonella* spp, *Enterococci* spp. etc. (Yadav, Grover, Batish, 1992).

Thus, contamination *i.e.* pathogenic bacterial flora is not observed in any of the branded samples while most contamination is observed in samples as obtained from Koradi and Khaperkheda. Contamination indicates danger of causing various diseases, toxicosis, poisoning etc. which can cause considerable harm to individuals consuming such dairy products (Mortality and Morbidity Weekly).

The common bacterial flora usually obtained in milk and milk products include Normal bacterial flora: *Lactobacillus* spp, *Streptococcus lactis, Leuconostoc, Pediococcus* etc.

Pathogenic bacterial flora: *S.aureus, Escherichia coli, Salmonella* spp, *Shigella* spp, *Streptococcal* spp, *Proteus* spp, *Klebsiella* spp, *Enterobacter* spp. also observed in some areas (Anmika Mayra, 1998)

Conclusion

It is concluded that a number of disease organism are present in the local dairy products samples specially those procured from Koradi and Khaperkheda which could be due to carelessness in handling and preservation. Also the water supplies are not suitably checked which is also a reason for this condition. Unmindful consumption of such dairy products can lead to a number of diseases and toxicosis cases and food poisoning targeting a large number of individuals causing food poisoning and disease outbreaks. An awareness must be brought among the common people about the hygiene concerned in the use of such dairy products.

Also cleanliness and safety should be on top priority in production and marketing of dairy products.

Acknowledgement

Thanks are due to the Head of the Department, Post Graduate Department of Microbiology, R.T.M. Nagpur University, Nagpur and the Complete faculty of the Department for providing me the facilities and for providing me encouragement and support.

References

Ali, Mohammad. *Food Nutrition in India: Interactions of Lactic Acid Nacteria with Human System.*

Ananthanarayan, R. and Jayram Paniker, C.K., 1999. *Textbook of Microbiology, Vol. 2: Systemic Bacteriology.*

Anmika Mayra Makinen Marc Bigret, 1998. Industrial use and production of lactic acid bacteria. In: *Lactic Acid Bacteria: Microbiology and Functional Aspects*, 2nd edn. (Eds.) Salminen Seppo, Atte Von Wright.

Axellson, Lans, 1989. *Lactic acid bacteria: Classification and physiology*. In: *Lactic Acid Bacteria: Microbiology and Functional Aspects*, 2nd edn. (Eds.) Salminen Seppo, Atte Von Wright.

Batish, V.K., Chander, H. and Ranganathan, B., 1982. *J. Food Proteins.*

Bergeys Manual of Determinative Bacteriology, 8th edn. 1974. Williams and Wilkins Co., Baltimore, U.S.A. Classification and Morphology of Bacteria.

Bergeys Manual of Systemic Bacteriology, 9th edn., 1986. Staley, Bryant Williams and Wilkins Co., Baltimoore, London.

Bradshaw L. Jack, 1979. Key to sps. of genus *Lactobacilli* in lactic acid bacteria microbiology.

Carpenter, Philip. *Microbiology of Foods and Dairy Products Morphology and Physiology of Lactic Acid Bacteria.*

Centre for Disease and Prevention Outbreak of *Salmonellae enteriditis* and *E. coli* Poisoning as associated with Home Made Icecream, Florida. Mortality and Morbidity Weekly Report 43: 599–571.

Centre for Disease Control and Prevention Update *Salmonella enteriditis* Infection and Grade A Shell Eggs-US 1989 Morbidity and Mortality E Weekly Report.

Krausker, M.M. and Hunschner, Martha. *Food Nutrition and Diet Therapy*, 5th edn. Philadelphia and W.L.Saunders, Lactic Acid Bacteria.

Prescott, S.C. and Dunn, C.G., 1957. *Industrial Microbiology*, New York Mc Graw Hill.

Shermann-Bacteriol, 1937, Reviews, 1997, Weekly, 56.

Yadav, J.S., Grover, Sunita and Batish, V.K., 1992. *A Comprehensive Dairy Microbiology Introduction on Dairy Products, Cheese, Butter, Icecream.*

Chapter 30

Ephedrine, Saponin and Ginsenoside from Sidabiyai: A Traditional Home Remedy Herbal Medicine for Leucorrhoea and Leucorrhoea Associated Pelvic Inflammatory Disease

S.R. Singh and M. Neshwari Devi

Post Graduate Studies Centre, HRDRI, Canchipur, Imphal – 795 003, Manipur

ABSTRACT

Sidabiyai, a traditional home remedy herbal medicine that therapeutically prescribed in Leucorrhoea by herbal medicineman have been practiced since immemorial time in Manipur. The present chapter explored the therapeutic organic compounds of Ephedrine, Saponin and Ginsenoside from the indigenous herbal medicine sidabiyai by using TLC and revitalize the laurel of the traditional technology and knowledge of home remedy in health care system. The therapeutic chemistry of the identified compounds, their effect in metabolism, immune system, organ functions etc. have been described with emphatic emphasize on urgent needs of hi-tech exploration on unmatched traditional wealth of indigenous therapeutic knowledge.

Keywords: *Home remedy herbal medicine, Therapeutic chemistry, Sidabiyai, Leucorrhoea, Indigenous technological knowledge.*

Introduction

Leucorrhoea a common gynecological problems faced by the gynecologist and are often difficult to treat, is an abnormal excessive vaginal discharge often associated with irritation and pruritus. Leucorrhoea could be physiological when associated with various phases of menstrual cycle or due to cervical vaginal inflammation or diseases. It can be due to infection with *Tricomonas vaginalis. Candida albicans* or mixed bacterial infections chronic cervicitis, cervical dysphasia malignancy or due to senile vaginitis (Bose and Bose, 1996). Broad symptoms including pain in the lumber region and the calves and a dragging sensation in the abdomen, symptoms of constipation and frequent headaches and intense itching leucorrhoea associated with Pelvic (Chidambaranathan, 2006). In chronic leucorrhea patient feels disruptive disturbance irritable and develops black patches under the eyes inflammatory diseases Pelvic Inflammatory diseases (PID) an upper genital tract infections encompasses with endometritis salpingitis-pentonitis, caused more complexity.

Materials and Methods

Sidabiyai, a widely used common safe phytomedicine for women, which is presented by Traditional medicineman, to Leucorrhea and Leucorrhea associative PID. It is crude plant drug composed of premium phytonutrients non-hormonal rational combination of valuable indigenous formulation from the bark and wood of *Terminalia arjuna, Santalum album* and *Adenanthera paronina* at the ratio of 4:2:1. The herbal extract drug is known to have direct action on the utchine musculature, improve its circulation and tone up the female genital tract. However, in depth therapeutic chemicals based on scientific research have not yet been undertaken in the indigenous herbal drugs in Manipur. It is worthwhile to detect the therapeutic organic compounds to enhance the good forth to traditional herbal drugs and to revitalize the indigenous technological knowledge of life saving herbal drugs and thereof health care system and its tradition of life saving drugs and thereof health care system and its tradition. For the present therapeutic chemical investigation of organic compounds in the medicine, freshly prepared medicine were made from the constituent bark and woods in the manner that all the bark of *Terminalia arjuna, Santalum album* and *Adenanthera paronina* collected from the Imphal during February/ March 2007, then, washed thoroughly with tap water and with distilled water, then they were chopping and dried under shade and ground into mixture to obtained powder of the bark or wood. The powder was mixed in the ratio of 4:2:1 of *Terminalia arjuna: Santalum album: Adenanthera paronina.* The mixture was then subjected to extraction in soxhlet by using different organic solvents, *viz.* EtOH, MeOH, $CHCl_3$, and Petroleum ether. Then 5 ml of concentrated was carried on using TLC on Silica Gel G Chromatoplate and Ninhydrine reagent, Anisaldehyde sulphuric acid and Vanilline phosphoric acid as spraying agent.

Results and Discussion

The TLC photo drug atlas has an immediate clarity of representation that facilitates the learning of TLC drug analysis and photographic reproduction of thin layer separation has a large didactic advantage over mer graphic representation (Wagner, 1996). Regarding documentation of TLC, Equipments, general technique and special technique in TLC by TLC module (NT curriculum project, UW-Madison; 1995-96). The present experimentation on the therapeutic compounds scavenge the Rf value 0.5, 0.3, 0.5 and 0.55, in the thin layer chromatographs indicating an organic natural compounds *viz.* Ephedrine, Saponin, (Senegae radix and Hippocastani semen), Ginsenoside respectively. The chromatograms are presented in Figures 30.1–30.3. The colouration and the characteristic of the spot elucidate and put construction on getting the axplicatory confirmation of the compounds. The chemical

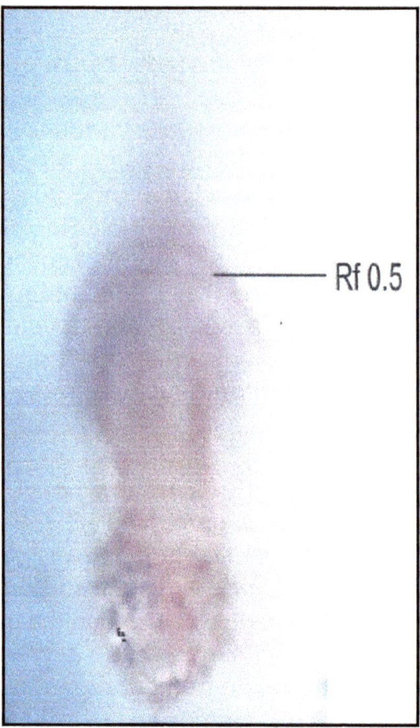

Figure 30.1: TLC Photograph of Ephedrine

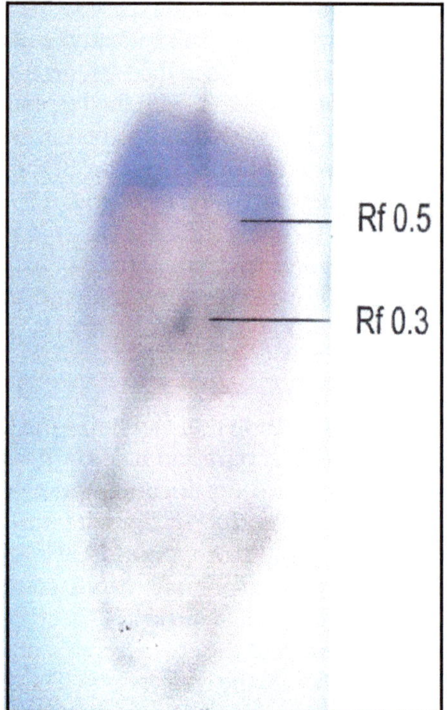

Figure 30.2: TLC Photograph of Saponin

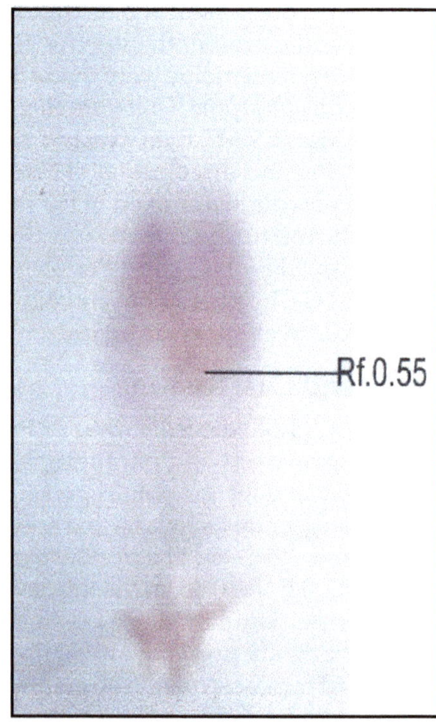

Figure 30.3: TLC Photograph of Ginsenoside

structure of the compounds Ephedrine, Saponin (*Senegae radix* and *Hippocastani semen*), Ginsenoside accords in Figures 30.4–30.7 respectively.

EtOH extract of whole indigenous traditional medicine, in test examination for analysis and purification of therapeutic compounds on TLC, the result assigned to distinct spot violet red with standard Rf 0.4 consequently verified and patented the compound Ephedrine. With the same technique and methods on herbal medicine, the violet red band with Rf 0.4-0.5 of Ephedrine was reported from *Aconiti tubera, Ephedrae herba*, by Wagner and Bladt (1996). In 1885 Nagayoshi Nagai first time isolate ephedrine from Vulgaris. Ephedrine is a sympathomimetic amine, that is its principal mechanism of action relies on its indirect action on the adrenergic receptor system, which is a part of sympathetic nervous system or SNS. Ephedrine's mechanism of action on neurotransmission in the brain is wide. Its action as against at most major norephedrine receptors and its ability to induce moderate stimulation of the release of both dopamine and to a lesser extent, serotonin, is presumed to have a major role in its mechanism of action. Ephedrine was once widely used as a topical decongestant and as a bronchodilator in the treatment for asthma. It continues to be used for these indications, although its popularity is waning due to the availability of more effective agents for these indication which exhibit

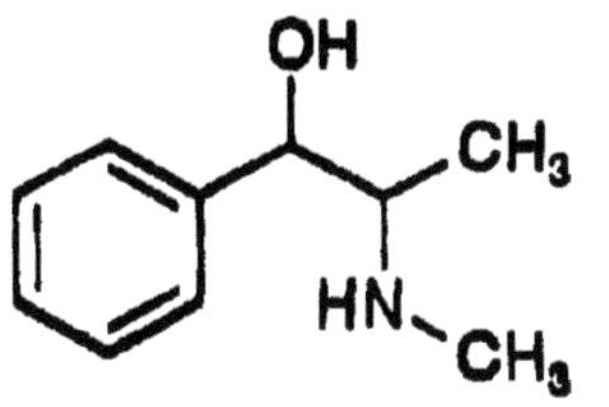

Figure 30.4: Structure of Ephedrine

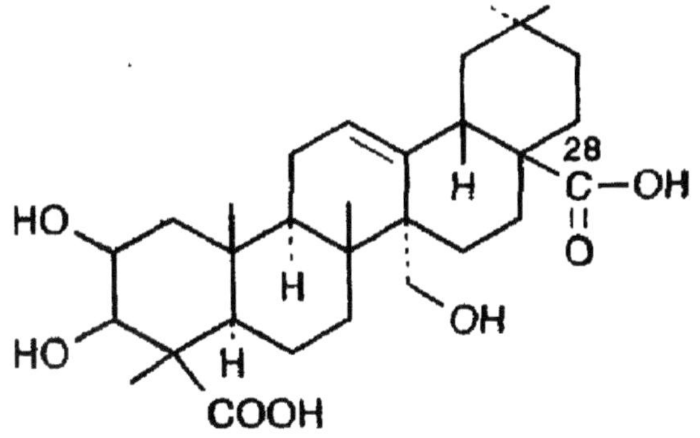

Figure 30.5: Structure of Saponin (*Senegae radix*)

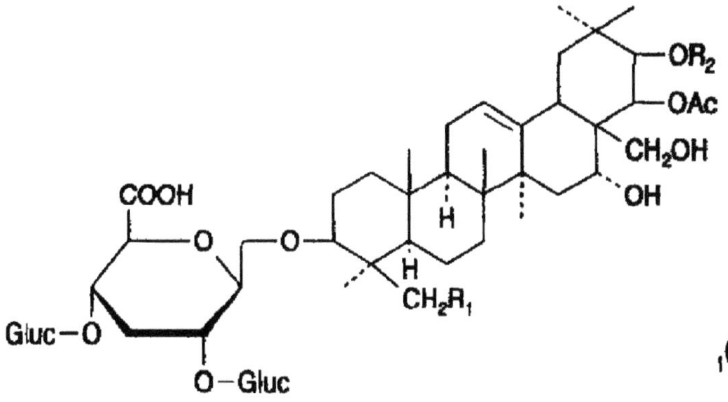

Figure 30.6. Structure of Saponin (*Hippacastani semen*)

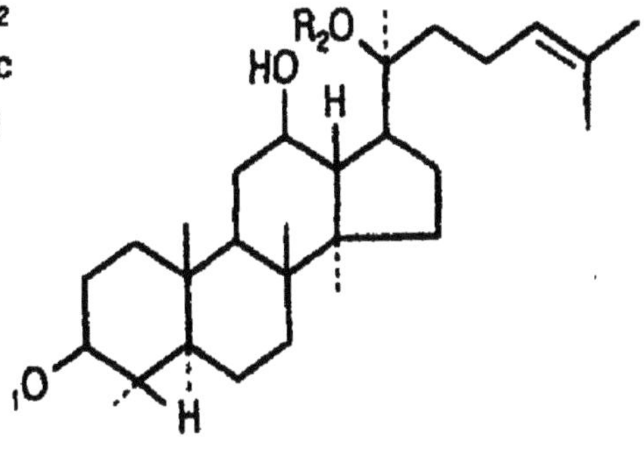

Figure 30.7: Structure of Ginsenoside

fewer adverse effects (BNF 2004)Ephedrine continues to be used intravenously in the reversal hypotension from spinal/epidural anesthesia (BNF 2004). It also used in hypotensive states, including overdose with ganglionic blocking agents, antiadrenergic agents, or other medications that lower blood pressure (Bicopoulus 2002).

In traditional Chinese medicine, ephedrine has been used in the treatment of asthma and bronchitis for centuries (Ford, *et al.*, 2001). Ephedrine therapy include (BNF 2004): Cardiovascular, Dermatological, Gastrointestinal, Genitourinary, nervous system, Respiratory etc. Ephedrine can also lead to damage of the brain receptors over a period of large usage this is because of its constant action on the neurochemicals. It also leads to high increase in blood pressure which over time can lead to damage in the blood vessels. Ephedrine has a reduced neurotoxic effect on dopamine than its amphetamine counterparts.

Ephedrine is used to treat low blood pressure and chronic asthma, it is more popularly used in diet pills (Anonymous a), Ephedrine mainly effects the cardiovascular and respiratory systems of the body. This means that it constricts blood vessels and enhance certain actions of the heart (Anonymous b). Capri Mara Fillmore, Lisa Bartoli, Richard Bach and Young Park (1999) stated that ephedrine stimulates the sympathomimetic and central nervous system. Ephedrine is usually mixed with other stimulants like caffeine and appetite suppressants to create a pill that companies claim help a person to lose weight without diet or exercise. In clinical review by Luke Bucci "lean mass is preserved better with ephedrine containing combinations during weight loss". Bucci also states that ephedrine is prohibited for sporting events due to its affect on physical enhancement (Bucci 2000).

Ephedrine is a central nervous system stimulant used to treat breathing problem (as bronchodilator), nasal congestion (as a decongestant) low blood pressure problems (orthostatic hypotension) or myasthenia gains. This alkaloid has also been used to treat certain sleep disorders (narcolepsy), menstrual problems (dysmenorrheal), urine control problem (incontinence or enuresis).

In another test on Ethanol extract of whole medicine it was made concentrated and wield to analysis and purification to TLC, the result assigned to distinct spot red in UV light, standard with Rf 0.55 pointing out the compound ginsenoside. The finding was in concordance with that of ginsenosides largely extend from 0.25 to 0.55 Rf from Ginseng radix, ethanolic extract (Wagner and Bladt, 1996). These gensenosides, the active ingredients of Ginseng, over-dominant as elixir of life and still act as miracle to medical science. In other words the active components of ginseng are considered to be ginsenosides, a group of steroidal Triterpene saponin (Gillis 1997). Ginsenosides are a class of steroid like compounds. It includes hypoglycemia, an increased risk of bleeding and a decreased anticoagulant effect of warfarin (Ang-Lee *et al.*, 2001). Ginsenoside in human are by Cui and Colleagues (Cui *et al.*, 1996, 1997). They showed that ginsenoside present in urine (in human) after oral ingestion. Different ginsenoside might have different or even antagonistic actions (Corthout *et al.*, 1999). It is used to lower cholesterol, increase energy and endurance, reduce fatigue and effects of stress, as well as prevent infections. It can alleviate some major effects of aging, such as degeneration of the blood system and increase mental and physical capacity. It even appears to help people with diabetes type 2. Ginsenoside benefits are increase memory, helps with radiation damage, acts as an antioxidant, Insomnia and sleep disorder, inhibits blood coagulation, increase overall citality, improves vision and hearing, strengthen nervous system, menopausal stages, works as general stimulant, removes toxin from body, slows degeneration of cells, increase longevity.

Ginseng or its ginsenoside component showed positive results on the central nervous system and gonadal tissues. In males, ginsenosides can facilitate penile erection by directly inducing the vasodilatation and relaxation of penile corpus cavernosum. Moreover, the effects of ginseng on the corpus cavernosum appear to be mediated by the release and or modification of release of nitric oxide from endothelial cells and pervascular nerves. Animal studies lend growing support for the use of ginseng in the treatment of sexual dysfunction and provide increasing evidence for a role of nitric oxide in the mechanism of ginsenoside action (McElhaney, 2004).

Ethanol extract of whole medicine with anisaldehyde sulphuric acid reagent detected unambiguous unique spot with Rf ~ 0.3 and ~ 0.5, pointing out the compounds Saponin (Senegae radix) and Saponin (*Hippocastani semen*). The finding substantiated the therapeutic compound. Saponin on TLC synopsis of saponin drugs under Senegae radix with Rf 0.2 to 0.4 and Hippocastani semen Rf 0.4 to 0.5 (Wagner and Bladt 1996). Further with solvent diisopropyl ether-acetone, Schmittmans *et al.* (1962) Sapogenin from Quillaja Rf value 0.2 to 0.93. Saponin are well known to have a wide spectrum of activities including lowering blood cholesterol, inhibiting cancer cell growth, and acting as antifungal and antibacterial agents. Phytochemist report that saponins can act by binding with bile acids and cholesterol. Saponins are glycosidic triterpenoids (Burnouf-Radosevich *et al.*, 1985; Miziui *et al.*, 1988, 1990; Mu *et al.*, 1989; Meyer *et al.*, 1990; Ridiout *et al.*, 1991).

Saponin is in their antibiotic, fungistatic, and pharmacological properties (Basu and Rastogi, 1967; Agarwal and Rastogi, 1974; Chandel and Rastogi, 1980; Nanaka, 1986). The pharmacological interest in saponin lies with their ability to induce changes in intestinal permeability (Gee *et al.*, 1989; Johnson *et al.*, 1986), which may aid the adsorption of particular drugs (Basu and Rastogi, 1967), and with their hypocholesterolenic effects (Oakenfull and Sidhu, 1990).

Glycosidic saponins were antifungal and molluscicidal in vitro and able to kill liver flukes both *in vitro* and *vivo* (Wran, 1988). More recently, biochemist have postulated that these saponins are responsible for the plant's expectorants actives. Saponins responses in gastric mucosa, which in turn activate mucous glands in the bronchi through parasympatholic signaling, to aid in the removal at mucous (Schulz *et al.*, 1998). Saponins comprising the hederagenin glycosides hederucoside C and a-hederin and the oleanolic acid glycosides hederacoside B and b-hederin (Wichtl and Bisset 1994). Saponins are glycosides with aglycone (the sapogenin), these rings are either steroidal and triterpenoidal (The Mark Index 1983). Among the associated properties of saponins are bitter testing compounds formation of stable foams when agitated in water, the formation of oil water emulsion and hemolytic activity (caution: injected into the blood stream causes dissolving of red corpuscles in extreme dilutions) (The Mark in Index 1983). Other properties of saponin include hormonal modulation, anti-inflammatory, diuretic, anti-microbial, stimulate mucosal secretion and emulsifier (Hoffman, 1-57).

Sidabiyai, is a traditionally prescribed indigenous herbal medicine for the Leucorrhea and PID which contains herbal known to useful to control white discharge like Ephedrine, Saponin, Ginsenoside. The present study detected the therapeutic organic compounds Ephedrine, Saponin, ginsenoside, presence in the herbal formulated medicine, Sidabiyai. The finding explored the gap missing between the present day modem hi-tech knowledge of therapeutic chemistry and the traditionally prescribed herbal medicines that practiced in health care system. Further, the gap missing being the present day need of the hour the Science and Technology has to filled up in depth research and development compelled through the subject and top in. priority so that renovation of traditional knowledge and innovation of the new vitas to frontline to the achievement of subject would come up enlighten.

References

Agarwal, S.K. and Rastogi, R.P., 1974. Triterpenoid saponins and their genins. *Phytochemistry*, 13: 2623–2654.

Ang–Lee, M.K., Moss, J. and Yuan, C.S., 2001. Herbal medicines and perioperative care. *JAMA*, 286: 208–216.

Anonymous, 2006a. Ephedrine and its weight loss. http://www.ravehard.com/doctor/ephidra.htm.

Anonymous, 2006b. Ephedrine and its weight loss. http://www.drbob.org/babble/20000401 /msgs/ 29022.html.

Basu, N. and Rastogi, R.P., 1967. Triterpenoid saponins and sepogenins. *Phytochemistry*, 6: 1249–1270.

Bicopoulus, D., 2002. *Aus. DI: Drug Information for the Healthcare Professional*, 2nd edn. Castle Hill: Pharmaceutical Care Information Services.

Bucci, Luke, R., 2000. Review: Selected herbals and human exercise performance. *American Journal of Clinical Nutrition*, 72(Suppl. 2): S624–S636.

Chandel, R.S. and Rastogi, R.P., 1980. Triterpenoid saponins and sapogenins: 1973–1978. *Phytochemistry*, 19: 1889–1908.

Chidambrathan, S., 2006. www.drcheena.com.

Corthout, J., Naessen, T., Apers, S. and Vlientinck, A.J., 1999. Quantitative determination of ginsenosides from *Panax ginseng* roots and ginseng preparation by thin layer chromatography-densitometry. *J. Pharm. Biomed. Anal.*, 21: 187–192.

Cui, J.F., Bjorkhen, I. and Eneroth, P., 1997. Gas chromatographic mass spectrometric determination of 20(S)-protopanaxadiol and 20(S)protopanaxatriol for study on human urinary excretion of ginsenoside after indestion of ginseng preparation. *J. Chromatogr. B. Biomed. Sci. Appl.*, 689: 349–355.

Cui, J.F., Garle, M., Bjorkhem, L. and Eneroth, P., 1996. Determination of aglycones of ginsenosides in ginseng preparations sold in Sweden and in urine samples from Swedish athletes consuming ginseng. *Scand. J. Clin. Lab. Invest.*, 56: 151–160.

Ford, M.D., Delaney, K.A., Ling, L.J., Erickson, T., 2001. *Clinical Toxicology*. WB Saunders Research Laboratories, Philadelphia.

Gee, J.M., Price, K.R., Ridout, C.L., Johnson, I.T. and Fenwick, G.R., 1989. Effects of some purified saponins on tranmural potential difference in mammalian small intestine. *Toxicol. In vitro*, 3: 85–90.

Gillis, C.N., 1997. Panax ginseng pharmacology: A nitric oxide link? *Biochem. Pharmacol.*, 54: 1–8.

Hoffman, D., 2002. *Therapeutic Herbalism*, pp. 1–57.

Johnson, I.T., Gee, J.M., Price, K.R., Curl, C.I. and Fenwick G.R., 1986. Influence of saponins on gut permeability and active nutrient permeability *in vitro J. Nutr.*, 116: 2270–2277.

Ma, W.W., Heinstein, P.F. and McLaughin, J.L., 1989. Additional toxic, bitter saponins from the seeds of *Chenopodium quinoa*. *J. Nat. Prod.*, 52: 1132–1135.

Madision, U.W., 1995–96. A Thin%20 Layer%20 Chromatography%20module.html.

McElhaney J.E. *et al.*, 2004. A placebo-controlled trial of a proprietary extract of North America ginseng (CVT–E002) to prevent acute respiratory illness in institutionalized older adults. *J. Am. Geriatr. Soc.*, 52(1): 13–19. PMID 14687309.

Meyer, B.N., Heinstein, P.F., Burnnouf-Radosevich, M., Delfel, N.E. and McLaughlin, J.L., 1990. Bioactivity directed isolation and characterization of quinoside a: One of the toxic/bitter principles of quinoa seeds (*Chenopodium quinoa* Wild.). *J. Agr. Food Chem.*, 38: 205–208.

Mizui, F., Kassai, R., Ohtani, K. and Tanaka, O., 1988. Saponins from brans of quinoa wild. II. *Chem. Pharm. But.*, 38: 375–377.

Nonaka, M., 1986. Variable sensitivity of *Trichoderma viride* to *Medicago sativa* saponins. *Phytochemistry*, 25: 73–75.

Oakenfull, D. and Sidhu, G.S., 1990. Could saponins be a useful treatment for hypercolesterolaemia? *Eur. J. Clin. Nutr.*, 44: 79–88.

Ridiout, C.L., Price, K.R., DuPont, M.S., Parker, M.L. and Fenwick, G.R., 1991. *Quinoa saponin*: Analysis and preliminory investigations into the effects of reduction by processing. *J. Sci. Food Agr.*, 54: 165–176.

Schmittmann, B. and Diplomarbit, B., 1962. In: *New Biochemical Separation*, (Eds.) A.T. James and L.J. Morris. Van Nostrand Company, London, 1964.

Schulz, V.R., Hnsel, V.E. and Tyler, 1998. *Rational Phytotherapy: A Physicians' Guide to Herbal Medicine*. Springer, New York.

The Merck Index (TMI), 1983. Tenth edition. Merck and Co. New Jersey, United States of America pp. 375/1204.

Wagner, Handbladt, S., 1996. *Plant Drug Analysis*. Springer Verlag, Berlin Heidelberg,pp. 384.

Wichtl, M. and Bisset, N.G., 1994. *Herbal Drugs and Phytopharmaceuticals*, Medpharm Scientific Publishers, Stuttgart.

Wren, R.C., 1988. *Potter's New Cyclopaedia of Botanical Drugs and Preparations*. The C.W. Daniel Company Ltd., Essex.

Chapter 31

First Report on the Effect of Effluents of Photo Colour Lab in Mayiladuthurai City

S. Ramalingam[1], T. Prabhu[1] and S.M. Nagarajan[2]

[1]Department of Physics, [2]Department of Botany
A.V.C. College (Autonomous), Mannampandal – 609 305, Tamil Nadu

ABSTRACT

Samples were collected from the municipality gutters in Mayiladuthurai city. The gutter effluents consist of major outlets of the color labs situated in and around the city. The samples are properly preserved and systematically analysed for physico-chemical parameters using APHA standard methods. The results revealed that the effluent had enormous amount of micro nutrients (zinc, copper, iron, manganese etc.) macro nutrients such as nitrogen and potassium and heavy metals like mercury, cadmium, lead, chromium, nickel and cobalt. From the results, it was concluded that the out-let of the photo printing color labs should be properly treated for chemical compounds before drainage.

Keywords: *Photo color lab, Gutter effluents, pH, Heavy metals.*

Introduction

In these modern days, a country is polluted by all means only because of cities. Most of the industries are factories situated in cities are the major pollution agents. The outlet of these industries and factories are drained in municipality gutters without any pretreatment. Usually in cities, the gutters are not properly maintained and they drain to the nearby ponds, lakes, pools and rivers etc. Generally, the sewage water is not treated properly for different chemical compounds before discharged. This result in the hazardous pollution of air, water and soil (Oblisami and Rajannan, 1978). Of the factories and industries, the photo color lab plays a major role in pollution, as the effluents carry the highly toxic chemicals to the gutters. All these above facts need a thorough monitoring of dangerous concentrations of hazardous materials, such as heavy metals and toxic compounds in the heart of the city, with reference to water and soil sediments. While doing so, attention should be paid to various physico-chemical parameters of water and soil in effluents of photo printing lab. Realizing this need, a detailed study of heavy metals distribution in water and soil sediments are made in this investigation in the Mayiladuthurai municipality area.

Materials and Methods

The photo printing color lab is situated in the municipality area of Mayiladuthurai city, Nagai District, Tamil Nadu. The soil and water samples were collected from the outlet of photo printing color lab in presterilized polythene bottles by dip method, avoiding bubbling during sampling process and its physico-chemical parameters were analysed for pH, temperature, conductivity, turbidity, color, organic matter, micronutrients, macronutrients and heavy metals. The temperature was measured at the collection site. The sample was preserved for the other parameters in accordance with standard methods (APHA, 1989).

Results and Discussion

The results of analysis of physico-chemical parameters of color lab effluent from outlets of color lab are shown in the Table 31.1 and 31.2.

In the present investigation, the physico-chemical characteristics of the color lab effluent from the area have been analyzed. The color of the effluent was black and odor was unpleasant. It may be due to presence of various chemicals being added for purifications. The pH of the effluent was neutral. The value of pH in the effluent discharged indicates that it is well within the permissible limits (CPCB, 1995).

The temperature was 33°C and the turbidity was 150 NTU. It will affect the oxygen combing capacity thereby indirectly affecting the dissolved oxygen in water. Result of the above study indicates color lab effluent has higher values compared to the standard (3000 ml/l) prescribed by CPCB (1995). Most of the values of inorganic substances and salt which show good conductivity (Manivasakam, 1984). In the present study high levels of BOD in color lab effluent was recorded which may be due to the presence of considerable amount of both organic and inorganic matter. The levels BOD were considerably higher than the CPCB (1995) standard prescribed for BOD (30–100 mg/l). Discharge of effluent into inland surface water and for irrigations purpose increases the BOD level and this increase in BOD is a reflection of microbial oxygen demand to depletion of BOD which may causes hypoxic condition with consent adverse effects in aquatic biota (Poole *et al.*, 1978)

The high BOD also creates septic conditions, generating foul-smelling hydrogen sulfide, which in turn can precipitate iron and any dissolved salts, turning the water black and highly toxic for aquatic life.

The presence of chloride in the outlet of color lab was mainly due to some chemical mixture and its value was found to be 385 mg/l. The presence of chloride concentrations in the wastewater is used as an indicator of organic pollution by industrial sewage (NEERI, 1979).The present investigation revealed that high levels of COD in the effluent sample from color lab do not meet the standard prescribed by CPCB (1995) for effluent discharge into inland surface water (COD levels of 250 mg/l). The high COD value may be due to significant amounts of biologically resistant organic matter (Nagarajan and Ramachandramoothy, 2002).

Table 31.1: Analytical Report of Soil Samples

Sl.No.	*Name of the Parameter*	*Analytical Value Color Lab*
1.	pH	7.10
2.	EC (dms^{-1})	0.80
3.	Colour	Yellow Black
4.	Texture	Sandy Clay Loam
5.	Lime Status	NIL
6.	Organic Matter	1.20
	Available Macronutrients	
7.	Available Nitrogen (Kg/ac)	134.48
8.	Available Phosphorus (Kg/ac)	6.0
9.	Available Potassium (Kg/ac)	289.0
	Available Macronutrients (in ppm)	
10.	Available Zinc	1.26
11.	Available Copper	1.56
12.	Available Iron	8.64
13.	Available Manganese	9.32
	Soil Fractions	
14.	Fine Sand (per cent)	48.69
15.	Coarse Sand (per cent)	26.12
16.	Silt (per cent)	10.36
17.	Clay (per cent)	14.83
18.	**Cat ion Exchange Capacity**	26.2
	(C. Mole Proton $^{+}$/kg)	
	Exchangeable Bases (c. Mole Proton+/kg)	
19.	Calcium	12.6
20.	Magnesium	10.2
21.	Sodium	2.15
22.	Potassium	0.69
	Heavy Metals (ppm)	
23.	Boron	0.09
24.	Molybdenum	0.10
25.	Cobalt	0.06
26.	Chromium	0.12
27.	Lead	0.19
28.	Mercury	0.01
29.	Cadmium	0.08

Table 31.2: Analytical Report of Water Samples

Sl.No.	*Detail of Parameter*	*Samples Details*
	Physical Parameter	
1.	Colour	>3 hue
2.	Odour	Unpleasant
3.	Turbidity	150 NTU
4.	Temperature	33 °C
	Chemical Parameter	
1.	pH	7.10
2.	EC (dms^{-1})	2.0
3.	Total Dissolved Solids (TDS) (mg/l)	1280
4.	Suspended Solids (mg/l)	412
5.	BOD (mg/l)	686
6.	COD (mg/l)	439
7.	DO (mg/l)	0.4
8.	Phosphorus (mg/l)	0.11
9.	Potassium (mg/l)	4.68
10.	Sodium (mg/l)	288
11.	Calcium (mg/l)	150
12.	Magnesium (mg/l)	72
13.	Carbonate (mg/l)	Nil
14.	Bi-Carbonate (mg/l)	427
15.	Chloride (mg/l)	385
16.	Sulphate (mg/l)	192
17.	Nitrate (mg/l)	2.12
18.	Nitrite (mg/l)	Nil
19.	Fluoride (mg/l)	1.20
20.	Zinc (mg/l)	0.16
21.	Copper (mg/l)	0.11
22.	Iron (mg/l)	2.19
23.	Manganese (mg/l)	0.58
24.	Chromium (mg/l)	0.53
25.	Nickel (mg/l)	0.14
26.	Mercury	Nil
27.	Cadmium	0.20

The maximum level of TDS was present in effluent which could be due to high salt content. The TDS values to be beyond the limits prescribed by ISI (1979) and CPCB (1995). The TDS in the effluents renders. Hence further treatment or dilutions is necessary before the effluent is being discharged. Level of suspended solids was only marginally higher in the effluent when compared to the permissible levels of 30 mg/l prescribed by the ISI (1979) for effluent discharge. The effluent sample showed slightly high values of TSS. The ions especially calcium, sulphate, potassium import hardness to water.

From the result of the above study it can be inferred that the physicochemical parameters such as pH, BOD, TDS, TSS of colour lab effluent were found to be high suppressed the permissible limits of CPCB (1995). Release of such effluent into water courses will cause water pollutions problem affecting aquatic organism. Hence the effluent should be treated before discharge.

According to the rules and regulations of environmental pollution control board (EPCB), the management of the industry should take steps to avoid the Industrial effluents being discharged into the gutter directly that is proper measures should be taken by the industry to treat the effluents properly before discharging.

References

APHA (American Public Health Association), 1989. *Standard Methods for the Examination of Water and Wastewater.* AWWA. Wat. Pollut. Control Fed., Washington, DC.

CPCB, 1995. *Pollution Control: Acts, Rules, and Modifications Issued there Under Central Pollutions control Board*, New Delhi.

ISI, 1979. *Guide for Treatment and Disposal of Effluents of Cane Sugar Industry*, IS: 4903 (1st revision) ISI, New Delhi.

Manivasakam, N., 1984. *Physico-chemical Examination of Water, Sewage and Industrial Effluents.* Pragati Prakashan, Meerut, p. 42.

Nagarajan, P. and Ramachandramoothy, T.R., 2002. Oil and grease removal from steel industry wastewater by chemical treatment. *J. Ecotoxicol. Environ. Monit.*, 12(2): 181–184.

NEERI, 1979. *Course Manual on Water and Wastewater Analysis.* NEERI, Nagpur, p. 134.

Oblisami, G. and Rajannan, G., 1978. Effect of industrial effluents on soil and crop plants In: *Abstracts, Int. Symposium on Environmental Agents and their Biological Effects*, Hyderabad, India.

Chapter 32

Rice Straw Utilization Pattern of the Farmers of Punjab

Harmeet Singh Saralch, Gur Reet Pal Singh Brar and Manpreet Singh

Department of Forestry and Natural Resources, PAU, Ludhiana

ABSTRACT

The survey was conducted to study the different aspects of rice straw utilization by Punjab farmers. For this purpose, a detailed questionnaire was prepared to collect information from 200 farmers across 20 districts of Punjab, randomly selecting 10 farmers per district. From the data collected, it was found that out of the total amount of the rice straw produced; 71 per cent was burnt on field, 11 per cent was incorporated in the soil, 10 per cent was fed to the cattle as fodder, 6 per cent was sold off by the farmers to the *gujjars,* one per cent was sold to pulpboard mills and one per cent was put to other uses mainly mulching in potato and other vegetables, animal bedding, making ropes for tying wheat and sugarcane crops after harvesting. Among the three agro-climatic zones of Punjab, maximum per cent was burnt in southern districts of the state, while maximum per cent was used as fodder in the submontane zone particularly by the gujjars.

Introduction

Rice, the world's most important food crop, has been grown for more than 6000 years in South Asia. More than 40 per cent of the world's population depends on rice as the major source of calories and 596 million metric tons of rice is produced annually, more than 90 per cent of which is produced and consumed in Asia (IRRI, 2001).

Due to the advent of modern high yielding varieties, chemical fertilizers, and agronomic management coupled with increase in area under paddy, the production of rice has increased many folds during the last three decades. Consequently, rice straw production has increased accordingly and disposal of straw, particularly in rice-wheat systems has been a major challenge. As estimated by

Sarkar *et al.* (1999), the rice-wheat system accounts for about one-fourth of the total crop residue production in India.

Each year more than 300 million tons (DRR Annual Report, 2000) of rice straw is generated in India. Out of this, the Punjab state is producing more than 15 million tons, annually. Maximum part of this is left unused and is disposed off by burning, while some part of it is used as cattle feed, or recycled for its beneficial effects on soil and plants. The burning of the straw is a serious environmental issue in many rice-producing countries. It is not only a source of atmospheric pollution but also leads to loss of a rich renewable soil rejuvenating organic resource. Alkaline ash left on the soil surface after the burning of crop residues causes an increase in urease activity and may cause N losses from soil and applied fertilizer because of ammonia volatilization (Bacon and Freney, 1989).

To stop this practice and to emphasize on the benefits of other means of disposal of rice straw, it was required that the present pattern of rice straw disposal by the farmers be studied. This survey was conducted to study the different aspects of means of disposal and to get an estimate of overall rice straw utilization pattern of Punjab farmers.

Materials and Methods

To study the means of disposal of rice straw by the farmers of Punjab, a survey was conducted across the state of Punjab. For this purpose, a detailed questionnaire was prepared. The questionnaire was filled from ten randomly selected respondents (farmers) from each district making the total number of surveyed farmers 200 from 20 districts of Punjab. This also gave an idea about disposal pattern for different zones of Punjab state. From the filled proformae, the total amount of rice straw produced by the surveyed farmers was worked out and the proportion of straw put to different means of disposal was calculated. The residue to product ratio was taken as 1.5:1.0 for calculating the amount of rice straw produced per unit grain production, as suggested by Thakur (2003).

Results and Discussion

The Overall Rice Straw Utilization Pattern for the Punjab State

Burning of Rice Straw

The Utilization pattern of rice straw by the farmers of Punjab is given in Table 32.1. It is clear from the Table that out of the total amount of the rice straw, 71 per cent was burnt on field. Most of the farmers stated that there is no such option for straw disposal which could be better as well as cost-effective alternative to open field burning. Furthermore, the time gap between rice harvesting and wheat sowing is short and thus, in order to vacate the fields in time rice straw burning turns out to be the easiest and most economical alternative.

Table 32.1: Rice Straw Utilization Alternatives for Different Zones of Punjab

Sl.No.	*Rice Straw Utilization Alternatives*	*Sub-montane zone (per cent)*	*Central Zone (per cent)*	*Southern Districts (per cent)*	*Punjab (per cent)*
1.	Open Field Burning	47	69	84	71
2.	Livestock Fodder	25	10	4	10
3.	Soil Incorporation	10	11	11	11
4.	Sold to Gujjars	16	8	0	6
5.	Sold to Board Mills	2	0	1	1
6.	Others	0	2	0	1
	Total	100	100	100	100

A study conducted by Singh (2003) also revealed similar trend of rice straw utilization, in which, thirty five farmers were surveyed in the 2 representative villages in Ludhiana district of Punjab and 56 farmers were surveyed in 6 villages in Karnal district of Haryana. It was found that 85 per cent of the farmers surveyed in the Ludhiana district of Punjab preferred to burn the rice straw in field. They do not feed it to animals for two reasons: (*i*) the availability of ample amount of wheat straw and (*ii*) animals do not prefer to eat the rice straw because of its high silica content. Some 5 per cent of farmers were also using rice straw as substrate for mushroom production or as fuel or bedding material for animals. In Kamal district of Haryana, farmers were growing basmati rice, the straw of which has less silica content thus they were using it for the purpose of animal feeding also to greater extent than the farmers in Punjab do. Here, 71 per cent of the surveyed farmers still preferred to bum the rice straw in field. Twenty six per cent of the surveyed farmers used rice straw as animal feed and 3 per cent of farmers were using rice straw for other purposes such as animal bedding material or fuel.

Similar results regarding rice straw burning were reported from a survey carried out in 1996 with 237 farmers selected by stratified random sampling from rice-wheat cropping area in four North Indian states. It revealed that 87 per cent of rice was combine harvested and the straw was burnt. Punjab which stood first in rice productivity was also found to be on first position in burning of rice straw while Haryana was second (Anonymous, 1999).

Other Major Uses of Rice Straw

Among the other uses of rice straw, incorporation in soil was found to be the second best alternative by the farmers. Out of the total amount of rice straw produced by the farmers under survey, 11 per cent was incorporated in the soil. The lower proportion of rice straw being incorporated in soil can mainly be attributed to the rising prices of diesel, as considerable quantity of diesel is required to incorporate the rice straw in the soil with help of the tractor driven disc harrow. It has been reported earlier that in most climates, removal or burning of crop residues deteriorated soil physical properties (Prasad and Power, 1991), while recycling of crop residues helped reducing the adverse effects of hardpan formation in rice-based cropping systems. It is further stated that trend of soil incorporation is decreasing steadily which is primarily due to rising prices of diesel. The other factors which account for the shift in this trend are: less time for vacating the field for wheat crop and slow decomposition of straw in the soil.

It was observed that 6 per cent of rice straw was sold off by the farmers to the *gujjars* and one per cent was sold to the pulpboard mills. The maximum amount of rice straw being sold to *gujjars* was found to be in the submountaneous districts of Hoshiarpur, Ropar and Gurdaspur. It was also observed that in these districts the rice straw fetched higher prices (Rs. 1000–1500 per acre) from *gujjars* as compared to other districts of Punjab (Rs. 200–500 per acre). This may be due to the high population of this community in sub mountaneous zone.

Use of Rice Straw as Animal Feed

The percentage of rice straw used as feed for animals was 10 per cent. It was again mainly used as feed in the sub mountaneous zone. This may be due to the fact that in the central and southern districts, there is plenty of wheat straw available for the cattle, therefore, rice straw is not preferred by cattle in these areas. The other uses of rice straw accounted for 1 per cent of the total rice straw produced by the farmers under survey. The other uses include mulching in potato and other vegetables, animal bedding, making ropes for tying wheat and sugarcane crops after harvesting.

Utilization Pattern for the Three Agro-climatic Zones of Punjab State

Utilization patterns for the three agro-climatic zones of Punjab state were worked out from the questionnaires and the data is presented in Figures 32.1–32.3, for Sub montane Zone, Central districts

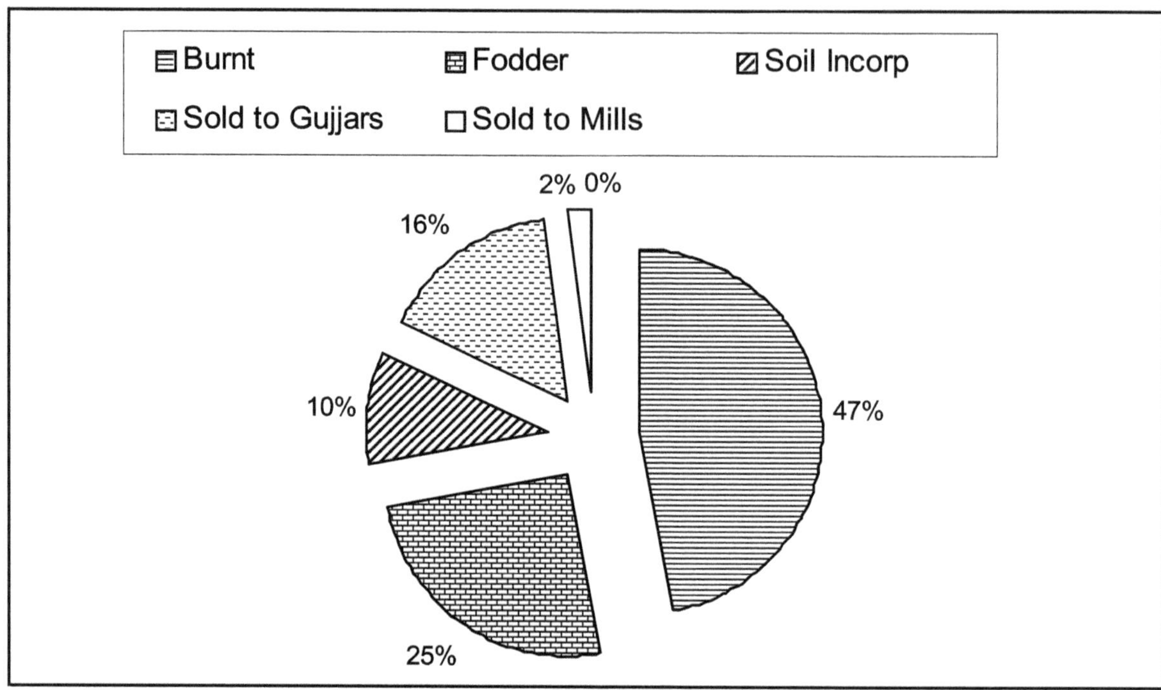

Figure 32.1: Rice Straw Utilization Pattern by the Farmers of Sub-Montane Zone of Punjab

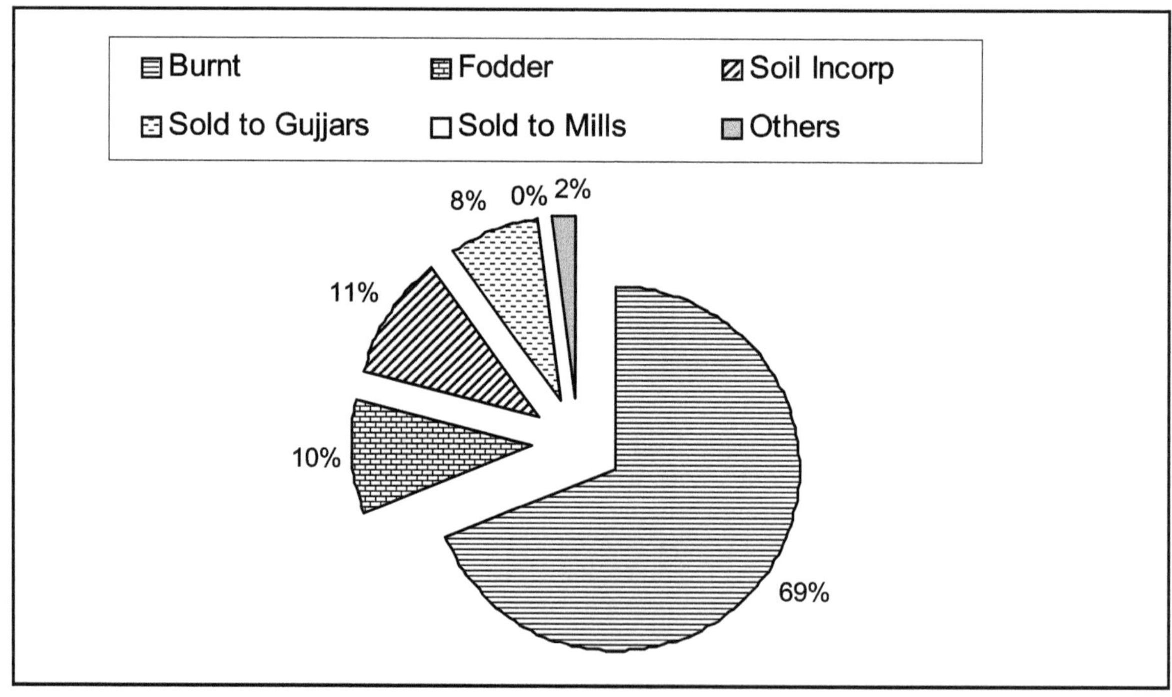

Figure 32.2: Rice Straw Utilization Pattern by the Farmers of Central Districts of Punjab

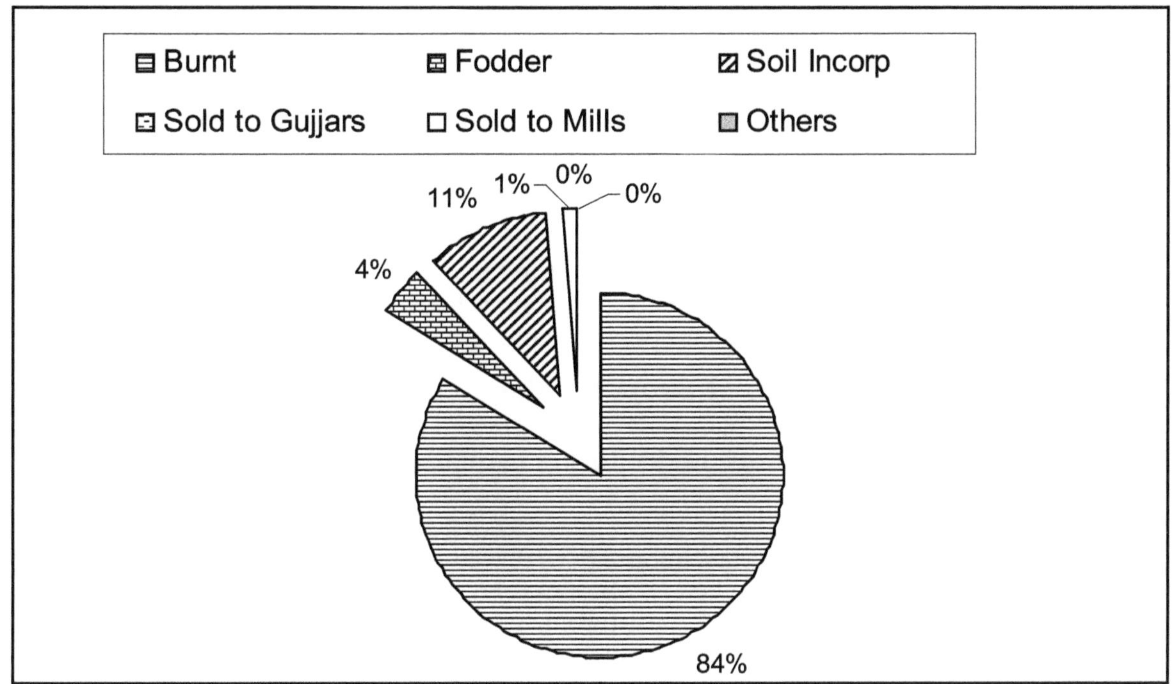

Figure 32.3: Rice Straw Utilization Pattern by the Farmers of Southern Districts of Punjab

and the southern zone, respectively. The salient observations made through perusal of data are discussed below.

Maximum open field burning of paddy straw was observed in southern districts (84 per cent) followed by central zone (69 per cent) and sub-montane zone (47 per cent). This may be attributed to the scarcity of fodder for cattles coupled with higher percent population of *gujjar* community in the sub montane zone of the state. This is also evident from difference in the proportion of paddy straw used as fodder as well as sold to *gujjars* in the three zones. The percent paddy straw used as fodder varied from 25 per cent in sub montane zone to only 4 per cent in Southern zone. It was also observed during the course of survey that there is huge scarcity of fodder during lean period in the sub montane zone while in southern districts the demand is fulfilled by regular supply of wheat straw from adjoining areas of Rajasthan and Haryana.

Sixteen per cent of paddy straw was sold to *gujjars* in sub montane zone while negligible amount of paddy straw was sold to *gujjars* in the southern zone. The sub montane zone has maximum percent population of *gujjars*, the nomadic tribe who are famous for rearing milch cattle, thus, increasing the demand for paddy straw as fodder. It is interesting to note that the trend of soil incorporation of paddy straw remains almost similar in all the three zones. This trend seems to be uniform all over the state because incorporating the paddy straw in the soil is not related to physical-edaphic, demographic or geographical characteristics of the zone, rather only knowledgable farmers were found to follow this practice to enrich soil health.

It is further stated that trend of soil incorporation is decreasing steadily which is primarily due to rising prices of diesel. The other factors which account for the shift in this trend are: less time for vacating the field for wheat crop and slow decomposition of straw in the soil. Regarding the other uses of paddy straw only farmers of central zone emerged in this segment with 2 per cent paddy straw being used for various purposes like mulching in potato, rope making, tying sugarcane bundles and as bedding material for animals.

Awareness Regarding Use of Rice Straw in Pulp and Paper Mills

All the farmers were aware of the use of rice straw in paper and pulp board industries, but majority of the farmers expressed that supplying the rice straw to these industries was not cost effective, as the costs thus incurred are more than the price fetched.

Awareness Regarding Hazards of Straw Burning

It was observed that all the farmers were aware of the environmental hazards of rice straw burning, regardless of the fact that whether they follow the practice or not.

Conclusion

The present study yielded important information regarding the trends of utilization of rice straw among Punjab farmers. It was observed that the burning of rice straw is the major method of disposal of rice straw being used by the farmers. This disturbing issue needs to be addressed with immediate effect. However, it was also revealed by the study that the percentage of the rice straw being burnt on-field is lower than the figures reported earlier. An in-depth study of the utilization pattern at micro level may also be initiated to further examine and promote other environment friendly uses of rice straw.

References

Anonymous, 1999. Enhancing the biodegradation of rice residue in rice based cropping systems to sustain soil health and productivity. *Terminal Proj. Rep. Indian Council of Agric. Res.*, New Delhi, India.

Bacon, P.E. and Freney, J.R., 1989. Nitrogen loss from different tillage systems and the effect on cereal grain yield. *Fert. Res.*, 20: 59–66.

DRR, 2000. *Annual Report*. Directorate of Rice Research, Rajendra Nagar, Hyderabad.

IRRI, 2001. *World Rice statistics*. IRRI, Los Banos, Philippines.

Lee, D.S. and Atkins, D.H.F., 1994. Atmospheric ammonia emissions from agricultural waste combustion. *Geophys. Res. Lett.*, 21: 281–284.

Prasad, R. and Power, J.F., 1991. Crop residue management. *Adv. Soil Sci.*, 15: 205–239.

Sarkar, S., Rathore, T.R. and Sachan, R.S., 1991. Influence of wheat straw on yield of rice and ammonium and nitrate contents of a soil. *J. Indian Soc. Soil Sci.*, 39: 377.

Singh, R., 2003. Recycling of rice straw for improving soil fertility and reducing environmental pollution. *M.Sc. Thesis*, Division of Environmental Sciences, Indian Agricultural Research Institute, New Delhi, p. 60.

Thakur, T.C., 2003. *Crop Residues as Animal Feed.* Addressing resource conservation issues in rice-wheat systems of South Asia: A resource book. Rice wheat consortium for Indo Gangetic Plains (CIMMYT).

Chapter 33

Scanning Electron Microscopic Studies of Cuticular Sensory Structures of the Legs of *Anisops* sp. (Notonectidae)

S. Gupta

Department of Ecology and Environmental Science, Assam University, Silchar – 788 011, Assam

ABSTRACT

Scanning electron microscopic studies on sensory structures of the foreleg, midleg and hindleg of *Anisops* sp. are reported. Study revealed that most of the sensory structures are sensilla trichoidea and the hairplate morphology of the coxa of the foreleg differed with that of the coxa-trochanter joint of the hindleg. The structure and function of eight varieties of sensilla trichoidea and one variety of sensilla basiconica were discussed in the light of their habitat, feeding, mating and defense.

Keywords: *Anisops sp., Sensilla trichoidea, Hairplate morphology, Scanning electron microscopy.*

Introduction

The legs of surface and subaquatic water dwellers play a most vital role in moving around, roaming, capturing food and mating. Moreover, the trichoid sensilla or hairplates located in the joint regions of the legs, neck and antennae are useful for taxonomic demarcation. Hence an attempt has been made to study the cuticular morphology of sensory structures of the foreleg, midleg and hindleg of *Anisops* sp. by scanning electron microscopy as micrographs of sensory structures reveal their orientation, structure and arrangement in detail (Schmidt and Schmidt, 1987).

* Corresponding Author; E-mail: susmita_au@rediffmail.com.

Materials and Methods

A few adults of *Anisops* sp. were collected by hand net from Ward Lake, a small artificial lake in Shillong (25°34'N, 91°52'E), Meghalaya State, India. Their forelegs, midlegs and hindlegs were fixed for 2hrs. in 2.5 per cent glutaraldehyde and post fixed for 2hrs. in 2 per cent osmium tetroxide, both fixatives buffered with 0.1M cacodylate. They were dehydrated in graded concentration of acetone, dried in a critical point drying apparatus, mounted on brass stubs and coated with gold in a fine coat ion sputter JFC 1100 (Gupta *et al.*, 2000). The specimens were examined and photographed in JSM 35CF Scanning Electron Microscope.

Results and Discussion

Forelegs of the *Anisops* sp. exhibit some typical cuticular patterns. On the dorsal side of the coxa, one hairplate consisting of 25 to 30 prominent socketed, cylindrical and hollow sensilla trichoidea (st1) with conically pointed tips were found (Figure 33.1). Their sockets have round collars (Figure 33.2). Nearer to the hairplate, a patch of thin, pointed tapered and pliable sensory hairs (sh) was present (Figure 33.3). On the outer side of the femur longitudinally arrayed socketed spine like sensilla trichoidea (st 2) and a cluster of short blade like sensilla basiconica (sb) were found. The slightly curved tips of the latter pop out from the thick collar of the sockets (Figure 33.4). The outer side of the tarsus of the forelegs is densly covered with flattened, slightly curved, socketed and pointed tipped sensilla trichoidea (st 3) (Figure 33.5). The most prominent feature of the fore leg is the presence of five finger like pliable socketed and pointed tipped sensilla trichoidea (st 4) in the tarsus just opposite to the claw. The inner side of the claw exhibit several layers of denticles usually arranged in nearly parallel rows while the outer surface is smooth with a curved tip (Figure 33.6).

In the midleg, the typical feature is the presence of a hairplate in the trochanter formed of a group of 20 to 22 sensilla trichoidea (st 5). They are socketed having serration at the broad flat end forming a spatula like structure (Figure 33.7). Stiff, socketed, unidirectional spine like sensilla trichoidea (st 6) arranged in a single row throughout the anterior portion of the tarsus is another characteristic cuticular feature of the midleg. Unlike the claw of the foreleg, the claw of the midleg looks like two horns set together (Figure 33.8).

In the hind leg, a hairplate in the coxa-trochanter joint is formed by sensilla trichoidea (st 7) which are socketed, pliable and hooked at the tip (Figure 33.9). The tarsus terminates in a pair of claw. The long unidirectional, socketed, pliable sensilla trichoidea (st 8) along the anterior side of tarsus resembles swimming hairs (Figure 33.10).

The variety of trichoid sensilla forming hairplates in the coxa and trochanter explain its function as position detectors providing information about the load and relative angles between body parts and joints (Graham, 1985; Schmidt and Smith,1987). The hairplate of the coxa of the foreleg formed of long cylindrical hollow sensilla trichoidea (st1) with conically pointed tip is similar to the fossa spongiosa (pad made of tens of thousands of individual hairs) of the Assassin bug which give the bug a good grip (McGavin, 1993). These long cylindrical sensilla trichoidea are known to secret a sticky fluid through the hollow spine and spread the secretion over the ventral side as recorded in the heteropteran families like Berytidae, Meridae and certain Notonectidae (McGavin, 1993). Further being located in a larger socket which allows movement in all directions these sensilla trichoidea (st1) have another function of cleaning the cuticle (Gaffal and Theiss, 1978; Mc.Iver, 1985; Schmidt and Smith,1987; McGavin, 1993). The presence of a patch of noninnervated, unsocketed hairs (sh) near the coxal hairplate serves the purpose of catching small insects by applying sticky secretions from the

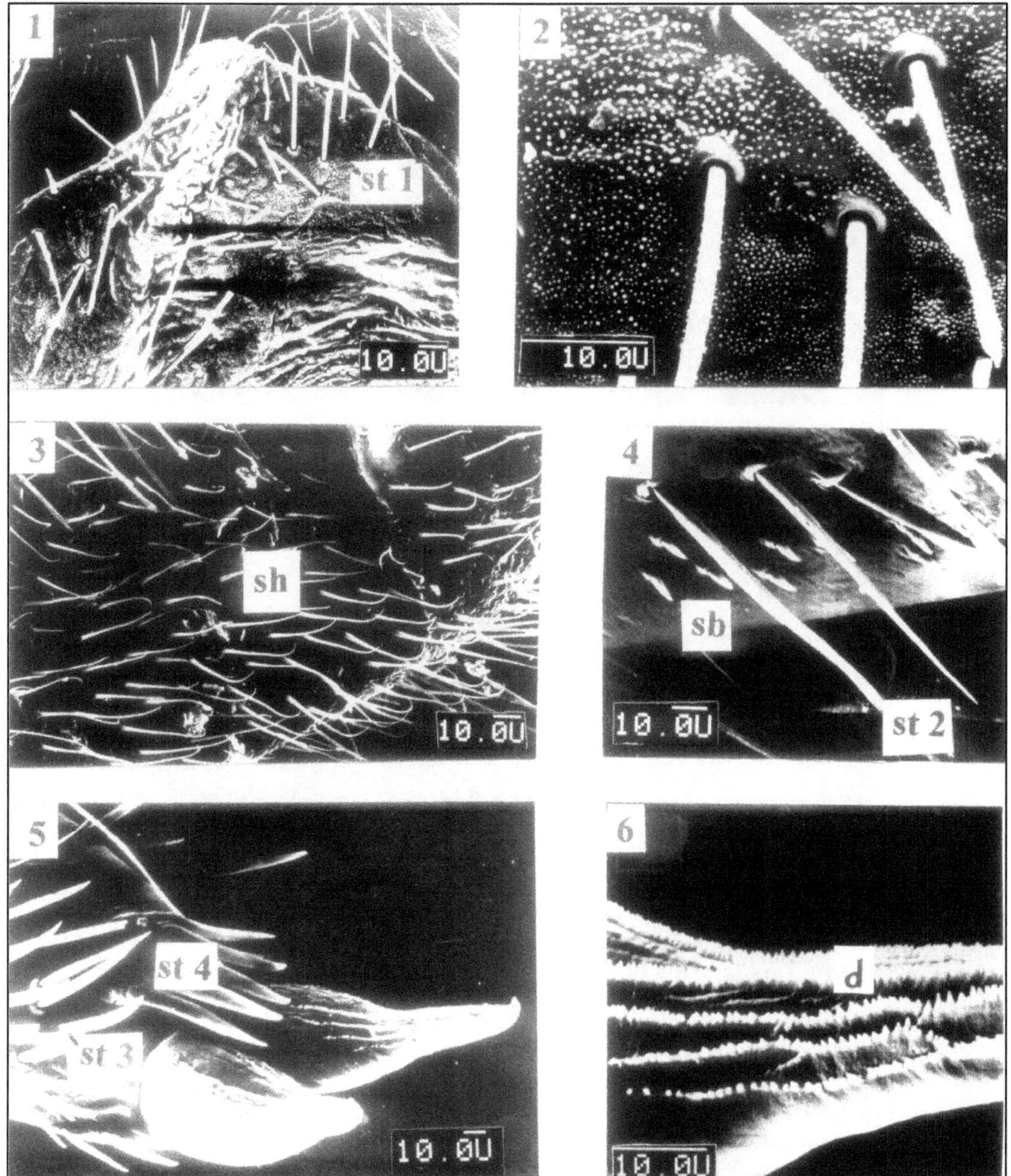

Figure 33.1: Coxal Hairplate of the Foreleg, st 1–*Sensilla trichoidea* Type 1

Figure 33.2: st1–Magnified

Figure 33.3: A Patch of Sensory Hairs in the Coxa of the Foreleg. Sh: Sensory hair.

Figure 33.4: Femur of the Foreleg. st 2: *Sensilla trichoidea*, spine shaped. Sb: *Sensilla basiconica*.

Figure 33.5: Tarsus of the Foreleg, st 3: *Sensilla trichoidea*, Flattened and Pointed Tipped

Figure 33.6: Claw of the Foreleg Magnified, d: denticles

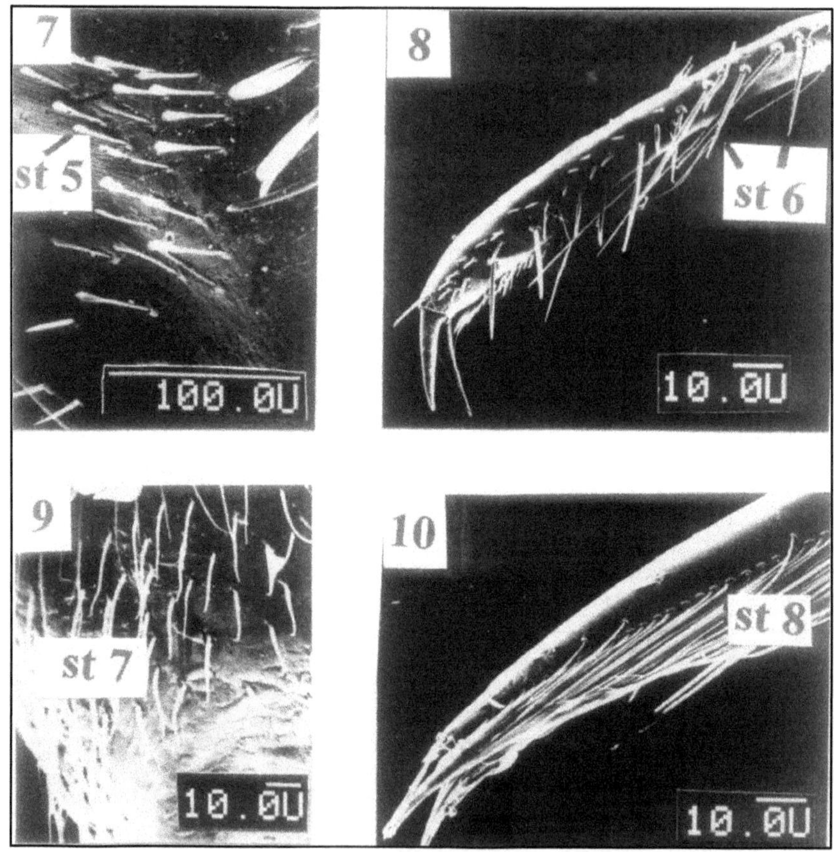

Figure 33.7: Hairplate in the Trochanter of the Midleg. st 5: *Sensilla trichoidea*, Spatula Shaped

Figure 33.8: Tarsus in the Midleg. st 8: *Sensilla trichoidea*, Stiff, Spine Like

Figure 33.9: Hairplate in the Coxa-trochanter Joint of the Hindleg. st7: *Sensilla trichoidea*, Hook Tipped

Figure 33.10: Tarsus of the Hindleg. st 8: *Sensilla trichoidea*, Swimming Hair Like

coxal hairplate beside preventing the growth of microorganism among the dense field of water repellant hairs (McGavin, 1993). The spine like unidirectional structure of the sensilla trichoidea (st 2) of the femur of the foreleg and their close fitting sockets indicate the strength of the spine which is used for holding the prey or for anchoring during feeding or prey capture. The broad based sensilla basiconica (sb) of the femur is olfactory or gustatory in function. The dense growth of the flattened slightly curved or pointed tipped sensilla trichoidea (st 3) in the outer side of the tarsus serve the purpose of a smooth brush (McGavin, 1993). The unique finger like pliable socketed sensilla trichoidea (st 4) along with denticles of the claw aid in grasping during mating and feeding. The hairplates of the trochanter of the midleg formed of spatula shaped sensilla trichoidea (st 5) are position detectors (Wright, 1976; Schmidt and Smith, 1987). The spined tarsus of the midleg is used for anchoring the prey while feeding or for kicking out the rivals for defending a territory (McGavin, 1993). In the hindleg, the hairplate of the coxa-trochanter joint formed of sensilla trichoidea with hooked tip (st 7) have hydrofuge properties and act as physical gill. As *Anisops* are expert swimmers (Ward, 1992), the long, unidirectional, socketed, pliable sensilla trichoidea (st 8) regarded as swimming hairs are used mainly for rowing and for enhancing swimming efficiency. The majority of the sensilla are mechanosensory. Therefore, *Anisops* depends mostly on tactile cues for grooming, prey capture and mating.

Acknowledgement

The author wishes to thank Professor D.T. Khathing, Director, Regional Sophisticated Instrumentation Centre, North Eastern Hill University, Shillong for his help and encouragement.

References

Gaffal, K.P. and Theiss, J., 1978. The tibial thread-hairs of *Acheta domestica* L (Saltatoria : Gryllidae). *Zoomorph.*, 90: 41–51.

Graham, D., 1985. Pattern and control of walking in insects. *Adv. Insect Physiol.*, 18: 31–140.

Gupta, S., Gupta, A. and Meyer-Rochow, V.B., 2000. Post-embryonic development of the lateral eye of *Cloeon* sp. (Ephemeroptera : Baetidae) as revealed by scanning electron microscopy. *Entomologica fennica*, 11: 89–96.

McGavin, G.C., 1993. *Bugs of the World.* Bladford, London, pp. 192.

McIver, S.B., 1995. Mechanoreception. In: *Comprehensive Insect Physiology, Biochemistry and Pharmacology, Vol. 6: Nervous System: Sensory*, (Eds.) G.A. Kerkut and L.I. Gilbert. Pergamon Press, Oxford, p. 71–132.

Schmidt, J.M. and Smith, J.J.B., 1987. The external sensory morphology of the legs and hairplate system of female *Trichogama minutum* Riley (Hymenoptera : Trichogrammatidae). *Proc. R. Soc. Lond.*, B 232:323–366

Ward, J.V., 1992. *Aquatic Insect: Ecology, Biology and Habitat*. John Wiley and Sons Inc., New York, pp. 438.

Wright, B.R., 1976. Limb and wing receptor in insects, chelicerates and myriapods. In: *Structure and Function of Proprioceptors in the Invertebrates*, (Ed.) P.J. Mill. Chapman and Hall, London, p. 323–386.

Chapter 34

Aeromycological Studies and Seasonal Variations of Ascomycotina Spore Types over Potato Agro Environs near Satara in Maharashtra State

Avinash V. Karne

Department of Botany, Shahajiraje Mahavidyalaya, Khatav – 415 505, District Satara, Maharashtra

ABSTRACT

Aeromycological studies were conducted with the help of Continuous Volumetric Tilak Air Sampler for consecutive Kharif season and a Rabi season. Though all the types of spores were encountered, only the Ascomycotina spore types have been dealt here. Out of 21 Ascomycotina spore types recorded during Kharif season of 2004, *Leptosphaeria* (3.38 per cent) contributed highest and *Amphisphaerella* (0.01 per cent) was lowest. During Rabi season of 2003 total 17 Ascomycotina spore types were recorded with *Chaetomium* (1.38 per cent) was the highest and *Rosellinia* (0.06 per cent) was the lowest in percentage contribution to the total airspora over potato agro environs. Seasonal variations and incidence of these spores in the air and changes in the meteorological conditions and the results are discussed in this chapter.

Keywords: *Aeromycogical study, Ascomytina, Airborne spore types, Aeroallergens, Potato fields.*

Introduction

In Indian agriculture, due to agro-economic and population factors, potato deserves a much more important place than it occupies. It is unique crop which can supplement the food needs of the country is a testimony of the fact that, whenever there has been scarcity of food grains, the potato has come to the rescue of people. Thus potato is most promising and productive crop. It is most important vegetable and cash crop grown in this region.

Potato crop is subjected to various diseases caused by fungi, bacteria and viruses which reduces the yield and production. Present investigations were conducted for first time in this unexplored locality and were aimed at analysis of the biopollutants and airspora over potato fields. It was an attempt to under stand variation in the airborne ascospores in different seasonal conditions. A correlation has been made between the presence of ascospores in air and meteorological parameters like temperature, relative humidity and rainfall. The significance of ascospores as aeroallergens and its relevance to allergic disorders is also considered.

Materials and Methods

Potato (*Solanum tuberosum* L.) variety Kufri Jyoti was sown in two acres field at Diskal, District Satara (M.S.). The air monitoring was initiated eight days prior to the sowing of potato in the field and continued for eight days more after harvesting of the same crop. Continuous Volumetric Tilak Air Sampler (Tilak and Kulkarni, 1970) was located in the middle of the experimental field with its orifice facing the west at a constant height of 2.5 feet above the ground level. The study site was completely exposed on all sides having a free display of air currents in all directions.

The studies were conducted for two seasons *i.e.* Rabi Season (from 13th October 2003 to 28th January 2004) and Kharif Season (*i.e.* 15th June 2004 to 30th September 2004). Slides were scanned for estimating the total number of spores and their percentage contribution. The identification of different spore types was based mainly on comparative spore morphology, spore description and microscopic characters. The spore number trapped in the sampler was expressed as number of spores per cubic meter of air. The daily meteorological data of temperature, relative humidity and rainfall was maintained.

Results and Discussion

During the Kharif Season of 2004 in the period of present investigation, 72 types of aerobiopollutants were trapped of which 65 belonged to spore type origin, while remaining 7 types belonged to group 'other types' comprising of algal filaments, hyphal fragments, insect scales (parts), pollen grains, protozoan cysts, trichomes (hairs) and unidentified fungal spores. Out of 67 fungal spore types in general, 3 belonged to Zygomycotina, 21 to Ascomycotina, 3 to Basidiomycotina and 46 to Deuteromycotina. Total 76 spore types were recorded during Rabi Season of 2003 of which 69 belongs to spore type origin, while remaining 7 types belonged to group 'other types'. Of the 69 fungal spore types in general, 3 belongs to Zygomycotina, 17 to Ascomycotina, 4 to Basidiomycotina and 45 to Deuteromycotina. Of the various groups, spore types belonging to the Deuteromycotina (69.99 per cent and 74.30 per cent) contributed highest catches to the total airspora followed by Ascomycotina (15.62 per cent and 9.34 per cent), other types (11.24 per cent and 10.49 per cent), Zygomycotina (2.12 per cent and 3.07 per cent) and Besidiomycotina (1.03 per cent and 2.80 per cent) respectively during Kharif Season of 2004 and Rabi Season of 2003. Though all the types of spores were encountered, only the ascospore types have been dealt here.

Ascomycotina group was found to contribute 15.62 per cent with 21 ascopore types during kharif season of 2004 and 9.34 per cent with 17 ascospore types during Rabi season of 2003, to the total airspora over potato fields. Among 21 ascospore types, spores of *Leptosphaeria* contributed 3.38 per cent followed by *Chaetomium* 2.20 per cent, *Lophiostoma* 1.64 per cent, *Didymosphaeria* 1.32 per cent, the lowest percentage was contributed by *Amphishaerella* 0.01 per cent. During Rabi season of 2003 among 17 ascospore spore types *Chaetomium* contributed highest percentage of 1.38 per cent followed by *Leptospheria* 1.05 per cent, *Hypoxylon* 0.91 per cent, *Pleospora* 0.79 per cent, *Lacanidion* 0.79 per cent, the lowest was *Rosellinia* 0.06 per cent (Table 34.1). Similar results were obtained by Pande (1974), Tilak and Bhalke (1979), Dhaware (1981) and Karne (2006).

Table 34.1: Percentage Contribution of Each Ascomycotina in the Spore Type over Potato Agro Environs

Sl.No.	*Spore Type*	*Kharif Season*		*Rabi Season*	
		Spore Concentration (m^3 of Air)	*Percentage Contribution*	*Spore Concentration (m^3 of Air)*	*Percentage Contribution*
1.	Leptosphaeria	51072	3.38	11620	1.05
2.	Chaetomium	33250	2.20	15386	1.38
3.	Lophiostoma	24780	1.64	1932	0.17
4.	Pleospora	23114	1.52	8820	0.79
5.	Didymosphaeria	19950	1.32	7532	0.68
6.	Pringsheimia	14056	0.93	5166	0.47
7.	Hypoxylon	12544	0.83	10108	0.91
8.	Otthia	11788	0.78	1722	0.16
9.	Hysterium	8008	0.53	6664	0.60
10.	Sordaria	7252	0.48	–	–
11.	Valsaria	4690	0.32	1400	0.13
12.	Parodiella	4382	0.29	2044	0.18
13.	Hysterographium	4382	0.29	5810	0.52
14.	Sporormia	3626	0.24	–	–
15.	Meliola	3318	0.22	5166	0.46
16.	Massarina	2870	0.19	–	–
17.	Rosellinia	2422	0.16	644	0.06
18.	Lacanidion	2114	0.14	8610	0.78
19.	Melanospora	1960	0.13	8386	0.76
20.	Xylaria	308	0.02	–	–
21.	Amphisphaerella	154	0.01	–	–
22.	Nobulosphaeria	–	–	2688	0.24

The occurrence and seasonal variation of the ascospores has been studied over potato fields, which clearly point out the majority of the spore types encountered in the air during the months of June, July, August and September which correspond with the rainy season. The concentration of ascospores gradually increases according to the rainfall. The average monthwise percentage of ascospores was 16.30 per cent in June, 15.21 per cent in July, 16.67 per cent in August and 14.25 per cent in month of September. It reaches peak in month of August followed by June. Similarly during Rabi season of 2003 it was 10.58 per cent in October, 9.36 per cent in November, 8.67 per cent in

December and 9.22 per cent in month of January. The lowest percentage was recorded in month of December. From the above results it is evident that during the rainy season most of the ascospores get released and constitute the dominant type in the composition of airspora. It also becomes very clear that there is a close relation between the rainfall and release of ascospores which is evident from the lowest concentration in December. The increase in the ascospore concentration in the rainy season is in accordance with the results of Tilak and Srinivasulu (1971), Tilak and Bhalke (1979) and Karne (2006). Besides rainfall, other meteorological factors like relative humidity and temperature have a profound effect in determining the concentration of the different ascospore spore types in the air.

The spores of *leptosphaeria* occurred frequently. The maximum monthly mean concentration (16590/m^3 of air and 3822/m^3 of air) was recorded in the month of August and November. The minimum monthly mean concentration (6930/m^3 of air and 2226/m^3 of air) was recorded in month of June and December. The maximum daily mean concentration (952/m^3 of air and 448/m^3 of air) was recorded on 11th and 23rd August 2004 and 14th November 2003 during Kharif season of 2004 and Rabi season of 2003 respectively.

Chaetomium spore type occurred continuously. Its maximum monthly mean concentration (11144/m^3 of air and 4788/m^3 of air) was recorded in the month of August and in the month of December. The minimum monthly mean concentration (976/m^3 of air and 758/m^3 of air) was recorded in month of June and October. The maximum daily mean concentration (672/m^3 of air and 336/m^3 of air) was recorded on 19th July 2004 and on 13th December 2003 during Kharif season of 2004 and Rabi season of 2003 respectively.

Spores of *Rosellinia* occurred frequently. The maximum mean concentration (1162/m^3 of air and 210/m^3 of air) was recorded during the month of August and November. The maximum daily mean concentration (126/m^3 of air and 56/m^3 of air) was recorded on 11th August 2004 and 26th–27th October 2003 during Kharif season of 2004 and Rabi season of 2003 respectively.

Amphispehaerella spore type was rare in occurrence. The maximum monthly mean (98/m^3 of air) was recorded in month of August and minimum monthly mean concentration of (56/m^3 of air) was recorded in month of July and maximum daily mean concentration (42/m^3 of air) was recorded on 10th August 2004 during Kharif season of 2004. This spore type was recorded only during the Kharif season.

All the ascospore types occurred with high concentration in air only when there was congenial weather conditions for sporogenesis and release of the spores into the atmosphere. Spores types of *Sordaria, Sporormia, Massarina,* and *Xyleria,* were restricted to Kharif season only, while spores of *Nodulosphaeria* were recorded only in Rabi season. The ascospore types reported during present study shows a tendency to occur and increase in the concentration after the onset of rain or high humid conditions and low temperature. The results obtained in the present investigation clearly showed that either the rainfall or high humid conditions were having their immediate impact on the spore release of Ascomycotina. These spore types were with "nocturnal pattern' and few were encountered during day time, forming the "wet spora" types. The maximum number of spores were observed in forenoon period fallowed by afternoon and night periods (Tilak, 1980). Ascospores in general were found practically throughout the period of investigation but none was found to be pathogenic (parasitic) on potato plants.

The maximum incidence of these spores was recorded in month of August and November, when there was a record of 22.69°C and 22.1°C temperature, 76.93 per cent and 56.93 per cent relative humidity and 126 mm and 4.00 mm rainfall during Kharif Season of 2004 and Rabi season of 2003 respectively. Role of ascospores like *Chaetomium*, *Leptosphaeria*, *Pleospora* etc. is reported in causing allergy and acute asthma by Frankland and Gregory (1953) and Karne and Pande (2006). It is well documented by earlier workers that exposure to outdoor bio-pollutants increase the airway responsiveness to the aeroallergens and causes allergy or may be responsible for inducing allergenic reactions to sensitive individuals. People working in agricultural fields often suffer from various allergic ailments like repeated cold, breathlessness and sneezing (Chanda and Mandal, 1980).

Acknowledgement

Author is thankful to Prof. Dr. B.N. Pande, Head, Department of Environmental Science, Dr. B.A.M. University, Aurangabad for suggestions and encouragement. He is also grateful to Shri Sanjay Patil, Principal, Shahajiraje Mahavidyalaya, Khatav for motivation and providing library and laboratory facilities.

References

Chanda, S. and Mandal, S., 1980, Aerobiology in India With reference to upper respiratory tract allergy and organic environment pollution. In: *Proc. Int. Aerobio. Conf.*, Munich, pp. 288–306.

Dhaware, A.S., 1981. Aeromycology at Udgir–I: Ascospores. In: *Proc. Nat. Conf. Env. Bio.*, pp. 55–57.

Frankland, A.M. and Gregory, P.H., 1953. Allergenic and agricultural implementation of airborne ascospores concentration. *Nature*, 245: 336–337.

Karne, A.V., 2006. Concentration of airborne ascospores in the atmosphere over potato fields. *Nat. Env. Poll. Tech.*, 5(4): 645–650.

Karne, A.V. and Pande, B.N., 2006. Aeromycological study of allergenic fungal aerobiopollutants over potato fields. *Poll. Res.*, 25(3): 525–530.

Pande, B.N., 1974. Studies in airspora over some fields at Nanded. *Ph.D. Thesis*, Marathwada University, Aurangabad.

Tilak, S.T., 1980, Aeromycology at Aurangabad–I: Ascospores. In: *Proc. Int. Conf. Aerobio. Conf.*, Munich, pp. 145–147.

Tilak, S.T. and Bhalke, S.P., 1979. Aeromycology at Aurangabad–III: Ascospores. *Biovigyanam*, 5: 175–177.

Tilak, S.T. and Kulkarni, R.L., 1970. A new air sampler. *Experientia*, 26: 443–444.

Tilak, S.T. and Srinivasulu, B.V., 1971. Airspora of Aurangabad–II: Ascospores. *Indian Phytopath..*, 24: 740–743

Chapter 35

Optimization of β-glucosidase Assay and Protein Estimation from Various Parts of *Rauvolfia vomitoria*

Om Prakash Verma*, Ankit Kumar, Nirmala Singh, A.K. Gupta and Subhash Chandra Verma

Allahabad Agricultural Institute-Deemed University, Allahabad – 211 007

ABSTRACT

Rauvolfia vomitoria is one of the medicinally most important plant and intensive efforts are required to understand the plant metabolism with respect to its phyto-chemical alkaloids. The activity of crude enzyme in *Rauvolfia vomitoria* was found highest in mature leaf-1 (100 per cent) followed by flower stem (78.25 per cent) whereas lowest in root (4.45 per cent). It was noticed that flowers of *Rauvolfia vomitoria* have highest protein content (3.497 mg/ml) and lowest protein was found in flower stem (1.168 mg/ml).

Keywords: *Rauvolfia vomitoria, Protein, Activity, Alkaloid, β-glucosidase.*

Introduction

Rauvolfia vomitoria is found in Africa, Upper Guinea, Uganda to Angola and Nyasaland, Democratic Republic of Congo and Ivory Coast. The common names include African snakeroot, poison devil's pepper and swizzle-stick. The natural components are ajmaline, serpentine, yohimbine and rescinnamine (Sahu, 1983). The botanical aspects include: tree type, whitish green tiny flower. Shrubs

---

* F.I.S.E.C. (Fellow of International Society for Ecological Communications).

are 5m tall, glabrous, stems erect and stiff. Leaves whorled; leaf blade broadly ovate or ovate-elliptic, 5–12 × 3–6 cm; lateral veins 8–17 pairs. Cymes usually 4 together. Corolla greenish or pale green, tube subcylindric, 6–12 mm, inflated at throat, pubescent inside; lobes dolabriform, 1–2 mm. Stamens inserted at corolla throat; disc ringlike, shorter than ovaries (Kumar *et al.*, 2003). Ovaries distinct, Style filiform, pubescent at base; pistil head fleshy, base membranous. Drupes 2, distinct, ovoid or ellipsoid, 0.8–1.4 cm × 6–9 mm.

Materials and Methods

Plant Material

Rauvolfia vomitoria was used for the β-glucosidase extraction and protein estimation (Esen, 1978). *Rauvolfia* was grown and maintained in the field, at Allahabad Agricultural Institute-Deemed University, Allahabad. The plants were grown following standard agronomic practices. The plant material was freshly harvested for use and processed immediately after harvest to avoid tissue breakage and loss in enzyme activity.

Chemicals

All chemical were of high analytical grade and purchased from Hi-media.

β-Glucosidase Assay

The plants parts of *Rauvolfia vomitoria* was ground to fine powder in a chilled mortar and pestle using liquid N_2 and extraction buffer (1 ml buffer/1 g tissue) was mixed to it. This was centrifuged at 12000 rpm/4°C/30 min. The supernatant was taken and it was used for activity and protein estimation. For activity estimation, para nitophenyl-β-D-glucopyranoside (pNPG) was used as a substrate. Optical density was taken using spectrophotometer at 405 nm for experimental and control reaction. For experimental, 165 μl citrate phosphate buffer (0.1 M, pH 4.8), 10 μl crude enzyme and 25 μl substrate (pNPG) was mixed at room temperature. After 15 min 800 μl, 1 M Na_2CO_3 was mixed to stop the reaction and O.D. was taken. For control, 165 μl citrate phosphate buffer (0.1 M, pH 4.8, Appendix 13), 10 μl enzyme, 800 μl Na_2CO_3 (1M) and 25 μl substrate (pNPG) was mixed at from temperature and after 15 min, O.D. was taken (Mahadaven and Sridhar, 1986).

Protein Estimation

Soluble protein estimation was performed calorimetrically with BSA as standard, using Lowry's method (Lowry *et al.*, 1951). The protein/enzyme extract (100 μl) was precipitated with 100 μl T.C.A. (12 per cent). After 30 min centrifugation was done at 10,000 RPM, 4°C for 5 min. Thus, obtained pellet was dissolved in 200 μl NaOH (0.1 N). Using these samples, protein was estimated. Taking replicates in different volumes such as, 50 μl and 100 μl, and 950 μl and 900 μl of 0.1 N NaOH was added. There after, 5 ml alkaline Cu-reagent was mixed to it. After 10 min, 0.5 ml F.C.C. reagent (IN) was added to it. After 30 min., absorbance was taken at 660 nm. Alkaline Cu-reagent and Folin Ciocalteus phenol dilution was done at the time of estimation only. All the samples including blank and standard BSA are done in triplicate.

Results and Discussion

Optimization of Extraction and Assay of β-glucosidase Enzyme

The enzyme from each part of *Rauvolfia vomitoria* was extracted with procedure described in Materials and Methods. The volume of supernatant and weight of respective tissues taken for different part of plants were tabulated in Table 35.1.

Table 35.1: Volume of Crude Enzyme of *Rauvolfia vomitoria*

Sl.No.	Sample	Wt. of Tissue Taken (g)	Volume of Supernatant (ml)
1.	Fruit	1.341	1.5
2.	Flower	7.381	5.2
3.	Flower stem	1.26	1.0
4.	Young leaf	0.7592	0.2
5.	Mature leaf–1	1.94	2.0
6.	Mature leaf–2	4.514	3.961
7.	Old leaf	2.741	1.1
8.	Stem	2.018	1.0
9.	Root	1. 041	0.74

Activity Estimation

The activity was estimated using spectrophotometer at wave length 405 nm. For the reaction, pNPG was used as a substrate. The reaction mixture was incubated for 15 min at room temperature. Changes in optical density were calculated and the activity was estimated in Table 35.2. The activity of crude enzyme in *Rauvolfia vomitoria* was found highest in mature leaf–1 (100 per cent) followed by flower stem (78.25 per cent) whereas lowest in root (4.45 per cent) as shown as Table 35.2.

Table 35.2: β-glucosidase Activity Profile from Various Parts of the Plant in *Rauvolfia vomitoria*

Sl.No.	Sample	Change in O.D./10 l	Activity Units/ml/min	Activity (per cent)
1.	Fruit	0.030	200	9.50
2.	Flower	0.137	913	43.7
3.	Flower stem	0.245	1633	78.25
4.	Young leaf	0.152	1031	48.53
5.	Mature leaf–1	0.313	2087	100
6	Mature leaf–2	0.191	1273	60.99
7.	Old leaf	0.127	847	40.58
8.	Stem	0.061	407	19.50
9.	Root	0.014	93	4.45

Total Soluble Protein Estimation from Extracts of *Rauvolfia vomitoria*

Total soluble proteins estimated as per details in Materials and Methods section and are represented in Table 35.3. From protein estimation data, it was noticed that flowers of *Rauvolfia vomitoria* have highest protein content (3.497 mg/ml) and lowest protein was found in flower stem (1.168 mg/ml) as shown in Table 35.3.

Specific Activity

Specific Activity was calculated by using the value of activity (International unit) and total protein. The data are presented in Table 34.4 and also represented through Figure 35.1. In *Rauvolfia vomitoria* the glucosidase activity was highest in mature leaf–1 (117.0 IU) followed by mature leaf–2 (60.0 IU) and flower stem (70.0 IU) while specific activity was highest in mature leaf–1 (71.297) and followed by flower stem (59.93) where as lowest glucosidase activity was found in root sample (4.0 IU). Also the specific activity was found lowest in root (1.705) as shown in Table 35.4. And the maximum numbers of bands of proteins were found in young leaves and mature leaf–l of *Rauvolfia vomitoria* (Laemmli, 1970).

Table 35.3: Protein Estimation in *Rauvolfia vomitoria*

Sl.No.	Sample	Protein (mg/ml)	Protein (mg/g)
1.	Fruit	2.167	2.4239
2.	Flower	3.497	2.4636
3.	Flower stem	1.168	0.9269
4.	Young leaf	1.581	0.4165
5.	Mature leaf–1	1.641	1.6917
6.	Mature leaf–2	1.381	1.2118
7.	Old leaf	2.679	1.0751
8.	Stem	1.612	0.7988
9.	Root	2.345	1.6669

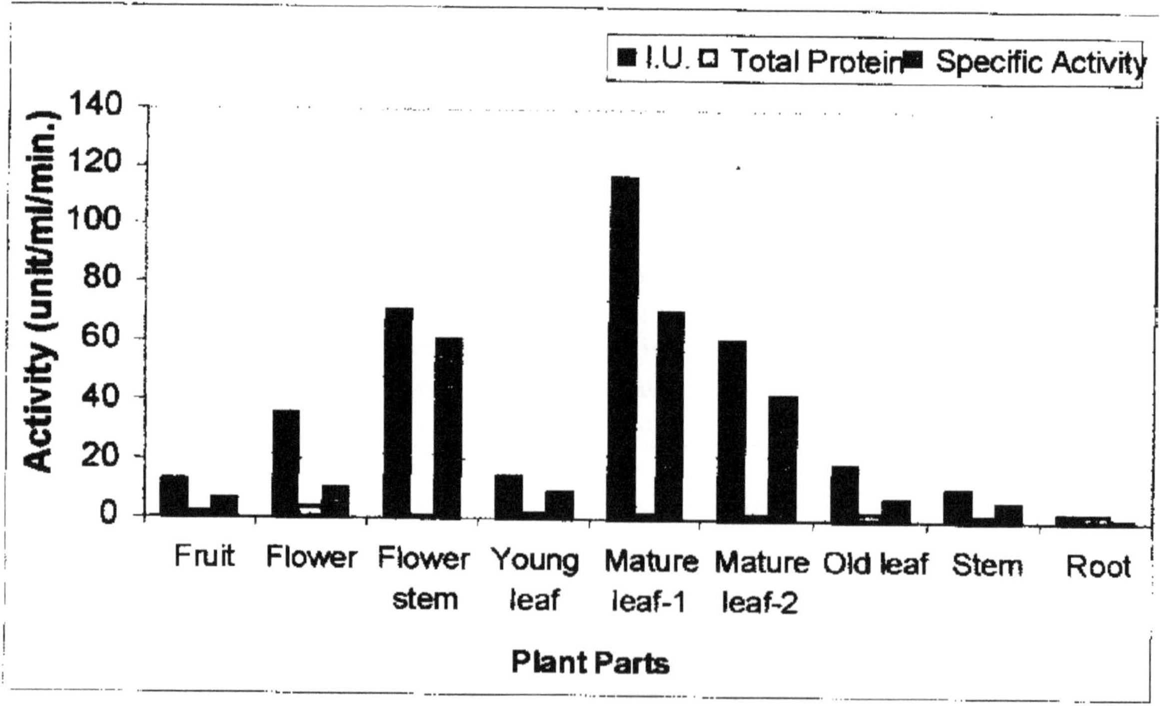

Figure 35.1: IU/g, Total Protein and Specific Activity of β-glucosidase from Various Parts of the Plant in *Rauvolfia vomitoria*

Table 35.4: Estimation of Specific Activity of β-glucosidase in *Rauvolfia vomitoria*

Sl.No.	*Sample*	*Activity Units/ml/min*	*Activity Units/g*	*I.U./g*	*Protein (mg/ml)*	*Specific Activity*
1.	Fruit	200	223	12.0	2.167	5.537
2.	Flower	913	643	35.0	3.497	10.008
3.	Flower stem	1633	1296	70.0	1.168	59.93
4.	Young leaf	1031	266	14.0	1.581	8.855
5.	Mature leaf–l	2087	2151	117.0	1.641	71.297
6.	Mature leaf–2	1273	1117	60.0	1.381	43.446
7.	Old leaf	847	339	18.0	2.679	6.718
8	Stem	407	201	11.0	1.612	6.823
9.	Root	93	66	4.0	2.345	1.705

Results of Native Gels Versus *Rauvolfia vomitoria* Plant Parts

The patterns of band over native gels of *Rauvolfia vomitoria* are given in Table 35.5. From the table, it is apparent that in all the three species glucosidase exhibited two zones of activities in flowers. Root had no detectable bands under the conditions of gel running and its development (Figure 35.2).

Table 35.5: Number of Isoforms of β-glucosidase found in Native PAGE

Plant Parts	*Rauvolfia vomitoria*
Fruit	–
Flower	2
Flower stem	1
Very young leaf	–
Young leaf	2
Mature leaf–1	2
Mature leaf–2	1
Mature leaf–3	1
Mature leaf–4	1
Old leaf	–
Stem	1
Root	0

During the native PAGE study, it was found that in *Rauvolfia vomitoria,* flower, young leaf and mature leaf–1 were found to have two isoform of β-glucosidase as shown in Table 35.5. These results are of physiological significance as they reveal requirement of more than one isoform (root) for the tissue to perform the deconjugation. Interestingly enough, their enhanced amount and presence in flowers and pedicel has not been reported earlier in the literature. This is for the first time we have observed tissue specific activity measurement of β-glucosidase. From native gel developed of crude enzyme, obtained from flowers of *Rauvolfia,* isoforms were obtained. These isoforms were also separated by column chromatography. These isoforms may be RG and SG or some other more isoform, which

Figure 35.2: Native PAGE of *Rauvolfia vomitoria*

Lane 1: Young leaf; Lane 2: Mature leaf–1; Lane 3: Mature leaf–2; Lane 4: Mature leaf–3; Lane 5: Mature leaf-4; Lane 6: Stem; Lane 7: Root; Lane 8: Flower and Lane 9: Flower stem.

may be concluded by further study of β-glucosidase (Kutchan, 1998). These isoforms play an important role in vomiline and ajmaline biosynthesis or may have some new role. Looking at the alkaloid biosynthetic pathway, these two isoforms of enzyme may be strictosidine glucosidase (SG) and raucaffricine glucosidase (RG) or it may be some other isoform of enzyme. If these are RG and SG, then these play a major role in the pathway of ajmaline and vomiline biosynthesis. If this enzyme has some other isoform, then it may have a new role to play. In *Rauvolfia,* particular flower it could have even some participation in anthocyanin development. It would be thus interesting to see the enzyme versus flower colour and development.

References

Esen, A., 1978. A simple method for quantitative, semi quantitative and qualitative assay of protein. *Anal Biochem.*, 89: 264–273.

Kumar, N., Khader, J.M.A., Rangaswami, P. and Irulappan, I., 2003. Sarpagandha (*Rauvolfia vomitoria* benth). In: *Commercial Medicinal Plants,* 3rd edn. Oxford and IBH Publishing Co. Pvt. Ltd., 19.14–19.17.

Kutchan, T.M., 1998. Molecular genetics of plant alkaloid biosynthesis. In: *The Alkaloids: Chemistry and Biology*, (Ed.) G.A. Cordell. Academic Press., San Diego, 50: 257–316.

Laemmli, U.K., 1970. Cleavage of structural proteins during the assembly of the head of bacteriophage T4. *Nature* (London), 227: 680–685.

Lowry, O.H., Rosebrough, N.J., Fano, A.L. and Randall, R.J., 1951. Measurement with the Folin phenol reagent. *J. Boil. Chem.*, 193: 265–275.

Mahadaven, A. and Sridhar, R., 1986. In: *Methods in Physiological Plant Pathology*, 3rd edn. Sivakami Publications, Chennai, pp. 61.

Sahu, B.N., 1983. *Rauvolfia vomitoria: Chemistry and Pharmacology.* Today and Tomorrow's Printers, 2: 595.

Chapter 36

Social Forestry vis-à-vis Joint Forest Management: A Case Study in Orissa

***T.L. Mohanty*[1], *A.N. Dey*[2] *and S. Sindhi*[3]**

Department of Forestry, Orissa University of Agriculture and Technology, Bhubaneswar – 751 003

[1]SMS, Forestry, KVK, Dhenkanal, OUAT

[2]Assistant Professor, Department of Forestry, UBKV, Pundibari, Cooch Behar – 736 165, West Bengal

[3]Fellow Programme in Management at XIM, Bhubaneswar

ABSTRACT

The study was undertaken in some selected village of Khurda district of Orissa namely Kusupalla, Mandalsingh, Sawnpur, Rolasingh and Gopalipada for comparative assessment between social forestry and joint forest management. It was observed that the concept of JFM is more meaningful to the local people than social forestry. The reason may be attributed to the participation of the local people in the management activities, through creation of Van Samrakshana Samiti (VSS) and the direct involvement and bonding of forest staff with the forest protection committees. The institutional arrangement in the JFM also clearly states the rights of the forest users and provides them with a sense of belongingness.

Keywords: *Social forestry, JFM, Van Samrakshana Samiti.*

Introduction

Forest occupies a substantial portion of India's geographical area. The forest cover in country is 63.7 million ha or 19.39 per cent of total geographical area, with average productivity of 0.7 m^3 per hectare per year which is far behind the world average of 2.1 m^3. On this basis the country needs a minimum of 0.4 7 ha of forest land for every individual against the actual availability of 0.11 ha (FSI, 1999). About 80 per cent of our population lives in rural area. According to Asian Survey of Agrarian Reforms and Rural Development sponsored by FAO, 70 per cent Indian farmers are marginal *i.e.* more than 200 million people have an average holding of less than 2 ha (Singh and Tiwari, 1996). They are living much below the poverty line and are residing in and around the forest areas. Forests serve as safety nets during lean periods for most of the poor rural population. In villages adjacent to the forests, forests also provide a source of livelihood and a means of income generation. But with widespread degradation and loss of forest cover, the situation is getting precarious. If this situation will be continued for longer time then it will indicate pervasive threats to our forests even if it will be completely decimated. This has generated an urgent need for sustainable management of forests not only for the planners and policy makers but also the forest dependent communities have felt the urge to protect the forests from further degradation. At this juncture it is highly essential to find the ways for sustainable management and for bridging the gap between demand and supply of forest products and timber.

In this context several attempts have been made to integrate people's need with forest protection, of which the most significant programme was "Social Forestry Programme". The term was coined by Westby during 9th Commonwealth Conference, held in Delhi in 1968. As the name suggests it was forestry of the people, by the people and for the people in the rural areas and was felt as an alternative to check the pressure of populace on the fast depleting forests In Orissa, Social forestry programme started in the year 1983 with financial assistance from Sweden International Development Agency (SIDA). After a decade of its implementation it was felt that social forestry couldn't fulfill the objectives because of lack of participation of the people and failure of implementation policy. This led to the initiation of JFM programme at Arabari in West Bengal in the late seventies. By now it was clear that forests can only be protected with the active participation of the local masses. With the success of Arabari experiment, JFM was adopted in many other states of India. JFM gave an opportunity to the people to have a say in forest management and also protect the forests for the larger and long term benefits. In this context, the Government of India circular of June 1990 formalized and endorsed Joint Forest Management system. It was laid out to develop to ride on the lacunae of social forestry programme and augurs well for the future. At the end of 1998, over 10.24 m ha of degraded forests in the country were being managed and protected by about 36,075 forest protection committees (FPC) or Van Samrakshana Samiti (GOI, 2000). Orissa has the highest average area under forest protection committees–about 597 ha/VFC. Joint Forest Management practice was adopted in Budhikhamari village of Mayurbhanja district of Orissa in 1987 to rehabilitate the degraded Sal forest (Tewari, 1993). This present chapter has focused on the intangible flow of benefits from the forest without affecting the density of standing crop by comparing social forestry vs. joint forest management.

Materials and Methods

The methodology of present study was, one of triangulation, which means a variety of technique and perspectives of data collection and analysis to approach a set of research question and objective. The data in the field were collected by adopting Participatory Rural Appraisal (PRA) method. About 125 people from 5 villages (25 persons per village) were interacted to provide information. Unstructured and semi-structured interviews were also conducted with users, non users, Van Samrakshana Samiti and forest officials. Secondary source of information such as published or printed materials, and available documents were reviewed in order to reconstruct the background and establish working hypothesis Relevant statistical information on population and the data required to assess the demand and supply of forest resources at the household level were collected. Simultaneously, the knowledge of elderly people was taken into consideration for the histories of the condition of forest in this area.

Results and Discussion

From the demographic study of the villages, it was revealed that collection of minor forest produce is one of the significant sources of livelihood of the target population. Besides agriculture, dairy, poultry and goattery are the subsistence type of farming system of the locality. Education level is poor. Proportion of service holder is negligible. The features of the study area of different villages were presented in Table 36.1.

Table 36.1: Features of Study Area of Different Villages

Categories/Features	*Kusupalla*	*Mandalsingh*	*Sawnpur*	*Rolasingh*	*Gopalipada*
Name of Forest	Panchgrah	Mandalsingh	Brahmanideo	Duguri DPF	Duguri DPF
Area	200 ha	200 ha	100 ha	100 ha	80 ha
Number of household	102	689	720	401	123
Population	553	2519	3588	2154	653
Literacy per cent	20.07	42.41	42.58	32.77	32.92
Per capita forest	0.81 ha	0.07 ha	0.02 ha	0.04 ha	0.12 ha
Legal status of forest	Reserve	DPF	DPF	DPF	DPF

Large chunks of land outside forests, located around almost all villages are lying as village common land, panchayat land, degraded revenue forest land, revenue waste land which remain as uncultivated, unculturable land and grazing grounds etc. The social forestry plantations were taken up in these common lands with fast growing local species for the benefit of rural people. The species as *Acacia auriculiformis, Cassia siamea* and *Eucalyptus* sp. were planted during the year 1983–84. The management of these plantations lay with the government. The plantation programme did not meet the objectives and expectations as desired to fulfill the fuel and fodder needs of the people. The major drawback was the non involvement of the local people in the management and protection of the plantations. This also failed to create a sense of belongingness amongst people. The local people were unsure about the procurement and distribution of the products from the plantations. The plantation were not handed over to any organization of that locality for its management as a result the concept of "Everyone's property becomes no one's property" was realized. The distribution of products from the plantation was mostly restricted to stronger section of the society which led to social conflict and lack of mass participation.

The JFM programme was also initiated in the year 1990 in above said villages in the respective parts of degraded protected forests. Vana Samaraksana Samiti (VSS) was set up for different villages as management unit and each household being a member of the VSS. A female and a male representative were essentially a part of each household and members of the VSS. The forester is the ex-officio-secretary of the VSS. The degraded lands under JFM programme were well protected and natural vegetation started coming. The major species as *Shorea robusta, Terminalia bellerica, Dalbergia sissoo, Anogeissus latifolia, Bridelia retusa, Emblica officinalis, Dendrocalamus stricuts, Diospyros melanoxylon* with the under growths of *Hollarhena antidysenterica, Strychnos nux-vomica, Zizyphus oenoplia* and *Phoenix acaulis* etc. came up naturally.

The success of JFM programme was due to the mass involvement and participation of local people with financial as well as technical assistance from the Government. A sense of belongingness amongst the local people, and definitive rules and regulations under JFM provided people a sense of security. The forest products like fruit, fodder, grass, bamboo, cane, fuel wood and minor forest product were distributed equitably among VSS member in the formative years of the forest growth. The members of VSS protected these areas for some tangible benefits but also understood the intangible benefits of the forests and therefore laid much importance on indigenous species. The earlier social forestry plantations in these villages were incorporated under JFM programme after assuring 100 per cent benefits both tangible and intangible to the local people. This extraordinary approach by Government has catalyzed the people's participation in the JFM programme.

The following inferences are drawn from the study area:

1. Failure in social forestry programme has not only created realization but also created awareness among the peoples about the importance of forests which paved the way for JFM.
2. The involvement of women as member of VSS is also an important reason which provides impetus to the programme.
3. Perception of the concept of JFM among VSS members is directly proportional to the level of awareness in the village. Most of the VSS member in Kusupalla village does not have a clear understanding of the concept of JFM while most of the VSS members are well aware of the provisions and the concept of JFM in the village of Mandalsingh and Sawnpur.
4. Co-operation among villagers in the forest protection and carrying out forestry operations is observed to be more effective with smaller population size as seen in village Kusupalla and Gopalipada with population of 553 and 653 respectively. But their forest protection activities are usually marred by the neighbouring villages which are not protecting the same piece of degraded land.
5. Villagers show greater interest in protecting and, managing the forest area available under VSS in Kusupalla and Gopalipada where per capita forest area (under VSS) are 0.81 ha and 0.12 ha respectively while in Mandalsingh, Sawnpur and Rolasingh people show less interest in forest management as per capita forest area are 0.07 ha, 0.02 ha and 0.04 ha respectively. This trend is perhaps directly related with the quantum of usufruct expected per household from the forest. In order to evoke greater interest in JFM, larger area should be allotted to VSS so that more usufructs will be available to local people.
6. Micro plan should be prepared scientifically to achieve maximum productivity in the form of different forest produce.
7. The technical expertise in the fields of Agroforestry, Forestry and Agrostrology etc. should be incorporated by involving scientists in JFM programme.
8. By analyzing the ground realities, all the goals set by VSS can be achieved by involving committed NGOs with capabilities of imparting training and sensitization programmes in field of JFM.
9. The state forest department personal should extend their co-operation to VSS members to make the programme successful.

The pressure on forest land and demand on forest produce are increasing day by day which has aggravated the drastic reduction of forest areas. So the country is facing acute shortage of forest produce and multifarious environmental problems. The prime objective is to prevent continuing degradation of forests and to reduce the gap between the demand and supply by augmenting productivity of forests. Joint Forest Management subsequently proves as viable alternative for conserving our forest at this juncture which needs an atmosphere of creativity, flexibility, mutuality and humanitarianism for its progress, growth development and success. Successful models of JFM are already available in various parts of the country. The people have slowly recognized its importance for their needs as well as forest ecosystem. The study of this approach in different places is likely to provide inputs for perfection of the strategy for rehabilitation of degraded forest areas with active participation of the people.

References

GOI, 2000. *Guidelines for Strengthening of Joint Forest Management Programme.* No. 22-8/2000-JFM (FPD). Ministry of Environment and Forest (Forest Protection Division), New Delhi.

Mishra, V.K. and Khare, A., 1998. *Joint Forest Management Update.* SPWD, New Delhi, India.

Patnaik, L.K., Egneus, H. and Das, S.S., 1989. *Social Forestry Handbook for Orissa.* Social Forestry Project, Orissa.

Singh, M.P. and Tewari, D.N., 1996. *Agroforestry and Wastelands.* Anmol Publication Pvt. Ltd., New Delhi, 299 pp.

Tiwari, D.N., 1993. Joint (participatory) forest management. *Ind. For.*, 119(9): 687–698.

Chapter 37

Ecological Use-Planning of Inputs and Climatic-Fertility for Agricultural Sustainability in India

S. Venkataraman

Retired Director, India Meteorological Department and
Former WMO/FAO Agromet Expert, 59/19, Nav Sahyadri Society, Pune – 411 052, Mh.

ABSTRACT

The terms sustainability of crop production and climatic-fertility are elucidated. Agricultural developments relating to use of inputs and production of irrigated, dryland and industrial crops are examined. Neglect of climatic influences on production and protection of crops leading to (*i*) non-sustainable rates of crop production due to environmental degradation resulting from over use of production inputs and (*ii*) imbalances in crop production are highlighted. Agro-climatic analyses for aiding a quick increase in gross crop production are detailed.

Keywords: *Ecocidal use of water and agrochemicals, GM crops, Quick increase in gross crop production, Agro-climatic outputs.*

Introduction

Sustainability of crop production is the "planned identification, location and efficient use of input-resources and climate to meet the agricultural crop needs of the expected increase in population without quantitative depletion of natural resources and qualitative degradation of the edaphic, ariel and hydro environments" (Venkataraman, 2003), The term "Climate Fertility" has been used (Bernard, 1992) to stress the direct link between climatic variability in weather parameters and variations in the yield-potential of crop cultivars. Thus, climate also constitutes a crop production input. A brief evaluation relating to irrigated, dryland and industrial cropping and use of input resources is presented below.

Developments and Current Scenario

Irrigated Cropping

Irrigation projects in independent India were started as a panacea for all agricultural problems. However, the practice of charging the farmer for area irrigated and not quantum of water supplied have led to farmers over-irrigating their crops and growing of water guzzling crops in unsuitable soils and in areas and periods of high water duty. The above have resulted in virgin soils opened up for irrigation becoming saline and the irrigation projects not realising the envisaged potential in gross irrigated acreage. Lives and capacity of constructed dams have steadily declined due to excessive sedimentation on account of deforestation in catchment areas of dams. Use of high yielding cultivars in the Green Revolution resulted in quantum jumps of unit area yields of irrigated crops. The last 15 years have, however, been characterized by the setting-in of a decline in rates of increase in crop yields even under irrigation in the erstwhile green revolution areas (Gadgil and Rao, 2000). A very high fraction of groundwater has been and is being used as an exclusive source of irrigation in raising commercial crops in high water duty areas and seasons leading to alarming drops in groundwater levels at all locations in the country.

Dryland Cropping

The new cultivars that played a major role in the green revolution lacked physiological senescence (Venkataraman, 1985). Therefore, their peak water consumption covered the maturity period during which dryland crops mostly suffer from soil moisture stress. The old varieties had the ability to curtail transpiration under inadequate soil moisture (Venkataraman, 1981). The tallness of old cultivars gave them fodder value. The new hybrid cultivars were, prima facie, not suitable for use under rainfed conditions.

The differences in yields of rainfed crops between the farmers' fields and the nearby research stations have been reported to be widening with increase in rainfall (Sivakumar *et al.*, 1983). One can understand the below par performance of Kharif crops in years of inadequate rains but not during good rainfall years. This shows that even available dry farming, agronomic technology to take advantage of good rains has not been transferred from lab to land. However, collection of surface runoff from rains in percolation tanks and use of the harvested rainwater for drought-alleviation of crops have been successful at many locations.

Industrial Cropping

Agro-industrial crops have been and are being grown ubiquitously irrespective of the climatic suitability of the areas and periods for their economic production. The policy of one and same pricing to ensure profitability in areas and/or seasons of very low productivity have, for any given crop, led to (*a*) a widening of the disparities in income amongst of farmers in various regions and even in the same region amongst farmers supplying their produce to the factories (*b*) non-exportability of surpluses and (*c*) the danger of import-dumping of produce at a price lesser than what the Indian consumer pays domestically.

Agrochemicals

Currently pesticides and fungicides are being applied in a routine manner at periodic intervals. Incidence of any crop pest or disease depends on pre-disposing, congenial weather conditions and the juxtapositioning of the infective stage of the organism and susceptible stage of the crop. Thus such routine applications are unnecessary and wasteful. Because of climatic influences, the distribution of weed flora across regions and their composition in a region vary greatly. There is no broad-spectrum weedicide effective against all weeds. So weedicide prescription *i.e.* selection of weedicide, its timing, concentration and frequency of application is a specialized job, which places the farmer at the mercy of MNCs and look up to expert advice. In the absence of a policy for basing fertiliser subsidies on an basis of N, P and K elements, our fertiliser usage has become lopsided with excessive applications of nitrogenous fertilizers.

Effects of Over-Use of Agrochemicals

Excessive use of biocides increases the cost of cultivation in a highly disproportionate way related to the quantum of produce saved. Overuse of pesticides and fungicides leads to the development of strains of pests and disease organisms resistant to chemo-control. Again, the indication is that over-use of weedicides for a long period of time will lead to chemo-resistance in weeds (Nandal *et al.*, 1999).

Biocides (pesticides, fungicides and weedicides) directly applied to the soil or entering the soil as leaf-drips and stem-flow from foliar sprays and incorporation of crop residues treated with systemic. biocides lead to a build up of concentrations of the biocides in top soil. Water seeping through or flowing over soils containing concentrated biocide residues pollute groundwater and surface water reservoirs respectively. Kler *et al.* (2005) have drawn attention to the accumulation of systemic biocides causing damage to the animals and humans consuming such sections of the crop. The decline in rates of increase in production of irrigated crops mentioned at the outset is due to over-use of agrochemicals leading to pollution of (*i*) soil and groundwater (Kler *et al.*, 2005) and (*ii*) air and surface water resources (Sahoo, 2000). Excessive nitrogenous fertilizers are either leached to contaminate groundwater or get denitrified to produce Dinitrogen oxide, which has the potential to destroy the Ozone layer (Kler *et al.*, 2005).

Neglect of Climatic Effects

Weather factors exert predominant, direct or indirect, influences on growth, development and yields of crops, distribution in time and space of weed flora, outbreak of pests and diseases, water needs and fertilizer requirements of crops, recharge of groundwater, regeneration and erosion of top soil. Thus, the above mentioned ecocidal effects of over-use of water and agrochemicals and non-realisation of potential (*i*) crop yields and (*ii*) acreage of assured crop production are traceable to the neglect of effects of climatic factors on several aspects of production and production of crops

Present Needs, Constraints and Feasibilities

Unit Area Productivity

The Prime Minister has called for a doubling of crop production in the country in 10 years' time. From considerations of conservation of our valuable biological diversity, afforestation, development of social forestry and growing of bio-fuel crops, additional land cannot be opened up for agriculture. Thus, realisation of the above target calls for an increase in unit area productivity of crops by a compounded rate of 7 per cent every year in the next 10 years. With existing varieties a plateau in unit area yields of crops even with adequate irrigation and other inputs have been reached. Higher unit area yields of crops can come only from new crop cultivars with higher net dry matter production per day. Genetically Modified, GM, crops are held out as panacea for our agricultural problems. The reasons advocated for use of GM cultivars, namely their resistance to weedicides and pests are not compulsive enough for their widespread adoption. The misgivings relating to GM crops such as (*i*) risks of escape of transgenes into natural ecosystems (*ii*) use of GM varieties of crops with indigenous inter-special genetic diversity (*iii*) environmental damage (*iv*) placing the farmers totally at the mercy of seed companies and (*v*) adverse effects of agricultural exports are genuine. Also, no clear-cut data is available relating to the superiority of GM or other cultivars in production of higher total dry matter per day or better Harvest Index vis-à-vis the crop varieties in vogue. In view of the above any significant increase in the near future of unit area yields of crops even in irrigated areas from the current plateau levels does not look feasible.

Increase in Irrigated Acreage

The other alternative is to increase the net area under irrigation. Creation of additional capacities for irrigation is time-consuming.. Completing all ongoing major and medium irrigation projects (Swaminathan, 2006) quickly will increase the net irrigated acreage by only 40 per cent (Garg and Hassan, 2007). Again, Projects like rivers interlinking will not solve water shortage problems of elevated regions.

However, there is considerable scope for drastic savings in use of surface irrigation water. Charging the farmer for the amount of irrigation water with a seasonal weightage for the charges as is being tried out in Maharashtra and irrigating all crops optimally and effectively will help save considerable amounts of irrigation water. Again, paddled rice is one crop in which water loss by way of percolation and run-off substantially exceed the crop's true water need. The widespread acceptance by rice farmers of System of Rice Intensification, SRI for conserving water in the water scarcity areas with an enhancement of yield potential and water use efficiency vis-à-vis flooded rice augurs well for drastic savings in amount of irrigation water at present used in rice culture. Crops that have their establishment phase in the hot weather period or have to pass through summer in their established phase pose problems of non-affordable luxury consumption of irrigation water. However, climate-adapted rationalization of agronomic practices of such irrigated crops like Sugarcane, Cotton etc. can lead to considerable savings in irrigation water.

The quantum of irrigation water saved as above can be used to increase the net irrigated area substantially even with available potential. However, in soils that have been subject to water logging for long periods, detection and correction of deficiencies of micro and trace elements, particularly Iron, need to be taken up. Similarly over-irrigated soils would need to be treated for salinisation. Considering the benefits derivable the expenses on soil amelioration would be worth the investment.

Need for and Role of Agro-Climatic Analysis

Agricultural Sustainability

Agro-ecology has not received as much attention as Industrial Ecology as environmental mismanagement in industrial production manifests itself perceptibly in air and water pollution in urban areas while neglect of ecological considerations in crop production leads to widespread but slow environmental degradation in an imperceptible way in rural areas. However, as neglect of ecological considerations emerges as a significant factor in the current unsatisfactory status and ecocidal nature of production and protection of crops, the need for and role of agro-climatic analyses for micro-level planning of agronomic systems for ensuring sustainability of crop production with environmental conservation becomes obvious.

Maximisation of Irrigation Potential, Net Income and Water Use-Efficiency

In light of the possibility that utilisable water resources might have been overestimated (Garg and Hassan, 2007), optimal irrigation of crops emerges as the single most important factor in maximisation of gross irrigated acreage with the available irrigation potential. Water needs of crops is principally weather-controlled due its direct effect on transpiratory needs of crops and indirect effects on life-duration and growth-rhythm of crops. It is agro-climatologically possible to determine for a given location, soil type and crop cultivar of specified life-duration and developmental rhythm not only its total crop water needs (Allen *et al.*, 1998) but also the optimal uniform interval of irrigation for distributing the total water need (Venkataraman, 2004). An uniform irrigation interval is necessary for proper distribution of crops in command area and convenience of reservoir operations. In irrigated relay cropping because of the dominant role of weather on water needs of crops, at a given location differences in composition and sequence of relay crops will only marginally affect the total crop water needs.

Because of agro-ecological reasons, in irrigated relay cropping even for a given crop its rotational crops cannot be the same in all regions. Under irrigation, Yield per day, YPD (Swaminathan, 1968) is conservative for a given crop cultivar on a seasonal basis. However YPD of a given cultivar can vary under irrigation with seasons (Tanaka, 1964; Oldeman *et al.*, 1987). Life-duration and developmental rhythm of crop cultivars are weather-influenced. Thus, the economic yield and hence water use-efficiency and net income per unit area will show wide a real variations even under irrigation in a season and seasonal variations at a location. The aim of climate-based planning for relay cropping must be to ensure that at a given place and season a crop is so chosen that its phasic weather requirements mesh with the temporal march of the concerned weather parameters. Such data can then be used for comparing the cropping patterns at a location or across locations and select the ones that will ensure maximal net income per hectare (Swaminathan, 2006) and water use-efficiency (Venkataraman, 2005a).

Thus, agro-climatological analyses for planning of optimal irrigation scheduling can assist not only in the determination of the net and gross irrigated acreage possible with existing irrigation potential but also in selection of crops and their sequencing to ensure maximal net profit to farmer and water use-efficiency. The advantage of agro-climatic analyses is that a cafeteria of cropping sequences and irrigation scheduling for such relay cropping can be generated even for virgin areas contemplated for provision of irrigation.

Groundwater Usage

On an overall national scale the point of overexploitation of groundwater appears to have been reached in 1997 itself (Garg and Hassan, 2007). The need is for augmentation of groundwater through artificial recharge. Groundwater recharge is highly erratic in time and space. So its development cannot be specifically planned. Social ownership of groundwater reserves is not possible. Groundwater cannot be conserved by not using it. However, over a period of time, say a decade, its use can be rationalized and restricted to the annual recharge from rains, which can be agro-climatically determined region-wise. In view, of the above, groundwater should not be allowed to be used as a sole source of irrigation and its drawal should be restricted to average annual recharge from rains and for critical irrigations for (*i*) saving rainfed crops during droughts and (*ii*) marginally extending the maturity periods of an irrigated relay crop nearing warm weather.

Dryland Crop Productivity

The crucial factor affecting dryland crop prospects, namely, the temporal distribution of rainfall over short-time periods varies extremely in time and space. Therefore, planning for an assured doubling of dryland crop production in the next ten years is practically impossible. The concept of climatic-fertility is eminently applicable to dryland crop productivity and the emphasis should be on realising the yield potential for a given rainfall distribution every year in all regions of the dry farming tract. Methodology for determination of times and duration of crop-life period for a given distribution of rainfall are available (Venkataraman, 2005b) and can be used to gauge on a location-specific basis, the frequencies and times of occurrence of specified crop-life periods. Such probabilistic assessments of crop life durations can be used to plan the dryland cropping system most suited to local rainfall climatology. Technology developed at research stations for (*i*) coping with inadequate rains and (*ii*) maximizing benefits from good rains should be disseminated to dryland farmers in neighbouring areas for adoption on receipt of medium range forecasts of likely rainfall situation.

Contrary to popular notion the water use efficiency of crops is 10 to 30 per cent lower under irrigation compared to unirrigated conditions (NCOF, 2004). Unlike irrigation rainfall is free. Thus, if farmers are charged for irrigation on the basis of amounts of water supplied as mooted above the net income per hectare of a crop under dry farming and irrigation in a region can be considerably narrowed down. By encouraging harvest of surface runoff from rains and re-use of the same in drought situations to cover all dry land areas much of the dry land crop losses that now occur, due to mid-seasonal droughts, can be avoided. Rainfall harvesting may be done on a watershed basis or even on an individual basis by construction of on-farm-reservoirs, OFRs. The area of OFR will be crop free and agro-climatological analyses can assist in arriving at the times and quantum of availability of surface runoff from rains and hence help in determining the optimal fraction of field to be used for construction of OFR. By augmenting recharge of groundwater and strategically using the same to combat drought, especially during maturity, the cost of water used in dryland crop production can be kept minimal.

Industrial Crops

Agro-industrial crops are mostly irrigated. In agro-climatological anslyses on delineation of homogenous climate zones for all major irrigated crops, yield levels for optimal crop period across locations and for early and late sowings in terms of normal date at a location can also be indicated. Such analyses can help overcome the problems posed by ubiquitous growing of industrial crops by going in for regionalized, specialized and economic production of specific agro-industrial crops for fully realizing the climatic-fertility potential. However, in all the regions the plantings/sowings of industrial crops have to be staggered to ensure steady daily feed of raw material to the mills/factories. The staggered plantings/sowings invariably mean differences in unit area yields of the crop and profits amongst farmers supplying, as per a time-schedule, their produce to the factories. The advantage of specialised and regionalised production of industrial crops is that pricing mechanism for payment to farmers can be meaningfully limited to minimise regional and seasonal inequalities in net profit to farmers.

Minimisation of Use of Agrochemicals

Weedicides

Difficulties and drawbacks in chemo-control of weeds has been mentioned earlier. Indian agriculture is labour-intensive and will continue to be so for many years to come. The effectiveness of traditional practice of hand weeding vis-à-vis the recommended weedicide applications is well established (Reshi and Sapru, 2003). The successful adoption of the System of Rice Intensification, SRI, involves keeping the fields weed-free manually and/or mechanically is a case in point. Thus, manual or mechanical means only need to be adopted for removal of weeds.

Pesticides and Fungicides

Determination of the pre-disposing weather conditions for incidence outbreaks of major pests and diseases can be used for weather-based demarcation of susceptible areas and periods for incidence of specific pests and diseases. Now a crop is susceptible to a pest or disease at a certain growth stage only and the pest or disease organism does damage at a certain development stage only. Thus in case of specific pests/or disease even in areas and periods endemic to it, simple measures like a change in sowing dates from the optimum (Rao *et al.*, 2000) can be used to avoid or at least minimise the setbacks due to pests and diseases. Again, undertaking control operations against pests and diseases only when weather conditions are expected to be pre-disposing for their incidence will result in drastic reductions in applications of pesticides and fungicides. Use of organic insecticides and fungicides of plant origin (Pandey *et al.*, 2004; Dwivedi and Shekhawat, 2005) and non-chemical control of pests and diseases such as (*i*) mixed cropping (to deny a pest or disease organism the ideal micro-climate for its proliferation) and biological control through local introduction of exotic natural predators and parasites of pests and diseases will make a very big reduction in the quantum or inorganic chemical pesticides and fungicides at present used.

Inorganic Fertilizers

Split applications of Nitrogen, deep placement of volatile nitrogenous fertilizers and use of (*i*) biocides of low solubility, high binding ability to soil and quick breakdown and (*ii*) slow release fertilizers have been suggested (Kler *et al.*, 2005) to minimise soil and groundwater pollution by inorganic fertilizers. We are dependant on import of Liquefied Natural Gas, LNG for production of inorganic fertilizers. World supply of LNG, fossil fuels and their by-products are dwindling and becoming increasingly not affordable. It has been reported that Gobar Gas Plants have the potential to meet the entire nutritional requirements of two annual crops on all arable areas in India (Firodia, 2004). Therefore, a big push needs tope given for the subsidized promotion of Gobar gas plants in villages.

Agro-climatic Output Requirements

For implementing the strategies outlined above agro-climatic analyses on a crop-wise, region-wise and season-wise basis for irrigate crops and rainfall zone-wise for dryland crops would be required. Almost all State Agricultural Universities, SAUs have agromet divisions. Thus the following agroclimatic outputs need to be generated by agromet divisions of SAUs for crops in their state/zone of jurisdiction with technical assistance from the agrimet divisions of central institutions like IMD and ICAR.

1. Delineated crop-climate zones for all irrigated crops indicating in each crop-climate zones the times and duration of crop growing periods for early, normal and late sowings.
2. Demarcated homogenous rainfall zones of the dry farming tract on a state-wise basis and delineation of times and duration of soil moisture availability periods in each of the above zones.

3. Assessed (*i*) times and quantum of surface runoff from rains and (*ii*) times and duration of soil moisture availability periods on a probability basis, in each of the rainfall zones. Demarcated endemic areas and periods of major crop pests/diseases based on pre-disposing weather and crop conditions for their outbreak Climatologically values on a Standard Week basis of ratios of Reference crop Evapotranspiration (ETO) of FAO Pan Evaporation for all evaporation stations.
4. Guidelines for determination of optimal irrigation scheduling from pan evaporation and crop phenological data.
5. Agro-climatologically determined optimal irrigation scheduling for all crops in their climate zones.

References

Allen, R.R., Pereira, L.S., Raes, D. and Smith, M., 1998. *Crop Evapotranspiration: Guidelines for Computing Crop Water Requirements*. Food and Agriculture Organisation (FAO) Irrigation and Drainage Paper, 56: 300.

Bernard, E.A., 1992. 'L' intensification de la production agricole par I 'agrometeorologie". *FAO Agrometeorology Working Papers Series*, FAQ, Rome, 1: 35.

Dwivedi, S.C. and Shekhawat, N.B., 2005. Studies on the efficacy of five botanical extracts as pupicidal against *Trogoderma granarinum* (Everts.). *Indian J. Environ. and Ecoplan.*, 10: 31–34.

Firodia, A., 2004. Cows are for ever: Methane gas-the answer to oil imports. *Times of India*, 8-12-04, p. 10.

Gadgil, S. and Seshagiri Rao, P.R., 2000. Farming strategies for a variable climate: A challenge. *Current Sci.*, 78: 1203–1215.

Garg, K.N. and Hassan, Q., 2007. Alarming scarcity of water in India. *Current Science*, 93: 932–941.

Kler, D.S., Navneet Kaur and Uppal, R.S., 2005. Soil and groundwater pollution by agrochemicals: A review. *Ind. J. Environ. and Ecoplan.*, 7: 285–294.

Nandal, D.P., Om, H. and Dhiman, S.D., 1969. Efficiency of herbicides applied alone and in combination against weeds in transplanted rice. *Ind. J. Weed Sci.*, 31: 239–242.

National Commission on Farmers (NCOF), 2004. *First Report: Serving Farmers and Saving Farming*. Private Communication, p. 28.

Oldeman, L.R., Seshu, D.V. and Caddy, F.B., 1987. Response of rice to weather variables. In: *Weather and Rice*. Publication IRRI, pp. 5–39.

Pandey, U.K., Pandy, V. and Singh, P., 2004. Response of some plant origin insecticides against *Spodoptera litura* (Tobacco) caterpillar infesting some food plants. *J. Curr. Sci.*, 5: 737–739.

Rao, K.N., Gadgil, S., Seshagiri Rao, P.R. and Savithri, K., 2000. Tailoring strategies to rainfall variability the choice of the sowing window. *Curr. Sci.*, 78: 1216–1230.

Reshi, Z. and Sapru, B.L., 2003. Weed control efficiency of different herbicides in transplanted paddy under temperate conditions of Kashmir Himalaya. *J. Curr. Sci.*, 3: 35–39.

Sahoo, S., 2000. Sustainable agriculture vs. Biodiversity. *Indian J. Environ. and Ecoplan.*, 3: 745–748.

Sivakumar, M.V.K., Singh, P. and Williams, J.H., 1983. In: Alfisols in the semi arid tropics. A Consultants' Workshop, ICRISAT, Hyderabad, pp. 15–30.

Swaminathan, M.S., 1968. Genetic manipulation of productivity per day. Special Lecture. *ICAR Symposium on Cropping patterns in India.*

Swaminathan, M.S., 2006, 2006–07. Year of Agricultural Renewal. Public Lecture, 93rd Indian Science Congress, Hyderabad, January, pp. 32.

Tanaka, A., 1964. Examples of plant performance. In: *Proc. Symp. on Mineral Nutrition of the Rice Plant,* International Rice Research Institute (IRRI), John Hopkins Univ. Press, London, pp. 37–49.

Venkataraman, S., 1981. Lysimetric observations on moisture accretion for and use by M-35-1 jowar at Solapur. *J. Maharashtra Agric. Universities,* 6: 36–40.

Venkataraman, S., 1985. Agrometeorology as a link in the technology transfer for the stabilization of crop outputs. *Int. J. Ecol. Environ. Sci.,* 11: 91–103.

Venkataraman, S., 2003. Impact of climatic factors on sustainability of crop production with environmental conservation in developing countries. *Int. J. Ecol. Environ. Sci.,* 29: 151–156.

Venkataraman, S., 2004. Agro-climatic determination of water needs of crops: A case study for wheat in India. *Indian J. Environ. and Ecoplan.,* 8: 565–568.

Venkataraman, S., 2005a. Climatic influences on water use-efficiencies in irrigated wheat in India. *J. Curro Sci.,* 7: 277–280.

Venkataraman, S. 2005b. A crop-weather model for water balance studies on Kharif crops through, rainfall budgeting. *Indian J. Environ. and Ecoplan.,* 10: 285–290.

Chapter 38

A Study of the Role of Plerouts and Phosphobacteria in Composting the Municipal of Jayankondam

M. Kalaivani and P. Kennadi

Department of Zoology, Government Arts College,
Ariyalur – 621 713, Tamil Nadu

ABSTRACT

The municipal waste was decomposed with the help of *Plerotus* and *Phosphobacteria.* The composting process was completed in 60 days. The waste materials undergo intensive decomposition under thermophilic and mesophilic conditions in the open place with adequate moisture and finally yield a brown to dark colour to the humified material, which is more stable in form, valuable for replenishment of plant nutrients. In the final compost the N, P, K content was analyzed. Composting maintains the soil organic matter and improve the soil condition.

Introduction

Composting is one of the best ways of managing the solid (Kiyonori Haga 1990). It defines composting as a process by which organic wastes are converted into organic fertilizers by means of biological activity under controlled conditions. Goleuka *et al.* (1977) reported that solid waste management whereby the organic component of the solid waste stream is biologically decomposed under controlled conditions to a state in which it can be handled, stored or applied to the land without adversely affecting the environment. Zadrazil (1977) reported that solid-state fermentation of the wastes through ignolytic fungi and other microbes is a good method of pollution free disposal as agro waters. Devarajan *et al.* (1993) reported that when the soil as irrigated with the effluent it increased the pH and organic carbon content of the soil. Rajarathnam *et al.* (1982) reported that the *Pleurotus* spp., are effective primary degraders of the lignocelluloses waste. Kaviyarasu and Natarajan (1997) reported that during the degradation *Pleurotus* spp., increases the amount of free sugars, amino acids, organic acids were also increased.

Materials and Methods

The garbage was collected from Jayankondam municipal area and segregated into compost able and non-compost able wastes. This was done manually. The compost able waste was taken from a windrow. The Municipal waste used for windrow is 60 kg. The effluent used for to windrow is 400 lts. *Plerotus* and Phosphobacteria were used for microbial inoculum. The preprocessed municipal wastes were used to make a windrow. The dimensions are as follows: 12 feet length, 6 feet width and 1.5 meters height.

Windrow was done to increase the efficiency of compost. The municipal waste was used to make a bed. After a week the effluent was added to the windrow as a moistening agent. Tilling was done for about 3 days. Ureas, Gipsum, Rock phosphate were added in 4 : 3 : 1 ratio. Then the fungal inoculums were added and windrow was undisturbed for a week, then the temperature and the moisture were checked.

Tilling was continued to increase the operation and to give the congenital temperature for fungal growth. Similarly the second bed was made with the help of phosphobacteria and was placed above the first bed. The third bed was prepared with the help of *Pleurotus* and was added above the second bed. Then the three beds were mixed well by means of tilling. Now the windrow height reached 1.5 mts. The temperature was observed everyday. Within 50 days the composting of the material was completed. Then the compost was cured in shallow piles during the period of two weeks.

Results and Discussion

The composting is completed within 50 days. The dark brown humus was observed with agreeable smell. The characteristic of the humus was observed and the results are recorded in the Table 38.1.

Table 38.1

Parameter	Compost
Color	Tea brown
Smell	Agreeable smell
pH	6.5
C/N ratio	15
Temperature	35C
Nitrogen	4.5 per cent
Phosphorus	2.3 per cent
Potassium	3 per cent

The municipal refuse contains Cellulose, Lignin, Nitrogen, Phosphorus and other nutrients essential for growth of microorganisms. *Pleurotus* spp., are effective primary degraders of lignocelluloses waste. During degradation the amount of free sugars, amino acids, organic acids are also increased. The free-living fixing bacteria grows very well in this condition, which improves the nitrogen contents of the final product. This final product increases the phosphorus availability, cationic exchange capacity and water retention.

In addition Rock phosphate was solobilized by the phosphobacteria, which increases the phosphorus content of the final product. The microbial population changes during aerobic composting. The initial stage of composting will occur in the mesophilic stage. When the degradation occurs the temperature of the microorganism rises and attains the termophilic stage. During the stage, the pathogenic organisms and resistant forms of *Ascaris* eggs and cyst of *Entamoeba histolytica* are destroyed. After that the temperature falls and attains the mesophilic temperature. When the temperature increases the pH of the compost turns slightly acidic pH 6 in nature.

The Carbon/Nitrogen ration of good compost is about 15–25 per cent. The Organics in combination of the organic fertilizer were for superior to the sole chemical fertilizer for agronomic efficiency (Sharma, 1991).

The composting is used in 2 ways, which are solid waste management and distillery effluent disposal.

Acknowledgement

The authors thank the Principal, Meenaakshi Ramasamy Arts and Science College, Thathanur for the facilities provided to carry out this work on their college premises.

References

Deverajan, L., Rajannan, G., Ramanathan, G. and Oblisami, B., 1993. Studies on the effluent on alcohol, distillery effluent on soil fertility status, yield and quality of crops produced in scheme report submitted to Shakti Sugars Ltd., Shakti Nagar, pp. 96–234.

Goleuke, C.G., Card, B.J. and McGanntly, P.H.A., 1977. Clinical evaluation of inoculums in composting. *Appl. Microbiol.*, 2: 45.

Kaviyarasan, C.A. and Nataraja, D.E., 1977. Organic acids in soil during decomposition of organic residues. *Indian Nat. Sci. Acad.*, 40: 68.

Kiyonori Haga, T., 1990. Reclamation of organic refuse by composting. *Tech. Bull. 9: Sanitary Engineering Project*, University of California, Berkeley.

Rajarathanam, N., 1982. Bioconversion of cellulosic residues to microbial biomass protein by Lignocellulolytic fungus. *Ph.D., Thesis*, IIT Delhi.

Sharma, R.C., 1991. Nitrogen management in potato. *Tech. Bull No.32* CPRI, Shimla, p.74

Zadrazil, F., 1997. The ecology and industrial production of *Pleurotus ostreatus, P. Florida, P. cornacopiae* and *P. erygil. Mush. Sci.*, 9: 621–652.

Chapter 39

Callus Initiation and Plantlet Regeneration from Cotyledon Explants of *Cucumis melo* var. Utilissimus

M. Venkateshwarlu

Department of Botany, Kakatiya University,
Warangal – 506 009, Andhra Pradesh

ABSTRACT

Callus induction from a suitable explants the present investigation callus induction in *Cucumis melo* L. for callus induction studies. Cotyledon explants excited from six days aseptically grown seedlings NAA 2mg/l + BAP 2 mg/l was found best among various combinations used for induction of callus. For differentiation of callus BAP 1 mg/l + NAA 0.5 mg/l combination was found to be the best that resulted in maximum number of shoot lets. For differentiation followed by profuse rotating to the regenerated shoots. In addition to auxin and cytokinin there are reports involving the possible roles of other growth regulators in the induction of organogenesis. Studies involving the transformation of protoplasts would be of little value unless the genetically altered plant material could be regenerated into a plantlet. Their observation was conformed and extended, who also proved the ability of auxins to stimulate root formation and inhibit shoot formations. A similar effect of the hormone in enhancing shoot induction has been reported in one of the Asclepiadaceae family members, (Beena *et al.*, 2003). The effect of benzyl amino purine in inducing shoot induction was already reported in some of the important medicinal plants (Komalavalli and Rao, 2000; Martin 2002).

Keywords: *Organogenesis, Leaf explants, Morphogenetic response Reproducible.*

Introduction

This variety is cultivated both as a hot weather crop and as a rainy season crop. The fruits are slender and elongated, the length varying from a few inches to about 3 ft. They are pale or dark green in colour, smooth or ridged, with soft downy hairs covering the skin when tender. The explants in cucumber tested include cotyledons (Cade *et al.*, 1987), roots (Trulson *et al.*, 1986) hypocotyls (Gadasi and Ziv, 1986) and leaves (Malopszy *et al.*, 1983). According to Karp 1991, several factors such as genotype tissue source and composition of the medium have been shown to influence regeneration frequency of plant tissue culture. The capacity to induce the formation of adventitious roots and shoots *in vitro* is of utmost importance in plant tissue culture methodology. Studies involving the transformation of protoplasts would be of little value unless the genetically altered plant material could be regenerated into a plantlet. Previous reports on members of cucurbitacea were focused on establishing procedures for regeneration of different cultivars of *Cucumis* (Punja *et al.*, 1990; Chee, 1991). Plantlet formation in *Cucurbita pepo* has been reported by Jelaska (1974).

Materials and Methods

The purpose of this study is to establish an efficient and reproducible method for regeneration and the development of mass variable population of plants in short period of time. To reach this aim it was necessary to identity the most potential explants and the suitable nutrient media and additives that would elicit optimum responses. In the present studies, L-Glutamic acid with a combination of cytokinin or auxin proved to be suitable for morphogenesis and regeneration from cotyledon derived callus. MS basal medium containing 3.0 mg/l. Glutamic acid along with BAP at a concentration of 0.5 mg/l was the best combination for shoot and differentiation in *Cucumis melo* var. *utilissimus*, (Table 39.1, Plate 39.1). Lower concentration of L-Glutamic acid *i.e.* from 0.5 mg/l BAP was in effective in producing shoot buds but enhanced the growth of callus and greening of callus. Cytokinin (BAP) in combination with auxin (NAA) was also found to be suitable for initiating shot buds on the calli. The convey concentration of BAP with 1.0 mg/l NAA produced small shoot buds.

The cotyledon explant callus whenever concentration was increased. The combination of BAP and NAA induced not only small shoot bud formation but also induced profuse rooting and maximum percentage of frequency of growth response of callus was obtained at a concentration of 1.0 mg/l BAP and 1.0 mg/l NAA, and decreased with further increase in concentration while small green spots were observed on the callus of ctyledon. When placed or MS medium with same growth regulators (2.0 mg/l BAP + 1.0 mg/l NAA), regeneration from cotyledon derived callus with small shoot buds.

Results and Discussion

The cotyledon explants used for initiation of callus were obtained from *in vitro* growth seedling and were inoculated on MS medium supplemented with auxins, cytokinins, and auxin and cytokinin combinations and amino acids. The effect had evoked different morphogenetic responses (Table 39.1, Plate 39.1). The addition of 2.0 mg/l BAP and 1.0 mg/l IAA to MS medium resulted in white soft and hard compact callus. The percentage frequency of growth response was high and is 60 per cent at 2.0 mg/l and 0.5 mg/l IAA. Addition of Glutamic acid to the medium produced more friable callus and greening of callus with multiple shoots.

The effect of various growth regulators on cotyledon explants had evoked different morphogenetic responses in *Cucumis melo* var. *utilissimus*. The cotyledon explants used for callus induction were obtained from young seedlings grown *in vitro*. The cotyledons were inoculated on MS medium supplemented with BAP and 2,4-D in different concentrations. Callus was initiated on MS medium

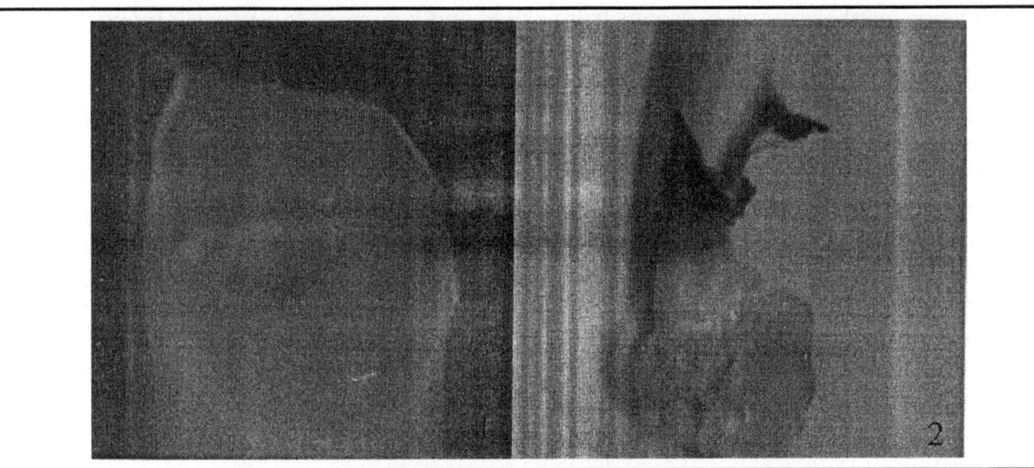

Plate 39.1: Callus Initiation and Plantlet Regeneration from Cotyledon Explants of *Cucumis melo.* var. *utilissimus*

fortified with 2.0 mg/l BAP and 1.0 mg/l 2, 4-D. Increase in the concentration of BAP and 2,4-D promoted production of triable callus BAP in combination with IAA or NAA yielded green friable callus BAP alone at 4.0 mg/l produced friable and compact callus with small shoot buds. NAA alone produced rooting of callus. 2,4-D at a concentration of 2.0 mg/l and BAP at 0.5 mg/l could yield highest amount of callus in fresh weight as well as dry weights. Regeneration from explant such as cotyledon is possible though direct regeneration or via a callus phase. In the present studies the induction of shoot buds was effective with addition of BAP, but kinetin did not yield positive results.

Table 39.1: Effect of L-Glutamic Acid + BAP + NAA on Differentiation of *Cucumis melo* var. *utilissimus* from Cotyledon Derived Callus

Amino Acids and Growth Regulators (mg/l)	*Cotyledon*	
	Per cent of Frequency of Growth Response	*Morphogenetic Response*
0.5 L-Glutanica acid + 0.5 BAP	40	Profused callus growth
1.0 L-Glutanica acid + 0.5 BAP	50	Greening callus
2.0 L-Glutanica acid + 0.5 BAP	65	Small nodules on growth
3.0 L-Glutanica acid +0.5 BAP	70	Small shoot buds (2-3)
4.0 L Glutanica acid +0.5 BAP	75	Small shoot dubs (4-8)
0.5 BAP + 1.0 NAA	40	Decreased in callus growth
1.0 BAP + 1.0 NAA	30	Greening of callus
2.0 BAP + 1.0 NAA	30	Shoot buds
3.0 BAP + 1.0 NAA	25	Shoot buds frooting
4.0 BAP+ 1.0 NAA	20	Profused rooting
5.0 BAP + 1.0 NAA	20	Rooting
6.0 BAP + 1.0 NAA	20	Rooting

The development of complete organs from callus and control of this process by phytohormones represent a major area of interest in plant tissue culture. Regeneration from explant such as cotyledon is possible through direct regeneration or viva callus phase (Cambley and Dodd 1990). Morphogenesis can be obtained directly from cultured explants and in directly from callus or suspension cultures (George and Sherington, 1984). In the present studies the induction of shoot buds was effective with addition of BAP, but kinetin did not yield positive results. NAA alone had inhibitory effect on shoot bud formation, while in combination with BAP has shown stimulatory effect. L-Glutamic acid at concentration of 2.0-2.5 mg/l along with 0.5 mg/l BAP proved to be the best combination for shoot formation in the four genera concerned. Few amino acids also have proved to be effective for the growth of excised embryos. Regenerated shoots are transferred to a root inducing medium. In many cases, auxin alone or in combination with a low level of cytokinin will enhance root primordial formation. In addition to auxin and cytokinin, there are reports involving the possible roles of other growth regulators in the induction of organogenesis.

Conclusion

In case of cotyledon calli the maximum number of shoot lets obtained on MS medium fortified with the combination of BAP, NAA and L-Glutamicacid. *In vitro* produced plants in any crop system have to finally reach to the field where *in vivo* conditions are totally different with practically no control on any of the environmental parameters.

References

Beena, M.R., Martin, K.P., Kirti, M.P. and Hariharan, M., 2003. Rapid *in vitro* propagation of medicinally important *Ceropegia Candelabrum. Plant Cell Tissue Organ Culture*, 72: 285–289.

Chee, P.P., 1991. Somatic embryogenesis and plant regeneration of squash *Cucurbita pepo* L, CVYC 60. *Plant Cell Rep.*, 9: 620–622.

Gadasi, G. and Ziv, M., 1986. Enhanced embryogenesis and plant regeneration from cucumber (*Cucumis sativus* L) callus by activated charcoal in solid/Liquid double layer cultures. *Plant Sci.*, 47: 115–122.

Gambey, R.L. and Dodd, W.A., 1990. An *in vitro* technique for the production *de novo* of multiple shoot in cotyledon explants of cucumber (*Cucumis sativas* L). *Plant Cell Tissue and Organ Culture*, 20: 177–183.

George, E.F. and Sherington, 1984. Plant propagation by tissue culture. In: *Handbook and Directory of Commercial Laboratories Exegetics Limited*, Eversley, Pasingstoke, U.K.

Karp, A., 1991). Oxford surveys of plant molecular and cell biology In: *On the Current Understanding of Somaclonal Variation*, (Ed.) B.J. Migglin. Oxford University Press in cooperation crith the International Society for Plant Molecular Biology, (15 PMB), 7: 1–58.

Komalavalli, N. and Rao, M.N., 2000. *In vitro* micropropagation of *Gymneme sylvestre* a rare multipurpose medicinal plant. *Plant Tissue Organ Cult.*, 61(2): 97–105.

Malopszy, S. and Nadolska-Orczyk, A., 1983. *In vitro* culture of *Cucumis sativas* L. regeneration of plantlets from callus formed by leaf explants. S. Pflauzen *Physiol. Bd.*, 1115: 273–276.

Martin, K.P., 2002. Rapid micropropagation of *Holestemma ada-kodien* schult: A rare medicinal plant through auxiliary bud multiplication and indirect organogenesis. *Plant Cell Rep.*, 21(2): 112–117.

Punja, S.K., Tang, T.A. and Sarmento, G.G., 1990. Isolation, culture and plantlet regeneration from cotyledon and mesophyll protoplasts of two pickling cucumber *Cucumis sativus* L. genotypes. *Plant Cell Reports*, 9: 61–64.

R.M., Wehner, T.C. and Blazich, F.A., 1987. Organogenesis and embryogenesis from cucumber (*Cucumis sativas* L.) cotyledon-derived callus. *Hort. Sci.*, 22: 154 (Abstr).

Telaskas, 1974. Embryogenesis and organogenesis in pumpkin explants. *Physiol. Plant*, 31: 257–261.

Trylson, A.J. and Shahin, F.A., 1986. *In vitro* plant regeneration in the genus cucumis. *Plant Sci.*, 47: 35–43.

Chapter 40

Impact of Lead and Influence of Different Feeds on Carbohydrate Metabolism of Liver Tissue on Freshwater Fish, *Oreochromis mossambicus* (Peters)

K. Aruldoss[1], K. Saravanan[2] and N. Indra[3]

[1]PG and Research Department of Biotechnology,
[2]PG and Research Department of Nutrition and Dietetics,
Bharath College of Science and Management, Thanjavur – 613 005
[3]Professor of Zoology, Annamalai University, Annamalai Nagar – 608 002

ABSTRACT

The present study was to investigate the effect of lead and the influence of different feeds on carbohydrate metabolism in freshwater fish, *Oreochromis mossambicus* for 15 and 30 days. The glycogen content decrease in the liver tissue of lead exposed fish fed with feed 1 (group 4) when composed to control fish fed with feed 1 (group 6) showed a slight decrease in the glycogen liver content when compared to control fish fed with feed 2 (group 2) and feed 3 (group 3). The per cent decrease for group 5 over control group 2 was 20.84 and 16.18 for group 6 over control fish (group 3) was 27.17 and 21.94 for 15 and 30 days respectively.

Remarkable minimum decrease in the content of glycogen was found in the lead exposed fish fed with feed 2 (group 5) than feed 3 (group 6) when compared to corresponding control group 2 and 3. There was no noticeable changes in the control fish fed with feed 2 (group 2) and feed 3 (group 3) when compared to feed 1 (group 1) for 15 and 30 days exposure periods. The decreased level of glycogen liver content in groups 4, 5 and 6 for both 15 and 30 days exposure periods was statistically significant.

***Keywords**: Heavy metals, Lead, toxicity, Supplementary feeds, Glycogen, Liver, Oreochromis mossambicus.*

Introduction

Environmental pollution is a global problem and is common to both developed as well as developing countries (Agarwal, 1940). The developing and progressive country like India is also facing the problem of air, water, land and noise pollution (Lodha, 1991). Heavy metal pollutants cause direct toxicity both to humans and other living beings, due to their presence beyond specified limits (Rai *et al.*, 1998).

In the recent years, industrial development and agricultural process have resulted in the increased levels of toxic metals in the environment, although relatively high concentrations can also occur naturally (Lopez Alonso *et al.*, 2002). Heavy metals have been recognized as strong biological poisons because of their present nature, toxicity, tendency to accumulate in organisms and under go food chain amplification (Kamble and Muley, 2000; Dinodia *et al.*, 2002); they also damaging the aquatic fauna including fish.

The aquatic animals are susceptible to various pollutants like heavy metals, pesticides and industrial effluents, but they have to adjust to these new circumstances by changing their metabolic activities (More *et at.*, 2003). Fish become one of the immediate targets of various pollutants as they are comparatively more susceptible to such pollutants (Prasad Nanda *et al.*, 2000). Lead poisoning is the most important environmental health problem (Committee on Environmental Health, 1993). Dietary constituents have been pointed out both as causative agents in oxidative stress and as protective agents in the antioxidant defense mechanisms against stress (Thomas, 1994). Studies on the elimination of metals from animal are most important from the point of view of human and animal health. Medicinal plants have been used in various traditional system, as they have immune potential against numerous diseases (Kottai Muthu *et al.*, 2005).

Plant products have been shown to have a good therapeutic potential as anti-inflammatory agent and promoters of wound healing due to the presence of active alkaloids terpenes and flavonoids (Porras-Reyes *et al.*, 1993; Mensah *et al.*, 2001). Many plant extracts and phytochemicals have been shown to have antioxidant free radical scavenging properties (Larson, 1986; Bagal *et al.*, 2003). Role of many plant extracts and Ayurvedic formulations in scavenging free radicals has been reported (Vani *et al.*, 1997; Bangul *et al.*, 2003). Experimental work on several plants has been carried out to evaluate their efficacy against chemically induced toxicity. Experimental studies conducted with the *Emblica officinalis* indicate the they have significant cytoprotective against radiation and heavy metal induced toxicities (Sharma, 1978). Juice and extract of the fruit of the plant *Emblica officinalis* are used as Rosayanas, which are claimed to have preventive, curative and health restorative functions (Satyavathi *et al.*, 1976).

Supplementary feeding in tilapia culture is an important one. Among the feeds employed Asia are rice bran, broken rice, oil cakes, flour, corn meal, kitchen refuse, rotten fruit coffee pulp and a variety of aquatic and terrestrial plants. Plants are a rich source of phytochemical. Vallarai leaves (*Centella asiatica*) are widely used in traditional medicines to treat wounds, leprosy, venous insufficiently and disorders of the central nervous system. In recent studies, Vallarai has been shown to improve antioxidant levels which inhibit lipid peroxidation. Freshwater fish *Oreochromis mossambicus*, showed retardation in growth under a sub-lethal gallium exposure (Lin and Hwong, 1998). The use of a biochemical approach has been advocated to provide an early warning of potentially damaging changes in stressed fish (Jee-Lee Yang and Hong-Cheng Chen, 2003).

The biochemical parameters in fish are sensitive for detecting potential adverse effects (Almeida *et al.*, 2002). Carbohydrate metabolism plays an important role in energy yielding process. Glycogen, being the chief source of energy for the fish, is the metabolite to be affected by an stress. Glucose, a reliable source of energy, is present in almost all tissues. Glycogen, the food reserve is utilized more to meet the extra demand of energy during stress conditions which leads to the decrement. Decreased glycogen synthesis is attributed to inhibition of enzyme glycogen synthesis (Stamp and Lesker, 1967). Alteration of carbohydrate metabolism is observed in *Tilapia mossambicus* exposed to arsenic toxicity (Sobha Rani, 2000) in *Labeo rohita* exposed to arsenic trioxide (Pazhanisamy, 2002) and in *Mystal guili* exposed to lead (Kasthuri and Chandran, 1997). Dietary intake of ascorbic acid to protect the body against toxic substances has been studied in Albino rats (Schlegal *et al.*, 1970 and Kamm *et al.*, 1973). Alteration in the level of glucose and lactic acid levels in blood, liver and muscle has been reported in the fish, *Heteropenustes fossilis* (Singhal, 1994). The present study has been undertaken to investigate the influence of different feeds (Vallarai leaves mixed feed *Centella asiatica* and *Emblica officinalis* fruit mixed feed on the changes in the glycogen contents in the liver tissue of freshwater fish, *Oreochromiss mossambicus* for 15 and 30 days.

Materials and Methods

Adult healthy fishes weighing 25030 gm and 13–14 cm length were collected from Annamalai Nagar, India and acclimatized to laboratory conditions for at least 2 weeks. The LC_{50} values were determined by following the method of Finney, (1971). Sub-lethal studies are helpful to assess the response of test organisms under augmented stress caused by metals. According to Konar, (1969) and Sprague, (1971), one tenth of 96 h LC_{50} values represent the lower sub-lethal concentration. 96 h LC_{50} value of lead was found to be 20.03 ppm. Hence the 1/10 of LC_{50} value (2.06ppm) was selected for the period of 30 days. The fishes were divided into 2 batches of 10 each. The fist batch was exposed to sub-lethal concentration of lead acetate (2.06 ppm) for 15 and 30 days. The second group was maintained as control. In the present investigation, sub-lethal concentration of lead acetate was selected and no mortality of fish was recorded throughout the experiment. The dechlorinated water was changed daily for both control and lead acetate stressed fishes. At the end of 15 and 30 days, the tissue was taken out form the experimental and control groups for Carbohydrate estimation.

The calorimetric micro method of Kemp and Kits Van Heijininger (1954) was employed for the quantitative estimation of glucose and glycogen. The mixture was cooled and made upto 5.0 ml with deprotainizing solution. Once again and later centrifuged at 2,000 rpm for 10 minutes. The mixture was heated in a boiling water bath for 6.5 minutes and subsequently cooled and developed colour was measured in greating spectrophotometer (Cecil, Model CE 3313) against the reagent black (3.0 ml conc. Sulphuric acid at 520nm). The glycogen values were expressed as mg/g wet weight of tissue.

Results and Discussion

In the present investigation, lead exposed fish fed with feed 1 shows a decrease in the content of glycogen which compared to control fish fed with feed 1. The decreased level of glycogen liver content in group 4,5 and 6 for both 15 days and 30 days exposure periods was statistically significant (Table 40.1). The present study, *Oreochromis mossambicus* exposed to sub-lethal concentration of lead (group 2 fed with control feed 1) shows a significant decrease in the glycogen level of liver, tissue of 15 and 30 days of exposure periods. Several investigations have been made on the effect of heavy metals on the glycogen and glucose levels of fishes (Karuppasamy, 2000; Datil and Dhande, 2000).

Table 40.1: Glycogen Level of Liver in Control and Experimental Groups

Groups	*Exposure Period in Days*	
	15–Days	*30–Days*
Group 1 (control–Feed 1)	12.75 ± 0.352	12.78 ± 0.427
Group 2 (control–Feed 2)	12.81 ± 0.278	12.79 ± 0.363
Group 3 (control–Feed 3)	12.73 ± 0.408	12.76± 0.216
Group 4 (Lead–Feed 1)	9.11 ± 0.326	8.01 ± 0.276
(Per cent COC–Group 1)	(–28.549)	(–38.289)
Group 5 (Lead–Feed 2)	10.14± 0.279	10.72 ±0.318
(Per cent COC–Group 2)	(–20.843)	(–16.184)
Group 6 (Lead–Feed 3)	9.27 ± 0.277	9.96 ± 0.378
(Per cent COC–Group 3)	(–27.179)	(–21.943)

Mean ± S.E. indicates the mean of six individual observations.

±: Indicates the per cent increase/decrease over the control. Values expresses in mg/g wet wt. of tissue.

Values which are not sharing common superscript differ significantly at 5 per cent level ($P<0.05$)

Duncan's Multiple Range Test (DMRT).

Vincent *et al.* (1995) have reported a decrease of carbohydrate content in liver, gill and intestine after exposure to chromium in Indian major carp *Catla catla*. They have further reported that increased glycogenolysis in fish under the stress of heavy metals may be expressed as depleted levels of tissue carbohydrates. Shoba Rani (2000) has reported a significant depletion of glucose content in the liver, muscle and gill of *Tilapia mossambica* exposed to arsenic. Sunita Shailajan *et al.* (2005) have reported the significant variation in rat tissue glycogen after treatment with CCl_4 impairment of liver metabolism. Since CCl_4 treatment causes reduction in tissue glycogen levels. Reduced glycogen level has been reported in *Channa punctatus* exposed to monocrotophos (Samuel and Sastry, 1989), in freshwater crab *Barytelphusa guerhi* exposed to monocrotophos (Venkateswarlu and Sunitha, 1995). Koundinya and Ramamurthy, (1978) have reported that stepped up glycogenolysis leads to a decrease in glycogen content.

Alkakhera *et al.* (2005) have reported the depletion of glycogen reserves of liver in the atrazine administered animal. Shakoori and Ali (1987) and Towry *et al.* (1985) have reported decreased glycogen content in the liver and muscle of fish treated with mercury. Decreased level of carbohydrate content in muscle, liver and gill tissues has been reported in larvivorous fish, *Gambusia affinis* exposed to different sub-lethal concentration of tannery effluent collected from a tannery in Chromepet on the Madras Chingalpet National Highway (Revathi *et al.*, 2005).

Dietary intake of feed 2 mixed with Vallarai leaves and feed 3 mixed with *Emblica officinalis* fruit to lead exposed fish show a significant increase in tissue glycogen when compared with lead exposed fish fed with normal feed (feed 1). The glycogen content are significantly increased in lead exposed fish fed with feed 2 than fish fed with feed 3 and it indicate that Vallarai mixed feed (feed 2) is more effective than Emblica fruit mixed feed (feed 3). Lead exposed fish fed with feed 2 brings back tissue glycogen values to near normal. In the control fish fed with feed 2 and feed 3 the level of tissue glycogen shows no significant changes when compared to control fish fed with basal feed (feed 1). Muthulingam (2002) has reported oral administration of chloroform and ethyl acetate extract of *Astercantha longifolia* and also *Silymarin* to carbon tetra chloride treated rats shows significant increase in glycogen content in liver and kidney when compared to CCl_4 alone treated rats.

Mohamed Salahy and Abd Allah Mohamoud, (2003) have reported hypoglycemia, hypolepidaemia, hypocholesteraemia, hypotriglyceridaemia, promotion of glycogenesis and lipogenesis in white muscle of carnivorous fish, *Chrysichthyl auratus* after oral administration of garlic Juice (*Allium sativum*). A similar decreasing trend of glycogen has been reported in (*Tenopharyngodone idellus,* Tilak *et al.*, 2000). Ascorbic acid alts as a protective agent against aldrin pollution in *Channa punctatus* in which mortality is reduce? after addition of ascorbic acid in diet (Agrawal and Mahajah, 2000). An improvement in the level of total protein and glucose level near to normal has been reported in the blood of lead intoxicated wistar rat supplemented with Vitamin C and silymarin (Shalan *et al.*, 2005). Thus feed 2 mixed with Vallarai leaves has more antioxidant properties than feed 3 (with *Emblica officinalis*) fruit that may minimize the deleterious effects generated by heavy metal lead thereby suggesting its use as a potent causative action.

Acknowledgement

The authors thank the authorities of Annamalai University and the Head, Department of Zoology for providing necessary Research facilities.

References

Agarwal, K.C., 1940. Environmental pollution and Law. In: *Environmental Pollution*. Agro-botanical Publishers, India, 4E–176, p. 1–2.

Agrawal, A.K. and Mahajan, R., 2000. Effect of piper longum, *Zingerber officinalis* and femia species on gastric ulceration and secretion in rats. *Indian J. Exp. Biol.*, 38: 994.

Almedia, J.A., Diniz, Y.S., Marques, S.F., Faine, L.A., Ribas, B.A., Burneiko R.C. and Novel, E.L., 2002. The use of the oxidative stress responses as biomakers in Nile tilapia (*Oreochromis niloticus*) exposed to *in vivo* cadmium contamination. *Environ. Int.*, 27(8): 673–379.

Committee on Environmental Health, 1993. *Pediatrics*, 92: 176.

Dinodia, G.S., Gupta, R.K. and Jain, K.L., 2002. Effect of cadmium toxicity on liver glycogen in some freshwater fishes. *Proc. 9th Natl., Symp. Environ.*, p. 236–238.

Jen-Lee Yang and Hon-Cheng Chen, 2003. Serum metabolic enzyme activities and Hepatocyte ultra structure of common carp after Gallium exposure. *Zoological Studies*, 42(3): 455–461.

Kamm, J.J., Dashman, T., Conney, A.B., and Barns, J.H., 1973. *Proc. Nact. Acad. Sci., USA.*, 70: 120.

Karuppasamy, R., 2000. Effect of phenyl mercuric acetate on carbohydrate content of *Channa punctatus* Uttar Pradesh. *J. Zool.*, 20(3): 219–225.

Kasthuri, J. and Chandran, M.R., 1997. Sublethal effect of lead on feeding energetic growth performance, biochemical composition and accumulation of the estuarine cat fish *Mystus galio* (Hamilton). *J. Environ. Biol.*, 18(1): 95–101.

Kemp, S.A. and Kitsvan Heijhinger, J.M., 1954. A colorimetric micro method for the determination of glycogen in tissues. *Biochem. J.*, 56: 646–648.

Kottaimuthu, A., Sethupathy, S., Maanavalan, R. and Karar, P.K., 2005. Hypolipidemic effect of methanlic extract of *Dolichos biflorus* Linn. in high fat diet fed rats. *Ind. J. Expl. Biol.*, 43: 522–525.

Koundiya, P.R. and Ramamurthu, R., 1978. Effect of Sumithion (feritrothion) on some selected enzyme systems in the fish *Tilapia mossambica* (peters). *Ind. J. Expt. Biol.* 18: 809–811.

Kumble, G.B. and Muley, D.V., 2000. Effect of acute exposure of endosulfan and chlorpysiphos on the biochemical composition of the freshwater fish, *Saratherodon mossambicus. Indian J. Environ. Sci.*, 4(1): 97–102.

Lin, H.C. and Hwang, P.P., 1998. Acute and chronic effects of gallium chloride (Oach) on Tilapia (*Oreochromis mossambicus*) larvae. *Bull. Environ. Contan. Toxical.*, 60: 931–935.

Lodha, R.M., 1991. Environmental essays. In: *Pesticides and Environmental Pollution.* Ashish Publishing House, New Delhi, India, p. 21–22.

Lopez Alonso, M., Benedito, J.L., Miranda, M., Castillo, C., Hernandez, J., and Shore, R.F., 2002. Interaction between toxic and essential trace metals in cattle from a region with low levels of pollution. *Arch. Environ. Contam. Toxicol.*, 42: 165–172.

Mohamed, B., Al-Salatry and Abd Allah, Mohmoud, B., 2003. Metabolic and histological studies on the effect of garlic administration on the carnivorous fish, *Chrysichthys auratus. Egyptian Journal of Botany*, 5: 94–107.

More, T.G., Rajput, R.A. and Bandela, N.N., 2003. Impact of industrial effluents on DNA contents in the whole body of freshwater bivalve, *Lamellidens marginalis. J. Industrial Pollution Control*, 19(2): 195–2002.

Muthulingam, M., 2002. Studies on the curative efficacy of *Astercantha longifolia* L. Nees (Acanthaceae) on carbon tetra chloride induced hepatoxicity in rats. *Ph.D. Thesis*, Annamalai University, India.

Patil and Dhande, 2000. Effect of mercurin chloride and cadmium chloride on haematological parameters of the freshwater fish, *Channa punctatus* (Bloch). *J. Ecotoxicol. Environ. Monit.*, 10(3): 177–181.

Prasanta Nanda, R.N., Panda and Behera, M.K., 2000. Nickel induced alterations in protein level of some tissues of *Heteropneustes fossilis. J. Environ. Biol.*, 21(2): 117–119.

Rai, A.K., Upadhyay, S.N., Kwnar, S. and Upadhya, Y.D., 1998. Heavy metal pollution and its control through a cheaper method. *J. IAEM*, 25: 22–51.

Revathi, K., Gulati, Suman and Sharief, Dawood, 2005. Tannery effluent induced biochemical changes in the larvivorous fish, *Gambusia affinis. Poll. Res.*, 24(4): 815–818.

Samuel, M. and Sastry, K.V., 1989. *In vivo* effect of monocrotophos on the carbohydrate metabolism of the freshwater snake head fish, *Channa punctatus* Pestic. *Biochem. Physiol.*, 34: 1–8.

Sathyavathi, G.V., Raina, M.K. and Sharma, M., 1976. In: *Medicinal Plants of India*. Indian Council of Medical Research, New Delhi, p. 277.

Schlegal, L.V., Pipkin, G.E., Nishmura, R. and Schultz, G.N., 1970. *J. Urol.*, 103: 155.

Shakoori, A.R. and Ali, S., 1987. Effects of cadmium and lead on common edible fish *Cirrhina mrigala*, of Pakistan. *Second Annual Technical Report of PSF Research Project*, p-pu/Bio (121) 63 p.

Shalan, M.G., Mostafa, M.S., Hossouna, M.M., Hassab, El-Nabi, S.E. and El-Refaie, A., 2005. Amelioration of lead toxicity on rat liver with Vitamic C and Silymarin supplements. *Toxicology*, 206: 1–15.

Sharma, P.V., 1978. *Dravyaguna Vijana.* Chaukhamba Sansthans, Varanasi.

Shoba Rani, A., Sudharson, R., Reddy, T.N., Reddy, P.V.M. and Raju, T.N., 2000. Effect of sodium arsenates on glucose and glycogen levels in freshwater teleost fish, *Tilapia mossambica. Poll. Res.*, 19(1): 129–131.

Singhal, K.C., 1994. Biochemical and enzymatic alterations due to chronic lead exposure in the freshwater cat fish, *Heteropheustes fossilis. J. Environ. Biol.*, 15(3): 185–191.

Stamp, W.D. and Lesker, P.A., 1967. Enzyme studies related to sex difference in mice with hereditary muscular dystrophy. *American Medical Journal Physio.*, 213: 587.

Sunitha Shailajan, Naresh Chandra, R.T., Sane and Sasikumar Menon, 2005. Effect of *Asteracantha longifolia* Nees against CCl_4 induced liver dysfunction in rat. *Indian Journal Exp. Biol.*, 43: 68–75.

Thomas, J.A., 1994. Oxidative stress, oxidant defense and dietary constituents In: *Modern Nutrition in Health and Diseases*, 8th edn., (Eds.) M. Shils, J.A. Olson, M. Shike. Lea Febiger, Philadelphia pp. 501–512.

Toury, R., Stelly, N., Biossonneau, E. and Dipuis, Y., 1985. Degenerative processes in skeletal muscle of cadmium treated rats and cadmium inhibition of mitochondrial calcium transport. *Toxicol. Appl. Pharmacol.*, 77: 19–35.

Venkateswarulu, M. and Sunitha, A., 1995. Impact of monocrotophos on tissues carbohydrates of the freshwater carp *Barytelphusa querini* (Mine Edwards). *Ind. Jour. Comp. Ani. Physio.*, 13(1): 35–36.

Verma and Tonk, L.P., 1983. Effect of a sub-lethal concentrations of mercury on the composition of liver, muscle and ovary of *Notopterus notopterus. Water, Air and Soil Pollution*, 29(3): 287–292.

Vincent, S., Arobrose, T., Cysil Arun Kumar and Selvanayagam, 1995. Biochemical response of the Indian Major Carp, *Catla catla* (Ham) to chromium toxicity. *Indian J. Environ. Hlth.*, 37(3): 192–196.

Chapter 41

Multiple Shoot Induction from Tendril Explants of *Cucumis melo* cv. *Bathasa*

M. Venkateshwarlu

Department of Botany, Kakatiya University, Warangal – 506 009, Andhra Pradesh

ABSTRACT

Multiple shoot induction was achieved in one of the important medicinal plant of cucurbitacae family, *Cucumis melo* cv. *Bathasa.* MS medium supplemented with 1.0 mg/l BAP + 2.0 mg/l NAA and 2.0 mg/l L-Glutamic acid was found to be optimum to induce shoots. The present study established reliable and reproducible protocol for rapid multiple shoot induction from tendril explants of *Cucumis melo* L. using different concentrations and combinations of cytokinins. Tissue culture techniques are now becoming popular as alternative means of vegetative propagation. Micropropagation involves multiplication of genetically identical individual by asexual reproduction within a short span of time with tremendous potential for the production of high quality plant based medicines (Murch *et al.,* 2000). The effect of benzyl amino purine in inducing shoot induction was already reported in some of the important medicinal plants (Komalavalli and Rao, 2000). The Tendril explants were inoculated on MS basal medium supplemented with various cytokinins *i.e.,* BAP and NAA. Coconut water also had a role in triggering the formation of multiple shoots. Addition of BAP at 2.0 mg/l concentration or NAA at 3.0 mg/l to the MS basal medium, induced regeneration from the Tendril explants.

Keywords: *Multiple shoots, Tendril explants, BAP, L-Glutamic acid, NAA.*

Introduction

In the present chapter, a simple and reproducible procedure was devised to obtain multiple shoots, from Tendril explants of *Cucumis melo* cv. *Bathasa* on MS medium fortified with plant growth regulators along with coconut milk and amino acids. The plants of Cucurbitaceae suffer from several diseases including the water melon mosoic virus (Greber, 1978), Cucumber green mottle mosoic virus

(Nijsden, 1984) and *Cucumis melo* cv. *Bathasa* also suffers from downy and powdery mildews which seriously limits the crop production. Georges Morel (1952) first demonstrated that virus free plants can be recovered from infected plants through shoot tip cultures. Axillary buds from pumpkin were reported by Jelaska (1972). In comparison to lactose for different straw lactose 3 per cent of lemon grass straw (580 gram) was proved to be less effective to others. The similar findings were also reported by Singh (2005). Growth of *in vitro* propagated plants is often stronger than in those cloned *in vivo*. This is mainly due to rejuvenation and the fact that they are disease free. Propagation is carried out in aseptic conditions, free from pathogens.

Materials and Methods

The Tendril segments of 2.0–3.0 cm long were cultured and surface sterilized with 0.1 per cent $HgCl_2$ for 5–7 minutes and rinsed with sterile distilled water. They were cultured on MS medium containing 2.5 per cent sucrose and 0.8 per cent Agar-Agar and different concentrations of BAP, NAA and L-Glutamic acid (Table 41.1). The pH of the medium was adjusted to 5.8 and later was autoclaved at 120°C for 17 minutes. Cultures were incubated under 16 hrs, illumination (250 lux) at 25±2°C temperature. Raising the level of BAP (0.5 to 2.0 mg/l) resulted in the an increase in the number of shoots from hypocotyls and cotyledon explants of Niger (Nikam and Shitole, 1993). The results from this study has shown that BAP induced the activation of totipotency at the Tendril explants, which resulted in the formation of multiple shoots.

Table 41.1: Effect of Growth Regulators on Multiple Shoot Induction from Tendril Explants of *Cucumis melo* cv. *Bathasa*

Growth Regulators	*Tendril*	
	Per cent Frequency of Shoots	*Mean No. of Shoots*
MS + 0.5 mg/l BAP + 0.5 L–Glutamic acid	35	Callus
MS + 1.0 mg/l BAP + 1.0 L–Glutamic acid	30	Callus
MS + 2.0 mg/l BAP + 2.0 L–Glutamic acid	25	Shoots (1–4)
MS + 3.0 mg/l BAP + 3.0 L–Glutamic acid	20	Shoots (2–4)
MS + 0.5 mg/l NAA + CM	30	Callus
MS + 1.0 mg/l NAA + CM	25	Callus
MS + 2.0 mg/l NAA + CM	22	Shoots (2–3)
MS + 3.0 Mg/l NAA + CM	20	Shoots (1–2)
MS + 4.0 mg/l NAA + CM	15	Shoots (2–3)

CM: Coconut milk water.

Results and Discussion

The mean number of shoots developed on the explants ranged from 1–4 to 2–3 by the addition of different concentrations of BAP and NAA. The Tendril cuttings were inoculated on MS basal medium fortified with various cytokinins *i.e.*, BAP and NAA. Coconut water also had a role in triggering the

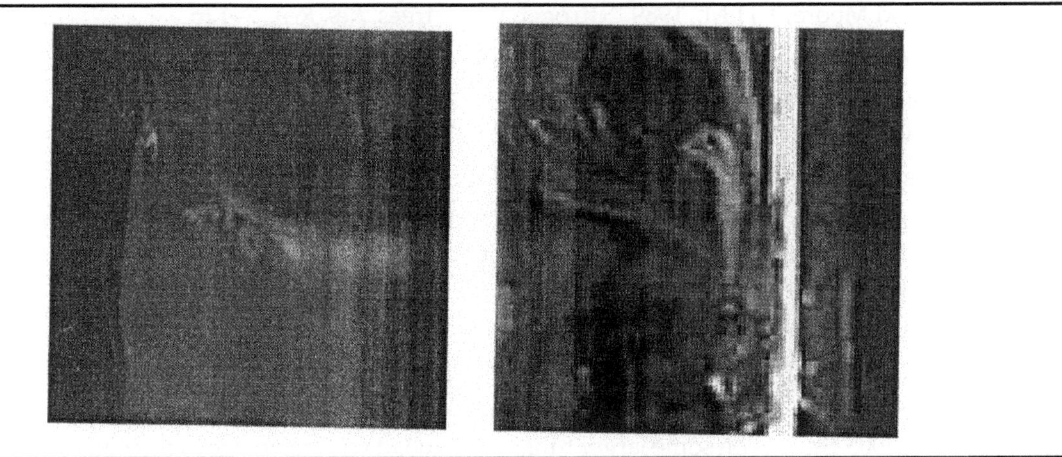

Plate 41.1: Induction of Multiple Shoots from Tendril Explants of *Cucumis melo* cv. *Bathasa*

formation of multiple shoots. Raising the level of BAP (3 mg/l to 4 mg/l) resulted in an increase in the percentage of shoots developed from Tendril cuttings. There was no significant increase in the number of shoots on NAA at low and high concentration (Plate 41.1). When MS medium supplemented with 10, 15, 20 per cent of coconut milk also triggered the induction of many multiple shoots. Low concentration of L-glutamic acid (0.5–1.0 mg/l, along with BAP (1.0 mg/l, has produced significant mean number of multiple shoots that ranged from 2–3 to 5–6 in both the explants.

Addition of NAA failed to produce many shoots but enlarged the Tendril segments. Lower levels of coconut milk (6, 12 per cent) induced callus formation. The results from study have shown the initiation of shoot buds and formation of multiple shoots from different explants *i.e.* Tendril cuttings of *Cucumis melo* cv. *Bathasa*. Among all explants used Tendril segments were the best for multiple shoot induction. With an increase in the level of BAP 2.0–3.0 mg/l the percentage of explants producing shoots also increased. The number of shoots developed on the explants ranged from 1–4 to 2–3 by the addition of BAP at a concentration of 1.0 mg/l or NAA at 2.0 mg/l. Among three concentrations used *i.e.* 10, 15 and 20 per cent, 15 per cent of coconut milk along with 0.5 mg/l BAP had proved to be ideal for multiple shoot induction. MS medium fortified with 1.0 mg/l BAP or 2.0 mg/l L-Glutamic acid also induced shoot buds on Tendril explants.

Conclusion

The method of repeated transfer of explant is considered to be useful for large scale production of plants, as it avoids isolation and culture of new explants. This is considered as one of the methods to increase the response in explants has suggested that repeated transfer of explant on multiplication media containing cytokinins succeeds in activating the plant materials.

References

Greber, R.S., 1978. Water melon mosoic virus 1 and 2 in Queensland Cucurbit crops. *Asst. J Agri. Res.*, 29: 1235–1245.

Halder, T.E. and Gadgil, V.N., 1987. Shoot bud differentiation in long term callus cultures of *Momordica* and *Cucumis*. *Ind. J. Expt. Biol.*, 20: 80–82.

Harris, P.J.C. and Moore, T.H.M., 1989. Nitrogen fixing trees. *Res. Rep.*, 7–12b.

Jelaska, S., 1974. Embryogenesis and organogenesis in pumpkin explants. *Physiol. Plant*, 31: 257–261.

Komalavalli, N. and Rao, M.N., 2000. *In vitro* micropropagation of *Gymneme sylvestre* a rare multipurpose medicinal plant. *Plant Tissue Organ Cult.*, 61(2): 97–105.

Morel Georges, 1952. Producing virus free cymbidium *Am. Orchid. Soc. Bull.*, 29: 495–497.

Murch, S.J., Krishna Raj, S. and Saxena, P.K., 2000. Tryptophan is a precursor for melatonin and serotonin biosynthesis in *in vitro* regenerated St. John's wort (*Hypericum perforatum* L. cv. *Anthos*) plants. *Plant Cell Rep.*, 19: 698.–704.

Nijsden, A.P.M. den, 1984. Attempts to introduce resistance from related species into the cucumber. In: *Proceedings of the Third Meeting on Breeding of Cucumber and Melons*, 2–5th July, Poldiv, Bulgaria.

Nikan, T.D. and Shitole, M.E., 1993. *Plant Tissue and Organ Culture*, 32: 345.

Singh, Rochica, 2005. Studies on some larger fungi of Faizabad with reference to their eco-physiological characteristics. *Ph.D. Thesis*, Dr. R.M.L. Avadh University, Faizabad, pp. 114–115.

Wehener, T.C. and Locky, R.D., 1981. *In vitro* adventitious shoot and root formation of cultivars and lines of *Cucumis sativus* L. *Hort. Sci.*, 16: 759–760.

Chapter 42

Effect of Arsenic on Free Amino Acid Levels in Freshwater Fish, *Labeo rohita* (Hamilton)

K. Pazhanisamy and K. Kannan

Department of Zoology, Government Arts College, Ariyalur – 621 713, Tamil Nadu

ABSTRACT

Arsenic forms a significant contribution to pesticides and industrial effluents released into different water bodies. It produces a variety of toxic effects in aquatic animals including fishes. The freshwater fish *Labeo rohita* treated with two sublethal concentrations ($1/10^{th}$–0.27ppm and $1/3^{rd}$–0.91 ppm of the 96 hr LC_{50}) of arsenic, have revealed a significant increase in the amino acid level in liver and muscle tissues throughout the period of 28th days of exposure. In the present study, an elevation of total free amino acids in the selected tissues of *Labeo rohita* might be due to intensive proteolysis in the respective tissues.

Keywords: *Labeo rohita, Free amino acid, Arsenic.*

Introduction

Environmental pollution with special reference to water pollution is a major problem in modern life. The domestic wastes and untreated or partially treated industrial effluents, supplemented with pollutants like heavy metals, pesticides and many organic compounds, have greatly contributed to massive fish death of aquatic ecosystems. These toxic chemicals and metals have changed the quality of waters that affect fish and other aquatic organisms (Dhasarathan *et al.*, 2000).

Almost all the heavy metals are toxic at higher concentrations and some are lethal even at a very low concentration. Among the heavy metals, arsenic is more significant from environmental and toxicological point of view (Newland, 1982). The determination of biochemical changes due to heavy metal toxicity could serve as an index of the test organism. In fishes, proteins are one of the most important and complete groups of biological materials comprising the nitrogenous constituents of the body and performing different biological functions. The estimation of protein and amino acids in fishes is considered important since proteins have diversified biological functions in animals (Jauncey, 1982).

Amino acids, which cannot be synthesized in the animal body, are called essential amino acids and must be supplemented through diet. The amino acids are building blocks of proteins (Ramesh Kumar, 1989). Moreover, estimation of proteins and amino acids of fishes is considered to be essential because the food value of fish directly relates to their protein concentration and contamination by toxic pollutants. An attempt has been made in present study to determine the extent of changes in the levels of amino acids in tissues of *Labeo rohita* exposed to lower and higher sublethal concentrations of arsenic trioxide at various time intervals.

Materials and Methods

Healthy *Labeo rohita* fish ranging from 10–12 cm in length and weighing 9–14g were collected from fish farm located in Puthur and acclimatized under laboratory conditions (29±1°C). The fish were fed daily on oil-less groundnut cake. The unused food was renewed after 2 hours and water was changed daily. Prior to experimentation the fish were acclimatized to experimental tanks for at least one week. The LC_{50} for arsenic trioxide for 96 hours was found using probit method (Finney, 1971) and $1/10^{th}$ of the LC_{50} value (0.27 ppm) and $1/3^{rd}$ of the LC_{50} value (0.91 ppm) were taken as lower and higher sublethal concentrations respectively. The fish were exposed to these sublethal concentrations for treated and control period of 7, 14, 21 and 28 days. A control group was maintained in an identical environment. The toxicant and normal waters were renewed every day. The fishes were sacrificed from both experimental and control groups on 7th, 14th, 21st and 28th days of exposure periods and subjected to analysis for the biochemical changes. The total free amino acids content was estimated by the method of Moore and Stein (1954).

Results and Discussion

The amino acid content in the liver and muscle tissues of control and arsenic treated fish, *Labeo rohita* are presented in Table 42.1.

In the present investigation, both lower and higher sublethal concentrations of arsenic exposed fish, *Labeo rohita* showed an increase in amino acids content of liver and muscle at all periods. It has been reported that the increased proteolysis becomes evident by the gradual increase in the magnitude of elevation of amino acid levels in the respective tissues when *Labeo rohita* was treated to arsenic toxicity. Similar results were reported by Sivaramakrishna and Radhakrishnaiah (1988) and Magendran (1990). The enhanced level of total free amino acids was found in the tissues of liver, muscle, intestine, kidney, brain and gill of *Channa punctatus* and *Oreochromis mossambicus* exposed to higher and lower concentration of phenyl mercuric acetate and ethofenprox after long periods respectively (Karuppasamy, 2000 and Muniyan, 1999).

Table 42.1: Changes in the Level of Amino Acid Content in Liver and Muscle Tissues of *Labeo rohita* Treated with Lower and Higher Sublethal Concentration of Arsenic

Tissues	*Group*	*Exposure Period in Days*				*F Value*
		7	*14*	*21*	*28*	
Liver	C	2.02±0.011	2.04±0.009	2.03±0.004	2.03±0.012	0.888^{NS}
	LC	2.25±0.020 (11.38)	2.50±0.014 (22.54)	2.79±0.007 (37.43)	3.00±0.043 (47.78)	173.56*
	HC	2.41±0.028 (19.30)	2.67±0.018 (30.88)	2.91±0.043 (43.35)	3.25±0.061 (60.10)	75.080*
Muscle	C	0.81±0.009	0.080±0.015	0.79±.0.018	0.080±0.008	0.377^{NS}
	LC	0.91±0.004 (12.34)	0.93±0.008 (16.25)	0.96±0.013 (21.59)	1.02±0.07 (27.50)	42.8*
	HC	0.94±0.002 (16.05)	0.99±0.013 (23.75)	1.05±0.022 (32.91)	1.10±0.004 (37.50)	40.14*

C: Control; LC: Lower sublethal concentration; HC: Higher sublethal concentration; Mean±S.E. indicates the mean of six individual observations.

(+/-) Indicates the per cent increase/ decrease over the control.

Values expressed in mg/g wet wt. of tissue

NS: Non significant.

*: Indicates significant at 5 per cent level of F test.

In the present study, arsenic stress severely affects the total free amino acids contents in liver and muscle tissues of *Labeo rohita*. Thus, the nutritive value of the fish could be altered during the arsenic exposure, Rangaswamy (1984) has reported that the decreased quantity of protein may be due to its conversion to ammoniated residues in order to increase amino acids pool. Meenakshi and Indra (1998) have noticed a depletion in the level of total protein in liver and muscle and increase in the total free amino acids in blood, liver and muscle of distillery effluent treated *Mystus vittatus*.

The remarkable increase in the free amino acids may represent proteolysis in the tissues to meet the demands for energy requisites in addition to the carbohydrate and fat. An increase in amino acid content in liver was observed in *Mystus vittatus* exposed to median lethal concentration of mercuric chloride (Jagadeesan, 1994) and in *Mystus vittatus* exposed to sublethal and median lethal concentration of copper (Rajamanickam, 1992).

An increased level of total free amino acids was found in liver tissues of *Mystus vittatus* when exposed to zinc sulphate (Ramesh Kumar, 1989). An increase in amino acid content in liver tissue may be due to enhanced proteolysis (Bass, 1962) and decreased utilization of amino acids for protein synthesis. The increased free amino acids levels in liver, muscle and brain tissues of *Cyprinus carpio* exposed to sublethal concentration of mercury was reported by Sivaramakrishnan and Radhakrishnaiah (1998). Thus, the increase in amino acids content is due to enhanced proteolysis and an inverse relationship between protein and amino acid. In the present study, an elevation of total free amino acids in the liver and muscle tissues of *Labeo rohita* might be due to intensive proteolysis in the respective tissues, which would have lead to increased deamination and oxidation of amino acid.

Acknowledgement

The author is thankful to Head of the Department of Zoology, Annamalai University, Annamalai Nagar, Tamil Nadu for providing laboratory facilities. The authors are also expresses his hearty acknowledgement to Dr. M. Vasanthy, Prof and Head, PG and Research Department of Environmental Sciences, Govt. Arts College Ariyalur, for the encouragement.

References

Bass, A., 1962. In: *The Denervated Muscle*, (Ed.) E. Gutrnann. Czec. Acad. Sci., Prague.

Dhasarathan, P., Palaniappan, R. and Ranjit Singh, A.J.A., 2000. Effect and endosulfan and butachlor on the digestive enzyme and proximate composition of the fish, *Cyprinus carpio. Indian J. Environ. and Ecoplan.*, 3(3): 611–614.

Finney, D.J., 1971. *Probit Analysis*, University Press, Cambridge, pp. 333.

Jauncey, K., 1982. Carp (*Cyprinus carpio* L.) nutrition. In: *Recent Advances in Aquaculture*, (Eds.) J.F. Muir and R.J. Roberts. Westview Press, INC, 5500, Colarado, pp. 22.

Jagadeesan, G., 1994. Studies on the toxic effects of mercuric chloride and the influence of antidote dimercaprol on selected tissues in *Labeo rohita* (Hamilton) fingerlings. *Ph.D. Thesis*, Annamalai University.

Karuppasamy, R., 2000. Short and long-term effects of phenyl mercuric acetate on protein metabolism in *Channa punctatus* (Bloch.). *J. Natcon.*, 12(1): 83–93.

Magendran, A., 1990. Studies of heavy metal (Cu, Cd and Hg) stress on pearl spot *Etroplus suratensis* (Bloch). *Ph.D. Thesis*, Annamalai University.

Meenakshi, V. and Indra, N., 1998. Sublethal toxicity of distillery effluent on the protein and free amino acids of the freshwater fish, *Mystus vittatus* (Bloch). *J. Natcon.*, 10(1): 87–91.

Moore, S. and Stein, W.H., 1954. A modified ninhydrin reagent for the photometric related com pounds. *J. Biol. Chem.*, 211: 907–813.

Muniyan, M., 1999. Effects of ethofenprox (trepon) on the biochemical and histological changes in selected organs of the freshwater fish, *Oreochromis mossambicus* (Peters). *Ph.D. Thesis*, Annamalai University, India.

Newland, L.W., 1982. Arsenic, beryllium, selenium and vanadium. In: *Handbook of Environmental Chemistry*, (Ed.) O. Hutzinger, 3(B): 27–67.

Rajamanickam, C., 1992. Effects of heavy metal copper on the biochemical contents, bioaccumulation and histology of the selected organs in the freshwater fish, *Mystus vittatus* (Bloch) *Ph.D. Thesis*, Annamalai University.

Ramesh Kumar, B., 1989. Histological and biochemical studies on the effects of zinc sulphate on the freshwater fish, *Mystus vittatus* (Bloch). *M.Phil. Thesis*, Annamalai University.

Rangaswamy, C.P., 1984. Impact of endosulfan toxicity on some physiological properties of the blood and aspects of energy metabolism of a freshwater fish, *Tilapia mossambica. Ph.D. Thesis*, Sri Venkateswara University, India.

Sivaramakrishna, B. and Radhakrishnaia, K., 1988. Impact of sublethal concentration of mercury on nitrogen metabolism of the freshwater fish, *Cyprinus carpio* (Linnaeus). *J. Environ. Biol.*, 19(2): 111–117.

Chapter 43

Subsurface Water Quality in Selected Areas of Visakhapatnam City

V. Saritha and K. Swapna Vahini

Department of Environmental Studies, GITAM Institute of Science, GITAM University, Visakhapatnam – 530 040, Andhra Pradesh

ABSTRACT

Assessment of some physico-chemical parameters of under groundwater samples from seven sampling sites of various urban and suburban areas of Visakhapatnam were collected and analyzed to evaluate its suitability for domestic purposes. Analysis of some physico-chemical characteristics like temperature, pH, EC, TS, TDS, Total Hardness, Alkalinity, DO, sulphates, phosphates and nitrates has been done and the groundwater samples were well within the pelmissible limits set by BIS, ISI, ICMR and WHO.

Keywords: *Groundwater, Physico-chemical parameters, Visakhapatnam.*

Introduction

Groundwater accumulates chiefly from the precipitation that percolates through the soil strata. The quality of groundwater of any area is of great importance for human beings and irrigation. All the groundwaters irrespective of their source of origin contain mineral salts and these constituents depend upon various geological and physical factors (Prasanthi and Murali, 2004). Groundwater extraction through dug and bore wells is resulting in the emergence of water claimants and it serves for different purposes *viz.*, domestic, industrial etc. However, the alarming salt content in groundwater due to local pollutants has biological and chemical affect on the groundwater quality adversely (Patel, 2006).

Increasing living standards, growing population, rapid industrialization and wide sphere of human activities have bought greater stress on land and water, which in turn results in steady increase in the demand for water resources. According to the World Health Organization about 80 per cent of all the diseases of human beings are caused by water therefore, water that is supplied for drinking and various other purposes must be of good quality (Venkatasubramani *et al.*, 2006).

Pollution of groundwater resources has become a major problem today. The pollution of air, water and land has an affect on the pollution and contamination of groundwater. Some 45 million people the world over are affected by water pollution marked by excess fluoride, arsenic, iron or the ingress of salt water (Nalini, 2006). The poor quality of drinking water in our country is more due to contamination than due to the inferiority of the source (Gibbons, 1984). Therefore, a continuous periodical monitoring of water quality is necessary so that appropriate steps may be taken for water resource management practices (Rajvaidya, 1998; Hariharan, 2007).

Keeping the above mentioned facts in view the present study has been taken up to assess the quality and suitability of groundwaters in the study area and to depict any groundwater contamination and the possible causes of this contamination.

Methodology

Study Area

Visakhapatnam is coastal station situated on the east coast of India. The extent of GVMC study area is 515 Sq. km and geographically lies between 17°32′ N to 17°51′N latitudes and 83°07′E to 83°21′E longitude. Visakhapatnam, described as the "City of Destiny" is the fastest growing city of the Asian Continent and is rated as one of the top twenty five cities in the world that have a high rate of expansion, both population wise and in the direction of industrial development. This phenomenal growth of the city is due to its resource potential and the aesthetic gift of the city provided by the nature in the form of long stretch of coast line and beautifully spanning hill ranges (Suvarna Lakshmi and Reddy, 2007). The city receives an average around 1000mm mean rainfall. The June–September and October–November are the main monsoon of the city during which the south-west, north-east monsoons being are average of 1000mm of mean annual rainfall.

The present study has been made on the bore-waters of selected areas of Visakhapatnam, at five selected sites situated at Rushikonda (S–1), Pendurthi (S–2), Bheemili (S–3), M.V.P. Colony (S–4), BHPV Area (S–5), Auto Nagar (S–6) and Aarilova (S–7). The water samples were collected in sterile glass bottles of 100 to 1000ml capacity. The collected water samples were analyzed for various physico-chemical parameters following standard methods of (APHA, 1975).

Results and Discussion

The physico-chemical parameters of the groundwater samples of the investigated area are listed in Table 43.1. Drinking water standards as prescribed by the Bureau of Indian Standards, Indian Council of Medical Research, Indian Standards Institution and World Health Organization are given in Table 43.2.

Colour

All the water samples were clear and colorless. Color in water may be aesthetically objectionable.

Table 43.1: The Physico-Chemical Parameters of Subsurface Waters of Selected Areas of Visakhapatnam

Sl.No.	*Parameters*	*Units*	*Sampling Site*						
			S-I	*S-2*	*S-3*	*S-4*	*S-5*	*S-6*	*S-7*
1.	Temperature	°C	25	26	28	25	26	27	25
2.	pH Value	–	7.23	7.35	7.33	7.53	7.5	7.29	7.7
3.	EC	S/cm	420	350	410	730	520	580	350
4.	Total Solids	mg/l	600	700	800	1400	900	1100	500
5.	TDS	mg/l	500	200	300	700	400	600	200
6.	Total Hardness	mg/l	130	110	250	300	310	360	370
7.	Calcium	mg/l	70	80	180	200	180	200	200
8.	Magnesium	mg/l	60	30	70	100	130	160	170
9.	Total Alkalinity	mg/l	170	420	800	470	490	420	280
10.	Chlorides	mg/l	95.56	150.05	175.19	130.14	125.13	190.2	155.17
11.	Dissolved Oxygen	mg/l	5.2	6.2	5.6	5.0	6	4.7	3.7
12.	Sulfates	mg/l	115	85	112	96	120	165	134
13.	Phosphates	mg/l	0.03	0.09	0.09	0.07	0.08	0.05	0.05
14.	Nitrates	mg/l	7.97	3.77	42.45	29.15	9.68	11.9	24.75

Note: S/cm = micro siemens per centimeter.

Table 43.2: Standards for Drinking Water

Parameter	*BIS*		*ICMR*		*ISI*		*WHO*	
	P	*E*	*P*	*E*	*P*	*E*	*P*	*E*
Colour	–	–	5	25	10	50	5	25
Taste and Odour	UO	UO	UO	UO	UO	UO	UO	UO
pH value	6.5–8.5	6.5–9.2	7 – 8.5	6.5–9.2	6.5–8.5	6.5–9.2	7 –8.5	6.5 – 9.2
Total Dissolved Solids	500	1000	–	–	–	–	500	1500
Total Hardness	300	600	300	600	300	600	–	–
Calcium	75	200	75	200	75	200	75	200
Magnesium	30	100	50	150	30	100	50	150
Chlorides	–	–	250	1000	250	1000	200	600
Sulfate	200	400	200	400	150	400	200	400
Nitrate	45	–	20	50	45	–	–	100

P: Permissible Limit; E: Excessive Limit; UO: Unobjectionable.

Note: All Units except pH are in mg/l, otherwise stated.

Taste and Odor

All the water samples were with in permissible limits with out objectionable tastes and odors.

Where, presence of tastes and odours can be attributed to contamination of water by various organic substances from industrial and domestic effluents.

Temperature

The temperature of groundwater remains relatively constant throughout the year. The ground temperature for an area is approximately equal to an area's annual average air temperature. The earth and groundwater temperatures are much more stable than the highly variable seasonal air temperature. The temperature of the water samples in the present area has varied from 25–28°C. However, when the temperature increases it enhances the dissolution of various chemicals and reduces the dissolved gases.

pH

Groundwater pH is a fundamental property that describes the acidity and alkalinity of groundwater and largely controls the amount and chemical form of many organic and inorganic substances dissolved in groundwater. It is known that pH of water does not cause any severe health hazard, however high pH induces the formation of trihalomethanes which are causing cancers in human beings. In the present study pH values are within the ICMR standards (6.5–8.5) and ISI standards (6.5–9.2). The pH of the water samples analyzed ranged from 7.23 to 7.7.

Electrical Conductivity

Electrical conductivity (EC) estimates the amount of total dissolved salts (TDS), or the total amount of dissolved ions in the water. EC is controlled by geology (rock types), the size of the watershed and "other" sources of ions to lakes. There are a number of sources of pollutants which may be signaled by increased EC. The electrical conductivity value is an index to represent the total concentration of soluble salts in water. The gradual increase of conductance shows entry of greater quantities of ionic matter into the source. Electrical conductance values of the present samples ranged from 350–730 µS/cm.

Total Dissolved Solids

Total Dissolved Solids values beyond the prescribed limit impart taste to water and reduce its palatability. The TDS values in the present study ranged within 200–700 mg/l. Most of the water samples in the present study are well within the safe limit of drinking water standards. Rambabu *et al.* (1996) and Joseph (2001) have expressed that the dissolution of soil particles containing minerals under slightly alkaline conditions favors in increasing the TDS concentration in groundwaters. The TDS values above the 1500 mg/l cause gastrointestinal irritation according to Bhavani Shankar and Muthu Krishnan (1994). Higher concentration of dissolved solids may produce distress in cattle and livestock and a salty taste to water. In the present study TDS values are within the WHO standards (500 to 1500mg/l) and BIS standards (500mg/l–1000mg/ l).

Total Hardness

The values of total hardness for all water samples fall within permissible limit. Hardness of water is important in determining the suitability of water for domestic and industrial uses. Absolutely soft waters are tasteless. On the other hand, hardness up to 500mg/l can be relished if got acclimatized to. More cases of cardiovascular diseases are reported in soft water areas. Hard water is useful for the growth of children due to presence of calcium. Magnesium hardness, particularly associated with sulphate ions has a laxative effect on persons unaccustomed to it. It makes food tasteless.

It also precipitates protein of meat and make tasteless. Hardness values were recorded between 110 to 370mg/l. The scale of hardness from the consumers' point of view may be taken as below:

0–50	Soft
50–100	Moderately soft
100–150	Slightly hard
150–250	Moderately hard
Over 250	Hard

Calcium

Calcium occurs in water mainly due to the presence of limestone, gypsum, dolomite and gypsiferous minerals. The determination of calcium is usually required for potable water. The values ranged from 70–200 mg/l. High calcium contents in water are undesirable for washing, bathing and laundering. It tends to create scales on utensils. The permissible limit of calcium is 75mg/l (ISI).

Magnesium

The values varied within the range of 30–170 mg/l in the sample waters. The permissible limit of magnesium is 100 mg/l (ISI). Magnesium is an essential element for man. However, at higher levels, magnesium salts have a laxative effect.

Alkalinitiy

Most of the alkalinity in natural water is formed due to dissolution of carbon dioxide in water. Large amounts of alkalinity impart bitter taste to water. Even though the carbonate alkalinity is absent, the total alkalinity is found which may be due to the accumulation of bicarbonates. In the present investigation the total alkalinity of water samples is found in the range 170–800.

Chloride

The standards published for chlorides are 250mg/l (permissible) and 1000mg/l (Excessive). Chlorides associated with sodium exert salty taste, when its concentration is more than 250 mg/l. Although chlorides are not harmful as such, their concentration over 250mg/l imparts a peculiar taste to the water, thus rendering the water unacceptable for drinking purposes from aesthetic point of view. The values in the present study lie within limits. Concentrations of chloride were found to vary within 95.56–190.2 mg/l.

Dissolved Oxygen

Dissolved Oxygen as a parameter is very useful for assessing the quality of water and providing a check on pollution. The drinking water should be rich in dissolved oxygen (DO) for good taste and the values of DO are well within limit and found suitable for appropriate taste. It is needed for living organisms to maintain their biological processes. The values of DO vary from 3.7–6.2 mg/l.

Sulphates

The values of sulphate ranged within 85 to 165mg/l. Sulphate may have laxative effect if magnesium is present at an equivalent concentration (Chatterjee and Raziuddin, 2002). Excess concentration of sodium sulphate in water can cause disfunctioning of the alimentary canal. So the recommended upper limit is 250mg/l in waters intended for human consumption. In the present study, the values fall within limit.

Phosphates

Phosphates are not toxic and do not represent a direct health threat to human health or other organism. But excess of phosphorus promotes the abundant growth of algae, and leads to eutrophication of water bodies. The phosphate values of the present study varied from 0.03–0.09 mg/l.

Nitrate

Nitrates are the end products of decomposition of organic matter present in fully oxidized waters and harmful above 45mg/l. In the present study nitrate in groundwater ranged between 3.77–42.45 mg/l which is below permissible limit. A few minerals contribute nitrates to groundwater. The most important source of nitrate is biological oxidation of nitrogenous substances which come in sewage, industrial wastes, chemical fertilizers, decayed vegetables, leaches from refuse dumps, septic tank effluent, etc.

Conclusion

The quality parameters were estimated through physico-chemical study. The study states that all drinking water quality parameters such as colour, taste, odour, turbidity and chemical parameters were found well within limit for all studied groundwater samples as prescribed by ICMR, BIS, ISI and WHO. Therefore the water from all groundwater samples is suitable for drinking purposes as specifications.

References

APHA, 1975. *Standard Methods for the Examination of Water and Wastewater*, 16th edn. American Public Health Association, American Water Works Assoc., Water Pollution Control Federation, Washington DC.

Bhavani, S. and Muthu Krishnan, N., 1995. Is the groundwater in Madras city potable. *Indian J. Env. Prot.*, 14(3): 176–179.

BIS, 1998. *Specification for Drinking Water*. Bureau of Indian Standard Institution, New Delhi.

Chatterjee, C. and Raziuddin, M., 2002. *Determination of Water Quality Index (WQI) of a Degraded River in Asansol Industrial Area*.

Gibbons, J.H., 1984. Protecting the nations groundwater from contamination. Congress of United States Office of Tech. Assessment, Washington, DC.

Hariharan, A.V.N.S.H., 2007. Some studies on the water quality parameters of Vuda (Mithilapuri) Colony, Visakhapatnam. *J. Ind. Poll. Cont.*, 23(1): 113–117.

ICMR, 1975. *Manual of Standards of Quality for Drinking Water Supplies*. Spl. Rep. Sl.No. 44. ICMR, New Delhi.

ISI, 1983. *Indian Standards Specifications for Drinking Water*. Indian Standard Institution, New Delhi.

Joseph, W.T., 2001. Groundwater chemistry in the valley De yabucoa alluvial aquifer south Eastern Puerto Rico. In: 3rd *International Symposium on Tropical Hydrology*, San Juan, USA.

Nalina, E. and Puttaiah, E.J., 2006. Characteristics of groundwater quality of Kadur Taluk: A preliminary observation. *Ecol. Env. Cons.*, 12(4): 617–620.

Patel, S.R. and Desai, K.K., 2006. Correlation on groundwater parameters of some villages of Surat District, Gujarat (India). *Poll. Res.*, 25(3): 659–660.

Prasanthi, A. and Murali, M., 2004. Some studies on correlation of physico-chemical characteristics of groundwater at Totaguruvu and Arilova, Visakhaptnam. *M.Sc. Project* submitted to Department of Environmental Studies, College of Science, GITAM, Visakhapatnam.

Rajvaidya, N. and Dilip Kumar, M., 1988. Advances in environmental sciences and tech: Water char. and properties. *APHB*, (1).

Rambabu, C. and Somashekara Rao, K., 1996. Studies on the quality of bore water of Nuzvid. *Indian J. Env. Prot.*, 16(7): 138–142.

Suvarna Lakshmi, P. and Reddy, E.U.B., 2007. Environmental concerns on greater Visakhapatnam municipal corporation by using remote sensing and GIS techniques. *M.Phil. Thesis*, Department of Environmental Sciences, Andhra University, Visakhapatnam.

Venakatasubramani, R., Meenambal, T., Livingston, P. and Peter Goldwyn, 2006. A correlation study on physico-chemical characteristics of groundwaters in Coimbatore district. *Poll. Res.*, 25(2): 371–374.

WHO, 1984. *Guidelines for Drinking Water Quality, Vol. 1: Recommendations.* General, p. 81–84.

Chapter 44

Effect of Weed Management Practices on Crop Growth in Chilli (*Capsicum annuum* L.)

M.J. Praveena Swamy, H.R. Bhoomika, S.L. Biradar, K.G. Santosh and N.S. Mavarkar

University of Agricultural Sciences, GKVK, Bangalore – 560 065

ABSTRACT

An experiment was conducted to study the effect of different weed management practices on crop growth in chilli at GKVK, UAS, Bangalore. The study comprised of 16v treatments including both herbicidal chemicals and cultivation practices laid out with three replications in RCBD. Chilli variety Byadagi Dabba was taken for the study. The results revealed that frequent hand weeding at 60, 90 and 120 DAT and at harvest results in good growth of crop in terms of plant height (86.33 cm, 98.73 cm and 104.47 cm respectively), number of branches per plant (22.40, 25.27 and 26.20 respectively), Leaf Area (6911.00 cm^2) and Leaf Area Index (2.560). Whereas root parameters like number of roots per plant (45.26) was maximum in treatment metachlor 59 EC @ 0.75 kg a.i./ha, while root length was maximum (25.41 cm) in treatment alachlor 50EC @ 1.5 kg a.i./ha. Hand weeding resulted in highest total dry matter production (128.60g/plant).

Keywords: *DAT (Days after Transplanting), Chilli, Byadagi dabba, Alachlor, Metachlor.*

Introduction

Vegetables constitute an important component of a balanced diet for man. They add nutrients, palatability, roughage, taste and flavor to the food. Among major vegetables chilli (*capsicum annuum* L.)

is an important commercial vegetable cum spice crop on India. It belongs to the family solanaceae and is a native of Peru or Mexico.

Adoption of wider spacing, adequate moisture supply, liberal use of fertilizers and intensive cropping systems encourage luxuriant weed growth and predispose the chilli crop to heavy weed infestation, leading to severe crop-weed competition, particularly during early stages of crop growth. Hence there is a greater need for developing sound weed management strategies for the crop.

Materials and Methods

The present study was undertaken at division of Horticulture, GKVK, UAS, Bangalore with chilli cv. Byadagi Dabba. The present experiment consisted of totally 16 treatments including un-weeded control. The experiment was laid out in Randamized Complete Block Design, with all 16 treatments in three replications. The details of the treatments are furnished below:

T_1: Alachlor 50EC @ 1.00 kg a.i./ha.
T_2: Alachlor 50EC @ 1.25 kg a.i./ha.
T_3: Alachlor 50EC @ 1.50 kg a.i./ha.
T_4: Pendimethalin 30EC @0.75 kg a.i./ha.
T_5: Pendimethalin 30EC @1.00 kg a.i./ha.
T_6: Pendimethalin 30EC @1.25 kg a.i./ha.
T_7: Metalachlor 50EC @ 0.75 kg a.i./ha.
T_8: Metalachlor 50EC @ 1.00 kg a.i./ha.
T_9: Metalachlor 50EC @ 1.25 kg a.i./ha.
T_{10}: Trifluralin 48EC @ 0.75 kg a.i./ha.
T_{11}: Trifluralin 48EC @ 1.00 kg a.i./ha.
T_{12}: Trifluralin 48EC @ 1.25 kg a.i./ha.
T_{13}: Hand weeding at 30, 60 and 90 DAT.
T_{14}: Intercultivation by light digging at 45 and 90 DAT.
T_{15}: Paddy straw mulching.
T_{16}: Unweeded control.

The herbicidal spray solution was prepared by using following formula:

$$\text{Chemical required/ha} = \frac{\text{Recommended a.i./ha}}{\text{EC (per cent)}} \times 100$$

Observations on plant height (cm) and number of branches per plant were recorded on 30, 60, 90 and 120 DAT and at final picking.

Leaf area (cm^2) was measured by disc method and is expressed as cm^2 per plant. At 100 DAT, a plant from each plot was uprooted and 50 discs of known size were taken through a cork borer. Both discs and remaining leaf blades were oven dried at 65°C–70°C and leaf area was calculated by using following formula:

$$LA = \frac{Wa \times A}{Wd}$$

where,

LA: Leaf area (cm^2)

Wa: Weight of all leaves (inclusive of 50 discs weight) in grams.

Wd: Weight of 50 discs in grams.

A: Area of 50 discs (cm^2).

Leaf Area Index (LAI)

LAI was worked out by using the following formula:

$$LAI = \frac{\text{Leaf area per plant } (cm^2)}{\text{Unit land area } (cm^2)}$$

Shoot and root dry matter production per plant were estimated and expressed in grams. Other observations made were root length (cm) and number of roots per plant. The results of the study are presented below.

Results and Discussion

There was significant difference among different weed management treatments for the trait plant height at all stages of crop growth (Table 44.1).

At 60 DAT, alachlor @ 1.50 kg/ha and hand weeding recorded significantly the higher plant height *i.e.* 58.53 cm and 58.00 cm respectively, and were *at par* with each other. At 90, 120 DAT and at harvest hand weeding recorded significantly the higher plant height of 86.33 cm, 98.73 cm and 104.47 cm respectively.

Number of branches per plant was highest (13.46) in the treatment alachlor @ 1.25 kg/ha, at 60 DAT, whereas at 90, 120 DAT and at harvest hand weeding recorded maximum number of branches per plant (22.40, 25.27 and 26.40 respectively) (Table 44.2).

The higher plant height and number of branches with hand weeding might be due to frequent removal of weeds throughout the crop life cycle, both in intra as well as inter crop row spacing, which resulted in lower weed density and lower weed biomass.

Higher crop weed competition in un-weeded check resulted in the lowest plant height (34.60, 59.13, 68.83 cm at 60, 90, 120 DAT and at harvest respectively) and number of branches per plant (7.33, 11.93, 16.73 and 17.47 at 60, 90, 120 DAT and at harvest respectively). This was because of better root development, which out grew that of crop plants, depleted soil moisture and occupied the space of crop plants. It might also be due to allelopathic effect of weeds on chilli growth (Hazarika and Sannigrahi, 2001). Similar results were reported by Rajagopal *et al.*. (1976) in chilli and Reddy *et al.* (1999) in tomato.

Table 44.1: Plant Height as Influenced by Weed Management Practices in Chilli cv. Byadai Dabba

Treatments	*Plant Height (cm)*			
	60 DAT	*90 DAT*	*DAT*	*At Harvest*
T_1: Alachlor 50EC @ 1.00 kg a.i./ha.	50.33	77.63	83.17	89.60
T_2: Alachlor 50EC @ 1.25 kg a.i./ha.	54.37	82.57	89.37	95.30
T_3: Alachlor 50EC @ 1.50 kg a.i./ha.	58.53	85.83	97.53	100.20
T_4: Pendimethalin 30EC @ 0.75 kg a.i./ha.	51.17	76.80	81.67	89.93
T_5: Pendimethalin 30EC @ 1.00 kg a.i./ha.	56.73	84.27	96.33	101.60
T_6: Pendimethalin 30EC @ 1.25 kg a.i./ha.	52.43	78.50	85.43	92.03
T_7: Metalachlor 50EC @ 0.75 kg a.i./ha.	48.03	75.23	81.63	83.03
T_8: Metalachlor 50EC @ 1.00 kg a.i./ha.	52.83	78.83	86.07	91.67
T_9: Metalachlor 50EC @ 1.25 kg a.i./ha.	56.23	84.00	91.07	96.37
T_{10}: Trifluralin 48EC @ 0.75 kg a.i./ha.	47.37	75.00	80.70	85.60
T_{11}: Trifluralin 48EC @ 1.00 kg a.i./ha.	51.70	77.90	84.97	89.17
T_{12}: Trifluralin 48EC @ 1.25 kg a.i./ha.	53.80	80.13	86.57	90.83
T_{13}: Hand weeding at 30, 60 and 90 DAT.	58.00	86.33	98.73	104.47
T_{14}: Intercultivation by light digging at 45 and 90 DAT.	51.40	77.97	85.47	88.00
T_{15}: Paddy straw mulching.	54.77	80.67	90.20	91.20
T_{16}: Unweeded control.	34.60	59.13	68.83	70.33
F-test	*	*	*	*
SEm±	3.734	4.022	3.679	3.761
CD at 5 per cent	10.78	11.62	10.62	10.86

Note: *: Significant at 5 per cent, DAT: Days after Transplanting.

Table 44.2: Number of Branches per Plant as Influenced by Weed Management Practices in Chilli cv. Byadai Dabba

Treatments	*Number of Branches per Plant*			
	60 DAT	*90 DAT*	*DAT*	*At Harvest*
T_1: Alachlor 50EC @ 1.00 kg a.i./ha.	11.00	18.33	21.73	22.40
T_2: Alachlor 50EC @ 1.25 kg a.i./ha.	11.86	19.93	23.87	24.53
T_3: Alachlor 50EC @ 1.50 kg a.i./ha.	13.46	20.87	25.13	26.40
T_4: Pendimethalin 30EC @ 0.75 kg a.i./ha.	11.03	18.27	21.20	21.90
T_5: Pendimethalin 30EC @ 1.00 kg a.i./ha.	13.13	20.47	24.33	25.60
T_6: Pendimethalin 30EC @ 1.25 kg a.i./ha.	11.40	19.67	22.27	22.93
T_7: Metalachlor 50EC @ 0.75 kg a.i./ha.	10.00	16.87	20.20	21.87
T_8: Metalachlor 50EC @ 1.00 kg a.i./ha.	11.40	18.93	22.53	23.20
T_9: Metalachlor 50EC @ 1.25 kg a.i.lha.	12.87	20.40	24.27	24.93

Contd...

Table 44.2–Contd...

Treatments	Number of Branches per Plant			
	60 DAT	90 DAT	DAT	At Harvest
T_{10}: Trifluralin 48EC @ 0.75 kg a.i./ha.	9.80	16.93	20.13	20.87
T_{11}: Trifluralin 48EC @ 1.00 kg a.i./ha.	10.80	18.33	21.93	22.67
T_{12}: Trifluralin 48EC @ 1.25 kg a.i./ha.	12.20	19.67	23.00	23.53
T_{13}: Hand weeding at 30, 60 and 90 DAT.	11.93	22.40	25.27	26.20
T_{14}: Intercultivation by light digging at 45 and 90 DAT.	9.63	18.33	21.87	22.27
T_{15}: Paddy straw mulching.	12.30	20.53	23.13	23.53
T_{16}: Unweeded control.	7.33	11.93	16.93	17.47
F-test	*	*	*	*
SEm±	1.025	1.533	1.313	1.403
CD at 5 per cent	2.96	4.43	3.79	4.05

Note: *: Significant at 5 per cent, DAT: Days after Transplanting.

Table 44.3: Leaf Area and Leaf Area Index as Influenced by Weed Management Practices in Chilli cv. Byadagi Dabba

Treatments	LA (cm^2)	LAI
T_1: Alachlor 50EC @ 1.00 kg a.i./ha.	5752.33	2.131
T_2: Alachlor 50EC @ 1.25 kg a.i./ha.	6142.33	2.275
T_3: Alachlor 50EC @ 1.50 kg a.i./ha.	6687.67	2.477
T_4: Pendimethalin 30EC @ 0.75 kg a.i./ha.	5705.33	2.113
T_5: Pendimethalin 30EC @ 1.00 kg a.i./ha.	6571.67	2.434
T_6: Pendimethalin 30EC @ 1.25 kg a.i./ha.	6054.67	2.242
T_7: Metalachlor 50EC @ 0.75 kg a.i./ha.	5183.00	1.920
T_8: Metalachlor 50EC @ 1.00 kg a.i./ha.	6037.67	2.236
T_9: Metalachlor 50EC @ 1.25 kg a.i./ha.	6428.33	2.368
T_{10}: Trifluralin 48EC @ 0.75 kg a.i./ha.	5231.67	1.938
T_{11}: Trifluralin 48EC @ 1.00 kg a.i./ha.	5885.00	2.180
T_{12}: Trifluralin 48EC @ 1.25 kg a.i./ha.	6284.33	2.327
T_{13}: Hand weeding at 30, 60 and 90 DAT.	6911.00	2.560
T_{14}: Intercultivation by light digging at 45 and 90 DAT.	6024.00	2.225
T_{15}: Paddy straw mulching.	6154.00	2.279
T_{16}: Unweeded control.	4165.00	1.543
F-test	*	*
SEm±	262.43	0.096
CD at 5 per cent	757.87	0.28

Note: *: Significant at 5 per cent; LA: Leaf Area; LAI: Leaf Area Index.

Leaf Area per plant and Leaf Area Index (LAI) at 90 DAT were maximum in hand weeding (6911.00 cm^2 and 2.560 respectively), which were *at par* with alachlor @ 1.5 kg/ha (6687.67 cm^2 and 2.477 respectively), pendimethalin @ 1.00 kg/ha (6571.67 cm^2 and 2.434 respectively) and trifluralin at 1.25 kg/ha (6284.33 cm^2 and 2.327 respectively). It might be due to decreased crop weed competition, higher plant height and more number of branches per plant. While the minimum Leaf Area and Leaf Area Index were recorded in unweeded control (4165.00 cm^2 and 1.543 respectively) (Table 44.3).

Number of roots per plant varied significantly due to different weed management practices (Table 44.4). Highest number of roots (45.20) per plant were recorded in metachlor at 1.25 kg/ha. Maximum root length was observed in alachlor at 1.50 kg/ha (25.41 cm), whereas unweeded control recorded least number of roots per plant (25.00) as well as shortest root (13.80 cm). This was because of the better root system of weeds, which restricted the spread and growth of chilli roots.

Table 44.4: Number of Roots per Plant and Length of Roots as Influenced by Weed Management Practices in Chilli cv. Byadagi Dabba

Treatments	*No. of Roots/Plant*	*Length of Roots (cm)*
T_1: Alachlor 50EC @ 1.00 kg a.i./ha.	30.40	17.20
T_2: Alachlor 50EC @ 1.25 kg a.i./ha.	39.00	21.27
T_3: Alachlor 50EC @ 1.50 kg a.i./ha.	40.93	25.41
T_4: Pendimethalin 30EC @ 0.75 kg a.i./ha.	27.67	17.60
T: Pendimethalin 30EC @ 1.00 kg a.i./ha.	40.40	24.10
T_6: Pendimethalin 30EC @ 1.25 kg a.i./ha.	33.20	18.90
T_7: Metalachlor 50EC @ 0.75 kg a.i./ha.	25.93	16.30
T_8: Metalachlor 50EC @ 1.00 kg a.i./ha.	30.60	18.50
T_9: Metalachlor 50EC @ 1.25 kg a.i./ha.	45.20	23.07
T_{10}: Trifluralin 48EC @ 0.75 kg a.i./ha.	32.03	17.03
T_{11}: Trifluralin 48EC @ 1.00 kg a.i./ha.	29.20	17.80
T_{12}: Trifluralin 48EC @ 1.25 kg a.i./ha.	32.47	18.50
T_{13}: Hand weeding at 30, 60 and 90 DAT.	39.60	21.70
T_{14}: Intercultivation by light digging at 45 and 90 DAT.	28.33	20.50
T_{15}: Paddy straw mulching.	28.47	17.50
T_{16}: Unweeded control.	25.00	13.80
F-test	*	*
SEm±	3.157	1.500
CD at 5 per cent	9.188	4.333

Note: *: Significant at 5 per cent.

The data regarding dry root weight per plant showed significant difference and is presented in Table 44.5. Maximum dry root weight was recorded in treatment alachlor at 1.50 kg/ha (11.50 g/plant) and maximum shoot dry weight was observed in hand weeded plots (118.7 g/plant). This might be due to higher utilization of resources as a result of less crop-weed competition during most of the growing period. The total dry matter production per plant was observed in un-weeded control (52.13 g/plant).

Table 44.5: Plant Dry Matter as Influenced by Weed Management Practices in Chilli cv. Byadai Dabba

Treatments	Dry Root Weight/Plant (g)	Dry shoot Weight/Plant (g)	Total Dry[1] Matter/Plant (g)
T_1: Alachlor 50EC @ 1.00 kg a.i.lha.	7.37	80.60	87.97
T_2: Alachlor 50EC @ 1.25 kg a.i./ha.	9.33	101.30	110.63
T_3: Alachlor 50EC @ 1.50 kg a.i./ha.	11.50	113.40	124.90
T_4: Pendimethalin 30EC @ 0.75 kg a.i./ha.	7.40	72.13	79.53
T_5: Pendimethalin 30EC @ 1.00 kg a.i./ha.	11.00	111.07	112.07
T_6: Pendimethalin 30EC @ 1.25 kg a.i./ha.	9.27	84.20	93.47
T_7: Metalachlor 50EC @ 0.75 kg a.i./ha.	6.80	64.20	71.00
T_8: Metalachlor 50EC @ 1.00 kg a.i./ha.	8.00	80.00	88.00
T_9: Metalachlor 50EC @ 1.25 kg a.i./ha.	10.80	107.80	118.60
T_{10}: Trifluralin 48EC @ 0.75 kg a.i./ha.	7.07	61.93	68.00
T_{11}: Trifluralin 48EC @ 1.00 kg a.i./ha.	7.73	62.70	70.43
T_{12}: Trifluralin 48EC @ 1.25 kg a.i./ha.	7.80	88.60	96.40
T_{13}: Hand weeding at 30, 60 and 90 DAT.	9.90	118.70	128.60
T_{14}: Intercultivation by light digging at 45 and 90 DAT.	9.00	76.00	85.00
T_{15}: Paddy straw mulching.	7.20	86.50	93.70
T_{16}: Unweeded control.	5.10	47.03	52.13
F-test	*	*	*
SEm±	0.833	5.731	6.424
CD at 5 per cent	2.40	16.55	18.55

Note: *: Significant at 5 per cent; [1]: Dry matter of plant excluding the chilli fruits.

With these findings it can be concluded that maintaining weed free condition by frequent hand weeding at 30, 60 and 90 DAT in chilli results in good crop growth, in terms of plant height, number of branches per plant, leaf area and leaf area index. In situations when use of herbicides is inevitable, the chemicals: alachlor@1.50 kg a.i./ha or pendimethalin @ 1.00 kg a.i./ha will be beneficial.

References

Hazarika, B. and Sannigrahi, A.K., 2001. Allelopathic research in vegetable production: A review. *Environ. Ecol.*, 19(4): 799–806.

Rajagopal, A., Muthukrishnan, C.R. and Ashok Mehta, V., 1976. Weed control studies in chilli. *Madras Agric. J.*, 63: 470–472.

Reddy, M.S., Rao, P.G. and Babu, R.S.H., 1999. Integrated weed management studies in tomato. *J. Res., ANGRAU*, 27(1 and 2): 7–11.

Chapter 45

Studies on the Effect of Lead on the Histology of Spleen Tissue of the Male Tree Frog, *Polypedates maculatus* (Gray)

K. Aruldoss, N. Indra and K. Senthamizh Durai

P.G. and Research Department of Biotechnology,
Bharath College of Science and Management, Thanjavur – 613 005, Tamil Nadu

ABSTRACT

Studies on the effect of sublethal dose of lead on male tree frog, *Polypedates maculatus* was studied for 30 days. Spleen tissues were removed after 30 days experiment for histological examinations. The spleen showed marked histopathological changes like variable necrotic areas and hypertrophied lymphocytes in experimental animals.

Keywords: *Heavy metals, Histology, Lead toxicity, Polypedates maculates.*

Introduction

Heavy metal pollutants cause direct toxicity both to humans and other living beings, due to their presence of beyond specified limits (Rai *et al.*, 1998). Metals are biologically essential but become toxic with increasing dosage (Kaul, 1997). Lead is the most serious heavy metal contaminant of soil. Routes of exposure to lead are diverse, with the most common pathway being the intake of lead containg air, food, water, soil, or dust (Kyung-Hwa Beak, *et al.*, 2005). Some elements such as lead, cadmium, and mercury has harmful effects on biological tissues at any concentration (Sataka *et al.*, 1997). The effects of heavy metals on eggs, embryos and tadpoles of amphibians are widely reported (Brige and Black, 1977; Hemant Ghate, 1985; Collins, 1987; Abbassi and Soni, 1993). Spleen is one of the organs with a high accumulation of lead, which could modify the immunological processes that take place in this organ (Cesar Teijon *et al.*, 2000).

Lead affects multiple organ systems of animals, resulting in a general decrease in health susceptibility to decrease or pathogen exposure (Eisler, 1988), Dietary lead (100 ppm) exposure altered immune responses of chicken, including phagocytosis and antibody titres (Hill and Qureshi, 1998). Lead appears to impair antibody production reduce disease resistance and increase mortality in animals infected with bacterial and viral agents (WHO, 1995). The histopathological technique to investigate to study the nature of the cell damage in the organism caused in relation to the concentration of the pollutant. According to Warner (1967) histological and biochemical bioassay would also reveal minute changes in the aquatic environment.

Athikesavan *et al.* (2006) have reported the nickel induced histopathological changes in the different tissues of freshwater fish, *Hypophthalmichya moltrix* (Valenciennes). The present study has been undertaken to study the histopathology of spleen tissue of male tree frog, *Polypedates maculates* exposed to sublethal dose of lead for the period of 30 days.

Materials and Methods

Healthy, mature male tree frog *Polypedates maculatus* ranging in weight from 12 to 15 gms were collected from nearby fields, gardens and building walls of Annamalai nagar and Childambaram during the monsoon season (July to October). Immediately after collection, the frogs were acclimatized to the prevailing laboratory conditions in groups (usually 10 animals in a group) in wooden cages, measuring 24″ × 24″ × 18″. Considering the amphibian habitat and to provide sufficient moisture condition, water was provided in a trough inside the cage. Fresh small twigs with leaves were kept every day, in a crises cross manner, inside the cage to duplicate arboreal conditions where the frog could rest. All the cages were cleaned every alternate day. The water in the trough was changed every 24 hrs to maintain hygienic aquatic environment for the experimental animals. The frogs were feed daily with live earthworms and grasshoppers and were allowed to feed *adlibitum.*

Polypedates maculatus were treated to sublethal does of lead acetate (2.75 mg/gm body wt) for 30 days. After completion of the experiment, the tissue was taken out from the experiment and control groups from animal and fixed in aqueous Bouins fixative. They were dehydrated through graded alcohol, cleared in xylene and embedded in Glaxo's Paraffin Wax (58–60°C M.P.). Sections were cut at 5µ thickness. They were stained with haematoxylin and eosin. Permanent mounts were prepared in DPX (Gurr, 1959).

Results and Discussion

The histology of normal spleen had distinct showing the organization of lymphocytes. Treatment with sublethal dose of lead acetate for 30 days showed histopathological lessions as reduction in the number of lymphocytes and hypertrophid lymphocytes, degeneration and variable necrotic areas (Figures 45.2–45.4). Similar observation have also reported by (Sivaprakasam, 1990; Baskaran, 1990). Adbel Rahman, (1999) have reported significant extra medullarly haematopoiesis and active hyperplasic in spleen pulps of rats after exposure to the cumulative dose of gamma irradiation.

A reduction in the number of lymphocytes liquidation and hypertrophied lymphocytes were reported in the zinc phosphate treated spleen of *Millardia meltada* (Gray) (Neelanarayanan and Kanakasabai, 1995). Anithakumari and Sree Ramkumar, (1996) have studied the effects of water pollution on the spleen of *Channa punctatus* from Hussainsagar Lake, Hyderabad.

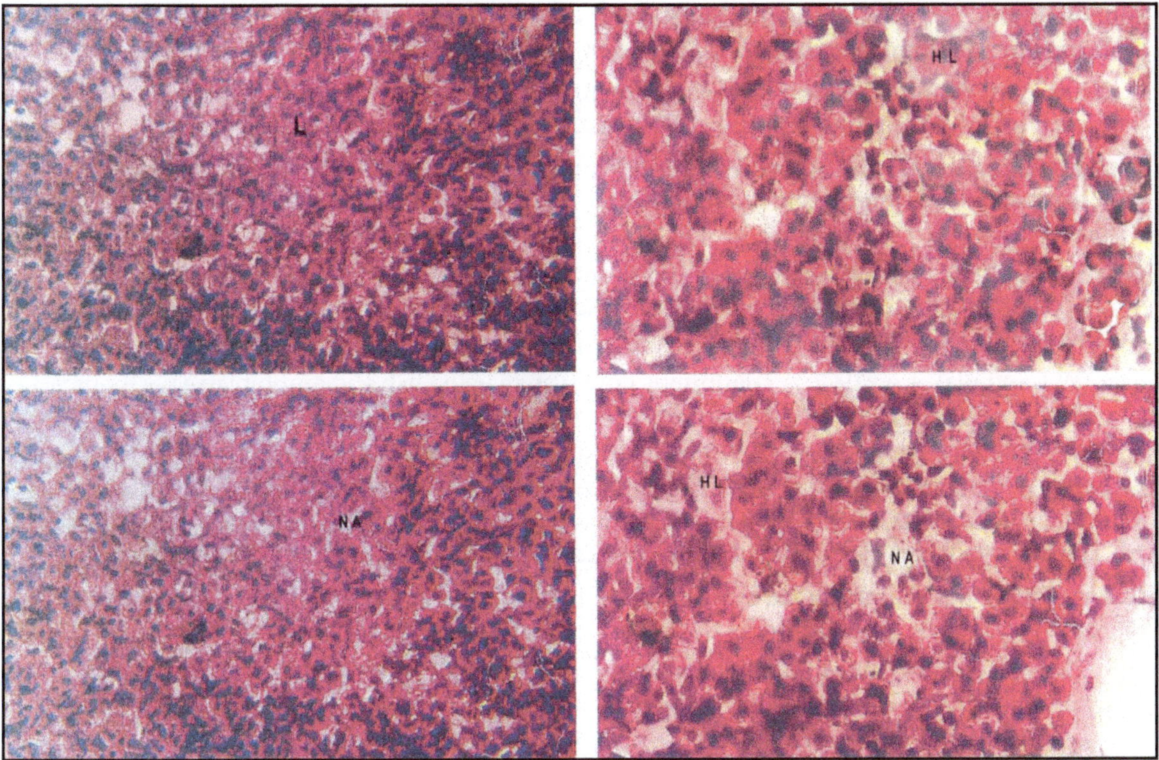

Figures 45.1–45.4: Light Micrograph of Spleen from (1) Normal frog showing Lymphocytes (Figure 45.1) Lead treated frog showing Hypertrophied Lymphocytes (HL) (Figure 45.2) Variable Necrotic Areas (NA), Hypertrophied Lymphocytes (HL) (Figures 45.3 and 45.4)

(Figures 45.1–45.4, Magnification x 100, x 200)

Spleen is found to possess Haematopoietic power only during early stages of development but resumes this function in the adult severe anaemia (Mitchell, 1956). Further they have revealed an increment in the number and decrease in the size of the melano-macrophage centers (MMC's) in the fish living in polluted water, when compared to spleen of control fish. Kranz (1989) has also reported an increased phagocytic activity of splenic macrophages under the influence of pollutants. Ellis *et al.* (1976) also supported the above findings. The results of the present study allowed a stimulation of mononuclear phogocytic system related to the presence of heavy metals. Thus the present study suggests that the virulence of the lead mainly depends upon the treatments periods.

Acknowledgement

The authors thank the authorities of Annamalai University and the Head, Department of Zoology for providing necessary research facilities.

References

Abbassi, S.A. and Soni, R., 1993. Computer aided studies on environmental impact assessment and management of seven heavy metals with respect to toxicity towards larvae of amphibian, *Rana tigrina*. *J. IPHE, India*, 2.

Abdal Rahman, N., 1999. Histopathology of the spleen of Adult male albino rats after whole body exposure to Gamma irradiation. *J. Egypt. Ger. Soc. Zool.*, 28(C): 13–26.

Anithakumari, S. and Sree Ramkumar, M., 1996. Effect of water pollution on the spleen of *Chnna punctatus* from Hussain Sagar Lake, Hydrabad, India. *J. Ecotoxicol. Environ. Monit.*, 6(1): 49–52.

Athikesavan, S., Vincent, S., Ambrose, T. and Velmurugan, B., 2006. Nickel induced histopathological changes in the different tissues of freshwater fish, *Hypophthalmicithays moltrix* (Valenciennes). *J. Environ. Biol.*, 27(2): 391–395.

Baskaran, I., 1990. Laboratory studies on efficacy and histopathological impact of bromadio lone (0.0005 per cent) against the soft furred field rat, *Millardia meltada* (Gray). *M.Sc. Dissertation*, Bharathidasan University, Trichy.

Birge, W.J. and Blak, J.A., 1977. A continuous flow system using fish and amphibian eggs for bioassay determinations on embryonic mortality and teratogenesis. *U.S. Environmental Protection Agency EPA*, 560: 15–77.

Cesar Teijon, Jesus M. Delsocorro, Juan. A. Martin, Mirian Lozano, Victoria Bernardo, M.A. and Dolores Blanco, M.A., 2000. Lead accumulation in rats at non acute doses and short periods of time: Hepatic, renal and haematological effects. *Ecotoxicology and Environmental Restoration*, 3(1).

Collins, T.F.X., 1987. Teratological research using in vitro system. V. Non mammalian model system. *Environ. Hlth. Persp.*, 72: 237–249.

Eisler, 1988. Exposure to lead can result in adverse effects of multiple tissues and organ that are critical to the viability and reproduction of wildlife. *Wildlife Resources*, 6–66, 2.1.

Ellis, A.E. Munro, A.L.S. and Roberts, R.J., 1976. Defence mechanisms in fish: A study of the phagocytic systems and the fate of intraperitoneally-injected particulate material in place plemonectes platessa. *J. Fish. Biol.*, 8: 67–78.

Ghate, Hemant, 1985. Embryopathic effects of lead in *Microhyla ornate* and its prevention of EDTA. *Poll. Res.*, 4(1): 711.

Gurr, E., 1959. *Methods for Analytical Histology and Histochemistry*. Leonard Hill (Books) Ltd., London.

Hill and Qureshi, 1998. Literature Review. Journal of the Australian College of Nutritional and Environmental Medicine, 15(2): 11–12.

Kaul, R.K., 1997. Metal pollution: Toxicology with special reference to factors influencing toxicity, and its treatment. *Chem. Environ. Res.*, 6 (3 and 4)

Kranz Heidemaire, 1989. Changes in splenic-melano mactophase centers *Dab limanda* during and after infection with ulcer diseases. *Dis. Aquai. Org.*, 6(3): 167–174.

Kyung-Hwa Beak, Hyun-Hee-Kim, Bum Han Bac, Yoon-Young Change and In-Sook Lee, 2005. EDTA-assisted phytoextraction lead from lead contaminated soils by *Echinochloa crusgalli* var. *framentacea*. *J. Environ. Biol.*, 26(1): 151–154.

Neela Narayanan, P. and Kanakasabai, K., 1995. Effect of zinc phosphate on intestine, liver, spleen and kidney of *millardia meltada*. *J. Ecobiol.*, 7(2): 85–90.

Rai, A.K., Upadhyay, S.N., Kumar, S. and Upadhya, Y.D., 1998. Heavy metal population and its control through a cheaper method. *Journal. IAEM*, 25: 22–51.

Sataka, M.Y., Mido, M.S., Sethi, S.A., Iqubal, H. Yasuhisa and Taguchi, S., 1997. *Environmental Toxicology.* Discovery Publishing House, New Delhi, (India), p. 96.

Sivaprakasam, C., 1990. Studies on determination of poison bait bases and laboratory evaluation of zinc phosphate in the Indian field mouse, *Musbooduga* (Gray). *M.Phil. Dissertation*, Bharathidasan University, Trichy.

Warner, R.E., 1967. Bioassay for micro chemical environment contaminants. *Bull. Wed. Hlths. Org.*, 36: 181–207.

WHO, 1995. Literature Review. Journal of the Australian College of Nutritional and Environmental Medicine, 15(2): 11–12.

Chapter 46

Forage Productivity in *Leucaena leucocephala* Lam. when Cultivated as Shrub Under Different Tree Environmental Conditions

***R.Y. Kulkarni*[1] *and D.V. Dev*[2]**

[1]*Department of Botany, Yogeshwari Mahavidyalaya, Ambajogai – 431 517, M.S.*
[2]*M.G.M. College of Biotechnology, Nanded – 431 602, M.S.*

ABSTRACT

In order to appreciate the magnitude of wasteland expanse of cultivation, the spread of shrub composition with grass species, coupled with browses showed varying productivity. Admittedly, the land cover retains a small portion, either in stunted form or barren areas, thus the biomass cover of wasteland is threatened by over utilization of resources.

There is need to mention here that the efforts have been directed to find out the "ways and means" to convent wasteland as source of biomass production of a good quality thereby, conforming the plant species × wasteland culturing are inseperable from each other.

Keywords: *Biomass, Dry matter, Crude protein, Tree environment, Harvest.*

Introduction

A combination of shrub land or grass species as animal feed is vital and can not be set-aside, since large areas are unsuited for crops but still can be used, justifiably to support "Fodder produce" may be of poor quality. Obviously, there exist no strategy for determining what contributions are desirable from shrub land species when associated with grasses as source of quality biomass. Rather, a relationship between the two sector (land × shrub species) can be established by taking into account the production level at local conditions. Wasteland utilization be focused on productivity, its establishment and efficient maintenance under existing conditions.

Materials and Methods

The land acquired was cleared one time, the clearance to extent of one hectare, minimum tillage practices were used during planning and planting of the shrub species. The trials revealed that tree species when converted to browse, sustained shaded condition, contributed high biomass, dry matter and crude protein yields.

The certified seeds of varieties were taken for the experimental trials on the field. The experiment was arranged in RBD with four replication. The experimental trials conducted for 3 years and the data is presented in the table. Freshly harvested plants were cut at (150 cm) HBH height, the samples were weighed and expressed as biomass into kg/u.a. The 100 gm. samples dried at 98°±2 in oven, the dried samples were powdered passed through fine (2 mm) mesh. The percent Nitrogen was determind by microkjeldhal's methods (Byers and Sturrock 1965). The crude protein was calculated as N per cent x 6. The result are presented in the Tables 46.1–46.3.

The harvest technology in different shrub land species employed to harness high biomass; revealed that high biomass yields at the height of 150 cm from ground level under varying tree environmental conditions.

Results

Biomass

The effect of environment on *Leucaena leucocephala* as shrub, the biomass yield is depicted in the Table 46.1. The results indicated that environment of *Leucaena leucocephala* effects were significant on biomass and between the harvests whereas environment × harvest were significant on biomass produce. *Leucaena leucocephala* being a slow growing yielded highest biomass when cultivated in the tree environment of *Leucaena leucocephala* in the 1st harvest of the 3rd year.

During 1st year, high biomass was recorded in the 2nd harvest while lowest in tree environment of *Sesbania grandiflora*. Interestingly, it was observed that *Leucaena leucocephala* as a shrub, when cultivated under tree environment of *Sesbania sesban* yielded 4765.3 Kg/u.a. On mean basis, highest biomass yields were recorded when cultivated with tree environment of *Leucaena leucocephala* while lowest in *Gliricidia maculata* tree. Consistently lowest yield were observed when cultivated in tree environment of *Sesbania grandiflora*. The low biomass is attributed to non-co-habiting relationship between legume to legume. Similar, were the results when cultivated under tree environment of *Gliricidia maculata*, *Leucaena leucocephala* as shrub species thrived well under tree environment of *Sesbania sesban*. The interaction effect revealed that their exist a significant relationship between harvests and different tree environment established.

Dry Matter

The character dry matter is important, it is photosynthetic accumulation. The data, presented (Table 46.2) notices that dry matter yields were significantly higher in the 3rd harvest of 1st year in the environment of *Leucaena leucocephala*. The harvest and environment × harvest effects were significant. When the data for biomass and dry matter is compared, a significant effect of harvest on biomass yield is noticed, similarly significant effects on dry matter yield were observed.

Table 46.1: Performance of *Leucaena leucocephala* for Biomass Yield Kg/u.a. Under Different Environmental Condition

Year	E_1T_1	E_2T_2	E_3T_3	E_4T_4	E_5T_5
1999–2000					
H_1	1269.7	4626.8	681.53	3385.6	800.3
H_2	1967.0	5250.7	614.88	3325.2	974.7
H_3	2149.2	4324.1	636.7	3812.3	1024.2
2000–2001					
H_4	2475.7	4414.8	967.48	3710.8	1053.2
H_5	2259.9	3519.2	1572.5	4200.6	986.6
H_6	2336.5	3769.2	1718.6	3459.3	1008..5
H_7	2935.3	4507.0	1980.8	3531.9	1006.4
2001–2002					
H_8	3672.7	11383.1	1807.9	2815.4	1612.4
H_9	4232.0	3405.2	1869.2	3015.4	1608.9
H_{10}	4156.5	753.7	2348.3	3605.6	1783.8
H_{11}	4765.3	815.6	2938.2	2850.3	1851.3
Mean	2929.1	4251.7	1557.8	3428.4	1245.4
ESE	317.06	HSE	470.27	E x HSE	1051.6
CD	877.46	CD	1301.5	CD	2910.2

H: Harvest; E_1T_1: Environment of tree *Sesbania sesban*; E_2T_2: Environment of tree *Leucaena leucocephala*; E_3T_3: Environment of tree *Sesbania grandiflora*; E_4T_4: Environment of tree *Moringa oleifera*; E_5T_5: Environment of tree *Gliricidia maculata*; ANOVA: Analysis of Variance; S_1: Indicates Shrub *Sesbania sesban.*

S_2: Indicates Shrub *Leucaena leucocephala*; S_3: Indicates Shrub *Sesbania grandiflora*; S_4: Indicates Shrub *Moringa oleifera*; S_5: Indicates Shrub *Gliricidia maculata*; G_1: Indicates Grass *Panicum maxicum*; G_2: Indicates Grass *Pennisetum typhoidum*; G_3: Indicates Grass *Euchleana maxicana*; Ap: Indicates Herb *Amaranthus paniculatus*

The results showed that *Leucaena leucocephala* has inbuilt mechanism for high dry matter yield. During 1st and 2nd year of harvest, the dry matter yield were significantly affected in all the harvests and tree environment. A significant high dry matter was observed in the 9th and 10th harvest of 3rd year. While, lowest in the 1st and 2nd harvest of the 1st year in tree environment of *Gliricidia maculata*. Invariably, the dry matter yields were high in shrub *Leucaena leucocephala* when cultivated under tree environment of *Sesbania grandiflora* and *Gliricidia maculata* resulted in low dry matter produce. On means basis the values for dry matter yields were highest in tree environment of *Leucaena leucocephala* followed by *Moringa oleifera* while lowest in tree environment of *Gliricidia maculata*. The biomass produce was significantly lowest in tree environment of *Gliricidia maculata*. The effect of tree environment practically did not have any impact on the shrub *Leucaena leucocephala*.

Crude Protein

The observations on crude protein yields (Table 46.3) showed that the yields significantly differ in *Leucaena leucocephala* when cultivated in different tree environments, the effect of harvests were significant and the interaction effects between environment × harvest as well. The crude protein yields showed an identical trend from highest to lowest as expressed in dry matter character. In the harvest of the 1st year, the crude protein yields were invariably low in tree environment of *Gliricidia maculata* and *Sesbania grandiflora* while, significantly higher crude protein yields were noticed in tree environment of *Leucaena leucocephala* and *Moringa oleifera*. As the age advances, the shrub *Leucaena leucocephala* established well in all the tree environments.

Table 46.2: Performance of *Leucaena leucocephala* for Dry Matter Yield Kg/u.a. Under Different Environmental Condition

Year	E_1T_1	E_2T_2	E_3T_3	E_4T_4	E_5T_5
1999–2000					
H_1	362.6	1166.0	179.9	925.4	161.2
H_2	504.5	1124.3	196.28	844.35	160.02
H_3	664.6	1309.0	205.1	932.88	179.55
2000–2001					
H_4	766.1	1087.2	194.92	899.48	194.03
H_5	689.2	1084.4	290.45	1047.5	204.82
H_6	633.7	880.8	397.05	869.8	194.92
H_7	767.3	967.8	534	867.55	288.05
2001–2002					
H_8	935.5	1123.9	613.1	707.77	415.05
H_9	1156.7	847.6	551.43	746.38	533.97
H_{10}	1156.7	387.3	507	899.23	610.85
H_{11}	1078.6	279.8	613.9	678.08	558.82
Mean	792.3	932.9	389.38	856.22	318.3
ESE	2.6051	HSE	3.8640	E x HSE	8.6401
CD	7.2096	CD	10.694	CD	23.912

Legend: Same as Table 46.1.

Discussion

The data presented in respective table reveals that the biomass, dry matter and crude protein yields showed significant difference under different tree environment. Similarly, significant differences between the harvest and years were observed in the biomass, dry matter and crude protein yields.

The interaction effects between harvest × year and harvest × year × environment were significant. As high as 5250.7 kg/bm/u.a., 304.9 and 119.7 kg of biomass, dry matter and crude protein yields were recorded in *Leuceana leucocephala* browse for first year (Tables 46.1–46.3). While, in the second year, although a significant differences were observed between the harvest and the tree environment. The shrub *Leuceana leucocephala* throughout maintained high biomass, dry matter and crude protein yields. On mean basis, highest yields were recorded in shrub *Leucaena leucocephala* under the environment of *Leucaena leucocephala*, followed 3428.4 kg/u.a. when cultivated in *Moringa* environment.

The results suggest that shrub *Leucaena leucocephala* yielded high biomass when grown in its own tree environment of *Moringa oleifera*. Hence, it is worthwhile its co-cultivation with *Moringa oleifera* tree. While, it did not harmoniously react when cultivated with tree *Gliricidia maculata*.

Table 46.3: Performance of *Leuceana lecocephala* for Crude Protein Yield Kg/u.a. Under Different Environmental Condition

Year	E_1T_1	E_2T_2	E_3T_3	E_4T_4	E_5T_5
1999–2000					
H_1	83.6	258.1	33.55	186.92	39.3
H_2	110.0	304.9	31.67	182.23	38.4
H_3	155.2	245.2	29.55	217.33	33.55
2000–2001					
H_4	171.7	242.5	66.97	206.22	31.67
H_5	164.6	213.7	88.42	244.05	29.55
H_6	143.1	223.2	124.2	196.2	29.57
H_7	181.6	260.9	138.45	192.12	62.77
2001–2002					
H_8	216.9	194.2	130.8	170.98	83.5
H_9	233.6	200.9	114.67	178.6	120.22
H_{10}	227.7	83.7	145.33	208.73	133.05
H_{11}	271.6	48.0	173.73	155.4	129.65
Mean	178.18	206.87	97.94	194.43	77.52
ESE	0.51195	HSE	0.75935	E x HSE	1.6979
CD	1.4168	CD	2.1015	CD	4.6991

Legend: Same as Table 46.1.

The increase dry matter yield is on account of synergistic effect under co-cultivation with tree and ecofriendly interaction between browse and tree species, our results are in confirmative with (Ghatnekar *et al.*, 1983) and reccomend that in shrub browse, biomass yields were and it is worthwhile of cultivation with *Moringa oleifera* tree.

The establishment of browse system in all is inexpensive, simple, developed into a fine coppice ensuring longevity and regeneration on repeated harvest.

All the converted shrub browse served as a valuable resources for forage produce besides the established tree system as source for fire and fuel. The results confirm the findings (Wilson *et al.*, 1980) and the shrub browse were source of high quality forage and the green leaves as source of manure (Jones and Jones, 1984). The variations in crude protein yields between the years were significant, so were between the harvests of the years under different tree environment.

The interaction between the harvest × years and harvest × year environment were significant (Table 46.3). Between the environment, dry matter yields were high, while crude protein yield showed negative co-relation when grown in association with trees (Table 46.3) [H7 x E2T2–260.9 kg/u.a.; H5 × E4T4–244.1 kg/u.a.; and H4 x E2T2–242.5 kg/u.a.]. Invariably, in the 3rd year, significantly, low crude protein yields were observed in *Leucaena leucocephala* with non legume. However, crude protein yields increased which confirms the co-cultivation with tree stepped up the crude protein yields. In the 3rd year (H9 × EITI–233.6 kg/u.a. and as low as H11 × E2T2–48.0 kg/u.a.). The studies reported by confirm the high biomass yields reported in this thesis with reference to *Leucaena leucocephala*.

The training and pruning to browse and its co-cultivation with tree stepped up forage yields, results confirm the finding reported (Semple and Pendlton, 1950), therefore, browse instead of tree is recommended for quality fodder (Narayanan and Sivagananam, 1967; Juliano and Carangal, 1979).

Acknowledgement

Authors are thankful to Dr. Suresh Khursale, The Principal Yogeshwari Mahavidyalaya, Ambajogai; Dr. S.W. Bhiogade, The Head and Dr. Acharya, Marathwada Agriculture University, Parbhani for the help.

References

Alferez, A.C., 1981. *Leucaena*: A renewable source of high protein animal feed and nitrogen-rich organic matter. *Scientia Fillipinas*, 1: 24.

Balsubramanian, T.N. and Robinson, J.G., 1984. Study on the inter cropping of *Leucaena leucocephala* with annual crops under dryland condition. *Mad. Agril. J.*, 71: 677.

Ghatnekar, S.D., Auti, D.G. and Kamat, V.S., 1983. Biomanagement of *Leucaena* plantations and by ion exchange. In: *Leucaena Res. on the Asian Pacific Region*; 109.

Hegde, N., 1983. *Leucaena* forage management in India. *Leucaena Research Asian Pacific Region*, Ottawa, p. 73.

Hill, G.D., 1970. Studies on the growth of *Leucaena leucocephala* production under grazing in the New Guinea lowlands papua and New Guinea. *Agril. J.*, 22: 73.

Jones, R.M. and Jones, R.J., 1984. The effect of *Leucaena leucocephala* on lime weight gain, thyroid size and thyroxine levels of steers in south-eastern Queensland. *Austr. J. Expt. agril. Anim. Husb.*, 24: 4.

Juliano, O.O. and Carangal, A.R. Jr., 1979. Crops for leaf protein and leaf meal production in the Philippines. In: *Proceedings of the Second International Green Crop Drying Congress*, (Ed.) R.E. Howarth. Saskatoon, Canada; University of Saskatchwqan, p. 5–10.

Kulkarni, R.Y. and Dev, D.V., 2007. Effect of harvest height on biomass productivity in different shrub land species. *Ecol, Env. and Cons.*, 13(3): 1–4.

Narayanan, S.S. and Sivagananam, L., 1962. A leguminous thornless quick hedge plant *Leucaena glance* Benth. *Madras Agril. J.*, 49: 110.

Partridge, I.J., and Ranacou, E., 1974. The effects of supplemental *Leucaena leucocephala* browse on steers grazing *Dichanthium caricosurn* in Fiji. *Tropical Grasslands*, 8(2): 107–112.

Rojoa, H., Naidu, S.N. and Owadlly, A.L., 1985. Effect of row spacing of Acacia plant *Leucaena leucocephala* (L) de wit variety peru on vegetative matter production. *Revue Agricole et sucriere de lle Maurice*, 64(1): 13–17.

Semple, A.T. and Pendlton, R.L., 1950. Woody legumes for the poor soils of humid equatorial lowlands. *Indian Fmg.*, 11: 223–225.

Wilson, L.L., Katsigianis, T.S., Greaves, Dorsett, A.A., Cathopoulis, T.E.A.G. and Bayor, J.E., 1980. Performance of native and Anglo-Nubian crosses and observations on improved pastures for goats in the Bahamas. *Tropical Agric.*, 57(2): 183–190.

Chapter 47

Floristic Distribution of an Invasive Weed *Hyptis suaveolens* Poit. in the Valley Districts of Manipur

Kh. Sandhyarani Devi, Y. Sanatombi Devi and P.K. Singh

Ethnobotany and Plant Physiology Laboratory, Department of Life Sciences, Manipur University, Canchipur, Imphal – 795 003, Manipur

ABSTRACT

A survey was conducted to study the distribution of *Hyptis suaveolens* Piot. of family Lamiaceae, which is becoming a potent invader in the valley districts of Manipur. During the survey, nine species were collected which were found to be associated with this plant. Habitat, taxonomic characters and its ethnobotanical uses and harmful effects were also studied. Phyto-sociological parameters for this plant were incorporated. Importance Value Index (IVI) for these plants were calculated by laying random quadrats of 1 metre square (1 quardrat = 1 m^2). The value of IVI is found to be maximum in *Hyptis suaveolens* (103.4) and minimum in *Sida acuta* (13.19) which shows that when this plant *Hyptis suaveolens* grows in an area, it starts dominating that particular area. Invasive plants are the second most important threat to our native plants after habitat destruction. As the plant is an invasive one, it is high time to understand this plant and preventive measures should be taken up to restrict further spread of this plant.

Keywords: *Hyptis suaveolens, Invasive plants, IVI, Dominant species.*

Introduction

The genus *Hyptis* consists of about 400 species which belongs to tribe Ocimeae of family Lamiaceae. Most ofthese species are widely distributed in the tropical and subtropical regions of new world.

Hyptis suaveolens Poit. commonly known as Tukma in Manipuri is native to tropical America, now become naturalized in almost all parts of the world including India. It has now established in Deccan Peninsula, Vindhyan Highlands, Andaman and Nicobar Islands and the North Eastern India (Yoganarasimhan, 2000). Tukma grows as a potent invader which disturbs the natural vegetation in the valley districts of Manipur. The state Manipur lies in the extreme north-eastern most part of India which is located at 23°50′N to 25°41′N latitude and 92°59′E to 94°4′E longitude with a total area of 22,327 km^2 (Vedaja, 1998). The valley districts of this state have an area of 2,067 km^2 which include four districts out of nine districts. The valley might be formed as a result of a lake being filled up by the river form sediments. The present Loktak lake is said to be the remnant of original lake in the past (Devi and Singh, 2005). The state which is described as "a little paradise on earth" is blessed with a rich floral diversity. However, due to human interferences and exploitation, the natural balance is degrading in an alarming level bringing an imbalance in natural and agricultural ecosystem. The intensification of agricultural land use and increasing urbanization leads to habitat destruction of many plants and animals. Thus, plants and animals adopt various mechanisms for their survival in the changing environment, as according to Darwin, "In the struggle of survival, the fittest win out at the expense of their rivals because they succeed in adopting themselves best to their environment". Some plants evolve themselves in their morphological, ecological and especially reproductive characteristics to adapt the disturbed geographical areas and too many plants turn out as invasive species and become weeds to many geographical areas. The invasive plants are non indigenous ones that adversely affect the habitats they invade economically, environmentally or ecologically. The magnitude and frequency of species introduction beyond their native range has increased dramatically with the expansion of global trait and passenger movements. Whether intentionally or unintentionally, the introduction of some invasive alien species have resulted in the degradation and destruction of natural ecosystem (Brown and Poorter, 2005). *Hyptis suaveolens* is also one such plant. It is becoming a potent invader and disturbing the natural vegetation of Manipur posing a threat to many endemic plants.

Materials and Methods

The survey was conducted in the four valley districts of Manipur (Bishnupur, Imphal East, Imphal West and Thoubal) during 2006–2007. Phyto-sociological parameters of *Hyptis suaveolens* with its associated species were studied in different sites of the valley districts where this plant is found to be grown abundantly using quadrat method taking the size of the quadrat as 1 metre square (1 quadrat = 1 m^2). Random quadrats were laid and in each quadrat the number of plant species and their total numbers were noted down for quantitative analysis for density (D), abundance (A) and frequency (F) of *Hyptis suaveolens* and its associated species and were calculated following the standard methods of Curtis and McIntosh (1950) and Oosting (1958). Importance Value Index (IVI) out of 300 was then calculated by adding the relative abundance (RA), relative frequency (RF) and relative density (RD) by adopting the method of Mishra (1968).

Results and Discussion

Hyptis suaveolens is a plant which is generally found to be grown in dry open locations like disturbed or overgrazed areas, wastelands, foothills, and roadsides. It is the plant which grows in hilly areas from where red hilly soils have been removed recently. However, other plant species are hardly found in such areas. The plant is found to be grown with other invasive plants like *Ageratum conyzoides* L., *Bidens pilosa* L, *Eupatorium odoratum* L., *Imperata cylindrica* (L.) P. Beauv., *Knoxia roxburghii* (Spreng) M.A. Rau, *Lantana camara* L., *Mikania cordata* (Burm.) B.L. Robinson and *Stachytarpheta cayennensis* (Rich.) Vahl.

Hyptis suaveolens is a strongly scented suffruticose herb attaining a height of about 1.8 m. Flowering of this plant occurs between September to January, and the plant propagated mainly through seed, although vegetative propagation through stocks also take place. Stems erect, quadrangular and hairy. Leaves ovate to broadly ovate (0.8–8 × 0.6–7) cm, base shallow cordate, margin serrulate, margin subacute to obtuse, pilosa in both adaxial and abaxial sides, 3–6 pedicellate blue flowers in small cymes, calyx persistent, 0.7–0.9 cm long, dilated upto 1.2 cm in fruit, hairy and glandular, teeth triangular, corolla blue, 0.8–1.0 cm long, nutlets 0.2–0.3 cm long with a brown notch at the tip when young. When dry seeds are soaked in water, a white gelatinous layer is formed.

Due to the presence of essential oils which is a characteristic of family Lamiaceae, the plant has some medicinal values. It has found to possess significant pharmacological (Kuhnt *et al.*, 1995), anti-cancerous (Mudgal *et al.*, 1997) and tumorigenic (Peerzada, 1997) properties. Locally, paste of this plant is applied on cutaneus infections. The decoction of this plant is taken as an appetizer, and dry seeds soaked in water for an hour is applied in boils and injuries for early suppuration. Although this plant possess medicinal properties, it is not efficiently utilized in this context. Further damage to the biodiversity of adjoining areas is much greater than its utilization as medicinal plants (Raizada, 2006). It has become an obnoxious weed and potential threat to our grazing grounds of wild animals and livestock. It forms dense stands over waste areas, threatening the native plant wealth (Murthy *et al.*, 2007). *Hyptis* also produces allelochemicals which disturb the growth of other plants leading to loss of many important palatable plants.

As per survey study, *Hyptis suaveolens* Poit. is found to be growing in association with *Imperata cylindrica* (L.) P. Beauv., *Bidens pilosa* L., *Stachytarpheta cayennensis* (Rich.) Vahl., *Knoxia roxburghii* (Spreng) M.A. Rau, *Mikania cordata* (Burm.) B.L. Robinson, *Lantana camara* L., *Euphorbia hirta* L., *Eupatorium odoratum* L. and *Sida acuta* Burm. F. Results of phyto-sociological studies in Table 47.1 showed the maximum values of density, abundance and frequency with their relative values are recorded in *Hyptis suaveolens* Poit. and minimum values in *Sida acuta* Burm. F.

Table 47.1: Phyto-sociological Studies of *Hyptis suaveolens* Poit. with its Associated Plants

Sl.No.	Name of Species	Density (1 m²)	RD (%)	Abundance (1 m²)	RA (%)	Frequency (%)	RF (%)	IVI Out of 300	A/F Ratio
1.	*Hyptis suaveolens* Poit.	16.2	48.80	16.2	39.3	100	15.3	103.4	016
2.	*Imperata cylindrica* (L.) P.Beauv.	5.20	15.7	5.77	13.8	90	13.8	43.30	0.06
3.	*Bidens pilosa* L.	3.80	11.5	5.43	13.1	70	10.76	35.36	0.07
4.	*Stachytarpheta cayennensis* (Rich.) Vahl.	1.80	5.42	2.57	6.23	70	10.76	22.41	0.04
5.	*Knoxia roxburghii* (Spreng) M.A.Rau	1.60	4.82	2.66	6.31	60	9.23	20.36	0.04
6.	*Mikania cordata* (Burm.) B.L. Robinson	1.10	3.3 I	1.83	4.37	60	9.23	16.91	0.03
7.	*Lantana camara* L.	1.10	3.3 I	2.20	5.33	50	7.60	16.24	0.04
8.	*Euphorbia hirta* L.	1.10	3.01	2.00	4.85	50	7.69	15.55	0.04
9.	*Eupatorium odoratum* L.	0.80	2.10	1.33	3.23	50	9.23	14.56	0.02
10.	*Sida acuta* Burm. F.	0.70	2.10	140	340	50	7.69	13.19	0.03
	Total	33.4		41.2		660			

The IVI value of the present investigation ranges from 13.9–103.4 and found to be maximum in *Hyptis suaveolens* Poit. (103.35) followed by *Imperata cylindrica* (L.) P. Beauv. (43.30). The minimum value of IVI is found in *Sida acuta* Burm.F. (13.19) followed by *Eupatorium odoratum* L. (14.56). From these values, it can be concluded that *Hyptis* is a dominant species in the areas where it is found. Also, the value of Abundance/Frequency ratio (A/F) of *Hyptis* was recorded more than 0.05 which is 0.16, this plant species was found to be distributed aggregately (Curtis, 1959).

Among the invasive weed species, *Parthenium hysterophorus* L. has been designated as worldwide troublesome noxious weed species, because of its prolific seed production and fast spreading ability, allelopathic effect on crops and health hazard to humans as well as animals (Shabbir and Bajwa, 2006). *Lantana camara* L. is also spreading fast due to the human disturbances such as cultivation, road construction and forest fragmentation and degradation (Raghubanshi *et al.*, 2005). In aquatic environment also the growth of aquatic weed water hyacinth, *Eicchornia crassipes* (Mart.) Solms-Laub. has led to loss of water in water bodies devoid of oxygen enhancing eutrophication (Rao, 2006). In the present scenario in the valley districts of Manipur, *Hyptis suaveolens* is a dominating species in areas where it is found and as it can grow in harsh environmental conditions with varied habitats, it turns out to be a potent invader disturbing the growth of other plants. The ability to reproduce vegetatively as well as sexually, rapid growth, high reproductive output, tolerance to a broad range of environmental conditions, and production of allelochemicals also help this plant in establishing and proliferating in a new environment. So preventive measures particularly awareness of this plant is necessary to restrict further spread in other areas as increasing anthropogenic disturbances, pollution levels, rapid industrialization and urbanization have put a stress on native vegetation.

Acknowledgements

We are grateful to CSIR/UGC for providing fellowship to Kh. Sandhyarani Devi for her research programme. We are also thankful to the Head, Department of Life Sciences, Manipur University, Canchipur, Imphal, Manipur, India for encouragement and facilities during the tenure of the programme. Sincere thanks to Nanda, Padmabati, Mona, Ksheroda and Research Scholars for their kind help and suggestions.

References

Brown, M. and Poorter, M.D., 2005. The Global Invasive Species Database (GISD) and International exchange of invasive information: using global expertise to help in the fight against invasive alien species. In: *Proceedings of International Workshop on Invasive Plants in Mediterranean Type Regions of the World*, 25–27th May.

Curtis, J.T., 1959. *The Vegetation of Wisconsin University, Wisconsin*. Wisconsin Press, Madison.

Curtis, J.T. and McIntosh, R.P., 1950. The interrelation of certain analytic and synthetic phyto-sociological characters. *Ecology*, 31: 434–455.

Devi, Y.N. and Singh, P.K., 2005. Species distribution of the family Verbenaceae in the valley districts of Manipur. *J. Curr. Sci.*, 7(1): 105–110.

Kuhnt, M., Probstle, A., Rimplex, H., Bauer, R. and Heinreich, M., 1995. Biological and pharmaceutical activities and further constituents of *Hyptis verticillata*. *Planta Medica*, 61(03): 227–232.

Mishra, R., 1968. *Ecology Workbook*. Oxford and IBH Publ. Co., New Delhi.

Mudgal, V., Khanna, K.K. and Hazra, P.K., 1997. *Flora of Madhya Pradesh, Vol. II*. Botanical Survey of India, p. 403–404.

Murthy, E.N., Raju, V.S. and Reddy, C.S., 2007. Occurrence of exotic *Hyptis suaveolens*. *Current Science*, 93(9): 1203.

Oosting, H.J., 1958. *The Studies of Plant Communities*. W.H. Freeman and Co., San Francisco, USA.

Peerzada, N., 1997. Chemical composition of the essential oil of *Hyptis suaveolens*. *Molecules*, 2: 165–168.

Raghubanshi, A.S., Rai, L.C. and Gaur, J.P., 2005. Invasive alien species and biodiversity in India. *Current Science*, 88(4): 539.

Raizada, P., 2006. Ecological and vegetative characteristics of a potent invader, *Hyptis suaveolens* Poit. from India. *Lyonia*, 11(92): 115–120.

Rao, A., 2006. *Invasive Species: Scourge of the Aliens*. World Prout Assembly, Proutist Universal Global Headquarters, Kolkata, India.

Shabbir, A. and Bajwa, R., 2006. Distribution of parthenium weed (*Parthenium hysterophorus* L.) an alien invasive weed species threatening the biodiversity of Islamabad. *Weed Biology and Management*, 6(2): 89–95.

Vedaja, S., 1998. *Manipur: Geography and Regional Development*. Rajesh Publications, New Delhi.

Yoganarasimhan, S.N., 2000. *Medicinal Plants of India*, (Eds.) V. Srinivasan and N. Kosal-Ram. Cyber Media, Bangalore, pp. 282.

Chapter 48

Impact of Acid Pollution on the Haemolymph Cholesterol Levels of a Freshwater Crab, *Paratelphusa hydrodromous* (Herbst)

***P. Usharani*[1], *N. Ezhili*[2], *A.A. Sivakumar*[3] *and K. Thirumathal*[4]**

[1]*Project Co-ordinator, Centre for Environmental Education, Coimbatore – 641 029*
[2]*Department of Zoology, PSGR Krishnammal College for Women, Coimbatore – 641 004*
[3]*PG and Research Department of Zoology, Kongunadu Arts and Science College, Coimbatore – 641 029*
[4]*PG Department of Zoology, A.P.A. College for Women, Palani – 624 614*

ABSTRACT

Both plasma and serum cholesterol levels were found to decrease insignificantly in *Paratelphusa hydrodromous* when they were exposed to sub-acute toxicity of formic and sulphuric acids. The formic acid was found to have more pronounced effect on the blood cholesterol levels. The decrease in blood cholesterol levels of crab under toxic stress in the present study may be probably due to cholesterol being converted into free fatty acids or reduced absorption of dietary cholesterol or inhibition of cholesterol biosynthesis.

Keywords: *Crab, Acid pollution, Blood cholesterol levels.*

Introduction

Acid rain is one of the two most global environmental problems associated with fossil fuel combustion (Carter, 1981). Acid rain containing oxides of sulphur (or) nitrogen has increased in pace with the use of fossil fuel especially from power plants, motor vehicles and melting of nonferrous metal throughout the world (Rosencrantz and Wetstone, 1980). Rastogi and Jayaraj (1984) observed a dangerous synergetic effect of sulphur oxide which is very unstable and oxidized to form sulphurdioxide mist. Such acid rain with huge amounts of sulphuric acid are reported from many of the industrially developed countries. The acidity of rainfall has been increasing in recent years (Krupa and Coscio, 1977). Most of the acidity in rain is attributed to SO_4 and NO_3 derived from anthropogenic activity. This is based on the correlation between these ions and pH of the rain. The effects of acid rain on the lake ecosystems are considerable (Bourdeau and Treshow, 1978). The acidification of freshwater ecosystem leads to many changes, most of which involve decrease in biological activity and important changes in nutrient cycling (Bourdeau and Treshow, 1978).

Materials and Methods

Specimens of *Paratelphusa hydrodromous* were collected from fresh water bodies in and around Coimbatore. The hydro biological features like dissolved oxygen, salinity, temperature and pH of the water used for the present study were determined for each set of experiment, as these factors have significant influence on the behaviour of toxicants. Healthy crabs with an average length of 2.5–3.0 cms and having an average weight of 13–16 gms were selected from the stock. Crabs belonging to both the sexes were used for each concentration 6 litres of water in circular plastic tubs was used. Small stones were kept to provide shelter for the crabs. 6 crabs were introduced in each tub, in which already different concentrations of formic and sulphuric acid solutions were added.

Stock solution of formic acid and sulphuric acid were prepared from analytical grade BDH acids. Preliminary tests were carried out to fmd out the median tolerance limits (LC_{50} of crabs to formic acid and sulphuric acid for 24 hrs. The LC_{50} concentration of formic acid and sulphuric acids were diluted by 5 times to get sub-lethal concentration respectively by trials. The particular concentration at which there was no mortality throughout the experimental period. Three sets of experiments were kept (control, Treatment–I, Treatment–II) with 6 litres of water and 6 crabs in each. Except the control tubs the rest received the respective sub-lethal concentration of acids.

The experimental tubs along with control were maintained for about 1½ months. At the end of every week 6 crabs were taken from control and experimental tubs. Haemolymph samples were drawn from the cut end of the dactylus region of the walking legs, using a pre-chilled calibrated micropipette.

Estimation of cholesterol was done by Ferric chloride method, using cholesterol clinical kit supplied by M/S New India Chemical Enterprises, Cochin. Haemolymph already collected was transferred to small test tubes containing sodium fluoride and ammonium oxalate (anti coagulant). The sample was mixed gently and centrifuged at 2500 rpm for 5 minutes and clean plasma was collected and used for the analysis. Haemolymph was transferred to a clean dry test tube and it was allowed to clot. The clots were loosened with a clean glass rod and centrifuged at 2500 rpm for 5 minutes and clean serum was collected and used for the analysis. To 0.5 ml of ferric chloride, 50ml of acetic acid solution was added. To 4.9 ml of working reagent, 0.1 ml of standard cholesterol and 3ml of sulphuric acid were added. 0.1 ml of plasma and serum, were taken in separate test tubes. To the sample, 9.9 ml of working reagent was added and kept for 15 minutes and centrifuged at 250 rpm for 15 minutes. The supernatant is a protein free solution which is used for analysis. To the 5 ml of supernatant, 3 ml of 36N sulphuric acid was added and were mixed well. The test tubes were kept for 30 minutes in the dark. The colour developed in the blank, standard and sample test tubes were read at 530µ using ERMA photoelectric colorimeter (Model AE-11N).

Results and Discussion

The medium lethal concentration for formic acid and sulphuric acid were 0.5ppm and 0.6ppm respectively. The sub-lethal concentration of both formic and sulphuric acids were arrived by trials, and were found to be 0.10 ml/L for the former and 0.012 ml/L for the latter. The plasma cholesterol data from formic acid indicate a gradual decrease starting from the 1st week upto 6th week both in the control and treated crabs. The treated crabs showed a decrease in their plasma cholesterol after every week when compared to that of their controls. The per cent decrease of plasma cholesterol in the treated crabs over the controls increased as the exposure time extended (Table 48.1). The maximum percent decrease was observed after the 6th week giving 17.84 per cent and the minimum value of 2.94 per cent after 1st week. The analysis revealed that per cent decrease at the end of each week was not significant. The plasma cholesterol data of the crab from the sulphuric acid also showed a similar pattern as that of the formic acid, the per cent decrease was least after the 1st week (2.63 per cent) and the maximum after the 6th week (39.06 per cent). The statistical treatment of the mean values of plasma cholesterol from sulphuric acid animals did not show any significant change.

Table 48.1: Changes in Plasma Cholesterol Levels of Crab *Paratelphusa hydrodromous* After Exposed to Sub-lethal Concentration of Formic Acid and Sulphuric Acid

Crabs	*Amount of Cholesterol (100 ml)*						
	Exposure of Formic Acid						
	Initial	*1st Week*	*2nd Week*	*3rd Week*	*4th Week*	*5th Week*	*6th Week*
Control	55.13 ±4.01	53.97 ±3.83	52.38 ±1.94	5.233 ±5.06	51.59 ±2.84	4600 ±7.24	44.44 ±7.33
Treated	53.33 ±4.85	52.38 ±1.94	50.79 ±2.29	46.03 ±7.24	42.86 ±3.72	38.08 ±5.06	36.51 ±5.22
Per cent change over control		–2.94	–3.04	–12.03	–16.92	–17.27	–17.84
Calculated 't' 0.05 level		1.890	0.233	0.070	2.170	0.912	0.330
Crabs	*Amount of Cholesterol (100 ml)*						
	Exposure of Sulphuric Acid						
	Initial	*1st Week*	*2nd Week*	*3rd Week*	*4th Week*	*5th Week*	*6th Week*
Control	55.13 ±4.01	53.97 ±3.83	52.38 ±1.94	52.33 ±5.06	51.59 ±2.84	46.00 ±7.24	44.44 ±7.33
Treated	53.96 ±3.30	52.55 ±6.74	49.21 ±5.22	44.45 ±7.67	42.86 ±5.38	28.57 ±3.18	27.08 ±262
Per cent change over control		–2.63	–5.96	–13.83	–18.17	–37.93	–39.06
Calculated 't' 0.05 level		0.2.55	0.444	0.063	0.270	0.216	1.424

The values are mean±SE of six individual observations.

–: Denotes per cent decrease over control.

The mean value of serum cholesterol with their standard error for the crabs exposed to formic acid are given in Table 48.2. The data indicate a decreasing trend in the serum cholesterol content both in the control and treated animals. However, certain amount of percent decrease in serum cholesterol was observed in the treated animals over their controls. After the 2nd week the per cent decrease was the least and maximum after the 6th week. The statistical analysis revealed that the 6th week value was significant and not the other values. The serum cholesterol concentration followed a decreasing trend as that of the formic acid. However in sulphuric acid, the per cent decrease in serum cholesterol was not gradual and steady but showed fluctuations after a week Table 48.2. The serum cholesterol decrease after 3rd week was highly significant and after 6th week it was significant as revealed by statistical analysis. Comparing the percent decrease in serum cholesterol between formic acid and sulphuric acid that the sulphuric acid is more effective than the formic acid.

Table 48.2: Changes in Serum Cholesterol Levels of Crab, *Paratelphusa hydrodromous* After Exposed to Sub-lethal Concentration of Formic Acid and Sulphuric Acid

Crabs	*Amount of Cholesterol (100 ml)*						
	Exposure of Formic Acid						
	Initial	*1st Week*	*2nd Week*	*3rd Week*	*4th Week*	*5th Week*	*6th Week*
Control	71.18 ±2.67	76.18 ±3.17	71.60 ±3.76	65.08 ±4.03	63.66 ±4.38	57.14 ±3.18	55.55 ±2.67
Treated	71.14 ±2.01	66.67 ±3.18	65.08 ±5.22	54.76 ±3.85	50.80 ±4.31	44.45 ±4.30	42.85 ±1.23
Per cent change over control		–12.48	–9.11	–15.86	–20.20	–22.21	–22.99
Calculated 't' 0.05 level		0.490	1.980	0.431	0.809	0.830	2.53*
Crabs	*Amount of Cholesterol (100 ml)*						
	Exposure of Sulphuric Acid						
	Initial	*1st Week*	*2nd Week*	*3rd Week*	*4th Week*	*5th Week*	*6th Week*
Control	71.18 ±2.67	76.18 ±3.17	71.60 ±3.76	65.03 ±4.03	63.66 ±4.38	57.14 ±3.18	55.55 ±2.67
Treated	74.60 ±3.17	60.32 ±3.67	57.80 ±4.30	47.62 ±3.89	44.45 ±4.30	42.08 ±4.89	41.27 ±4.29
Per cent change over control		–20.82	–19.27	–26.83	–30.18	–26.36	–25.71
Calculated 't' at 0.05 level		0.27	0.10	7.53**	1.21	1.29	2.57*

The values are the mean±SE of six individual observations; –: Denotes per cent decrease over the control; *: Values are significant at 0.05 level 't' = 2.23. **: Values are highly significant.

Perkins (1974) reported that the sub-lethal impairment of animals development or its capacity to perform and adapt can reduce the chance of survival and potential growth and reproduction. He also pointed out that an adaptive response doesn't necessarily increase the individual's chance of survival, but just increase the probability of survival in normal environmental conditions. Eisler and Edmunds

(1966) reported in fish an increase in cholesterol level after exposed to endrin toxicity and the authors suggested that endrin either enhances the production by liver or inhibits its excretion to the bile ducts. In *Saccobranchus fossilis* and *Labeo rohita*, a decrease level of cholesterol due to increased breakdown of cholesterol into free fatty acids was observed by Verma *et al.* (1979) and Bansal *et al.* (1979) when fishes were exposed to chlordane toxicity. Pant and Gill (1984) observed in *Punctius conchonius* that chronic exposure to chromium poisoning resulted in hypo-in cholestremia in blood and elevated levels of tissue cholesterol. In the present study, the plasma and serum cholesterol levels showed a reduction in its concentration over that of their controls. The percent increase was found to be dose and time dependent. The percent decrease in plasma and serum cholesterol levels was at minimum after the 1st week of exposure and maximum after the 6th week of treatment with regard to the effectiveness of acids on the cholesterol content, the sulphuric acid had pronounced effect than formic acid. According to Wasserman *et al.* (1970) that very little of serum cholesterol will be utilized for steroid hormones. In *Tilapia mossambica*, the reduction in serum cholesterol level of the fish exposed to phosphamidon revealed that the pesticide stress either affects the absorption of dietary cholesterol or the synthesis at the liver as it is the main source of cholesterol biosynthesis (Jayantha Rao *et al.*, 1984). In *Mystus vittatus*, Dalela *et al.* (1980) reported that reduction in blood cholesterol after pesticide exposure may be due to increased break down of cholesterol into tree fatty acids. The reduced cholesterol level of the haemolymph in the present study may be due to the inhibition of cholesterol biosynthesis or cholesterol being converted into free fatty acid or due to reduced absorption of dietary cholesterol recalling the situations reported by the earlier studies.

References

Bhagyalakshmi, A., Srenivasulu Reddy, P. and Ramamurthi, R., 1984. Subacute stress induced by sumithion on certain biochemical parameters in *Oxiotelphusa senex senex* the freshwater rice field crab. *Toxicol. Lett.*, 21: 127–132.

Bourdeau, P. and Treshow, M., 1978. Ecosystem response to pollution. In: *Principles of Ecotoxicology*, (Ed.) Gc. Butler. John Wiley and Sons, New York, pp. 318.

Carter, J., 1981. *The President's Environmental Program 1979*. The President's Council on Environmental Quality, Washington, DC, pp. 57.

Dalela, R.C., Rani, S., Kumar, V. and Verma, S.R., 1981. *In vivo* haematological alterations in a freshwater Teleost, *Mystus vittatus* following sub acute exposure to pesticides and their combinations. *J. Environ. Biol.*, 2(2): 79–86.

Eisler, R. and Edmunds, P.M., 1966. Effects of endrin on blood and tissue chemistry of a marine fish. *Trans. Am. Fish. Soc.*, 95: 153–159.

Hodges, L., 1973. *Technological Hazards in Environmental Pollution.* Holt, Rinehart and Winston, New York, pp. 216–217.

Jayantha Rao, K., Azhar Baig, M.D. and Ramamurthy, K., 1984. Effects of a systemic pesticide phosphamidon on some aspects of freshwater fish, *Tilapia mossambica. Indian J. Environ. Hlth.*, (1): 60–64.

Kallapur, V.L. and Yadwad, V.B., 1986. Changes in fuel reserves of the premoult field Crab, *Paratelphusa hydradromous* (Marine Edwards) following Endosulfon treatment. *Indian J. Exptal. Biology*, 20(10): 535–538.

Krupa, S.V. and Coscio, M.R., 1977. Impact of stock emission from NSP SHERCO Power Plant on terrestrial vegetation: The chemistry of rain 1976–1977. *Second Annual Report to Northern States Power Company*, Minneapolis, Minnesota.

Larsson, A., 1976. Some biochemical effects of cadmium on fish, from haematological and biochemical effects of cadmium on fish, from effects of pollutants on aquatic organisms. *D. Lockwood, A.P.M., Soc. for Expti. Biol. Seminar Series*, 2: 35 – 45.

Odum, E.P., 1971. Pollution and environmental health. In: *Fundamentals of Ecology*, 3rd edn. W.B. Saunders Company, Philadelphia, London, pp. 432.

Perkins, E.J., 1974. *The Biology of Estuaries and Coastal Waters*. Academic Press, London.

Rastogi, V. and Jeyaraj, M.S., 1981. *Textbook of Animal Ecology and Distribution of Animals.* Kedarnath, Meerut, Ramnath, Delhi, pp. 273.

Rosencranz, A. and Wetstone, G., 1980. Acid precipitation: National and international responses. *Environment*, 22(5): 6–8.

Sprague, J.B., 1970. Measurement of pollutant toxicity to fish. III. Sub-lethal effects and "safe" concentrations. *Water Res.*, 5: 245–266.

Verma, S.R., Sansal, S.K., Gupta, A.K. and Dalela, R.C., 1979. Pesticides induced haematological alterations in a freshwater fish, *Saccobranchus fossilis. Bull. Environ. Contam. Toxicol.*, 22: 467.

Wasserman, D., Michel, R. and Wasserman, M., 1970. Effect of organochlorine insecticides on serum PBI (Protein bound iodine) level in occupationally exposed people. *Bull. Environ. Contam. Toxicol.*, 5: 37.

Chapter 49

Control of Thrips (*Scirtothrips dorsalis*, Hood) in Tea [*Camellia sinensis* L (O) Kuntze] Plantations of Barak Valley, Assam (India)

P. Choudhury[1]* and B.K. Dutta[2]*

[1]Department of Ecology & Environmental Sciences,
[2]Microbial and Agricultural Ecology Laboratory,
Department of Ecology & Environmental Sciences,
Assam University, Silchar – 788 011, Assam

ABSTRACT

Thrips cause considerable loss of crop (25 per cent) in north-eastern tea fields. They cause damage to tender leaves and stems with the anchor like structures present on their legs, that cause laceration and thus juice oozes out from the damaged tissues. In all the five sub-areas of Barak valley, the pest infestation was more or less similar. The infestation caused by the pest was observed throughout the year. However, during the period of bud breaking (January/February), the infestation was more pronounced. Among the inorganic pesticides, Endosulphan, and of the two bio-pesticides, a Neem (*Azadirachta indica*) based product (*i.e.*, Econeem), while among the plant extracts, *Andrographis paniculata* was found to be more efficacious compared to others.

Keywords: *Bio-pesticides, Barak valley, Plant extracts, Seasonal variation, Thrips.*

* E-mail: parthankar@rediffmail.com; parthankar@yahoo.co.in; ** E-mail: bimankdutta@rediff.com.

Introduction

Thrips inflict heavy damage to tea throughout Asia and Africa. (Ananthakrishnan, 1973; Andrews, 1925; Chen, 1979; Kodomari, 1978; Lefroy, 1909; MKWaila, 1982; Muraleedharan and Kandasamy, 1980; Muraleedharan, 1983). Among all the pests occurring in the tea fields of Barak valley, Thrips (*Scirtothrips dorsalis,* Hood) was found to occur as one of the major damage causing pest occurring throughout the year. So far, comparatively very little work was done on the various aspects of the pest under the agro-climatic conditions of Barak valley, Assam. The present investigation was carried out to study the seasonal variation in infestation, seasonal variation in population of the pest and the influence of environmental factors, control aspects etc. under the agro-climatic conditions of Barak Valley, Assam.

Materials and Methods

An initial survey was carried out in five tea growing sub-areas of Barak valley (Assam), *i.e.,* Chatlabheel, North Cachar, Happy valley, Hailakandi, and Longai Valley/Karimganj sub-area. Young twigs of tea plantation areas showing various degrees of thrips (*Scirtothrips dorsalis,* Hood) infestation were categorized as follows:

Mild:	Less than 20 per cent young twigs infested
Mild-moderate:	21–40 per cent young twigs infested
Moderate:	41–60 per cent young twigs infested
Moderate-severe:	61–80 per cent young twigs infested
Severe:	81–100 per cent young twigs infested.

For estimating the population of thrips, the following method was adopted.

A young twig (Two and a bud) was plucked and kept inside a plastic vial (5.5 × 8.8cms) covered with lid. The collected specimens were brought to the laboratory and a piece of cotton wool soaked in ethyl alcohol was placed inside the vial to make the thrips senseless for counting. For sampling, 50 bushes (Ooman, 1982) were selected at random. All the samples were carefully transplanted to laboratory and counted under binocular microscope. Thrips population was expressed as the average number of individuals per twig (in the fifty twigs collected from fifty randomly selected bushes). Throughout the year long study, the sampling of twigs was done in the afternoon hours at fortnightly intervals. Care was taken against dislodging of insects while handling the plant parts (*i.e.,* young twigs) for sampling. Mean of the two observations taken in a month was calculated and it was then converted into percentages (*i.e.,* pest population/100 twigs). This was compared with the environmental factors, *i.e.,* temperature, humidity and rainfall to find if any correlation exists.

The thrips infested areas of Experimental site I [Arcuttipore Tea Estate (under Happy valley sub-area)] was taken as the experimental site. The experiment was laid out in randomized block design with three sets of replications. Sample blocks were prepared comprising 50 bushes. The insecticide Endosulphan (Thiodan 35 EC) was used at 1:200 (Thiodan: Water) concentrations. Approximately 500 ml of the diluted insecticide was sprayed on the top hamper of the individual bushes, while untreated plots were left for comparison. In parallel to the treatment with inorganic pesticides, treatment with some of the common plant extracts and bio-pesticides having known insecticidal/repellent properties were also done on the thrips infested areas of the same experimental site.

The bio-pesticides used were (I) a Neem (*Azadirachta indica*) based formulations (Econeem) and (II) Karanjin based formulations (Bio-2001). These were used at 1:100 and 1:200 concentrations respectively.

The different plant extracts were prepared following the modified method as described by Ghosh Hazra *et al.* (1994) and is as follows:

Lantena camara var aculeata

Green leaves of this plant species were taken. From one Kilogram of green leaves, about 400 ml of juice was extracted. The green leaves were grinded with a big size mortar and pastel to extract its juice and the same was then filtered. The filtrate was then diluted in water in the ratio of 1:10.

Chromolaena odorata

The method of preparation of leaf extract was same as that of *L. camara var aculeata.*

Andrographis paniculata

The method of preparation of leaf extract was same as that of *L. camara var aculeata.*

Curcuma sp.

Rhizome were shredded in a big sized mortar and pastel. From one kilogram of rhizome about 350 ml juice was extracted. This was then diluted in water in the ratio of 1:10 for spraying in the tea bushes.

All the above mentioned bio-pesticides/plant extracts were applied separately as foliar spray on the top hamper of the bushes with the hand sprayer at the rate of 250ml per bush. Untreated (control) plots were left for comparison. The efficacy of the treatments (*i.e.*, reduction in population/infestation) was evaluated by counting the average number of individuals per twig in the fifty twigs taken at random from each sample block and thus following the same method as followed in population estimation. The post treatment data were subjected to analysis of variance (ANOVA) to find out CV(per cent), test of significance (F value) and critical difference (CD at 5 per cent and 1 per cent) and also Paired 't' test to find out if any observable variation exists between the pre-treatment and post treatment.

Results and Discussion

In all the sub-areas of Barak valley the damage caused by the pest was almost uniform with little variation. Although, in all the sub-areas, mild and moderate degree of infestation persisted almost uniformly, but severe infestation was observed in some of the tea estates under Happy valley and Hailakandi sub-areas (Table 49.1).

Table 49.1: Thrips Infestation in Different Sub-areas of Barak Valley, Assam (Data based on observation in 15 tea estates in each sub-area)

	Percentage of Infestation				
	Mild	*Mild-moderate*	*Moderate*	*Moderate-severe*	*Severe*
Chutlabheel	14	14	43	29	0
North-Cachar	36.4	36.4	18.2	9	0
Happy Valley	36	29	7	21	7
Hailakandi	46	8	23	15	8
Longai Valley and Kxj	50	12.5	25	12.5	0

Seasonal variation in thrips population: Thrips population was observed almost throughout the year with a variation in their seasonal abundance. During the bud breaking season (*i.e.*, February and March), maximum population was observed (per 50 twigs 66 thrips in February and 60 thrips in March). While in April, there was little reduction (*i.e.*, 42 thrips per 50 twigs), followed by a rise in May (56 thrips per 50 twigs) and June (50 thrips per 50 twigs). From the month of July onwards, with the onset of heavy rain, the pest population in the studied areas started declining, giving an indication that the pests might have been washed down due to heavy shower. From November, till January, the population of thrips took an ascending trend, suggesting that cold weather favours their growth and multiplication (Figure 49.1B)

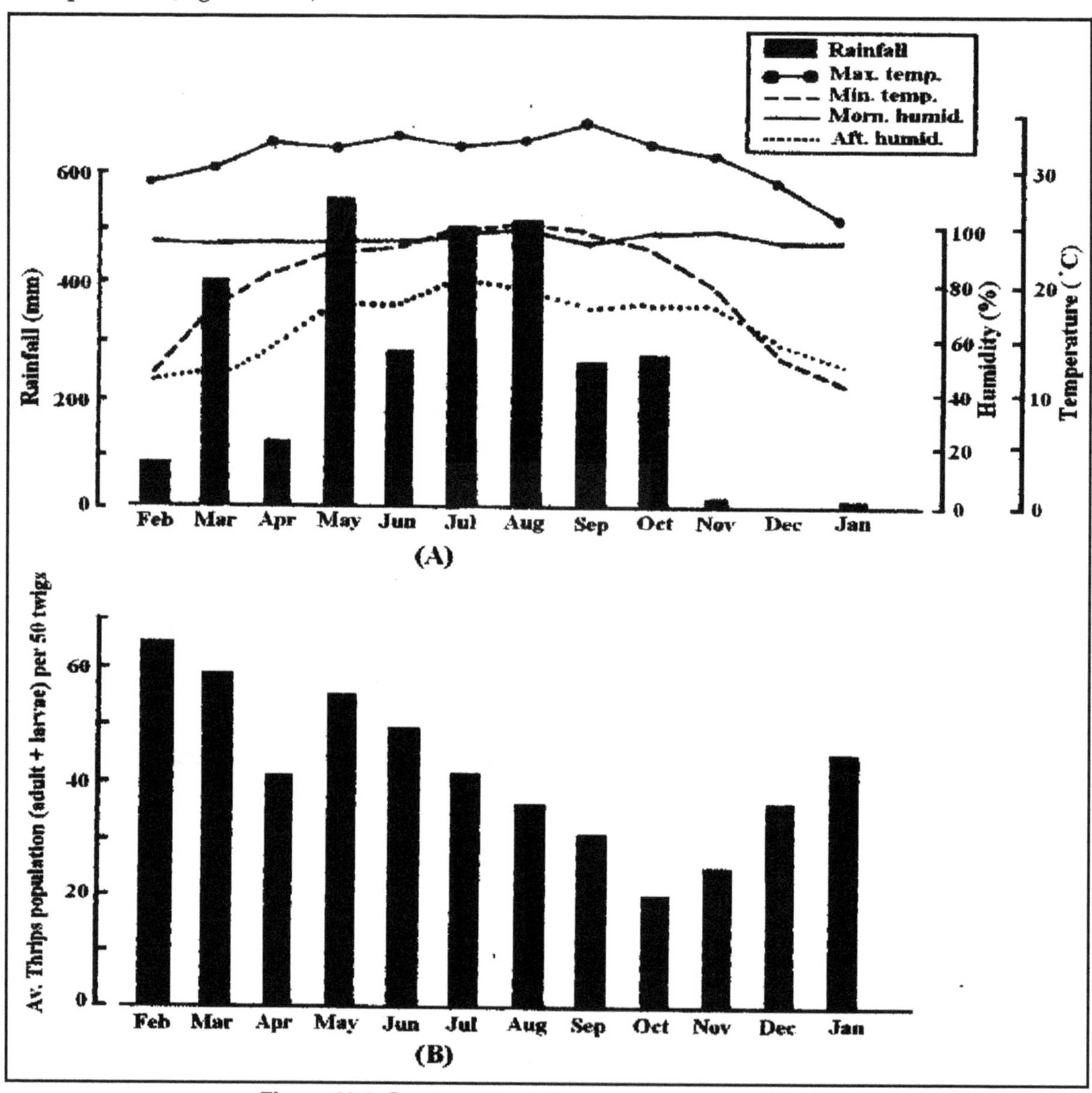

Figure 49.1: Seasonal Variation in Thrips Population in Relation to Environmental Conditions

Rainfall has been observed to have an inverse relationship with the incidence of thrips in tea. This was most prominent during the periods from May to October. From November onwards, with reduced monthly precipitation, the pest population had an increasing trend (Figures 49.1 A and B). The present investigation reveals that, thrips in general occur more during the months of reduced ambient temperature (*i.e.*, January to March) (Figures 49.1A and B)

Control measures: Treatment with the Endosulphan (Thiodan 35EC)/bio-pesticides/plant extracts against thrips revealed that after one month of treatment, the percentage of infestation declined considerably. Among the plant extracts, *Andrographis paniculata* was found to be most effective, and between the two biopesticides used, Econeem was found to be more efficacious over Bio-2001. The efficacy of plant extracts/biopesticides were observed to be at par with the inorganic pesticide (*i.e.*, Endosulphan) used. F value (For treatment) was higher than the table value, suggesting the treatment to be significant. (Table 49.2).

Table 49.2: ANOVA Test on Efficacy of Pesticides

	Months Following Treatments (March–August, 2003)					
	1	*2*	*3*	*4*	*5*	*6*
Control	80	81	81	82	85	85
Chromolaena odorata	32	30	26	16	10	4
Curcuma sp.	24	20	18	12	8	3
Lantena camara var aculeata	20	18	17	17	5	0
Andrographis paniculata	7	5	2	0	0	0
Econeem	16	10	7	5	0	0
Bio-2001	18	12	9	6	0	0
Endosulphan	15	7	5	0	0	0
CV (per cent)	17	18.9	22.8	21.6	19.9	23.7
F Value (for treatment)	77.1*	98.5*	88.9*	156.9*	354*	358.1*
CD at 5 per cent level	7.9	7.6	8.2	6.5	4.7	4.8
CD at 1 per cent level	10.9	10.5	11.5	9	6.5	6.6

Paired 't' test revealed the table value of $'t'_{0.05}$ for 6df for one tailed test is 1.94. The calculated value of 't' (*i.e.*, 20.65) was therefore found significant. It had been concluded that mean of the set of observation after application of pesticides (in March) was definitely significantly smaller than that of the set of observations before treatment (in February).This suggests that the application of pesticides had significant effect in reducing the infestation of tea caused by thrips on the tea bushes (Table 49.3).

Thrips incidence was observed almost throughout the year. However, during the bud breaking season, they were seen in sizable numbers. Gope (1991) reported that this pest causes major damage during early part of the season. In the present study, the pest population was observed to be higher during January–March, however, it took a declining trend from June/July onwards, due to heavy precipitation.

Among the inorganic pesticides, Endosulphan, and between the two bio-pesticides, a Neem (*Azadirachta indica*) based product (*i.e.*, Econeem), while, among the plant extracts, *Andrographis paniculata* was found to be more efficacious compared to others.

Table 49.3: Paired 't' Test

Sl.No.	Treatment done with	Period		d.f.	Calculated Value of 't'	Table Value of $t_{0.05}$
		Before Treatment	After Treatment			
1.	Endosulphan, 1:200	Feb, 2005	March, 2005	6	20.65*	1.94
2.	*Chromolaena odorata*					
3.	*Curcuma* sp.					
4.	*Lantena camara var aculeata*					
5.	*Andrographis paniculata*					
6.	Econeem					
7.	Bio-2001					

On an overview, the present investigation revealed that almost all the bio-pesticides and plant extracts were found to have various degrees of potentiality in bringing down the infestation under control. These observations suggest that they can be used in the tea fields as efficient alternatives to their inorganic counterparts. Being biodegradable, they can also be conveniently used in combination with traditional inorganic pesticides as an aid in IPM strategy, which should also help in reducing the total quantum of inorganic pesticide input in the tea agro-ecosystem.

Acknowledgements

The authors are thankful to the Managers and Staff of various tea estates of Barak valley for their kind help during field studies and during the field experiments. The authors are also thankful to the Tea Research Association (Cachar Advisory Branch, ASSAM) for supplying the meteorological data.

References

Ghosh Hazra, N., Kabir, S.E. and Bisen, J.S., 1994. Botanical pest control: An alternative to minimize residue toxicity in Darjeeling tea. *J. Darjeeling Planters' Club*, p. 58–74.

Gope, B., 1991. A few problematic pest of tea and their control. In: *Field Management in Tea*, (Ed.) J. Chakravartee. TRA Publication, Jorhat, India, p. 154–160.

Muraleedharan, N., 1983. *Tea Entomology: An Overview*. UPASI, TRI, Cinchona, India, Occasional Publication, 4: 1–11.

Muraleedharan, N. and Kandasamy, C., 1980. Tea thrips and their control. *Planters Chronicle*, 75(10): 447–448.

Ooman, P.A., 1982. Studies on population dynamics of the Scarlet mite, *Brevipalpus phoenicis*, a pest of tea in Indonesia. Mededelingen Landbouwhogeschool, Wageningen, Netherlands, 82(1): 556–557.

Waila, M.K.B., 1982. The occurrence of tea thrips: A review. *Q. Newsl. Tea Res. Found., Cent. Africa*, 66: 8–12.

Chapter 50

Double Sampling for Ratio Estimation and Hansen and Hurwitz Technique

Salam Opendra Singh

Department of Statistics, Manipur University, Canchipur, Imphal

ABSTRACT

In this chapter, Ratio Estimator is considered under double sampling where the samples are independent in presence of non-response when the population mean of the auxiliary variable is unknown. Under this condition, the first phase sample size and second phase sample size are obtained with the sub-sampling fraction for the proposed estimator.

Keywords: *Double sampling, Ratio estimator, Auxiliary variable non-response, Hansen and Hurwitz estimator.*

Introduction

It is a common experience in surveys that information cannot always be collected for all the units selected in the sample. Thus, the selected units may or may not be found at first attempt even after some call-backs and some may refuse to cooperate with the interviewer even if contacted. Many respondents do not reply and the available sample of the returns is incomplete. This incompleteness is called non-response. To face this non-response is to recontact the non-respondents and collect the information as much as possible. Hansen and Hurwitz (1946) were the first to deal with the problem of incomplete samples with the technique (*i*) select a random sample of respondents; (*ii*) send a mail questionnaire to all of them; (*iii*) after the deadline is over, identify the non-respondents and select a sub-sample of non-respondents; (*iv*) collect data from the selected non-respondents in the sub-sample by interview; and (*v*) combine data from the two parts of the survey to obtain unbiased estimates of the population values concerned. Using this technique, Cochran (1977) proposed the ratio and regression estimators of the population mean of the study variable in which information on the auxiliary variable is obtained from all the some sample units failed to supply information of the study variable. Further, Rao (1986, 1987) and Khare and Srivastava (1993, 1997) suggested the estimation procedure for population mean in presence of non-response using auxiliary character.

In this chapter, Ratio Estimator is considered under double sampling where the samples are independent in presence of non-response when the population mean of the auxiliary variable is unknown. Under this condition, the first phase sample size and second phase sample size are obtained with the sub-sampling fraction for the proposed estimator.

Sampling Scheme

To estimate the population mean $\overline{X}$ of the auxiliary variable x, a first phase sample of size n′ is drawn by simple random sampling without replacement from the population having N units and let us assume that at the first phase, all the units in the sample supplied the information on the auxiliary variable x. And a second phase sample of size n is drawn with the help of simple random sampling without replacement and both auxiliary variable x and main variable y are measured. If there is non-response in the second phase sample, draw a sub-sample of the non-respondents and re-contact them. Divide the second phase sample size n into two classes n_1 and n_2, n_1 units supply the information on both x and y and n_2 refuse to respond such that $n_1 + n_2 = n$. Draw a sub-sample of size $m\left(=\frac{n_2}{k}, k \succ 1\right)$ from n_2 at random and applied Hansen and Hurwitz approach (1946) assuming that all the units in m supply the information and population of size N is divided into two classes, those who will respond at the first attempt and those who will not, and call them the response class and non response classes with size N_1 and N_2 respectively.

This method of double sampling can be applied to estimate the family expenditure in a household survey where the household size is considered as auxiliary variable.

Ratio Estimate in Double Sampling in Presence of Non-response

The proposed Ratio estimate of the population mean $\overline{Y}$ is defined as

$$\overline{y}_R^* = \frac{\overline{y}^*}{\overline{x}^*}\overline{x}'$$

$$= R^*\overline{x}' \qquad (i)$$

where, $\overline{x}^* = w_1\overline{x}_1 + w_2\overline{x}_2$ and $\overline{y}^* = w_1\overline{y}_1 + w_2\overline{y}_2$ are the Hansen and Hurwitz estimators of $\overline{X}$ and $\overline{Y}$ respectively, $w_i = \frac{n_i}{n}$ and $w_i = \frac{N_i}{N}$, $i = 1, 2$.

Error in the estimate is given by

$$\overline{y}_R^* - \overline{Y} = \frac{\overline{y}^*}{\overline{x}^*}\overline{x}' - \overline{Y}$$

$$=\frac{\bar{y}^*}{\bar{x}^*}\bar{x}' - \frac{\bar{y}^*}{\bar{x}^*}\bar{X} + \frac{\bar{y}^*}{\bar{x}^*}\bar{X} - \bar{Y}$$

$$=\frac{\bar{X}}{\bar{x}^*}\left(\bar{y}^* - R\bar{x}^*\right) + R\left(\bar{x}' - \bar{X}\right)$$

If the second sample size n is large then $\frac{\bar{X}}{\bar{x}^*}$ and $\frac{\bar{X}}{\bar{x}'}$ will be equal to 1 and $\frac{\bar{y}^*}{\bar{x}^*} \cong R*$, then

$$\bar{y}_R^* - \bar{Y} \cong \left(\bar{y}^* - R\bar{x}^*\right) + R\left(\bar{x}' - \bar{X}\right) \text{ and } E\left(\bar{y}_R^*\right) = \bar{Y}$$

The approximate variance of $\bar{y}_R^*$ (for the large sample size) is given by

$$V\left(\bar{y}_R^*\right) = R^2\left(\frac{1}{n'} - \frac{1}{N}\right)S_x^2 + \left(\frac{1}{n} - \frac{1}{N}\right)S_p^2 + \frac{W_2(k-1)}{n}S_{2p}^2 \qquad (ii)$$

where,

$$S_p^2 = S_y^2 + R^2 S_x^2 - 2RS_{xy}^2$$

$$S_{2p}^2 = S_{2y}^2 + R^2 S_{2x}^2 - 2RS_{2xy}^2$$

S_x^2, S_y^2 are the population mean square and S_{2x}^2, S_{2y}^2 are population mean square for the non-respondents for the variable x and y respectively and R is the population ratio of $\bar{Y}$ to $X.S_{xy}^2, S_{2xy}^2$ are covariances for the whole population and population of non-respondents respectively.

Proof,

We know that

$$V\left(\bar{y}_R^*\right) = V\left(\bar{y}^* - R\bar{x}^*\right) + R^2 V\left(\bar{x}' - \bar{X}\right)$$

$$V\ \bar{y}^* \quad R^2 V\ \bar{x}^* \quad 2\mathrm{cov}\ \bar{y}^*, R\bar{x}^* \quad R^2 V\ \bar{x}'$$

Since $\bar{y}^*$ and $\bar{x}^*$ are the Hansen and Hurwitz estimators.

Using the theory of the Hansen and Hurwitz estimators, we get

$$V\left(\bar{y}_R^*\right) = R^2\left(\frac{1}{n'} - \frac{1}{N}\right)S_x^2 + \left(\frac{1}{n} - \frac{1}{N}\right)S_p^2 + \frac{W_2(k-1)}{n}S_{2p}^2$$

Optimum Value of n', n and k

Let us consider the cost function for the proposed estimator $\bar{y}_R^*$ as

$$C = c'n' + cn + c_1n_1 + c_2m \qquad (iii)$$

where,

- c': The unit cost associated with the first phase sample
- c: The cost of the first attempt on y with the second phase sample
- c_1: The cost of getting, editing and processing information per unit in the response class.
- c_2: The cost associated with the sub-sample.

The expected cost will be used in planning of the survey because the value of n_1 and m is not known until the first attempt is made. The expected value of $n_1 = W_1n$ and the expected value of m is

$$m = \frac{W_2 n}{k}$$

Thus the expected cost is

$$E(C) = C^* = c'n' + \left(c + c_1W_1 + \frac{c_2W_2}{k}\right)n \qquad (iii)$$

To get the optimum value of *k* and *n* such that cost is minimized for the fixed variance V_0, we have

$$\phi = C^* + \lambda\left[V\left(\bar{y}_R^*\right) - V_0\right]$$

$$= c'n' + \left(c + c_1W_1 + \frac{c_2W_2}{k}\right)n + \lambda\left[R^2\left(\frac{1}{n'} - \frac{1}{N}\right)S_x^2 + \left(\frac{1}{n} - \frac{1}{N}\right)S_p^2 + \frac{W_2(k-1)}{n}S_p^2 - V_0\right]$$

where, λ is Lagrange's multiplier.

Using the technique of Lagrange's multiplier we get:

(*i*) The optimum value of k is

$$k_{opt} \quad \sqrt{\frac{c_2(S_p^2 \quad W_2S_{2p}^2}{S_{2p}^2(c \quad c_1W_1)}}$$

(*ii*) The optimum value of n' is

$$n'_{opt} = \frac{\sqrt{\left(c + c_1W_1 + \frac{c_2W_2}{k_{opt}}\right)\left(R^2S_x^2\right)\left[S_p^2 + W_2\left(k_{opt} - 1\right)S_{2p}^2\right]} + \sqrt{c'\left[S_p^2 + W_2\left(k_{opt} - 1\right)S_{2p}^2\right]R^2S_x^2}}{\sqrt{c'\left[S_p^2 + W_2\left(k_{opt} - 1\right)S_{2p}^2\right]}\left\{V_0 + \frac{1}{N}\left(S_p^2 + R^2S_x^2\right)\right\}}$$

(*iii*) Optimum value of n is

$$n_{opt} \quad \frac{\sqrt{c \quad c_1W_1 \quad \frac{c_2W_2}{k_{opt}} \quad R^2S_x^2 \; S_p^2 \quad W_2\,k_{opt} \quad 1\,S_{2p}^2} \quad \sqrt{c \;\; S_p^2 \quad W_2\,k_{opt} \quad 1\,S_{2p}^2}\; R^2S_x^2}{\sqrt{c \quad c_1W_1 \quad \frac{c_2W_2}{k_{opt}}}\; \sqrt{R^2S_x^2} \quad V_0 \quad \frac{1}{N}\,S_p^2 \quad R^2S_x^2}$$

References

Cocran, W.G., 1977. *Sampling Techniques*, 3rd edn. John Wiley and Sons, New York.

Durbin, J., 1954. Non-response and call-backs in surveys. *Bull. Inter. Stat. Inst.*, 34: 72–86.

Hansen, M.H. and Hurwitz, W.N., 1946. The problem of non-response in sample surveys. *J Amer. Stat. Assoc.*, 41: 517–529.

Khare, B.B. and Srivastava, S., 1997. Transformed ratio type estimators for the population mean in the presence of non-response, *Comm. Stat. –Theory Methods*, 26(7): 1779–1791.

Khare, B.B. and Srivastava, S., 1993. Estimation of population mean using auxiliary character in presence of non-response. *Natl. Acad. Sci., India*, 16(3): 111–114.

Nargundkar, M.S. and Joshi, G.B., 1975. *Non-response in sample surveys. In: 40th Session of the ISI*, Warsaw, Contributed paper, p. 626–628.

Rao, P.S.R.S., 1986. Ratio estimation with sub-sampling the non-respondents. *Survey Methodology*, 12(2): 217–230.

Chapter 51

Effect of Pre-Emergence Herbicides in Relation to Cultural Practices on Weed Control in Chilli (*Capsicum annuum* L.)

M.J. Praveena Swamy, H.R. Bhoomika, S.L. Biradar, K.G. Santosh and N.S. Mavarkar

University of Agricultural Sciences, GKVK, Bangalore – 560 065

ABSTRACT

An experiment was conducted to study the effect of pre-emergence herbicides in relation to cultural practices on weed control in chilli cv. Byadagi Dabba, at GKVK, UAS, Bangalore. The study resulted in significant differences among treatments for the traits under investigation. The experiment consisted of sixteen treatments along with un-weeded control. Straw mulching resulted in least weed population at all the growth stages *viz.* 30, 60, 90, 120 DAT (Days After Transplanting) and at harvest–11.23, 21.33, 32.33, 41.67 and 68.33/m^2 respectively. Whereas hand weeding brought about efficient weed control with 84.95 per cent and 87.41 per cent WCE (Weed Control Efficiency) at 120 DAT and at harvest respectively with 0 per cent WI (Weed Index).

Keywords: *Chilli, Byadagi Dabba, Weed, WCE, WI, DAT.*

Introduction

Chilli botanically known as *Capsicum annuum* L. is an important commercial vegetable cum spice crop in India. It belongs to the family solanaceae and is a native of Peru or Mexico. Chilli is valued for its biting pungency attributed to the alkaloid capsaicin and captivating red colour due to pigments capsanthin and capsorubin (Verghese, 1999). Major chilli growing states in India are Andhra Pradesh,

Karnataka, Maharashtra, Orissa, Tamil Nadu and West Bengal. Andhra Pradesh tops the list with respect to area (2.261 ha.), production (4.03 1 t) and productivity (1790 kg/ha) of dry chillies, while Karnataka stands second in area (1.703 l ha) and production (1.465 1 t) (Anon, 2002).

Chilli exhibits greater susceptibility to weed competition owing to extremely slow growth at early stages of crop growth. Weeds reduce the crop yields and increase cost of cultivation. To ensure higher yields, effective control of weeds at right time with right method is a must. Keeping this in view the experiment was conducted to study the efficacy of pre-emergence herbicides in relation to cultural practices on control of weeds in chilli at GKVK, University of Agricultural Sciences, Bangalore.

Materials and Methods

There were totally 16 treatments in the experiment, the details of which are furnished below:

T_1: Alachlor 50EC @ 1.00 kg a.i./ha.

T_2: Alachlor 50EC @ 1.25 kg a.i./ha.

T_3: Alachlor 50EC @ 1.50 kg a.i./ha.

T_4: Pendimethalin 30EC @0.75 kg a.i./ha.

T_5: Pendimethalin 30EC @1.00 kg a.i./ha.

T_6: Pendimethalin 30EC @1.25 kg a.i./ha.

T_7: Metalachlor 50EC @ 0.75 kg a.i./ha.

T_8: Metalachlor 50EC @ 1.00 kg a.i./ha.

T_9: Metalachlor 50EC @ 1.25 kg a.i./ha.

T_{10}: Trifluralin 48EC @ 0.75 kg a.i./ha.

T_{11}: Trifluralin 48EC @ 1.00 kg a.i./ha.

T_{12}: Trifluralin 48EC @ 1.25 kg a.i./ha.

T_{13}: Hand weeding at 30, 60 and 90 DAT.

T_{14}: Intercultivation by light digging at 45 and 90 DAT.

T_{15}: Paddy straw mulching.

T_{16}: Unweeded control.

The experiment was laid out in RCBD with all 16 treatments in three replications. The gross plot size was 3m × 2.25m with a net plot of 1.8m × 1.35m size.

The chilli variety Byadagi Dabba was taken for the study. Above mentioned herbicides were sprayed as pre emergence, uniformly on soil surface after three days of transplanting at the rate of 750 liters of spray solution per hectare, using knapsack sprayer.

Hand weeding was done with kurfi at 30, 60 and 90 DAT. Mulching with paddy straw was done on the day of herbicide application after dipping it in insecticide solution to reduce termite infection.

Intercultivation by slight digging with hand guddali was done at 45 and 90 DAT.

Unweeded control plots were sprayed with water and weeds were not disturbed throughout the cropping season.

Observations on Weeds

Weed Population (No./m²)

Number of mono cot and dicot weeds in one square meter area was counted at 30, 60, 90 and 120 DAT and at harvest.

Dry Weight of Seeds (g/0.25 m²)

Monocot and dicot weeds present in 0.25 m² area were uprooted, washed, dried in hot air oven at 80°C and dry weight was recorded.

Weed Control Efficiency (WCE) (per cent)

$$\text{WCE (per cent)} = \frac{\begin{matrix}\text{Dry weight of weeds in} \\ \text{unweeded control plot}\end{matrix} - \begin{matrix}\text{Dry weight of weeds in treatment} \\ \text{plot for which WCE has to be worked}\end{matrix}}{\text{Dry weight of weeds in unweeded control plot}} \times 100$$

Weed Index (WI)

Is the reduction in yield due to presence of weeds in comparison to no weed plot.

$$\text{WI} = \frac{X - Y}{X} \times 100$$

where,

X: Total chilli fruit yield from hand weeding treatment.

Y: Chilli yield from the treatments for which WI has to be worked out.

Results and Discussion

The weed flora associated with chilli in the experimental plot could be classified into three classes *viz.* Sedges, grasses and broad leaved weeds. Observations were recorded separately under the following headings.

Weed Population (Number/m²)

Total weed population per square meter was found significant at all the intervals due to various weed management practices and the results are presented in Table 51.1.

At 30 DAT, straw mulching recorded significantly least number of weeds (11.33/m²). The maximum number of total weeds (143.00/m²) were found in unweeded control. Among herbicides alachlor @ 1.5 kg/ha recorded minimum number of weeds (29.33/m²). Similar trend was observed at all the intervals. Straw mulching at 60 DAT, recorded significantly the least total number of weeds (21.33/m²) and among herbicides alachlor @ 1.50 kg/ha proved better with 40.67 weeds/m². Similarly at 90 DAT, the minimum number of total weeds noticed with straw mulching was 32.33/m² and with alachlor @ 1.5 kg/ha it was 60.33/m². At 120 DAT and at harvest the least weed population observed with straw mulching is 41.67/m² and 68.33/m² respectively. While alachlor @ 1.5 kg/ha resulted in 251.33/m² of weeds at 120 DAT and 19.22/m² at harvest. Prakash *et al.* (1999a) and Saimbhi and Randhawa (1976) also reported that alachlor controls weeds effectively.

The lowest weed population in straw mulching was due to inhibition of weed seed germination and suppression of weed growth. Roe and Stoftella (1993) in chilli and Hundal *et al.* (2002) in tomato, reported similar results. All the weed management treatments had significantly reduced weed population compared to unweeded check at all the stages of crop growth.

Table 51.1: Total Weed Population (no./m²) as Influenced by Weed Management Practices in Chilli cv. Byadagi Dabba

Treatments	*Weed Population (no./m²)*				
	30 DAT	*60 DAT*	*90 DAT*	*120 DAT*	*At Harvest*
T_1: Alachlor 50EC @ 1.00 kg a.i.lha.	42.6	55.33	80.00	107.67	104.33
T_2: Alachlor 50EC @ 1.25 kg a.i./ha.	34.00	44.00	70.33	99.33	97.33
T_3: Alachlor 50EC @ 1.50 kg a.i./ha.	29.33	40.67	60.33	89.67	93.67
T_4: Pendimethalin 30EC @ 0.75 kg a.i./ha.	53.67	73.67	87.33	118.67	116.00
T_5: Pendimethalin 30EC @ 1.00 kg a.i./ha.	39.33	58.00	73.33	10 1.67	104.33
T_6: Pendimethalin 30EC @ 1.25 kg a.i./ha.	50.33	64.00	82.67	109.33	106.67
T_7: Metalachlor 50EC @ 0.75 kg a.i./ha.	59.67	62.00	90.33	125.00	120.33
T_8: Metalachlor 50EC @ 1.00 kg a.i./ha.	57.00	63.33	81.33	111.67	110.00
T_9: Metalachlor 50EC @ 1.25 kg a.i./ha.	40.00	58.00	77.00	107.00	110.67
T_{10}: Trifluralin 48EC @ 0.75 kg a.i./ha.	60.00	76.67	96.00	126.67	126.00
T_{11}: Trifluralin 48EC @ 1.00 kg a.i./ha.	63.33	81.00	104.33	130.67	127.67
T_{12}: Trifluralin 48EC @ 1.25 kg a.i./ha.	50.67	63.67	82.33	111.67	107.67
T_{13}: Hand weeding at 30,60 and 90 DAT.	124.33	4.00	43.33	51.00	70.33
T_{14}: Intercultivation by light digging at 45 and 90 DAT	136.33	37.67	61.33	60.00	78.33
T_{15}: Paddy straw mulching.	11.23	21.33	32.33	41.67	68.33
T_{16}: Unweeded control.	143.00	194.67	127.33	251.33	238.33
F-test	*	*	*	*	*
SE±	0.361	0.358	0.389	0.332	0.339
CD at 5 per cent	1.04	1.04	1.12	0.96	0.98

Note: *: Significant at 5 per cent, DAT: Days after Transplanting.

Total Weed Dry Matter (g/0.25m²)

The total dry weight of weeds (g/0.25m²) recorded at different intervals differed significantly and is presented in Table 51.2.

The lowest total weed dry weight at 30 DAT (0.46g/0.25 m²) was in straw mulching, but at 120 DAT and at harvest, the lowest total weed dry weight (6.11 and 7.10 g/0.25 m² respectively) was with hand weeding.

However, the weed population was lower but dry weight of weeds in mulching was higher than hand weeding which may be due to better growth as a result of less competition under lower plant population. In hand weeding frequent weeding caused less time for dry matter accumulation by weeds. Roe and Stofetella (1993) observed higher weed dry weight with organic mulch weed coverage was less. Similarly Prakash *et al.* (1999b) obtained higher weed biomass with mulching than with hand weeding. Similar lower weed dry weight with hand weeding was reported by Singh *et al*. (1995) in chilli.

Table 51.2: Total Dry Weight (g/0.25m²) as Influenced by Weed Management Practices in Chilli cv. Byadagi Dabba

Treatments	*Total Dry Weight (g/0.25m²)*		
	30 DAT	*120 DAT*	*At Harvest*
T_1: Alachlor 50EC @ 1.00 kg a.i./ha.	4.54	15.10	24.42
T_2: Alachlor 50EC @ 1.25 kg a.i./ha.	3.53	12.73	21.55
T_3: Alachlor 50EC @ 1.50 kg a.i.lha.	2.93	11.86	19.22
T_4: Pendimethalin 30EC @ 0.75 kg a.i./ha.	4.49	18.33	27.85
T_5: Pendimethalin 30EC @ 1.00 kg a.i./ha.	3.82	12.97	20.10
T_6: Pendimethalin 30EC @ 1.25 kg a.i./ha.	4.15	14.78	22.10
T_7: Metalachlor 50EC @ 0.75 kg a.i./ha.	4.97	18.83	30.73
T_8: Metalachlor 50EC @ 1.00 kg a.i./ha.	4.55	14.03	23.99
T_9: Metalachlor 50EC @ 1.25 kg a.i./ha.	4.09	13.84	22.92
T_{10}: Trifluralin 48EC @ 0.75 kg a.i./ha.	5.14	18.70	31. 00
T_{11}: Trifluralin 48EC @ 1.00 kg a.i./ha.	4.98	17.36	28.27
T_{12}: Trifluralin 48EC @ 1.25 kg a.i./ha.	4.36	14.32	22.20
T_{13}: Hand weeding at 30, 60 and 90 DAT.	19.92	6.11	7.10
T_{14}: Intercultivation by light digging at 45 and 90 DAT.	20.72	7.36	8.94
T_{15}: Paddy straw mulching.	0.46	9.03	10.61
T_{16}: Unweeded control.	21.43	41.13	57.10
F-test	*	*	*
SEm±	0.077	0.113	0.156
CD at 5 per cent	0.22	0.33	0.45

Note: *: Significant at 5 per cent, DAT: Days after Transplanting.

Among herbicide treatments, alachlor at 1.5 kg/ha recorded the lower total weed dry weight (2.93, 11.86 and 19.22 g/0.25 m² at 30 DAT, 120 DAT and at harvest respectively) and it was followed by alachlor @1.25 kg/ha. It might be due to lower weed population and increased alachlor absorption. These results are in confirmation with Bhalla (1980) and Joshi *et al*. (1995) in chilli. Alachlor kept maximum check on weed biomass production because it restricted the growth of emerged weeds.

Alachlor inhits seed germination and growth of young seedlings by inhibiting root germination, mitotic activity and enlargement of cortical cells. The protein synthesis is the primary site of action and also inhibits GA_3 induced α-amylase production by seeds during germination (Rao and Duke, 1976).

Weed Control Efficiency (WCE) (Per cent) and Weed Index (WI) (Per cent)

Higher WCE (84.95 per cent) was in hand weeding followed by inter-cultivation (81.85 per cent) and straw mulchingn (77.70 per cent). This was because of season long weed free conditions due to frequent weeding which resulted in lower weed density and weed biomass. Similar reports were made by Ram *et al.* (1994) confirming the present findings. Among herbicide treatments alachlor at 1.5 kg/ha, pendimethalin @1.00 kg/ha and alachlor @ 1.25 kg/ha recorded the maximum WCE (73.84 to 68.05 per cent) which was due to destruction of weeds early in the season. Lower WCE (53.64 to 55.05 per cent) was noticed in metalachlor @0.75 kg/ha, trifluralin at 0.75 kg/ha and pendimethalin at 0.75 kg/ha. As observed in the present study Kumar *et al.* (1995) in chilli and Prakash *et al.* (1999a) in tomato reported better weed control efficiency with alachlor and pendimethalin respectively (Table 51.3).

Table 51.3: Weed Control Efficiency (WCE) and Weed Index (WI) as Influenced by Weed Management Practices in Chilli cv. Byadagi Dabba

Treatments	*Weed Control Efficiency (WCE)*		*Weed Index (WI) (per cent)*
	120 DAT	*At Harvest*	
T_1: Alachlor 50EC @ 1.00 kg a.i.lha.	62.58	56.66	30.92
T_2: Alachlor 50EC @ 1.25 kg a.i./ha.	69.02	62.42	17.18
T_3: Alachlor 50EC @ 1.50 kg a.i./ha.	73.84	65.05	2.71
T_4: Pendimethalin 30EC @ 0.75 kg a.i./ha.	55.05	50.28	33.68
T_5: Pendimethalin 30EC @ 1.00 kg a.i./ha.	68.05	64.14	5.28
T_6: Pendimethalin 30EC @ 1.25 kg a.i./ha.	63.63	60.20	20.84
T_7: Metalachlor 50EC @ 0.75 kg a.i.lha.	53.64	45.45	44.44
T_8: Metalachlor 50EC @ 1.00 kg a.i./ha.	65.43	57.37	23.35
T_9: Metalachlor 50EC @ 1.25 kg a.i./ha.	65.82	58.97	23.29
T_{10}: Trifluralin 48EC @ 0.75 kg a.i./ha.	53.93	44.97	40.57
T_{11}: Trifluralin 48EC @ 1.00 kg a.i./ha.	57.32	49.86	25.38
T_{12}: Trifluralin 48EC @ 1.25 kg a.i./ha.	64.87	60.68	18.04
T_{13}: Hand weeding at 30, 60 and 90 DAT.	84.95	87.41	0.00
T_{14}: Intercultivation by light digging at 45 and 90 DAT.	81.85	84.27	25.4
T_{15}: Paddy straw mulching.	77.00	81.24	18.53
T_{16}: Unweeded control.	00.00	0.00	50.58
F-test	*	*	*
SEm±	2.383	3.753	4.667
CD at 5 per cent	6.88	10.84	13.48

Note: *: Significant at 5 per cent, DAT: Days after Transplanting.

Weed Index indicating yield reduction due to weed competition was highest in unweeded control (50.58 per cent). This result is in conformity with Schreoder (1992) and Morales Payan *et al.* (1998), while alachlor at 1.50 kg/ha (2.71 per cent) and pendimethalin at 1 kg/ha (5.28 per cent)recorded the lowest WI besides hand weeding where satisfactory weed control was observed which allowed plants to grow well and yield higher. Several workers (Ram *et al.*, 1994 with hand weeding, Prakash *et al.*,

1999b with alachlor in chilli and Prakash *et al.*, 1999a with pendimethalin in tomato) reported similar observations as noticed in the present study.

With these results it can be concluded that, frequent hand weeding at 30, 60 and 90 DAT is best to keep the weeds under check, causing no yield reduction. Intercultivation at 45 and 90 DAT will be the next best treatment followed by straw mulch.

Hence hand weeding could be very well adapted in areas where labour availability and incessant rains are not a problem. But in situations of labour scarcity and one has to go for herbicides, alachlor @ 1.5 kg/ha or pendimethalin @ 1.00 kg/ha would be the better options.

References

Anonymous, 2002. *Report on Area, Production, Productivity and Prices of Agricultural Crops in Karnataka, 1999–2000*. Directorate of Economics and Statistics, Bangalore, pp. 21.

Bhalla, P.L., 1985. Herbicides for transplanted bell pepper. In: *Proceedings of 39th Annual Meeting of the North-Eastern Weed Science Society*, pp. 188–190.

Prakash, V., Pandey, A.K., Singh, P., Chandra, S., Chauhan, V.S. and Singh, R.D., 1999a. Integrated weed management in tomato (*Lycopersicum esculentum* Mill.) under North-Western Himalayas. *Crop Res.*, 17(3): 345–350.

Prakash, V., Pandey, A.K., Singh, R.D. and Mani, U.P., 1999b. Integrated weed management in chilli (*Capsicum annuum* L.) under mild hill conditions of North-Western Himalayas. *Journal of Weed Sci.*, 31(3 and 4): 164–166.

Ram, B., Singh, V.P. and Singh, J.P., 1994. Weed management in kharif tomato in low hill/valley region in U.P. hills. *Veg. Sci.*, 21(1): 20–22.

Rao, V.S. and Duke, W.D., 1976. Effect of alachlor, propachlor and prynachlor on GA3 induced production of protease and amylase. *Weed Sci.*, 24: 616–618.

Roe, N.E. and Stoftella, P.J., 1994. Growth and yield of Bell pepper and Winter squash grown with organic and living mulches. *J. American Soc. Hort. Sci.*, 119(6): 1193–1199.

Saimbhi, M.S and Randhawa, K.S., 1976. Herbicidal control of weeds in transplanted chilli (*Capsicum annuum* L.). *Punjab Hort. J.*, 16: 146–148.

Schroeder, J., 1992. Oxyfluorofen for directed post-emergence weed control in chile pepper (*Capsicum annuum* L.). *Weed Tech.*, 6: 1010–1014.

Singh, R., Jangir, R.P. and Poonia, B.L., 1995. Evaluation of herbicides for control of weed in chilli. *Indian J. Agric. Sci.*, 65(10): 723–726.

Varghese, J., 1999. Scanning paprika *oleoresin* and chilli *colour* specifications/quality parameters. *Indian Spices*, 36(1): 7–11.

Chapter 52

Poisonous Snakes (Reptilia : Ophidia) of Mizoram, North East India

Daya Nand Harit*

Department of Zoology, Government Champhai College, Mizoram – 796 321, Assam

ABSTRACT

Mizoram is one of the North Eastern Hill states of India, with an area of 21,081 sq km and lies between 21°58′–24°30′N and 92°16′–93°25′E. The state roceives heavy rain fall and has rich faunal and floral diversity, which has not been explored much so far. Snakes constitute one of the important faunal components of the biodiversity. Studies on snakes of the state have been made by some workers. This chapter aimed to report poisonous snakes of the state based on local surveys and the literature available, which belong to three families like Elapidae, includes three species *Ophiophagus hannah, Naja naja kaouthia* and *Calliophis macclellandi.* Family Viperidae includes *Daboia russelli* (*Vipera russelii*) and Family Crotalidae includes *Trimeresurus monticola, Trimeresurus gramimeus, Trimeresurus eythurus* and *Trimeresurus albolabris.* Total eight species of the poisonous snakes are being reported herewith.

Keywords: *Poisonous snakes, Mizoram, North East India.*

Introduction

Reptiles are fascinating gift of nature, having verities of forms, living on land. These include crocodiles, alligators, tortoises, snakes and lizards. I shall be confining myself to the poisonous snakes of the state. Snakes are having the elongated body with no external limbs except some primitive forms

* E-mail: dnharit@yahoo.co.in.

having vestigial limbs. Snakes are able to make twisting movements of the body which is a characteristics feature of the serpentine motion, otherwise not possible by other vertebrates.

Mizoram is one of the North Eastern Hill states of India, with an area of 21,081 sq km and lies between 21°58′–24°30′N and 92°16–93°25′E. The state receives heavy rain full and has rich faunal and floral diversity, which has not been explored much so far. Snakes constitute one of the important faunal components of the biodiversity. Studies on snakes of the state have been made by some workers like Ramanujam and Harit (2002), Mathew (2007) and Harit (2007) etc. There has been no report on the poisonous snakes of the state so far, therefore it was decided to report and document the poisonous snakes of the area, based on local surveys and the literature available, which belong to three families like Elapidae, includes three species *Ophiophagus hannah, Naja naja kaouthia* and *Calliophis macclellandi.* Family Viperidae includes *Daboia russelli* (*Vipera russelii*) and Family Crotalidae includes *Trimeresurus monticola, Trimeresurus gramimeus, Trimeresurus eythurus* and *Trimeresurus albolabris.* Total eight species of the poisonous snakes are being reported herewith, listed in Table 52.1.

Methodology

Methodology adapted is the visual encounter surveys of the local area involving searching for animals in the area and recording all the viable specimens during the survey and the literature available. Road kills were also collected and studied for tile faunal study. Various published reports also have been taken into consideration. Identification of the encountered species are based on Gharpurey (2006), Daniel (2002), Shaw *et al.* (2000), Smith (2003), and Wall (2000).

Results

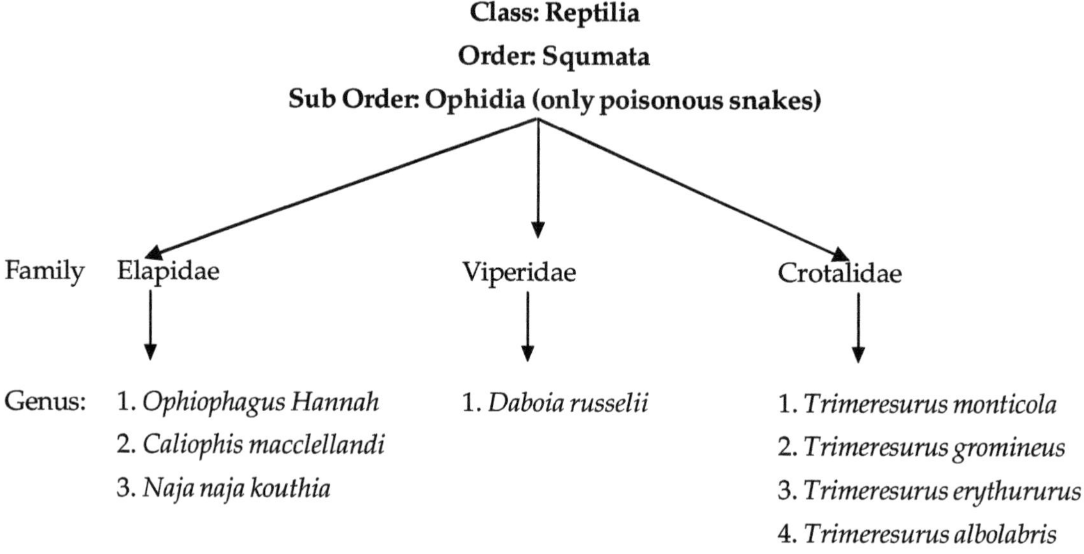

Family: Elapidae

Ophiophagus hannah (Cantor) (King Cobra)

Identification

Presence of a pair of occipital shield, costals 19 : 15 : 15, third supralabials touching the eyes and nasal, robust and glossy scaled snake, hood comparatively less dilated then cobra, eyes moderately round with round pupil.

Table 52.1: Poisonous Snakes of Mizoram

Sl.No.	*Family* *Scientific Name*	*Common Name*	*Local Name (in Mizo)*
1.	Elapidae		
	1. *Ophiophagus hannah*	King Cobra	Chawngkawrlian
	2. *Caliophis macclellandi*	Macclelland's Coral Snake	–
	3. *Naja naja kouthia*	Monocellate Cobra	Chawngkawr
2.	Viperidae		
	1. *Daboia russelii*	Russel's Viper	Rulngan
3.	Crotalldae		
	1. *Trimeresurus monticola*	Large Spotted Pit Viper	Rulmuk
	2. *Trimeresurus gramineus*	Green Bamboo Pit Viper	Rultuha
	3. *Trimeresurus erythururus*	Green Viper	Rultuha
	4. *Trimeresurus albolabris*	Green Viper	Rultuha

Size

55 meters (18.4 inches) recorded (Daniel, 2002).

Distribution

Dense forests of eastern ghats and the hills of forests, plains and estuaries of Orissa, Bengal, Indo Chinese regions, China, Malaya.

Remarks

This species has been reported from Kawnpui and Kolasib localities of Mizoram (Mathew, 2007).

Calliophis macclellandi (Reinherdt) Macclelland's Coral Snake

Identification

The broad enameled white band across head is distinctive, costals 13 across, seven supralabials with 5th and 6th touching the temporal shield, distinguish the species from other Indian snakes, head flattened, body cylindrical and moderately robust and uniform thickness throughout, tail short.

Size

Largest recorded 2 feet and 8 inches (Daniel, 2002).

Distribution

Western Himalayas to Southern China and Taiwan, common in the hills of Assam and eastwards. Typically a jungle, as well as hill species (Daniel, 2002).

Remarks

This species has been reported by Ramanujam and Harit (2002).

Naja naja kouthia (Lesson) (Monocellate Cobra)

Identification

Presence of a small 'cuneate' scale between 4th and 5th infralabials, preocular touching the nasal, 3rd supralabials touching the eye, presence of a hood which can be expanded and relaxed, presence of a angle "O" duped mark on the hood.

Distribution

Eastern India and eastwards of India.

Remarks

This species has been reported by Ramanujam and Harit (2002).

Daboia russelii Shaw = *Vipera russelli* (Russel's Viper)

Identification

Head covered with small scales, costals 27–33 at midbody, body massive cylindrical, narrowing at both ends, head flat with triangular with short snout, large gold flecked eyes with vertical pupil and large open nostrils, tail short, three series of large oval spot, one vertebral and two costals/laterals.

Size

150 cm recorded.

Distribution

Throughout India, Myanmar, Thailand, Indo Chinese region, prefers open county.

Remarks

This is considered to be deadly poisonous snake in Mizoram and has been reported from Mizoram (Government of Mizoram, 2005).

Trimeresurus monticola Gunther (Large Spotted Pit Viper)

Identification

Supralabials 10, first supralabials separated from nasals, third supralabials largest, no sub ocular shield present, two series of small scales present between eyes and supralabials, head scales smaller, internasals comparatively larger then others, loreal pit present, body scalation 23 : 23 : 19, smooth, V-142, C-51, paired, A-1, eyes small, head scales unequal.

Size

115 cm (female) recorded (Wall, 2000).

Colouration

Large brownish black square patches are present on the middle of the back in two rows, small 2–3 series of small patches are also present below the series of dorsal square patches on the lateral side of the body, a creamy streak extends from the top of the snout to both dorso-lateral sides of the head, joining at the posterior part of the head on back, giving a broad arrow shaped view.

Distribution

Himalayan region includes hills of Assam, Burma, Meghalaya, and eastern Himalayas Wall (2000).

Remarks

This species has been reported for the first time from Champhai District of Mizoram (Harit, 2007).

Trimeresurus gramineus Shaw (Green Bamboo Pit Viper)

Identification

Supralabials 11, head scales small, unequal, smooth supraocular shield present, first supralabials separated from nasals, third supralabials largest, two series of small scales per cent between supra labials and sub ocular, temporals small, loreal pit present, body scalation 21 : 21 : 15, posterior scales feebly keeled, V-147, C-60, paired, A-1.

Size

76 cm.

Colouration

Foliage grass green, a well defined white flank line extends from neck to tail region, belly greenish white, tail posterior part greenish brown or dirty reddish brown.

Distribution

Most widely distributed Pit viper, western ghat, Himalayan region, Neelgiri hills and other hills of India, don't find in plains.

Remarks

This species has been reported for the first time from Clamphai District of Mizoram (Harit, 2007).

Trimeresurus erythurus Cantor

Identification

Head scales small, supraocular narrow, intemasals larger then adjacent scales separated with one another by a single scale, supralabials 9–13, first supralabials united with nasal, third supralabials largest, two rows of scales between supralabials and elongated subocular, temporals small scales keeled, V 151-180, C 49-79, green in colour, tail usually spotted with brown colour.

Size

120 cm recorded.

Colouration

Green above, pale green or yellowish below, upper lip whitish or pale green, tail usually spotted with brown colour.

Distribution

Bengal, Assam, Eastern Himalayas and Bunna (Wall, 2000).

Remarks

This species has been reported from Mizoram (Ramanujam and Harit, 2002) and from Vairengte of Mizoram (Mathew, 2007) also found in Champhai District of Mizoram.

Trimeresurus albolabris Gray

Identification

Intranasal larger and separated by a single scale, supralabials 10–12, first supralabials united with nasal, third supralabials largest, two rows of scales between supralabials and suboccular, temporals small smooth or feebly keeled, 19 or 21 scales round the body, tail not spotted with brown colour.

Size

94 cm recorded (Smith, 2003).

Distribution

Snake of plain and hills (Wall, 2000).

Remarks

This species has been reported from Bilkhawthlir of Mizoram (Mathew, 2007) and also has been encountered among bamboo forests of Champhai District of Mizoram.

Acknowledgements

Author is grateful to the Principal, Government Champhai College. Mizoram for providing facilities. He is also thankful to Dr. B. Prasad, Professor of Zoology (Retired), Manipur University and Dr. G.R. Shukla, Reader in Zoology, D.A.V. College, Muzaffarnagar for encouragements. Smt. Suman Rakesh Harit also deserves thanks for her overall help rendered during the surveys and study. Author is also grateful to the University Grants Commission for financial assistance.

References

Daniel, J.C., 2002. *The Book of Indian Reptiles and Amphibians.* Bombay Natural History Society. Bombay.

Gharpurey, C.K.G., 2006. The Snakes of India. Asiatic Publishing Hosue,. Delhi.

Govermnent of Mizoram, 2005. Working plan of Champhai District.

Harit, D.N., 2007. Pit Vipers *Trimeresurus monticola* Gunther and *Trimeresurus gramineus* Show (Crotalidae : Reptilia) in Mizoram, North East India. *Cobra.* (Communicated).

Mathew, R., 2007. Addition to the snakes fauna of Mizoram. *Cobra,* 1(1): 5–9.

Ramanujam, S.N. and Harit, D.N., 2002. Report on Ophidian fauna of Mizoram, India. *Indian Journal of Environment and Ecoplanning,* 6(2): 335–337.

Shaw, Shebbeare and Barker, 2000. *The Snakes of Sikkim and Bengal.* Asiatic Publishing House, Delhi.

Smith, M.A., 2003. *Handbook of Indian Snakes.* Cosmo Publications, New Delhi.

Wall, F., 2000. *The Poisonous Terrestrial Snakes.* Asiatic Publishing House, Delhi.

Chapter 53

Evaluation of Basmati Rice Based Crop Sequences for Productivity, Economic Returns and Soil Health

M.S. Sidhu* and R. Sikka**

**Senior Agronomist and **Soil Chemist*
Department of Agronomy and Agrometerology, Punjab Agricultural University, Ludhiana – 141 004, Punjab

ABSTRACT

A field experiment was conducted consecutively for three years (1996–97 to 1998–99) at research farm of Punjab Agricultural University, Ludhiana, to evaluate the productivity of different basmati rice based crop sequences and their effect on soil health. Maximum mean grain yield of basmati rice was recorded in fallow-basmati rice sequence (22.9 q/ha) and minimum grain yield was recorded in green manure-basmati rice sequence (14.3 q/ha). However, the yield of subsequent crops, wheat, celery, field peas and sunflower increased substantially where green manure was practiced. Maximum basmati rice equivalent yield was recorded in *bajra* (f)-basmati rice-sunflower crop sequence (53.8 q/ha) and minimum was obtained in green manure-basmati rice-celery sequence (25.7 q/ha). The gross returns also followed the same trend. The mean net returns were, however, maximum in *bajra*-basmati rice crop sequence (Rs. 19900 /ha) and minimum in green manure-basmati rice sequence (Rs. 8800 /ha) but there was not much variation in the fallow-basmati and *bajra (f)*-basmati rice crop sequence. Incorporation of *Sesbania aculeata* increased the organic carbon from 0.33 per cent as its initial value to 0.49 per cent and available N from 110 kg/ha to 127 kg/ha after three crop cycles. The content of available P increased than its initial status under all the treatments. However, a significant decrease in available K content of the soil was observed after three crop cycles under all treatments.

Keywords: *Diversification, Green manure, Basmati rice, Sunflower productivity, economic returns, soil fertility.*

Introduction

Growing demand for food in the country due to burgeoning human population on one hand and progressive shrinking of agricultural land available per capita on the other hand led to intensification of time and space dimensions in Indian agriculture. This has been made possible through adoption of multiple cropping where more than two crops are grown one after another during a year on the same piece of land.

Rice-wheat crop sequence is the backbone of India's food security. This system occupies an area of 10.5 m ha in the Indo-Gangetic plains of India. There are reports of grain yield stagnation and decline in soil productivity (Yadav *et al.*, 1998), due to continuous adoption of this system. Rapid spread of rice-wheat system in the post-green revolution era in this region caused an eclipse on sustainability of soil productivity (Rattan and Singh, 1997) and low economic returns.

The decline in crop yields and economic returns could be minimized by introduction of alternate crops and crop sequences and inclusion of green manures (Chauhan, 2000). Basmati rice is one such alternative known for its aroma and cooking quality and has great export potential. Generally wheat is grown in sequence with basmati rice but wheat gives poor production due to its delayed sowing. Moreover, both the crops are cereals, which create problem of mining of nutrients from soil. Sunflower another such crop has also been found remunerative than late sown wheat.

Therefore, a need is felt to diversify the rice-wheat system by introducing celery, field peas sunflower and *bajra* in rotation with basmati rice and including organic source of fertilizer like green manure crop May and June before planting basmati rice. Accordingly, different basmati rice based crop sequences were tried to evaluate their productivity, economic returns and their effect on soil health.

Materials and Methods

A field experiment was conducted at the research farm of Punjab Agricultural University, Ludhiana for 3 consecutive years. The experiment was conducted on a sandy loam soil, low in organic carbon (0.33 per cent) and available N (110.2 kg/ha), high in available P (42.8 kg/ha) and low in available K (132.6 kg/ha) content. The pH and EC of the soil (1 : 2) were 7.62 and 0.23 dS/m, respectively. The experiment was laid out in split plot design. There were three main treatments *viz.*, fallow (where no crop was grown before basmati), *bajra* (a fodder grown before basmati) and green manure (*Sesbania aculeata* grown before basmati) which were kept as main plots and four crop sequences followed thereafter *viz.*, basmati rice (*Oryza sativa*)-wheat (*Triticum aestivum*), basmati rice-field peas (*Pisum sativum*) basmati rice-celery (*Apium graveolens*) and basmati-sunflower (*Helianthus annuus*) which were kept as sub plots. *Bajra* was grown as fodder and harvested after 8 to 7 weeks and green manure dhaincha (*Sesbania aculeata*) was grown for about 60 days and incorporated in soil a day before transplanting basmati rice in the respective treatments.

In all there were twelve treatments, which were replicated thrice. All recommended package of practices were followed to grow different crops. The varieties and fertilizer applied to different crops are given in Table 53.1. The crops were harvested at proper maturity and yield data were recorded and the yield of basmati rice, wheat, celery, field peas and sunflower were adjusted at 18, 10, 10, 10 and 14 per cent moisture respectively.

The productivity of different crop sequences were compared by converting the yield of all the component crops of a crop sequence into basmati equivalent yield on the basis of minimum support price and gross and net returns were also worked out. After the completion of three cycles of crop rotations surface soil samples (0-15 cm) were taken and analyzed for pH (1 : 2), EC (1 : 2), OC (Walkley and Black 1934), available N (Subbiah and Asija 1956) available P (Olsen *et al.*, 1956) and available K (Meernin and Peach 1950) to determine the changes in soil fertility.

Table 53.1: Varieties of Crops and Nutrients Applied to Different Crops

Crop	*Variety*	*Nutrient Applied (kg/ha)*		
		N	P_2O_5	K_2O
Sesbania aculeate (GM)	Local	0	0	0
Bajra	Local	50	0	0
Basmati rice	Basmati–370/386	60	30	0
Field peas	Field peas-48	30	40	0
Wheat	PBW–373	120	60	30
Celery	Local	100	40	0
Sunflower	MSFH–8	60	30	30

Results and Discussion

Crop Productivity

The data on yield of component crops of different crop sequences are given in Table 53.2. The grain yield of basmati rice ranged from 22.0 to 24.1 q/ha. in fallow-basmati rice sequence and from 18.2 to 19.4 q/ha in different *bajra (f)*-basmati rice sequence and the corresponding values ranged from 14.2 to 14.4 q/ha in green manure-basmati rice crop sequences. When the mean values were compared it was found that maximum mean grain yield of basmati was recorded in fallow-basmati rice sequence (22.9 q/ha) and minimum mean grain yield of basmati was recorded in green manure-basmati rice sequence (14.3 q/ha).

The reduction in grain yield in green manure-basmati rice sequence may be due to severe lodging which occurred in two out of the three years of experimentation. As basmati rice grown after green manured crop also received recommended N which might have made the crop more succulent and, therefore, resulted in lodging. In the *bajra (f)*-basmati rice crop sequences the mining of nutrients by *bajra* (fodder) might have resulted in the low yield of following basmati rice crop as is also evident from nutrient status of soils particularly N and K after harvest of the crop (Table 53.4).

Earlier also it has been reported by Sidhu *et al.* (2004) that where green manuring was practiced along with application of recommended level of N the basmati crop suffered from severe lodging which resulted in low grain yield. On the other hand the yield of wheat, celery, field peas and sunflower succeeding basmati rice increased substantially where green manure was practiced. The mean grain yield of wheat was 44.9, 43.9 and 47.1 q/ha, respectively, in fallow-basmati rice-wheat, *bajra (f)*-basmati rice-wheat and green manure basmati rice-wheat crop sequence. The seed yield of celery, field peas and sunflower were also higher in crop sequences with green manuring, indicating there by that there was considerable residual effect of green manure in the form of increasing the available nutrients in the soil (Table 53.4) and the other associated beneficial effects.

There was not much variation in the seed yield of celery and sunflower in fallow-basmati rice and *bajra* (fodder)-basmati rice crop sequences. However, the seed yield of field peas was high in fallow-basmati rice sequence as compared to *bajra (f)*-basmati rice-field pea sequence. Green manure of *dhaincha* not only benefit the crop before which incorporated but also leave sufficient residual effect to benefit the succeeding crop.

Table 53.2: Grain Yield of Component Crops and BREY of Basmati Rice Based Crop Sequences (3 year mean)

Crop Sequence	*Grain Yield (q/ha)*		*BREY (q/ha)*
	I	*II*	
Fallow			
Basmati rice-wheat	22.0	44.9	41.4
Basmati rice-celery	24.1	5.6	34.2
Basmati rice-field peas	22.2	14.6	35.5
Basmati rice-sunflower	23.4	27.0	46.7
Mean	22.9		39.5
Basmati rice-wheat	(472) 18.5	43.9	50.3
Basmati rice-celery	(455) 19.4	5.7	42.1
Basmati rice-field peas	(455) 18.7	11.7	41.7
Basmati rice-sunflower	(443) 18.2	27.2	53.8
Mean	18.7		47.0
Green manure			
Basmati rice-wheat	14.2	47.1	34.6
Basmati rice-celery	14.2	6.3	25.7
Basmati rice-field peas	14.3	16.2	29.1
Basmati rice-sunflower	14.4	28.2	38.8
Mean	14.3		32.1

Figures in parentheses are fresh biomass of *bajra* (fodder) q/ha.

Basmati Rice Equivalent Yield (BREY) and Economic Analysis

The grain yield of component crops of different crop sequences was converted into basmati rice equivalent yield (BREY) and the data are presented in Table 53.2. The maximum BREY was obtained in *bajra (f)*-basmati rice-sunflower crop sequence (53.8 q/ha) and minimum was obtained in green manure-basmati rice-celery system (25.7 q/ha) which could be due to inclusion of *bajra* crop in the former and in the later crop sequence it could be because of low grain yield of basmati rice. Maximum mean BREY was obtained in *bajra (f)* based crop sequence (47.0 q/ha) followed by fallow (39.0) and minimum was obtained in green manure based crop sequence (32.1).

As expected gross returns (Table 53.3) also followed the same trend and the mean gross returns were maximum (Rs. 51,700 /ha) in *bajra*-basmati rice sequence and minimum in green manure-basmati rice sequence, which could also be due to low yields of component crops in these sequence (Table 53.2).

On the other hand when net returns (Table 53.4) were compared it was found that the mean net returns obtained from green manure-basmati rice were significantly below as compared to fallow-basmati rice and *bajra (f)*-basmati rice which may be due to poor yield of basmati rice obtained in this crop sequence. However there was not much variation in the mean net returns obtained from fallow-basmati rice and *bajra* (*f*)-basmati rice (Rs. 19200 and 19900/ha). But when sub crop sequences were compared fallow-basmati rice-sunflower and *bajra* (*f*)-basmati rice-sunflower gave the highest net returns, which were significantly better than fallow-basmati rice-wheat and *bajra* (*f*)-basmati rice-wheat sequence.

Table 53.3: Gross Returns from Basmati Rice Based Crop Sequences (mean of 3 years)

Crop Sequence	*Fallow*	*Bajra (fodder)*	*Green Manure*	*Mean*
Basmati rice-wheat	45.6	55.3	38.0	46.6
Basmati rice-celery	37.6	46.3	28.3	37.4
Basmati rice-field peas	39.0	45.9	31.9	38.9
Basmati rice-sunflower	51.5	59.2	42.7	51.1
Mean	43.4	51.7	35.2	

CD at 5% Main plot: NS, Sub plot: 5.7, Interaction: NS.

Table 53.4: Net Returns from Basmati Rice Based Crop Sequences (mean of 3 years)

Crop Sequence	*Fallow*	*Bajra (fodder)*	*Green Manure*	*Mean*
Basmati rice-wheat	20.8	23.0	11.0	18.3
Basmati rice-celery	15.2	16.4	3.7	11.8
Basmati rice-field peas	14.4	13.8	5.1	11.1
Basmati rice-sunflower	26.3	26.5	15.3	22.7
Mean	19.2	19.9	8.8	

CD at 5% Main plot: 4.7, Sub plot: 2.8, Interaction: NS.

Nitrogen Addition by Incorporating Green Manure

About 60 days old dhaincha was incorporated for green manure during all the three years of cropping which produced 51.6, 54.5 and 62.0 q/ha of dry weight during 1996, 1997 and 1998 respectively. The N content of green manure during the three years was 2.2, 1.97 and 1,87 per cent, respectively and that added 113.5,107.4 and 115.9 kg N/ha during different years. Earlier, Meelu *et al.* (1994) had also reported that incorporation of green manure resulted in addition of 75–100 kg N/ha. Therefore, the improvement in soil fertility and BREY was observed in green manure compared to other treatments due to addition of biomass, nitrogen and other associated beneficial effects.

Soil Health

After three cycles of cropping there was a reduction in the soil pH than the initial value (Table 53.5). The mean soil pH decreased from 7.62 to 7.45, 7.48 and 7.46 in fallow-basmati rice, *bajra* (*f*)-basmati rice and green manure-basmati rice crop sequence. Although within the mean values there were marginal changes but when compared with initial pH (7.62) there was considerable decline in soil pH. This could be due to addition of urea which is known to be acid forming fertilizer and release of some organic acids produced during the decomposition of roots of different crops remaining in the soil.

Table 53.5: Influence of Basmati Rice Based Crop Sequences on Soil Health

Crop Sequence		*pH (1 : 2)*	*OC (%)*	*Available N*	*Available P (kg/ha)*	*Available K*
Fallow						
Basmati rice-wheat		7.52	0.34	114.4	47.6	119.0
Basmati rice-celery		7.56	0.43	125.4	55.0	108.9
Basmati rice-field peas		7.36	0.36	122.3	55.0	119.7
Basmati rice-sunflower		7.36	0.35	101.8	52.0	88.6
Mean		7.45	0.37	115.9	52.4	109.0
Bajra (fodder)						
Basmati rice-wheat		7.50	0.36	112.9	47.6	89.9
Basmati rice-celery		7.46	0.38	109.7	51.7	99.7
Basmati rice-field peas		7.49	0.39	106.6	55.0	59.3
Basmati rice-sunflower		7.49	0.41	103.4	54.6	106.1
Mean		7.48	0.38	108.1	52.0	95.0
Green manure						
Basmati rice-wheat		7.54	0.51	130.1	64.4	119.7
Basmati rice-celery		7.53	0.52	126.9	57.2	115.2
Basmati rice-field peas		7.39	0.49	128.5	59.5	123.2
Basmati rice-sunflower		7.38	0.45	122.2	58.2	116.6
Mean		7.46	0.49	127.0	59.2	118.7
CD at 5%	Main plot	0.05	0.06	NS	5.6	7.9
	Sub plot	NS	NS	NS	NS	NS
	Interaction	NS	NS	NS	NS	NS
Initial value		7.62	0.33	110.2	42.8	132.6

The mean organic carbon content of the soil increased over its initial content but the extent of increase was higher where green manure was incorporated and it increased from its initial status of 0.33 per cent to 0.49. Mahapatra *et al.* (2002) reported beneficial effects of *Sesbania aculeate* green manuring in rice-wheat sequence on soil organic carbon and available N. There was not much variation in the available N content of the soil than its initial content except in green manuring treatment where it increased to 127 kg/ha over its initial content of 110 kg/ha.

On the other hand mean available P of the soil increased than its initial status under all the treatments but again the extent of increase was maximum under green manure incorporation. This may be due to addition of P fertilizers to crops and because of lesser P uptake by the crops than the amounts applied to different crops. Earlier Sikka and Sidhu (1997) had reported that lesser P uptake in spite of much higher P application to different crop sequences resulted in build up of available P in soils.

However, the available K content of the soil decreased significantly than its initial status in all the crop sequences but the magnitude of reduction was little less in green manure incorporation. The decrease in available K status may be due to exceedingly higher K uptake by the crops and no addition of K fertilizer to the crops as is evident from the data in Table 53.1. The mean available K content was minimum under *bajra*-basmati rice sequence, which may be due to greater removal by *bajra* crop, which received no K application.

It may be concluded that fallow basmati rice-sunflower sequence was more viable both economically and biologically as it produced substantially higher rice equivalent yield and higher returns over variable cost than the prevalent basmati rice-wheat sequence and also sustained the nutrient availability in the soil except K. It may also be mentioned that green manuring did not prove beneficial for basmati rice as it resulted in severe lodging of basmati causing a reduction in its grain yield.

References

Chauhan, D.S., Sharma, R.K. and Kharub, A.S. 2000. Diversification of rice-wheat system for sustainability and profitability. In: *Extended Summary, International Conference on Managing Natural Resources,* held during 14–19 Feburary at IARI, New Delhi.

Mahapatra, B.S., Kumar, Ajay, Shukla, D.K. and Shukla, R K. 2002. Summer legumes in relation to productivity and fertility in rice-wheat cropping system. In: *Paper Presented at the International Conference on Managing Natural Resources,* held during 14–19 Feburary at IARI, New Delhi, 1: 155–156.

Meelu, O.P., Singh, B. and Singh, Y. 1994. Effect of green manure and crop residue cycling on N economy, organic matter and physical properties of soil in rice-wheat cropping system. In: *Proceeding of Symposium on Sustainability of Rice-wheat System in India,* held during 7–8 May at Regional Station, Karnal, p. 15–24.

Merwin, H.D. and Peech, M. 1950. Exchangeability of soil potassium in the sand, silt and clay fractions as influenced by the nature of the complementary exchangeable cations. *Soil Sci. Soc. Am. Proc.,* 15: 125–128.

Olsen, S.R., Cole, C.V., Watanabe, F.S. and Dean, L.A. 1954. Estimation of available phosphorus by extraction with sodium bicarbonate. *USDA Circular,* 939: 19.

Rattan, R.K. and Singh, A.K. 1997. Role of balanced fertilization in rice-wheat cropping system. *Fertilizer News,* 42(4): 592–600.

Sidhu, M.S., Thakar Singh and R.K. Sikka. 2002. Response of basmati rice to nitrogen in different cropping systems. In: *Extended Summary of 2nd International Agronomy Congress,* New Delhi, p. 402–403.

Sidhu, M.S., R. Sikka and Thakar Singh. 2004. Performance of transplanted Basmati rice in different cropping systems as affected by nitrogen application. *IRRI Notes,* p. 62–65.

Sikka, R. and M.S. Sidhu. 1997. Evaluation of summer groundnut based cropping systems for nutrient uptake and changes in soil fertility. In: *Proceedings of the 62nd Annual Convention of the Indian Society of Soil Science* held from 18–21 October at Kolkata, pp. 209.

Subbiah, B.V. and Asija, G.L. 1956. A rapid procedure for estimation of available nitrogen in soils. *Curr Sci.,* 25: 259–260.

Walkley, A. and Black, I.A. 1934. An estimation of degtareff method for determining soil organic matter and a proposed modification of the chromic acid titration method. *Soil Sci.,* 4: 51–56.

Yadav, R.L., Yadav, D.S., Singh, R.M. and Kumar, A. 1998. Long term effects of inorganic-inputs and crop productivity in rice-wheat cropping system. *Nutrient Cycling in Agroecosystems,* 51: 193–200.

Chapter 54

Analytical Study of Molecular Interactions in Binary and Ternary Liquid Mixtures

D.K. Gupta and G. Meenakshi

Department of Physics, Kanchi Mamunivar Centre for Post Graduate Studies, Lawspet, Pondicherry – 605 008

ABSTRACT

Ultrasonic velocity measurements are extensively used to study the physico-chemical behaviour of liquid mixtures. In recent years, a good deal of work has been done to correlate the experimental ultra sonic velocities with those of computed theoretically using empirical and semi-empirical relations. The ultrasonic velocities have been evaluated in the binary systems of Bromobenzene with Methanol, Ethanol, Propan-1-ol and Butan-1-ol at 303.15 K. Sound velocity has also been evaluated in ternary mixtures of Aniline + Benzene + Acetic acid, Aniline + Benzene + Propionic acid and Aniline + Benzene + Butryic acid at a temperature of 303.15 K. Alpha, excess molar volume, excess velocity are also evaluated and are used to explain the interaction in the mixtures.

Keywords: *Binary and ternary liquid mixtures, Molecular interactions, Sound velocity.*

Introduction

Liquids and liquid mixtures have found wide applications in the chemical textiles, leather and nuclear industries (Rowlinson and Swinson, 1982; Acree, 1984 and Prausnitz *et al.*, 1986). The

* E-mail: ashwinikdixit@yahoo.co.in.

measurement of ultrasonic velocity in the liquid mixtures has been found to be an important tool to study the physico-chemical properties of the mixtures. The ultrasonic velocity is also useful to study the molecular interactions in variety of liquid mixtures. The main objective of the present investigation is to explore molecular interactions in binary and ternary liquids, because there is no much data available in such systems in literature. For this purpose four different alcoholic liquids with Bromobenzene for binary system and three different acids with Aniline and Benzene for ternary system are selected. The theoretical evaluation of sound velocity and excess parameters are important as the deviation of experimental values from the ideal values may yield the information on the nature of molecular interactions. The calculations are based on the experimental results of Nikam *et al.*, 2000 for binary mixtures and Kannappan *et al.*, 1992 in the case of ternary mixtures.

Materials and Methods

Theoretical evaluation of sound velocity for binary and ternary liquid mixtures are done using the following empirical formulae (Pandey *et al.*, 1991; Gupta and Pandey, 1988; Srivastava, 1992 and Ali *et al.*, 2000):

$$U_{mix} = \left[\frac{X_1R_1 + X_2R_2}{X_1V_1 + X_2V_2}\right]^3 \tag{1}$$

where,

R_1, R_2 are the molar sound velocities, X_1, X_2 are mole fractions and V_1, V_2 are the molar volumes.

$$\left[\frac{1}{X_1M_1 + X_2M_2}\right]\left[\frac{1}{U^2{}_{im}}\right] = \left[\frac{X_1}{M_1V_1{}^2} + \frac{X_2}{M_2V_2{}^2}\right] \tag{2}$$

$$U_{mix} = \left[\frac{K}{L_{f(mix)}\rho^{1/2}{}_{mix}}\right] \tag{3}$$

where,

K is the temperature constant and ρ_{mix} is the density of the mixture.

$$U_{mix} = \left[\frac{U_{\infty}(X_1S_1 + X_2S_2)(X_1B_1 + X_2B_2)}{V_m}\right] \tag{4}$$

where,

S_1, S_2 are collision factors, $U_{\infty} = 1600\ ms^{-1}$, B_1, B_2 are actual volumes of the molecule per mole and U_m is molar volume.

Results and Discussion

Bromobenzene in Methanol, Ethanol and Propan-1-ol and Butan-1-ol at the temperature 303.15 K are discussed. The experimental value of ultrasonic velocity and the percentage deviation of the theoretical velocity from the experimental values in the above mixtures of different mole fractions of Bromobenzene are reported in Table 54.1. The calculated values of alpha (α), excess velocity (U^E) and excess volume (V^E) in different alcohols with mole fraction of Bromobenzene are also given in the same table.

Table 54.1: Percentage Deviation of the Calculated Value of Ultrasonic Velocity from the Experimental Value and Excess Parameters of Binary Liquid Mixtures at 303.15 K

Name of Compounds	Mole Fraction of Bromobenzene	Experimental Value of 'U' ms^{-1}	Alpha (α)	Percentage Deviation $\left(\frac{\Delta U}{U}\right)\%$				Excess Velocity (UE) ms^{-1}	Excess Volume (V^E) $10^{-3}m^3mol^{-1}$
				Nomoto	Van-Dael	FLT	CFT		
Bromo	0.0222	1070.7	−0.0425	−2.33	1.16	−3.16	−3.67	−23.51	−0.4212
benzene	0.0805	1068.3	−0.0514	−3.11	7.66	−6.75	−7.30	−28.55	−0.1913
in Methanol	0.1697	1066.4	−0.0617	−3.97	14.36	−10.50	−11.08	−34.49	−0.0459
	0.3226	1076.0	−0.0566	−3.91	21.31	−12.07	−13.08	−31.81	−0.0245
	0.6473	1096.1	−0.1468	−3.14	22.88	−9.77	−10.15	−27.42	−0.1812
Bromo	0.0316	1134.2	−0.0008	−0.05	2.48	−0.57	−0.81	−0.43	−0.1691
benzene in	0.1118	1126.4	−0.0156	−0.78	6.83	−2.47	−3.15	−8.55	−0.3135
Ethanol	0.2269	1120.3	−0.0264	−1.39	11.11	−4.60	−5.26	−15.11	−0.3962
	0.4064	1115.0	−0.0368	−1.95	14.28	−5.33	−6.72	−21.13	−0.4128
	0.7255	1121.6	−0.0276	−1.44	12.45	−3.97	−4.73	−15.80	−0.2068
Bromo	0.0407	1197.8	0.0095	−0.55	2.25	−1.50	−0.06	−5.68	−0.3627
benzene in	0.1410	1172.2	−0.0240	−0.99	4.04	−2.76	−2.48	−14.32	−0.6305
Propan-1-ol	0.2768	1155.5	−0.0393	−1.66	6.24	−3.56	−4.03	−23.43	−0.6283
	0.5173	1142.2	−0.0396	−1.61	7.78	−4.05	−4.36	−23.28	0.3287
	0.7751	1131.0	−0.0346	−1.50	5.21	−3.28	−3.28	−20.07	0.3774
Bromo	0.1683	1191.2	−0.0366	−1.70	1.56	−1.86	−2.50	−22.40	−0.1048
benzene in	0.3208	1172.2	−0.0455	−2.08	2.78	−2.59	−3.31	−27.63	−0.2699
Butan-1-ol	0.4145	1162.2	−0.0484	−2.20	3.10	−3.10	−3.57	−29.17	−0.3276
	0.5241	1149.2	−0.0539	−2.49	2.87	−3.87	−3.89	−32.27	−0.2939
	0.6538	1137.1	−0.0551	−2.59	2.23	−2.95	−3.85	−32.66	−0.2100

From the Table 54.1, it is observed that the percentage deviations of computed ultrasonic velocity using various theories, like Nomoto's method, Van Dael ideal mixing relation, collision factor theory and FLT for the mixture of Bromobenzene with different alcoholic group show that the Nomoto's method is most suitable for the calculation of ultrasonic velocity. The interaction effect can be explained from the calculated excess thermodynamic parameters. Both Bromobenzene and alcohols are polar liquids which can interact by forming hydrogen bonding through dipole-dipole interaction. Alcohols are liquids, associated through hydrogen bonding and in the pure state exhibit equilibrium between monomers and multimer species. In the pure state the halogenated compounds present weak dipole-dipole interaction. When Bromobenzene is mixed with alcohols the halogenated solvents can interact with or groups (Arul and Palaniappan, 2001; Kannappan and Rajendran, 1992). The aromatic derivatives set up an interaction between the electron cloud and the hydroxyl groups. Of course this interaction is of minor intensity compared with hydrogen bonding but they can lead to the formation of intermolecular complexes.

In the present study α, excess velocity and excess volume are found to be negative. The negative values of excess parameters indicate that the strong interaction in the liquid mixtures as well as interstitial accommodation of Bromobenzene molecules is in aggregates of alcohol. The negative excess volume indicates the formation of molecular clusters and complexes. This may involve in charge transfer complexes. The negative excess sound velocity vary with mole fraction may be attributed to the breaking of both the hydrogen bond and dipolar interaction in mixing. The average percentage deviation of ultrasonic velocity in ideal mixing relation decreases in the order Methanol > Ethanol > Propan-1-ol > Butan-1-ol (13.78 > 9.84 > 5.33 > 1.77). This trend suggests that the hetro association and homo association of molecules decreases with increase of chain length of carbon atoms in alcohols. This is probably due to less proton donating tendency of higher alcohols. The decrease of average percentage deviation in the ideal mixing relation with the increasing chain length of the carbon atoms in alcohols suggests that the ideal mixing relation can give good results with higher alcohols.

The ternary systems chosen for this study are Aniline and Benzene with carboxylic acids namely Acetic, Propionic and Butryic. The percentage deviation of ultrasonic velocity for the systems Aniline and Benzene in Acetic acid, Propionic acid and Butryic acid and the calculated values for α, excess volume and excess velocity are listed in Table 54.2. It is observed that the collision factor theory agrees better with the experimental values.

Table 54.2: Percentage Deviation of the Calculated Values of Ultrasonic Velocity from the Experimental Values and Excess Parameters of Ternary Liquid Mixtures at 303.15 K

X_1 and X_3 Refers to	Name of Compounds	Mole Fraction X_1 (U^2)	Mole Fraction X_3	Experimental Value of ms^{-1}	Alpha (α)	Percentage Deviation $\left(\frac{\Delta U}{U}\right)\%$				Excess Velocity (U^E) ms^{-1}	Excess Volume (V^E) $10^{-3}\,m^3\,mol^{-1}$
						Nomoto	Van-Dael	FLT	CFT		
Aniline	Aniline +	0.0980	0.5041	1234.1	–0.4167	–3.30	1.49	–3.69	–3.09	–297.58	–0.4795
and	Benzene +	0.2002	0.4014	1304.0	–0.4531	–1.35	4.80	–1.26	–0.98	–399.82	–0.743
Acetic	Acetic Acid	0.2988	0.3000	1352.2	–0.4890	–0.86	5.82	–0.76	–0.45	–284.53	–0.7433
Acid		0.3990	0.2003	1399.2	–0.5221	–0.34	6.07	0.24	0.00	–430.01	–0.6579
		0.4967	0.1023	1440.4	–0.5519	–0.02	5.26	0.15	0.17	–474.76	–0.5466
Aniline	Aniline +	0.0694	0.5266	1224.0	–0.4398	–3.24	–1.61	–3.58	–3.11	–317.03	–0.8575
and	Benzene +	0.1954	0.4100	1293.0	–0.4766	–1.75	1.35	–1.89	–1.54	–362.87	–0.8235
Propionic	Propionic	0.2911	0.3053	1344.3	–0.4983	–0.83	2.89	–0.80	–0.62	–394.59	–1.3683
acid	acid	0.3950	0.2058	1395.8	–0.5277	–0.08	3.91	0.47	0.08	–436.59	–1.0608
		0.4972	0.1044	1439.9	–0.5544	0.19	3.97	0.27	0.26	–478.25	–0.7915
Aniline	Aniline +	0.1007	0.4975	1249.0	–0.4624	–1.18	–0.14	–1.46	–1.28	–338.27	–0.7555
and	Benzene +	0.2015	0.3951	1295.7	–0.4921	–0.72	1.12	–0.98	–0.86	–376.81	–0.5751
Butryic	Butryic acid	0.3008	0.2989	1344.8	–0.5109	–0.10	2.34	–0.37	–0.26	–406.98	–1.0725
acid		0.3999	0.1979	1390.0	–0.5384	0.09	2.94	–0.07	–0.05	–447.61	–0.6365
		0.5058	0.1012	1436.9	–0.5567	0.19	3.26	0.26	0.08	–481.47	–0.9028

Benzene is a non-polar solvent and the carboxylic acids are polar in nature. Therefore they are associated with unsymmetrical molecules. In view of the high dipole moments, Aniline can be classified as dissociating solvent as it has the ability to dissociate the molecules of associative liquids like water, alcohols and carboxylic acids (Kannappan and Rajendran 1992) etc. The α, excess velocity and excess molar volume are found to be negative for all the concentrations studied. The negative value of a indicates the presence of complex formation. As explained in the case of binary mixtures, in ternary mixture also show negative values of excess sound velocity which may be attributed to the breaking of both hydrogen bond and bipolar interaction on mixing. The negative value of excess volume indicates the formation of molecular clusters and complexes and may involve in charge transfer complexes. The various results show the existence of strong hydrogen bonding interactions as well as interaction due to the formation of charge transfer complex between acids and benzene.

Conclusion

Generally in liquid mixtures the possible interactions are between like molecules as well as unlike molecules. These interactions are of two types; namely, long range and short range. The long range interaction includes electro static induction and dispersion forces. These forces arise without the overlap of the electron clouds due to the closer approach of interacting molecules. Normally the long range interactions are highly directional. On the other hand, short range interactions such as dipole-dipole, dipole-induced dipole, charge transfer, complex formation and hydrogen bonding interactions arise when the molecules come closer together, resulting in a significant overlapping of electron clouds. Therefore, when two or more liquids are mixed (in the case of binary and ternary mixtures), then the mixture is not ideal. Thus the deviation from the ideality is explained on the basis of molecular interactions between the components of the liquid mixtures.

Acknowledgements

The authors are thankful to Dr. S. Mayilavelan, Head of the Department of Physics, Dr. Ashwini Kumar Dixit, Department of Plant Science and the Director, Kanchi Mamunivar Centre for Post Graduate Studies, Lawspet, Pondicherry, for their constant encouragement and valuable suggestions while preparing this manuscript. The authors are also wish to thank the Staff members of Physics Department, for their constant encouragement during the preparation of this manuscript.

References

Acree, W.E., 1984. *Thermodynamic Properties of Non-electrolytic Solutions.* Academic Press, New York.

Ali, A., Tiwari, K., Nain, A.K. and Chakravortty, V., 2000. Ultrasonic study of molecular interaction in ternary mixtures of dimethylsulphonamide + corbomtetrachloride + aromatic hydrocarbons. *Ind J. Physics,* 74B(5): 351–355

Arul, G. and Palaniappan, C., 2001. Molecular studies in the ternary liquid mixture of cyclohexene + tolune + 2 proponol. *Ind. J. Pure and Appl. Phys.,* 39: 561.

Gupta, U. and Pandey, J.D., 1988. Sound velocity in reciprocal fused salt pairs. *Acoustic Letters,* 12(3): 55–58

Kannappan, A.N. and Rajendran, V., 1992. Acoustic parameters of some ternary liquid mixtures. *Ind. J. Pure and Appl. Phys.,* 30: 240.

Kannappan, A.N. and Rajendran, V., 1994. Molecular interaction studies in ternary liquid mixtures from ultrasonic data. *Ind. J. Pure and Appl. Phys.,* 29: 465.

Nikam, P.S., Kapade, V.M. and Hasen, M., 2000. Molecular interaction in binary mixtures of bromo benzene with normal alkanols: An ultrasonic study. *Ind. J. Pure and Appl. Phys.*, 38: 170–173.

Panday, J.D., Dubey, G.P., Shukla, B.P. and Dubey, S.N., 1991. Prediction of ultrasonic velocity and intermolecular tree length and their correlation with interaction in binary liquid mixtures. *Pranama J. Phy.* 37(6): 497–503.

Prausnitz, J.M., Linchenthalr and Azeved, E.G., 1986. *Molecular Thermodynamics of Fluid-phase Equilibrium*, 2nd edn. Eagle Wood Cliffs, Prentice Hall Inc.

Rowlinson, J.S. and Swinson, F.L., 1982. *Liquid and Liquid Mixtures*, 3rd edn. Buttor Worth Scientific, London.

Srivastava, A.P. 1992. Evaluation of sound velocity in binary systems of anisole with some aromatic compounds by different methods. *Ind. J. Chem.*, 31A: 577–580.

Chapter 55

Comparison of Energy Interaction Parameters for the Complexation of Pr(III) with 1,10-Phenanthroline and 2,2'-Bipyridyl in Different Organic Solvents using 4f-4f Transition Spectra as Probe: An Absorption Spectral Study

*N. Yaiphaba[1], Th. David Singh[2], Ch. Sumitra[2], M. Indira Devi[2] and N. Rajmuhon Singh[1]**

[1]Department of Chemistry, Manipur University, Imphal – 795 003, India
[2]Department of Chemistry, Nagaland University, Mokokchung – 798 601, India

ABSTRACT

Lanthanide co-ordination chemistry in solution is of great importance with the increase use of lanthanide as probes in the exploration of the structural functions of biomolecular reactions. Comparative absorption and spectral analysis of 4f-4f transitions of the Pr(III) with 1,10 Phenanthroline (phen) and 2,2'-Bipyridyl (bpy) has been explored in aqueous and different aquated organic solvents, like DMF, CH_3CN, CH_3OH, Dioxane and their equimolar binary mixtures. The variation in energy parameters like Slator Condon (F_k) Racah (E^k) and Lande Spin orbit coupling constant (ξ_{4f}), Nephelauxetic ratio (β), bonding parameter ($b^{1/2}$) and per cent covalency (δ) are discussed to explore

* Corresponding Author; E-mail: rajmuhon@yahoo.co.in.

the mode of interaction between Pr(III) and 1,10-Phenanthroline (phen) and 2,2′-Bipyridyl (bpy). The marginal decrease in the values of F_k, E^k, ξ_{4f} are correlated with the increase in the value of β. The difference in the energy parameters with respect to donor atoms and solvents, reveal that the chemical environment around the lanthanide ion has great impact on *f–f* transition and any change in the environment results in modifications of the spectra.

Keywords: *1,10-Phenanthroline, 2,2′-Bipyridyl, Absorption spectra, Nephelauxetic effect, Pseudohypersensitive transition.*

Introduction

The trivalent lanthanides Ln(III) constitute a series of hard Lewis acids with similar chemical behaviors and their complexes with anionic ligands containing oxygen as donar atoms have been the most studied (Sinha, 1966) and their co-ordination chemistry in solution is of great importance with the increase use of lanthanide as probes in the exploration of the structural functions of biomolecular reactions (Evans, 1990). The hard metal acceptors like lanthanide (III) prefer oxygen and nitrogen donor ligands for stable complexes with predominantly ionic character although the affinity for oxygen donar ligand is more. Shah and Shah (2001) studied comparative 4f–4f transition spectra of Pr(III) with lysozyme by using energy interaction parameters to explain the behaviour of binding between them. Studies on Ln(III) complexes derived from nitrogen donor ligands (Khan *et al.*, 2000) have shown that the Ln-N interaction is stronger than that previously believed. It has been found that the number of coordinated ligands to the lanthanides depends upon the ability of the anion to compete for the coordination sites and there relative size of the ligands.

The present work discusses the quantitative spectral analysis of energy interaction parameters of Pr(III) complexes with 1,10-phenanthroline and 2,2′-bipyridyl in MeOH, CH_3CN, DMF, Dioxane and their equimolar binary mixtures at 298 K. The ligand mediated pseudohypersensitive transitions of Pr(III), $^3H_4 \rightarrow {}^3P_0$, $^3H_4 \rightarrow {}^3P_1$, $^3H_4 \rightarrow {}^3P_2$ and $^3H_4 \rightarrow {}^1D_2$ which have been found to show substantial sensitivity with minor changes in the immediate coordination environment around it even in the presence of structurally related ligand have been utilised to calculate the magnitude and variation of energy interaction parameters *viz.*, Slater Condon (F_k), Lande spin orbit coupling (ξ_{4f}), Nephelauxetic ratio (β), bonding ($b^{1/2}$), Racah (E^k) and covalency (δ) parameters to discuss the interaction of Pr (III) with structurally related phen and bpy.

Materials and Methods

1,10-Phenanthroline and 2,2′-bipyridyl of analytical grade from Qualigens and Praseodymium(III) nitrate hexahydrate of 99.9 per cent purity from M/s Indian Rare Earths Ltd. are used for synthesis and spectral analysis. The solvents used are MeOH, CH_3CN, DMF and Dioxane which are of A/R grade from E. Merk. All the spectra are recorded on a temperature controlled Perkin Elmer Lambda-35 UV-Visible Spectrophotometer in the range 350–910 nm in the concentration of 0.01 M. The temperature of all the observations is maintained by using water circulating HAAKE DC 10 thermostat.

The Nephelauxetic effect, a measure of covalency (Peacock, 1964; Jorgensen and Judd, 1964; Henrie *et al.*, 1976) has been interpreted in terms of Slater-Condon and Racah parameters (inter electronic repulsion parameters) as well as by the ratio of the free ion and complex ion (Sinha and Schmidtke, 1965; Sinha, 1966; Sinha *et al.*, 1972; Condon and Shortley, 1963).

$$\beta = \frac{F_K^c}{F_K^f} \text{ or } \frac{E_c^K}{E_f^K} \tag{1}$$

where,

F_k (K = 2,4,6) is the Slater-Condon parameter and E^k is the Racah parameters for complex and free ions respectively. The bonding parameter ($b^{½}$) is inter-related to Nephelauxetic effect as:

$$\beta^{½} \quad \frac{1-\beta}{2}^{½} \tag{2}$$

The electrostatic term E_0 is expressed in terms of the product of Slater radial integral known as Slater-Condon parameter, F_k and is given by

$$E_0 \quad \sum_{\kappa\ 0}^{\kappa\ 6} K^{\kappa} F_{\kappa} \tag{3}$$

The Slater-Condon parameters are also known as direct-integral and is decreasing function of K as given by the relation:

$$F_1^{\kappa} \quad \int_0 \int_0 \frac{r^{\kappa}}{r^{\kappa\ 1}} R_i^2(r_i) R_j^2(r_j) r_i^2 r_j^2 dr_i dr_j \tag{3}$$

where,

R is the 4f-radial wave function; $r_<$ and $r_>$ are the radii of near and more distant electrons; i and j are the ith and j^{th} electrons under consideration. Condon and Shortley (1963) redefined F^k integrals in terms of reduced integral F_k related to each other and the relation is

$$F_K = \frac{F^K}{D_K} \tag{5}$$

Combining relation (4) and (5), the reduced Slater-Condon integral can be written as:

$$F_{\kappa} \quad \frac{1}{D_{\kappa}} \int_0 \int_0 r^{\kappa} r^{\kappa\ 1} R_i^2(r_i) R_j^2(r_j) T_i^2 r_j^2 dr_i dr_j \tag{6}$$

Here D_k is the denominator and F_k are coefficient of linear combination and represent the angular part of the interaction F_k is the expectation value of scalar product $(C_1^{(K)} C_2^{(K)})$.

The Racah energy interaction parameter E^{Ki}'s are linear combinations of F_K given by

$E^1 = (70 \times F_2 + 231\ F_4 + 20.02\ F_6)/9$

$E^2 = (F_2 - 3\ F_4 + 7\ F_6)/9$

$$E_3 = (5\ F_2 + 6\ F_4 - 9\ F_6)/3 \tag{7}$$

The energy, E_{so} arising from the most important magnetic interactions, which are spin orbit interactions may be written as

$$E_{so} = A_{so}\,\xi_{4f} \tag{8}$$

where,

A_{so} is the angular part of spin orbit interaction and ξ_{4f} is the radial integral and is known Lande's parameter by first order approximation the energy E_j of the jth level is given by Wong (1961; 1963) as:

$$E_j(F_\kappa, \xi_{4f}) \quad E_{oj}(F_\kappa^0, \xi_{4f}) \quad \frac{E_j}{F_\kappa}\Delta F_\kappa \quad \frac{E_j}{\xi_{4f}}\Delta\xi_{4f} \tag{9}$$

where,

E_{oj} is the zero order energy of the jth level. The value of F_k and ξ_{4f} are given by

$$F_k = F_k^{\,0} + \Delta F_k$$

$$\xi_{4f} = \xi^0_{\;4f} + \Delta\xi_{4f} \tag{10}$$

The difference between the observed E_j value and the zero order values, ΔE_j is evaluated by

$$\Delta E_j \qquad \underset{k\;2,4,6}{} \frac{E_j}{F_\kappa}\Delta F_\kappa \quad \frac{E_j}{\xi_{4f}}\Delta\xi_{4f} \tag{11}$$

By using the zero order energy and partial derivatives of Pr(III) ion given by Wong (1961; 1963), the above equation can be solved by least square technique and the value of F_2 and can be found out. From these the value of F_2 and are found out by using relation (10). The estimated values of F_4 and F_6 are calculated by the relations,

$$\frac{F_4}{F_2} = 0.1380 \text{ and } \frac{F_4}{F_2} = 0.0150 \tag{12}$$

The per cent covalency parameter (δ) representing the Nephelauxetic effect was given by the relation

$$\delta \quad \frac{1-\beta}{\beta} \quad 100 \tag{13}$$

Results and Discussion

The shape, energy and oscillator strength of hypersensitive or pseudohypersensitive transitions (Karraker, 1967) can be correlated with coordination number and diagnostic of immediate coordination environment around lanthanides ions. In Figure 55.1, red shift is observed when Phen/bpy is added to Pr(III). The variation of the magnitude of energy interaction parameters like Slater Condon (F_k), Lande factor (ξ_{4f}), Racah energy (E^K), Nephelauxetic ratio (β), bonding ($b^{½}$) and percentage covalency (δ) for Pr(III), Pr(III): bpy and Pr(III) : phen in DMF, CH_3CN, CH_3OH, Dioxane and their equimolar binary mixtures is shown in Table 55.1. The marginal decrease in the values of F^K, ξ_{4f} and E^K indicating complexation led to the increase in the values of Nephelauxetic ratio. Jorgensen and Ryan (1966) suggested that shortening in the metal ligand distance occurs with decrease in the co-ordination number. To interpret the correlation and analysis of the relationship between Nephelauxetic effect and geometry, energy parameters have been derived and evaluated as reported earlier (David *et al.*, 2003; 2004; 2005; 2006) for complex compound using the angular overlap model, the value of 'n' is proportional to the Nephelauxetic effect as

$$n \quad \frac{(1 \quad \beta^{1/2})}{\beta^{1/2}} \tag{14}$$

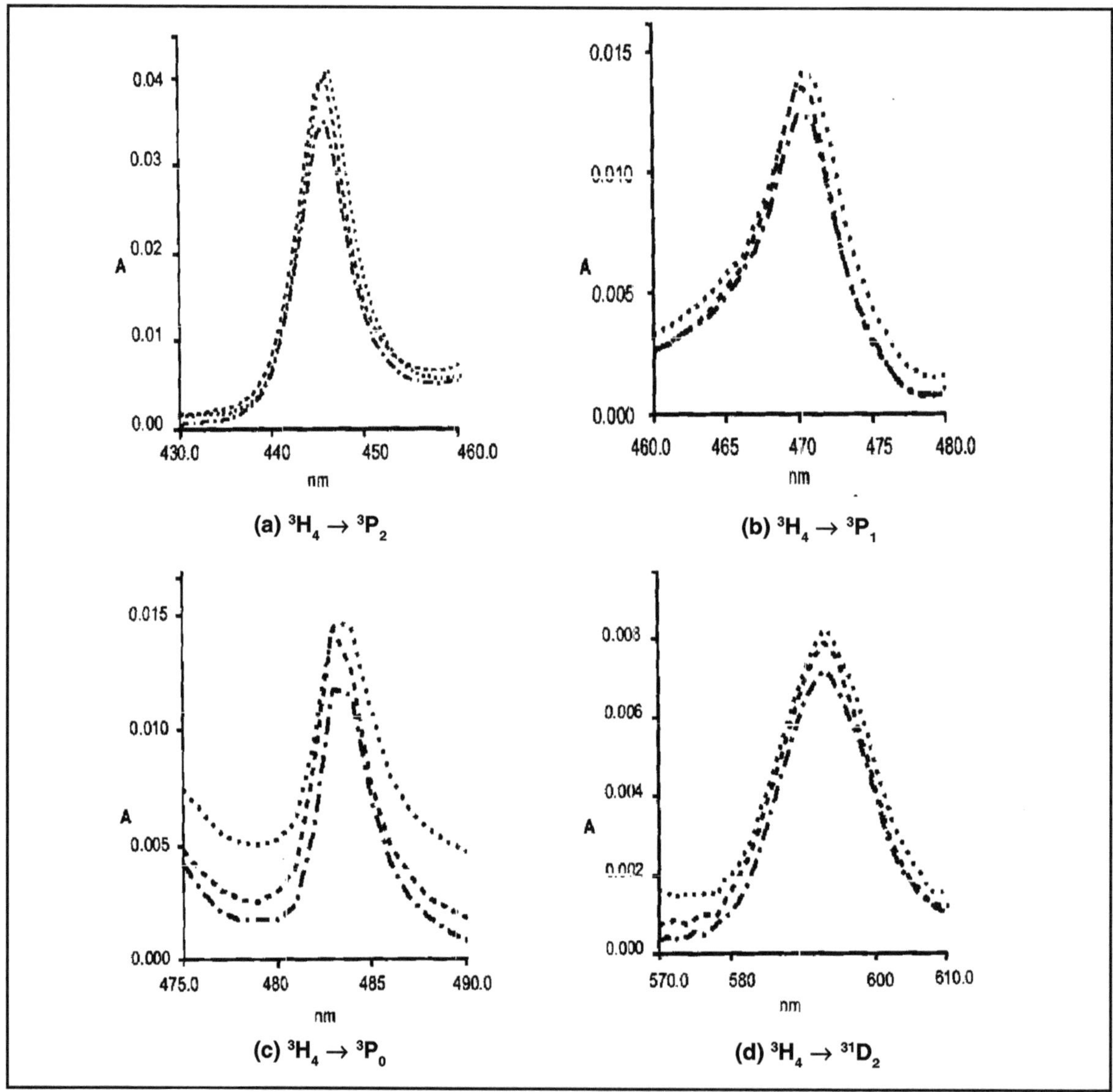

Figure 55.1: Comparative Absorption Spectra of Pr(III) –.–.–.–.–, Pr(III):bpy - - - - - - and Pr(III) : phen . . . . in DMF

Table 55.1: Computed Values of Energy Interaction Slator Condon (F_k), Racah (E^k) and Lande Spin Orbit Coupling (ξ_{4f}), Nephelauxetic Ratio (β) and Bonding Parameter ($b^{½}$) and Covalency Parameters (δ) of Pr(III), Pr(III) : bpy and Pr(III) : phen in DMF, CH_3CN, CH_3, OH, Dioxane and their Equimolar Binary Mixtures

Sl.No.	System	F_2	F_4	F_6	ξ_{4f}	E^1	E^2	E^3	β	$b^{½}$	δ
1.	Solvent–MeOH										
	Pr(III)	308.8467	42.6363	4.6636	719.6274	3506.8470	23.7314	586.0264	0.9444	0.1668	5.8904
	Pr(III) + Bpy	308.8586	42.6379	4.6638	719.8563	3506.9810	23.7323	586.0488	0.9445	0.1665	5.8709
	Pr(III) + Phen	308.8850	42.5638	4.6557	721.0336	3500.8850	23.6911	585.0302	0.9446	0.1665	5.8686
2.	Solvent–ACN										
	Pr(III)	308.3532	42.5682	4.6561	723.8837	3501.2430	23.6935	585.0900	0.9466	0.1635	5.6453
	Pr(III) + Bpy	307.2411	42.4146	4.6393	721.9964	3488.6150	23.6081	582.9798	0.9436	0.1679	5.9720
	Pr(III) + Phen	306.0091	42.2446	4.6207	722.3418	3474.6260	23.5134	580.6420	0.9421	0.1702	6. 1482
3.	Solvent–Dioxane										
	Pr(III)	308.1013	42.5334	4.6523	722.6680	3498.3830	23.6742	584.6120	0.9454	0.1653	5.7794
	Pr(III) + Bpy	307.9974	42.5190	4.6508	722.0624	3497.2030	23.6662	584.4149	0.9448	0.1661	5.8429
	Pr(III) + Phen	306.9069	42.3685	4.6343	720.7680	3484.8210	23.5824	582.3456	0.9432	0.1698	6.1216
4.	Solvent–DMF										
	Pr(III)	308.1064	42.5341	4.6524	719.3425	3498.4400	23.6746	584.6216	0.9431	0.1687	6.0339
	Pr(III) + Bpy	308.2119	42.5487	4.6540	718.9176	3499.6380	23.6827	584.8218	0.9430	0.1689	6.0493
	Pr(III) + Phen	308.0735	42.5296	4.6519	719.2725	3498.0670	23.6720	584.5593	0.9430	0.1688	6.0447
5.	Solvent–MeOH + ACN										
	Pr(III)	308.2608	42.5554	4.6547	720.0623	3500.1940	23.6864	584.9146	0.9438	0.1676	5.9532
	Pr(III) + Bpy	308.2599	42.5553	4.6547	720.9904	3500.1840	23.6863	584.9129	0.9444	0.1667	5.8821
	Pr(III) + Phen	307.0358	42.5863	4.6362	721.2242	3486.2840	23.5923	582.5903	0.9428	0.1691	6.0652

Contd...

Table 55.1–Contd...

Sl.No.	System	F_2	F_4	F_6	ξ_{4f}	E^1	E^2	E^3	β	$b^{1/2}$	δ
6.	Solvent–MeOH + Dioxane										
	Pr(III)	308.0788	42.5303	4.650	721.4075	3498.1270	23.6724	584.5693	0.9445	0.1666	5.8798
	Pr(III) + Bpy	308.2658	42.5561	4.6548	721.3876	3500.2500	23.6868	584.9240	0.947	0.1662	5.8506
	Pr(III) + Phen	307.2837	42.4205	4.6400	721.3444	3489.0990	23.6113	583.0605	0.9433	0.1684	6.0152
7.	Solvent–MeOH + DMF										
	Pr(III)	308.3668	42.5700	4.6563	719.6374	3501	23.6949	585.1157	0.9437	0.1678	5.9684
	Pr(III) + Bpy	308.3394	42.5663	4.6559	719.3452	3501.0860	23.6925	585.0637	0.9434	0.1682	5.9954
	Pr(III) + Phen	308.2668	42.5562	4.6548	719.0844	3500.2610	23.6869	584.9260	0.9432	0.1686	6.0274
8.	Solvent–DMF + ACN										
	Pr(III)	308.1657	42.5423	4.6533	719.5563	3499.1140	23.6791	584.7342	0.9433	0.1683	6.0077
	Pr(III) + Bpy	308.1514	42.5403	4.6531	719.3057	3498.9510	23.6780	584.7070	0.9431	0.1686	6.0294
	Pr(III) + Phen	307.8109	42.4933	4.6479	720.0342	3495.0850	23.6518	584.0609	0.9431	0.1686	6.0293
9.	Solvent–DMF + Dioxane										
	Pr(III)	308.2380	42.5523	4.6544	720.0548	3499.9350	23.6847	584.8713	0.9438	0.1677	5.9575
	Pr(III) + Bpy	308.2832	42.5585	4.6551	720.6835	35000.4480	23.6881	584.9572	0.9443	0.1669	5.9018
	Pr(III) + Phen	307.9737	42.5158	4.6504	721.0008	3496.9340	23.6644	584.3699	0.9440	0.1673	5.9282
10	Solvent–DX + ACN										
	Pr(III)	308.1142	42.5352	4.6525	723.3610	3498.5290	23.6752	584.6364	0.9459	0.1645	5.7243
	Pr(III) + Bpy	305.8562	42.2235	4.6184	719.5063	3472.8910	23.5017	580.3520	0.9399	0.1733	6.3926
	Pr(III) + Phen	305.8562	42.2235	4.6184	719.5063	3472.8910	23.5017	580.3520	0.9399	0.1733	6.3926

Table 55.2: Computed and Observed Values of Energies (cm^{-1}) and RMS Values for Pr(III), Pr(III) : bpy and Pr(III) : phen in DMF, CH_3CN, CH_3OH, Dioxane and their Equimolar Binary Mixtures

Sl.No.	System	$^3H_4$ Eobs	$^3P_2$ Ecal	$^3H_4$ Eobs	$^3P_1$ Ecal	$^3H_4$ Eobs	$^3P_0$ Ecal	$^3H_4$ Eobs	$^1D_2$ Ecal	R.M.S.
1.	Solvent–MeOH									
	Pr(III)	22490.10	22425.21	21311.51	21216.33	20737.42	20661.21	16893.89	17117.24	131.30
	Pr(III) + Bpy	22490.61	22427.16	21313.33	21218.07	20736.99	20662.48	16898.46	17118.45	129.47
	Pr(III) + Phen	22464.34	22396.81	21283.84	21185.13	20693.22	20627.05	16885.05	17097.19	126.18
2.	Solvent–ACN									
	Pr(III)	22519.99	22413.27	21311.96	21198.67	20681.67	20634.69	16890.75	17106.92	135.24
	Pr(III) + Bpy	22439.13	22328.65	21240.89	21113.24	20613.03	20553.06	16801.64	17050.31	136.25
	Pr(III) + Phen	22357.36	22247.14	21129.14	21028.28	20525.87	20467.26	16767.83	16994.68	138.94
3.	Solvent–Dioxane									
	Pr(III)	22508.33	22390.14	21316.51	21176.19	20669.70	20614.70	16837.00	17091.81	137.38
	Pr(III) + Bpy	22510.36	22380.07	21312.42	21166.50	20665.85	20606.26	16812.38	17085.27	135.51
	Pr(III) + Phen	22427.05	22299.89	21228.27	21084.94	20596.05	20527.27	16748.17	17031.38	136.89
4.	Solvent–DMF									
	Pr(III)	22432.09	22373.76	21255.34	21163.32	20686.80	20608.72	16862.27	17082.38	128.85
	Pr(III) + Bpy	22435.61	22378.75	21264.38	21169.03	20696.22	20615.32	16859.99	17085.99	130.75
	Pr(III) + Phen	22429.07	22370.97	21256.24	21160.54	20684.66	20606.12	16748.17	17031.38	130.96
5.	Solvent–MeOH + ACN									
	Pr(III)	22470.90	22387.81	21290.19	21177.01	20694.08	20620.93	16849.20	17091.57	137.75
	Pr(III) + Bpy	22473.43	22392.42	21297.44	21180.63	20688.94	20622.63	16860.85	17094.22	135.60
	Pr(III) + Phen	22416.50	22310.89	21244.50	21095.79	20601.57	20537.18	16770.64	17038.63	136.25

Contd...

Table 55.2–Contd...

Sl.No.	System	$^3H_4$ Eobs	$^3P_2$ Ecal	$^3H_4$ Eobs	$^3P_1$ Ecal	$^3H_4$ Eobs	$^3P_0$ Ecal	$^3H_4$ Eobs	$^1D_2$ Ecal	R.M.S.
6.	Solvent–MeOH + Dioxane									
	Pr(III)	22489.60	22382.28	21305.61	21169.60	20674.40	20610.72	16828.50	17087.11	138.86
	Pr(III) + Bpy	22492.13	22394.81	21296.99	21182.62	20688.09	20623.80	16857.72	17095.65	132.30
	Pr(III) + Phen	22454.25	22328.24	21257.60	21113.63	20619.41	20554.80	16773.74	17050.38	131.25
7.	Solvent–MeOH + DMF									
	Pr(III)	22451.22	22392.83	21275.69	21182.74	20703.50	20627.55	16877.64	17095.21	127.63
	Pr(III) + Bpy	22450.22	22389.51	21275.69	21179.66	20703.08	20625.08	16868.81	17093.10	131.62
	Pr(III) + Phen	22440.14	22383.29	21268345	21173.54	'20699.22	20619.48	16865.11	17089.00	131.08
8.	Solvent–DMF + ACN									
	Pr(III)	22434.60	22378.84	21258.96	21168.33	20689.80	20613.29	16870.52	17085.72	125.99
	Pr(III) + Bpy	22435.11	22376.61	21256.70	21166.33	20690.23	20611.81	16864.83	17084.34	128.38
	Pr(III) + Phen	22420.52	22357.27	21243.15	21145.37	20662.44	20589.30	16851.19	17070.80	129.56
9.	Solvent–DMF + Dioxane									
	Pr(III)	22454.75	22386.23	21277.05	21175.38	20692.37	20619.31	16865.11	17090.50	133.39
	Pr(III) + Bpy	22458.29	22392.45	21277.50	21181.05	20692.37	20623.69	16881.06	17064.40	132.07
	Pr(III) + Phen	22444.67	22373.13	21266.64	21160.62	20668.84	20602.57	16861.13	17081.10	13 1.47
10.	Solvent–Dioxane + ACN									
	Pr(III)	22520.49	22394.49	21293.81	21179.84	20668.42	20616.92	16859.71	17094.41	135.14
	Pr(III) + Bpy	22360.53	22278.50	21148.68	21075.32	20575.13	20483.72	16673.92	16993.63	133.08
	Pr(III) + Phen	22354.36	22222.56	21125.57	21006.31	20530.09	20451.14	16695.61	16979.41	132.03

The Nephelauxetic effect increases with the co-ordination number decreases. The Ln–N distance shortens in spite of the addition nature of β and decrease in the number of the co-ordinating ligand. The variation in the value of E^k (K = 2, 4, 6); corresponds to that in the value of F^k, since they are interrelated. Misra *et al.* (1979; 1981) observed a general decrease in the values of F_k, E^k and ξ_{4f} parameters as compared to the corresponding parameters of the free ion. The values of nephelauxetic effect (β) in all the system is less then unity and the values of bonding parameters ($b^{½}$) are positive which indicates covalent bonding. The comparative absorption spectra of pseudohypersensitive transitions of Pr(Ill) ion, Pr(III): bpy (in 1: 1 molar ratio) and Pr(III): phen (in 1 : 1 molar ratio) in different organic solvents are given in Figures 55.1a to 55.1d. The addition of bpy/phen in Pr(III) resulted in a marginal red shift of 4f-4f transition bands. The redshift is more in case of phen indicating a better ligand than bpy. The weak Ln-N interaction is enhanced by the presence of extra pi electrons as in the case of phen than in bpy, increasing the ligand character of the donor N-atom. The redshift is maximum in case when solvent is DMF or one of the solvent is DMF. Table 55.1 gives the computed values of Slator-Condon (F_k), Racah (E^k) and Lande (ξ_{4f}), Nephelauxetic parameter (β), bonding parameter ($b^{½}$) and covalency (δ) of Pr(III), Pr(III): bpy and Pr(III): phen in different organic solvents. Table 55.2 gives the computed and observed values of energies for the various transition bands. Further work on the evaluation of intensity parameters is going on.

References

Condon, S.U. and Shortley, G.H., 1963. *The Theory of Atomic Spectra.* University Press, Cambridge.

David Singh, Th., Singh, Mohondas N., Yaiphaba, N., Debecca Devi, H., Sumitra, Ch., Kriyananda, T., Indira Devi, M. and Rajmuhon Singh, N., 2003. *Chem. and Environ. Res.*, 12(3&4): 295.

David, Th., Sumitra, Ch., Bag, G.C., Indira Devi, M. and Rajmuhon Singh, N., 2006. *Spectrchimica Acta* (A), 63: 154.

David, Th., Sumitra, Ch., Yaiphaba, N., Debecca Devi, H., Indira Devi, M., Rajmuhon Singh, N., 2005. *Spectrchimica Acta,* (A), 61: 1219.

Debecca Devi, R., David Singh, Th., Yaiphaba, N., Sumitra, Ch., Indira Devi, M. and Rajmuhon Singh, N., 2004. *Asian Journal of Chemistry*, 16(1): 412.

Evan, C.H., 1990. *Biochemistry of Lanthanides.* Plenum Press, New York, Chapter 4.

Henrie, D.E., Fellow, R.L. and Choppin, G.R., 1976. *Coord. Chem. Rev.*, 18: 429.

Jorgensen, C.K. and Judd, B.R., 1964. *Molec. Phys.*, 8: 281.

Jorgensen, C.K. and Ryan, L., 1966. *J. Phys. Chem.*, 70: 2845.

Karrakar, D.G., 1967. *Inorg. Chem.*, 6: 1863.

Khan *et al.*, 2000. *Ind. Journal of Chemistry*, 39A: 1286.

Langhlin, R.D. and Conway, J.G., 1963. *J. Phys. Chem.*, 38: 1037.

Misra, S.N., Joshi, G.K. and Vaaishnav, P.P., 1979. *Indian J. Pure. App. Phys.*, 16: 553.

Misra, S.N., Megh Singh Joshi and G.K., 1981. *J. Inorg. Nucl. Chem.*, 43: 207.

Peacock, R.D., 1964. *Chem. Phys.*, 8: 281.

Shah, K.J. and Shah, M.K., 2001. *Bull. Pure Appl. Sci.*, C20(2): 81.

Sinha, S.P and Schmidtke, R.H., 1965. *Molec. Phys.*, 38: 2190.

Sinha, S.P. 1966. *Spectrochim. Acta.*, 22: 57.

Sinha, S.P., 1966. *Complexes of Rare Earths*. Pergamon Press, New York.

Sinha, S.P., Mehta, P.C. and Surana, S.S.L., 1972. *Molec. Phys.*, 23: 807.

Wong, E.Y., 1961, 1963. *J. Chem. Phys.*, 35: 544; 38: 976.

Chapter 56

Clinical Manifestation of Experimentally Induced Fish-Borne Trematode Infection in Experimental Animals

S.I. Shalaby[1], Olfat Anter[2], Nabila Hassan[2], M. El-Mahdy[2], Iman Mohamed[2] and Neelima Gupta[3]*

[1]Department of Complementary Medicine, National Research Center, Dokki, Cairo, Egypt
[2]Department of Parasitology, Faculty of Veterinary Medicine, Cairo University, Giza, Egypt
[3]Department of Animal Science, M.J.P. Rohilkhand University, Bareilly, India

ABSTRACT

Rats and dogs experimentally fed with viable encysted metacercariae from *Mugil cephalus* and *Tilapia nilotica* revealed the presence of seven species of trematodes. The pre-patent period was 3–10 days post-infection according to the type of trematode species. *Heterophyes hterophyes, H. heterophyes nocens, H. katsuradai, H. dispar* and *H. aequalis* were recovered from rats and dogs while *Haplorchis pumilio* and *Prohemistomum vivax* were only obtained from dogs. The number of recovered worms in rats was less as compared to dogs indicating that the former is a less favored host for the investigated trematodes. The experimentally infested animals showed signs of diarrhoea. The structural and functional mechanisms of diarrhoea induced by these parasites are discussed.

Keywords: *Metacercaria, Trematode, Mugil, Tilapia, Diarrhoea.*

* Corresponding Author; E-mail: guptagrawal@rediffmail.com.

Introduction

Fish is considered to be one of the important sources of parasitic infestations to man and fish eating mammals particularly after increased pollution of rivers and lakes. Ooi *et al.* (1999) reported mass infection of heterophyid metacercariae in cultured Japanese eels, *Anguilla japonica* in Taiwan. However, not all final host mammals might be favorable for parasitism with trematodes that are transmitted through eating fish infested with encysted metacercariae of those parasites. Boulos (1979) reported *Pygidiopsis genata* from dogs experimentally fed with *Tilapia* sp. from lake Edku, Maryut and Manzala. Ahmed (1980) found *Heterophyes heterophyes. Haplorchis yokogawai* and *Prohemistomum vivax* in stray dogs from Mansoura, Egypt. Riffat *et al.* (1980) observed *H. heterophyes.* in puppies fed on *Mugil* sp. and *Tilapia* sp. El-Mokaddem (1982) reported *H. heterophyes* from cats in Sallum and Fayoum, Egypt; Shalaby (1982) described 11 species of trematodes from experimentally infected dogs with encysted metacercariae in *Bagrus bayad, Syndontis schall* and *Tilapia nilotica.* Fahmy *et al.* (1984) examined 45 stray dogs, 29 stray cats and 673 different rodents and recorded haplorchid, prohemistomatid and cynodiplomatid trematodes from them. Noshy *et al.* (1987) collected *Pygidiopsis genata* and *Haplorchis pumilio* from experimentally infected hamsters, mice, rats, cats and puppies with *Tilapia* fishes having encysted metacercariae. El-Reid (1994) infested puppies, albino rats, chickens, ducks and pigeons by feeding them fish muscles and recorded *Heterophyes, Haplorchis, Prohemistomum* and *Mesostephanus* spp. Shalaby and Easa (1994) observed *Phagicola ornamentata* and *Pharyngostomum flapi* from dogs in Cairo. Moreover, Mahdy *et al.* (1995) fed kitten and puppies with infected tilapia fish in Manzala lake and obtained *Heterophyes, Ascocotyle, Haplorchis* and *Pygidiopsis* spp. Shalaby *et al.* (2003a) infected puppies and albino rats with heterophyid encysted metacercariae and could find extra-intestinal lesions due to the parasites. Also, Shalaby *et al.* (2003b) could trace the effect of parasitization by certain heterophyid species in dogs and rats.

Byong *et al.* (1980) reported that clinically. Echinostomes caused intermittent abdominal pain and nausea in a 60 years old woman. Byong *et al.* (1981) stated that *Heterophyes* spp caused eosinophilia in 7 per cent of human cases passing eggs in their stools. Line (1981) reported pain, gurgling and loose stools in a 40 years old farmer who ate improperly cooked fish. Byong *et al.* (1984) reported that the clinical picture in man as a result of heterophyid infection was represented by non-specific gastrointestinal symptoms such as epigastric pain, easy fatigability, weakness, indigestion and cardiac arrhythmia. Byong *et al.* (1985) reported that the major subjective symptoms in human volunteers and in experimental rats were diarrhoea and abdominal pain. Ana *et al.* (1991) recorded abdominal liver function tests, eosinophilia and obstructive jaundice complicated by sepsis and transhepatic cholangeogram demonstrated diffuse irregular dilatation of biliary tree with abrupt ending of terminal branches. Several small fusiform defects were noticed within small ducts. *Opisthorchis* eggs were found during aspiration of bile of this patient. Metaplasia of goblet cells and adenomatous hyperplasia of bile ducts epithelium with time followed by periductal fibrosis and thickening of ductal wall with irregular structure of its lumen were also reported. Jong *et al.* (1993) observed that *Metagonimus* spp. caused diarrhoea in infested persons. In Egypt, Shalaby (2002) stated that heterophyiasis constitutes a public health problem where people eat raw, inadequately salted or otherwise improperly prepared infested fish. This may result in chronic intermittent mucous diarrhoea which may occasionally be bloody, alternating with constipation. Moreover, Shalaby *et al.* (2002) reported that patients mostly had indigestion, abdominal pain or heart problems in the form of arrhythmia and.ventricular premature beats. Sometimes, obstructive jaundice, complicated by sepsis and disseminating intravascular coagulopathy-like syndrome was evident.

This investigation reports the suitability of the final host (mammal) for trematodes transmitted through fish, the effect of the numbers of parasites on the final host as well as the clinical picture in the final hosts.

Materials and Methods

Twenty, albino rats (110–150 gm B.W.) and 12 puppies (4 weeks old) were used for experimental infection. Prophylactic dose of Praziquantel (50 mg/10 kg B.W.) was given once a week before experimental infection. Each animal was given 150 encysted metacercariae collected from *Mugil cephalus* and *Tilapia* spp. Faecal examination was done daily to determine the pre-patent period. The clinical picture was reported and samples were taken. Finally animals were sacrificed and their intestines were examined.

Results

Experimental feeding of rats and dogs with viable encysted metacercariae revealed the presence of seven species of trematodes. Eggs of trematodes appeared after 3–10 days post infection according to the type of trematode species. Five species of digenetic trematodes were collected from experimentally infested rats and dogs, namely: *Heterophyes heterophyes, H. heterophyes nocens, H. katsuradai, H. dispar* and *H. aequalis*. However, *Haplorchis pumilio* and *Prohemistomum vivax* were obtained only from dogs. The characteristic features of the trematodes were studied.

It was noticed that the number of recovered worms in albino rats were less as compared to those obtained from dogs.

It was obvious from the clinical observation of the infested animals that they suffered from diarrhoea, passed abnormal fluidy faeces which was usually accompanied by increased volume and frequency of defecation. The animals were nervous and their blood films showed eosinophilia.

Discussion

Day to day, new records are coming up proving that fish is a source for serious parasitic infectious diseases for man, animals and birds. Now, it is believed that some trematodes other than *Heterophyes heterophyes* can be transmitted to man and animals through fishes.

In the present investigation, five species of heterophyid trematodes were recovered when rats and dogs were experimentally infected namely, *Heterophyes heterophyes, H. heterophyes nocens, H. katsuradai, H. aequalis* and *H. heterphyes dispar*. Moreover, *Haplorchis pumilio* and *Prohemistomum vivax* were obtained from the small intestine of dogs only.

The pre-patent period ranged between 3 to 10 days post-infection according to the type of trematode species. Nearly similar results were obtained by Dennis and Penner (1971) (4–6 days), Galaktionov (1980) (7–8 days), Pike (1980) (3 days) and Shalaby (1985), (4–5 days) post infection. The morphological characters of the trematodes in this study were identical with those of the original description given by Siebold (1853), Sonsino (1892), Looss (1902), Onji and Nishio (1915), Ozaki and Asada (1926), Witenberg (1929), Chen (1936), Fayek *el al.* (1986) and El-Reid (1994).

Heterophyes heterophyes was found more commonly in human beings in Egypt and in the middle east (Taraschewski, 1984) and later it was found in the Far East such as Japan and Korea (Yokogawa *et al.*, 1965; Chai *et al.*, 1985). *H. heterophyes* was detected from an Egyptian child (Fayek *et al.*, 1986). This species was reported repeatedly from man, birds and carnivorous mammals such as dogs, cats and foxes (Harford and Arlen, 1994).

Heterophyes heterophyes nocens is an intestinal trematode of dogs, cats and human beings of eastern countries like Japan and Korea (Chai and Lee, 1990), the parasite was considered as a rare species in Egypt as it was detected only by Abdel-Maksoud (1992) and El Reid (1994). However, this parasite may become endemic in Egypt, *Heterophyes dispar* was first found in dogs and cats from Egypt (Looss, 1902; Mahdy *et al.*, 1995). Also, it was reported from a variety of carnivorous mammals including foxes and wolves (Witenberg, 1929; Taraschewski, 1984). The parasite was found in 7 men from Korea (Eom *et al.*, 1985).

Heterophyes aequalis was recovered from the small intestine of house rats, domestic dogs, domestic and wild cats as well as jackal (Wells and Randell, 1956). Harford and Arlen (1994) mentioned that *H. heterophyes, H. dispar* and *H. aequalis* occur in man in the Mediterranean, Far East and Egypt, infestation with *H. pumilio* in dogs was recorded by Azim (1938), El Mokaddem (1982) and Shalaby (1982). This parasite has been shown to be of public health importance (Shalaby, 1982; Mahmoud, 1983).

Prohemistomum vivax infestation in dogs was reported by Azim (1938) and Ahmed (1980). In man, it was reported by Nasr (1941).

H. pumili and *P. vivax* were recovered from dogs only during the present study. This finding is in conformity with the results of Fahmy *et al.* (1976), Shalaby (1982) and Mahmoud (1983).

It was further noticed that the number of observed worms in albino rats were less as compared with those obtained from dogs. This may be attributed to the severe tissue reaction against the parasite in rats rendering them as less favorable hosts for these flukes and could be explained due to the difference in physiological status and acidity of the stomach of both hosts as well as other anatomical and physiological factors. These results are in conformity with those reported by Jong (1979), Shalaby *et al.* (1989) and Sung *et al.* (1990).

These flukes are responsible for causing gastrointestinal disturbances such as abdominal pain, gastric discomfort and/or diarrhoea in human hosts (Seo, 1978; Jong *et al.* (1986) and are thus of medical importance. However, certain kinds of heterophyids were reported to produce erratic parasitism in many other organs. In the present investigation, it was obvious from the clinical observation of infested animals that they suffered from diarrhoea with abnormal faeces, fluid in nature which was usually accompanied by Increased volume and increased frequency of defecation. This is a common sign occurring in many enteric malfunctions. The available data were insufficient bases for speculation on the structure and function of the intestine. Therefore, the diarrhoea produced by these helminths may be due to the mechanical and chemical stimulation of the epithelial layers of villi and/or crypts (secretory diarrhoea). A second factor in the pathogenesis of diarrhoea may be the decreased intestinal assimilation of nutrients (malabsorption diarrhoea). Malabsorption occurred mostly as a result of the reduced villous and microvillous surface (villous atrophy). The third mechanism for production of diarrhoea is the increased permeability of the mucosa which may contribute to diarrhoea by permitting increased retrograde movement of solute and fluid from the lateral intercellular space to the lumen or by facilitating transdaction of tissue fluid (filtration secretion). Effusion occurred in inflamed lamina propria due to increased vascular permeability and proprial oedema. Increased exfoliation of epithelium and erosions provide further potential sites of interstitial fluid effusion (effusive diarrhoea). These results were in agreement with Cheville (1983), Stuart and Bernard (1989) and Robert (1993).

References

Abdel-Maksoud, S., 1992. Zoonotic agents in marine fish marketed in Dumyat. *M.Sc. (Zoonoses)*, Fac. Vet. Med., Zagazig Univ.

Ahmed, F., 1980. Intestinal helminthes among stray dogs in Mansoura City, Egypt. *J. Egypt. Soc. Parasitol.*, 10: 289–294.

Ana, H., Sydney, F. and Benton, R., 1991. A case of Opisthorchiasis diagnosed by cholangiography and bile examination. *Amer. Surg. J.*, 57: 206–209.

Azim, M., 1938. Intestinal helminthes of dogs in Egypt. *J. Egypt. Med Assoc.*, 21: 105.

Boulos, L., 1979. Parasitological studies on the heterophyids in *Tilapia* fish from Edku and Maryut lakes and their transmission to laboratory animals. *Ph.D. Thesis (Parasitology)*, Fac. Med., Alex. University.

Byong, S., Sung, Y. and Jong, Y., 1980. Studies on intestinal trematodes in Korea: A human case of *Echinostoma cinetorchis* infection with an epidemiological investigation. *Seoul J. Med.*, 21: 21–29.

Byong, S., Sung, Y. and Jong, Y., 1981. Studies on intestinal trematodes in Korea. III. Natural human infections of *Pygidiopsis summa* and *H. h. nocens*. *Seoul J. Med.*, 22: 228–235.

Byong, S., Soon, H., Sung, Y. and Jong, Y., 1984. Studies on intestinal trematodes in Korea. III. Two cases of natural human infections by *H. continua* and the status of metacercarial infection in brackish water fishes. *Korean J. Parasitol.*, 22: 51–60.

Byong, S., Kwang, S., Soon, H., Sung, Y. and Jong, Y., 1985. Studies on intestinal trematodes in Korea. XVII. Development and egg laying capacity of *Echinstoma hortense* in albino rats and human experimental infection. *Korean J. Parasitol.*, 23: 24–32.

Chai, J., Hong, S., Sohn, W., Lee, S. and Seo, B., 1985. Further case of human *Heterophyes heterophyes nocens* infection in Korea. *Seoul J. Med.*, 26: 197–200.

Chai, J. and Lee, S., 1990. Intestinal trematodes of human in Korea: *Metagonimus. Heterophyes* and echinostomes. *Korean J. Parasitol.*, 28: 103–122.

Chen, H., 1936. A study of *Haplorchidae* (Loos, 1899). Poche (1926) (Trematoda : Heterophyidae). *Parasitology*, 28: 40–55.

Cheville, N., 1983. *Cell Pathology*, 2nd edn. Ames, Iwa, USA, 223 pp.

Dennis, E. and Penner, L., 1971. *Mesostephanus yedeae* sp.n, its life history and descriptions of the developmental stages. Occasional Papers, Univ. Connecticut, 2: 5–15.

El-Mokaddem, A., 1982. The prevalence of heterophyids in man and animals and various fish hosts in Edku area. *M.Sc. Thesis (Zoonoses)*, Public Health Institute, Alex University.

El Reid, A., 1994. Role of marine fish in transmission of some parasites to animals and birds. *Ph.D. Thesis (Parasitology)*, Fac. Vet. Med., Zagazig Univ.

Eom, K., Yoo, H., Chung, M., Lee, J. and Rim, H., 1985. Heterophyid trematodes human infection imported from Sudan to Korea. *Korean J. Parasitol.*, 23: 360–431.

Fahmy, M., Mandour, A. and El-Naffar, M., 1976. Successful infection of dogs and cats by *P. vivax* and *H. yokogawai*. *J. Egypt. Soc. Parasitol.*, 6: 77–82.

Fahmy, M., Arafa, M., Abdel Rahman, A. and Mounib, M., 1984. Studies on helminth parasites in small mammals in Assiut. 1-Trematode parasites. *Assiut Vet. Med. J.*, 11: 43–50.

Fayek, S., Nada, M. and Ahmed, B., 1986. Heterophyid parasites recovered from dogs with special reference to *Phagicola longus*. *Vet. Med. J.*, 34: 201–210.

Galaktionov, K., 1980. The life cycle of *M. appendiculatus* (Ciurea, 1916). Vestnik Leningradskego-Univ., Biologiya, 21: 27–34.

Harford, W. and Arlen, J., 1994. *Parasitic Worms of Fish*, 1st edn. Taylor and Francis, London, New York and Philadelphia.

Jong, Y., 1979. Study of *Metagonimus yokogawaj* in Korea. V. Intestinal pathology in experimentally infected albino rats. *Seoul J. Med.*, 20: 104–112.

Jong, Y., Byog, S. and Soon, H., 1986. Human infection by *H. heterophyes, H. dijpar* imported from Saudi Arabia. *Korean J. Parasitol.*, 24: 82–88.

Jong, Y., Sun, H., Jae, R., Jina, K. and Soon, H., 1993. An epidemiological study of *Metagonimus* along the upper reaches of Namhan river. *Korean J. Parasitol.*, 31: 99–108.

Line, O., 1981. *Echinochasmus perfoliatus* in man. A case report. *Natural Med. J. China*, 61: 249–252.

Looss, A., 1902. Notizen Zur Helminthologie Egyptens. V. Eine Revesion der Fasciolden-gattung Heterophyes. Cobb. Centralab. Bakt. Orig., 32: 886–891.

Mahdy, O., Manal, A., Essa, A. and Eaisa, M.El.S., 1995. Parasitological and pathological studies on heterophyid infections in *Tilapia* spp. from Manzala lake, Egypt. *Egypt. J. Comp. Pathol. and Clin. Pathol.*, 8: 131–145.

Mahmoud, N., 1983. Parasitic infestations of some native species of fishes in Cairo markets with special reference to parasites transmissible to man and animals. *M.V. Sc. (Hygiene and Food Control)*, Fac. Vet. Med., Cairo University, Cairo, Egypt.

Nasr, M., 1941. The occurrence of *Prohemistomum vivax* infection in man with description of the parasite. *Lab. Med. Prog.*, 2: 135–149.

Noshy, S., Magda, M., Helen, N., Nibal, A. and Laila, M., 1987. *Heterophyes metacercariae* in Maryut and Manzala lakes in Egypt. *J. Egypt. Soc. Parasitol.*, 17: 481–493.

Onji, Y. and Nishio, T., 1915. Studies on the metacercaria in *Mugil cephalus*. *Tokyo Tji Shinshi*, p. 2390–2395.

Ooi, H., Wang, W., Tu, C., Chang, H. and Chen, C., 1999. Natural mass infection by heterophyid metacercariae in aquacultured Japanese eel in Taiwan. *Dis. Aquat. Org.*, 35: 31–36.

Ozaki, Y. and Asada, J., 1926. A new human trematode *H. katsuradai*. *J. Parasitol.*, 12: 216–218.

Pike, A., 1980. Observations on the life cycle of *M. milvi* Yamaguti, 1939. *J. Nat. Hist.*, 14: 389–399.

Riffat, M., Salem, S., El-Kholy, S., Hegazi, M. and Youssef, M., 1980. Studies on the incidence of *Heterophyes heterophyes* in Dakahlia. *J. Egypt. Soc. Parasitol.*, 10: 369–373.

Robert, M., 1993. Diarrhoea in helminthic infections. *Clin. Inf. Dis.*, 16: 2120–2122.

Seo, B., 1978. *Clinical Parasitology*. Revised Ed. II Cho-kah Seoul, pp. 111–113.

Shalaby, S.I., 1982. Studies on the role played by some Nile fishes in transmitting trematodes. *M.Sc. Thesis (Parasitology)*, Cairo University, Cairo, Egypt.

Shalaby, S.I., 1985. Further studies on the role-played by catfish in transmitting some trematodes to fish eating mammals, with special reference to the morphobiology of *M. appendiculatous*. *Ph.D. Thesis (Parasitology)*, Cairo University, Cairo, Egypt.

Shalaby, S.I., 2002. Clinical picture of diseases transmitted to man through fish. 1st Annual International Congress of the Medical Division, N.R.C., Egypt, pp. 68.

Shalaby, S.I., Ibrahim, M., Mahmoud, N. and El Assaly, T., 1989. Parasitological and pathological studies on encysted metacercariae in the musculature and different organs of *T. nilotica. J. Comp. Pathol. and Clin. Pathol.*, 2: 186–212.

Shalaby, S.I. and Easa, M.El.S., 1994. Trematode parasites transmitted to man and fish-eating mammals through *Tilapia nilotica.* II. New trematode species. In: *Proc. 18th World Bur Cong.*, Bologna, Italy.

Shalaby, S.I., Mohamed Imam and Gupta Neelima, 2002. Clinical and parasitological studies on human fish borne trematodes with emphasis on intestinal histoenzymatic changes in Egypt: Comparative records from India. *Indian J. Environ. and Ecoplan.*, 6: 409–414.

Shalaby, S.I., Nabiha Hassan, El-Mahdy, M., Olfat Anter, Eman Mohamed and Gupta, Neelima, 2003a. Parasitological and pathological studies on extra-intestinal heterophyiasis. *J. Par. Dis.*, 27: 39–46.

Shalaby, S.I., Nabiha Hassan, Raafat, A. Olfat Anter, El-Mahdy, M. Iman Mohamed and Gupta, Neelima, 2003b. Effect of parasitization by certain heterophyid species in rats and dogs in Egypt. *Biol. Memoirs*, 29: 21–28.

Siebold, C., 1853. Uber Leucochoridium paradoxum. *Zeitschr. Wissensch. Zool.*, 5: 425–436.

Sonsino, P., 1892. Studi sui parasiti di molluschi di acqua dolle nei dinotorni di Cairo in Egitto. Festochr.Z. Gebutstag R. Leue Karts Leipsing, pp. 134–146.

Stuart, K. and Bernard, F., 1989. Pathological effects of *Echinostoma caproni* in domestic chicks. *J. Helminth.*, 63: 227–230.

Sung, T., Sun, H., Woon, G. and Soon, H., 1990. Changes in histopathological and serological findings of the liver after treatment in rabbit clonorchiasis. *Seoul J. Med.*, 31: 117–127.

Taraschewski, H., 1984. Die trematoden der gattung Heterophyes taxonomie, biologie, epidemiologie. *Ph.D. Dissertation*, Universitat Hohenheim.

Wells, W. and Randell, B., 1956. New hosts of trematodes of the genus *Heterophyes* in Egypt. *J. Parasitol.*, 42: 286–292

Witenberg, G., 1929. Studies on trematodes: Family Heterophyidae. *Annals Top. Med. and Parasitol.*, pp. 231–239.

Yokogawa, M., Sano, M., Itabashi, T. and Kachi, S., 1965. Studies on the intestinal flukes. II. Epidemiological studies on heterophyid trematodes of man in Chiba Prefecture. *Japan J. Parasitol.*, 14: 577–585.

Chapter 57

On a New Species of the Genus *Gryllotalpa* Latreille (Orthoptera : Gryllotalpidae) from India

R. Manisha Awate and T.V. Sathe

Department of Zoology, Shivaji University, Kolhapur – 416 004

ABSTRACT

A new species of Mole cricket, *Gryllotalpa kolhapurensis* sp. nov. (Orthoptera : Gryllotalpidae) has been described for the first time from India. The species runs close to *G. ornata* Walker by having following characters:

1. Size small (20–30 mm).
2. Colour dark brown.
3. Ocelli very small.

However, it differs from the above species by having following features:

1. Mirror elongately narrow.
2. Flagellar formula.
3. Hind tibia armed on the internal margin with 6 long and slender spines.
4. Fore tibial structure.

Keywords: *Mole cricket, Description, Gryllotalpa kolhapurensis sp. Nov, India.*

Introduction

Order Orthoptera of class insecta is very important from the view point of economic importance of insects. Grasshoppers, locusts, crickets and mole crickets are important groups of insect which are included under this order. Mole crickets have special importance since they are associated with human habitation and they contaminate our food and secondly, they produce monotonous noise, hence they are nuisance to humans. Therefore, present work was aimed to study the taxonomy of mole crickets.

The order Orthoptera comprises twelve families. Out of which family Gryllotalpidae contain mole crickets. The Gryllotalpidae is known by their trivial name as mole Crickets. They have elongate body and small and conical-head with two oceli. Pronotum shape is convex and ovipositor is completely aborted and a typical modification in fore tibia. The family contain two genera *viz. Scapteriscus* Scudder and *Gryllotalpa* Latr.

The genus *Gryllotalpa* was erected by Latreille in 1802 with the help of type species *Gryllus* (*Acheta gryllotalpa*) L. The genus is distributed in Europe, Asia, Africa, Australia and Newzeland. From this genus five species have been reported from India (Chopard, 1969). In past Chopard (1928, 1933, 1935, 1951, 1969), Pantel (1886), Sandrasagara (1954), etc. attempted taxonomical studies on Crickets.

Materials and Methods

Mole crickets have been collected from human habitations, specially hostels, hotels and in and around human dwellings by visiting frequently, 15 days interval. The mole crickets were collected with the help of insect net and killed in cyanide bottle or chloroform. The killed insects were pinned through mesothorax and kept for drying at 50°C in drying chamber for 2 to 4 days. The dried specimen were taken for taxonomical studies. Taxonomical studies were made on the various body parts, *i.e.* head, thorax and abdomen and their appendages. Slides of wings, antenna and cerci have been prepared for studies. The specimens are time beirg with T.V. Sathe, Zoology Department, Shivaji University, Kolhapur and will be deposited to ZSI Kolkata.

Results [Figure 57.1: (Male) *Gryllotalpa kolhapurensis* sp. nov.]

Male

30.00 mm long excluding anal appendages, head 5.00 mm long; thorax 14.00 mm long; elytra 14.00 mm long; hind wing 25.00 mm long; abdomen 11.00 mm long, yellowish brown; anal cerci 8.00 mm long, dark brown.

Head

5.00 mm long, 4.00 mm broad, triangular, occiput black, hairy, compound eyes dark black; ocular distance 0.72 mm ; interocular distance 1.83 mm ; 2 ocelli whitish brown; pronotum with regularly convex side, broad posteriorly.

Antenna

8.00 mm long, 0.16 mm broad, dark brown, black margin on antenal segment, hairs present on antenal segment; scape 0.50 mm long, 0.38 mm broad, yellowish brown; pedicel 0.29 mm long, 0.25 mm broad; flagellum 92 segmented, 1st segment 0.16 mm long, 0.19 mm broad, 15th segment 0.09 mm long, 0.16 mm broad, terminal segment 0.08 mm long, 0.09 mm broad.

Figure 57.1: (Male) *Gryllotalpa kolhapurensis* sp. nov

Flagellar Formula

I = L/W = 0.84, 15 = L/W = 0.56, Ts = L/W = 0.88, A= 0.76.

Thorax

14.00 mm long, 7.00 mm broad, metallic brownish golden, hairy.

Elytra

Long narrow, extended to the 3rd abdominal tergite; mirror elongately narrow, discoidal cell twice as long as wide; 13.00 mm long, 8.00 mm broad, dark brown.

Hind Wing

25.00 mm long, 12.00 mm broad, brown.

Legs

Legs lighter than body colour.

Fore Leg

16.58 mm long, 3.00 mm broad; coxa 4.00 mm long, 3.00 mm broad, yellowish brown; trochanter, reddish brown; 3.00 mm long, 3.00 mm broad; spine present at tip of trochanter, densely hairy, yellowish brown; femur stroughtly builded, reddish brown, densely hairy at inner margin, 6.00 mm long, 3.00 mm broad; tibia reddish brown; densely hairy at outer margin; 3.00 mm long, 2.00 mm broad; tibia modified into four tibial dactyls; dactyls short and wide, 1st tibial dactyl 0.62 mm long, 0.52 mm broad; 2nd tibial dactyl 1.00 mm long, 0.41 mm broad; 3rd tibial dactyl 0.75 mm long, 0.45 mm broad; 4th tibial dactyl 1.28 mm long, 0.41 mm broad, fourth tibial dactyl bifurcated, black at tip and reddish brown in colour, densely hairy; tarsus 0.58 mm long, 0.35 mm broad.

Mid Leg

9.00 mm long, 2.00 mm broad, coxa 2.00 mm long, 2.00 mm broad, yellowish brown; trochanter 2.00 mm long, 2.00 mm broad, yellowish brown, hairy; femur 6.00 mm long, 2.00 mm broad, dark brown at outer margin, sparsely hairy; tibia 5.00 mm long, 1.00 mm broad, 3 reddish brown spines present at base of tibia; tarsus 2 segmented, 4.00 mm long, 0.3 mm broad; 1st segment 3.00 mm long, 0.3 mm broad; 2nd segment 1.00 mm long, 0.3 mm broad, yellowish brown sparsely hairy; claws reddish brown.

Hind Leg

26.00 mm long, 2.00 mm broad; coxa 3.00 mm long, 2.00 mm broad, yellowish brown, hairy; trochanter 2.00 mm long, 2.00 mm broad, yellowish brown, sparsely hairy; femur 9.00 mm long, 3.00 mm broad, reddish brown tinge at outer margin, sparsely hairy; tibia 7.00 mm long, 1.00 mm broad, reddish brown tinge at margin, densely hairy, 7 tibial dark reddish brown spines present; tarsus 5.00 mm long, 0.5 mm broad, 2 segmented 1st segment 3.00 mm long, 0.5 mm broad, 2nd segment 2.00 mm long, 0.5 mm broad, yellowish brown; claws reddish brown.

Abdomen

11.00 mm long, 7.00 mm broad, tapering to posterior side, dark brown, 10 segmented; anal cerci 8.00 mm long, 0.5mm broad, yellowish brown, hairs present.

Colour

Yellowish Brown

Fore leg coxa, mid leg coxa, torchanter, tarsus, hind leg coxa, trochanter, tarsus.

Dark Brown

Fore wing, hind wing, mid leg femur, abdomen, anal cerci, antenna.

Dark Black

Compound eye, tibial dactyls, head, black margin on antennal segment.

Reddish Brown

Fore leg trochanter, femur, tibia, tibial dactyls, tibial spines of mid leg, tibia, femur margin of hind leg, tibial margin of hind leg, tibia of hind leg, fore leg tarsus, claws of mid leg, hind leg.

Holotype

Male, India, Maharashtra, Coll. 4-X-2006; Awate M.; Antenna, fore wing, hind wing, cerci on slide labelled as above; other parts with pined form, labelled as above.

Paratype

8 males, 0 females; Coll. From 10-X-2006 to 22-IX-2007. Same data as above.

Discussion

Gryllotalpa kolhapurensis runs close to *G. ornata* Walker by having following characters:

1. Size small (20–30 mm)
2. Colour dark brown
3. Ocelli very small

However, it differs from the above species by having following features:

1. Mirror elongately narrow.
2. Flagellar formular.
3. Hind tibia anned on the internal margin with 6 long and slender spines.
4. Fore tibial structure.

Acknowledgement

Authors are grateful to Shivaji University Kolhapur for providing facilities.

References

Chopard, L., 1928. Revision of the Indian Gryllidae. *Rec. Ind. Mus.*, 30: 1–36.

Chopard, L., 1933. New species of Indian Gryllidae (Orth.). *Stylops*, 11: 115–120.

Chopard, L., 1935. On a collection of Gryllidae form India made by Mr. B.M. Fletcher. *Ann. Mag. Nat. His.*, 16(10): 284–296.

Chopard, L., 1951. A revision of the Australian Grylloidea. *Rec. S. Austl. Mus., Adelatide*, 9: 397–533, 89 figs.

Chopard, L., 1969. *The Fauna of India and Adjacent Countries, Orthoptera, Vol. 2.* BMP Calcutta, pp. 1–421.

Pantel, J., 1886. Notes: *Orthopterolgiques. Ann. Soc. Esp. Hist. Nat.*, 25: 47–118.

Sandrasagara, T.R., 1954. Checklist of the Tridactylidae an Gryllidae (Insecta : Orthoptera) of Ceylon, with records of distribution. *J. Bombay Nat. Hist. Soc.*, 52: 540–562.

Chapter 58

Alterations in Kidney Transaminases After Combined Exposure to Sulphur Dioxide and Nitrogen Dioxide in Albino Rat

Asha Agarwal and Shaista Parveen

Department of Zoology, School of Life Sciences, Dr. B.R. Ambedkar University, Khandari Campus, Agra – 282 002, U.P.

ABSTRACT

The present study was undertaken in order to assess the impact of (20 + 20) ppm and (40 + 40) ppm combined exposure to sulphur dioxide and nitrogen dioxide gas on kidney transaminases in albino rat for 15 and 30 days for one hour per day. The results show that a highly significant increase in kidney glutamic oxaloacetic transaminase and glutamic pyruvic transaminase activities after (20 + 20) ppm and (40 + 40) ppm combined exposure to sulphur dioxide and nitrogen dioxide gas.

Keywords: *SO_2, NO_2, GOT, GPT, Albino rat, Kidney.*

Introduction

Air pollution evolved as a problem of regional or global dimension. The oxides of sulphur and nitrogen are major threat to social, economic development and even to man's survival. Sulphur dioxide and nitrogen dioxide are the highly toxic pollutants and their chief sources are automobile emission, power plants, combustion of fuel or coal etc. These toxic gases when absorbed leading various disorders in respiratory system and most noticeable biochemical changes observed in tissue. Health effects of air pollution has been suggested by Bernstein *et al.* (2004). Kidney is one of the complex organ of the five major organs of the body and it is finally regarded as a common target of toxic chemicals. Transaminases are important for diagnostic and prognostic purpose. The work of Glutamic Oxaloacetic Transaminase (GOT) and Glutamic Pyruvic Transaminase (GPT) in different tissues of rats and guinea pigs exposed to varying levels of altitude stress for different periods of time observed by Mukherjee and Ghosh (1987). There is no much data concerning the effect of air pollution on the activities of kidney enzymes.

Therefore, the object of present investigation is to observe the adverse consequences of combined sulphur dioxide and nitrogen dioxide on kidney enzymes of albino rat, *Rattus norvegicus* (Berkenhout).

Materials and Methods

Thirty adult and healthy albino rats of almost equal size and weight ranges from 90–120 gms were selected for the present study. Rats were bred at the animal house of Zoology Department. They were kept in polypropylene cages, provided with Gold Mohar Brand rat and mice feed, Hindustan Lever Ltd. India and, water *ad libitum*. Rats were acclimated for one month prior to experiment.

Nitrogen dioxide gas was generated in generator according to the method described by Saltzman (1954) and modified by Levaggi *et al.* (1972). The sulphur dioxide gas was generated in the generator according to the method described by Singh and Rao (1979).

Three sets one control (A) and two experimental sets (B and C) were taken of ten rats each. Experimental rats were exposed to (20 + 20) ppm and (40 + 40) ppm combined sulphur dioxide and nitrogen dioxide gas exposure for one hour per day, while control rats to ambient air in fumigation chamber for 15 and 30 days. Five rats from each set 'A', 'B' and 'C' were sacrificed after 15 days exposure and remaining five rats of each set after 30 days exposure to combined gas for kidney tissue collection.

Kidney glutamic oxaloacetic transaminase and glutamic pyruvic transaminase were determined by the method described by Reitman and Frankel (1957). The readings so obtained were subjected to the formulae for different statistical calculations (Fisher and Yates, 1950).

Results and Discussion

The results obtained for kidney glutamic oxaloacetic transaminase (GOT) and Glutamic Pyruvic Transaminase (GPT) after combined exposure to sulphur dioxide and nitrogen dioxide are shown in Table 58.1.

An elevation in the activities of kidney transaminases is the indication of impaired kidney function due to inflammatory action of the combined gas. An inflammatory process characterized by increased membrane permeability at the loss of soluble enzyme through the membrane might act as a stimulus to increase enzyme synthesis (Chatterjee and Shinde, 1994). In accordance to present findings, Rahman *et al.* (2001) have reported an elevation in kidney transaminases activity due to leakage of enzymes into blood following reservoir tissue damage or dysfunction. Mukherjee and Ghosh (1987) have also observed that both glutamic oxaloacctic transaminase and glutamic pyruvic transaminase activities were significantly increased in different tissues including kidney of both rats and guinea pigs during short period of exposure. A rise in the level of tissue glutamic pyruvic transaminase and glutamic oxaloacetic transaminase activities have also been observed by Whitehead *et at.* (1996) in men with increasing consumption of cigarette smoking.

Table 58.1: Values of Kidney GOT (Units/ml) and GPT (Units/ml) After Combined Exposure to Sulphur Dioxide and Nitrogen Dioxide Gas in Albino Rat

Parameters	*Exposure Days*	*Control Set (5)*	*Experimental Set (5)*	
			Concentrations of Combined Gas (Sulphur Dioxide + Nitrogen Dioxide	
			20 ppm + 20 ppm	*40 ppm + 40 ppm*
		Range Mean±S.Em	*Range Mean±S.Em*	*Range Mean±S.Em*
	15	50.00–54.50	53.80–56.10	57.70–60.20
Kidney GOT		53.00±0.79	55.24±0.42*†	60.062±0.87**†
(units/ml)	30	52.10–55.10	56.80–60.00	60.10–63.00
		53.94±0.57	58.24±0.57***†	61.72±0.47***†
	15	64.30–68.50	68.20–78.10	70.00–89.40
Kidney GPT		66.66±0.74	73.98±2.21*†	78.80±3.13**†
(units/ml)	30	66.50–69.00	69.00–72.00	79.00–83.40
		67.80±0.46	70.16±0.54**†	81.02±0.80***†

PPM: Parts per million; S.Em: Standard error of mean; 5: No. of rats; †: Increase; *: Significant; **: Highly significant; ***: Very highly significant.

Present findings are also supported by Gupta (1998) who reported an increase in the level of SGOT and SGPT activities in albino rats after combined exposure to sulphur dioxide and nitric oxide. Further, Mederios *et al.* (1983) in humans, Shimizu *et al.* (1986) in rats and Bree *et al.* (1992) in rats and guinea pigs have also reported an increase in the level of SGOT and SGPT activities after exposure to air pollutants.

In the present study, a very highly significant elevation in the level of kidney transaminases with increase in concentration and exposure time indicates that the tissue injury became more pronounced due to inflammatory response of toxic gases.

Present study reveals that combined action of sulphur dioxide and nitrogen dioxide gas could affect the renal tissues and impaired kidney functions.

References

Bernstein, J.A., Alexis, N., Barnes, C., Bernstein, I.L., Bernstein, J.A., Nel, A., Peden, D., Diaz-Sancher, D., Tarlo, S.M., Williams, P.B., 2004. Health effects of air pollution. *J. Allergy Clin. Immunol.*, 114: 1116–1123.

Bree, L.V., Marra, M. and Peter, J.A., 1992. Differences in pulmonary biochemical and inflammatory responses of rats and guinea pigs resulting from day time and night time, single and repeated exposure to ozone. *Toxicol. Appl. Pharmacol.*, 116: 209–216.

Chatterjee, C. and Shinde, R, 1994. *Textbook of Medical Biochemistry*. Arun Publishers, Medical Allied Agency, Kolkata.

Fisher, R.A. and Yates, F., 1950. *Statistical Methods for Research Workers*, 12th edition. Oliver and Boyd, Edinburgh, pp. 365.

Gupta, D., 1998. Combined effect of sulphur dioxide and nitric oxide on the serum enzymes in albino rat. *M.Phil. Dissertation*, Dr. B.R. Ambedkar University, Agra.

Levaggi, D.A., Wayman, S. and Feldstein, M., 1972. Method for the production of nitric oxide. *Environ. Sci. Technol.*, 6: 250.

Mederios, M., Bechana, H.G., Naoum, E.J.H. and Mourao, C.A., 1983. Oxygen toxicity and haemoglobinemia in subjects from highly polluted town. *Arch. Environ. Hlth.*, 38: 11–16.

Mukherjee, K. and Ghosh, N.C., 1987. GOT and GPT activities in different tissues of rats and guinea pigs exposed to varying levels of altitude stress for different periods of time. *Aviat. Space Environ. Med.*, 58(1): 18–23.

Rahman, M.F., Siddiqui, M.K.J. and Jamil, K., 2001. Effects of Vepacide (*Azadiracta indica*) on aspartate and alanine aminotransferase profiles in a sub-chronic study with rats. *Human Exp. Toxicol.*, 20: 243–249.

Reitman, S. and Frankel, F., 1957. A colorimetric method for the determination of glutamic oxaloacetic and glutamic pyruvic transaminases. *Am. J. Clin. Path.*, 28: 56–63.

Saltzman, B.E., 1954. Colorimetric microdetermination of nitrogen dioxide in the atmosphere. *Anal. Chem.*, 26: 1949–1955.

Shimizu, T., Sotokawa, H., Hatano, M., Izumiyama, M., Otomo, H. and Kogure, K., 1986. Effects of 50 ppm nitrogen dioxide gas exposure on physiological functions of rats. *Toxicol. Hlth. Camst.*, 32(1–2): 29–36.

Singh, N. and Rao, D.N., 1979. Studies on the effects of sulphur dioxide on Alfa red plants especially under condition of natural precipitation. *Ind. Air Poll. Contr.*, 2(2): 55–59.

Whitehead, T.P., Robinson, D. and Allaway, S.L., 1996. The effects of cigarette smoking and alcohol consumption on serum liver enzyme activities: A dose related study in men. *Ann. Clin. Biochem.*, 33: 530–535.

Chapter 59

Fern and Fern-allies of Kumbhalgarh Wildlife Sanctuary Rajasthan, India

B.L. Chaudhary and Y.S. Khichi*

Bryology Laboratory, Department of Botany, College of Science,
Mohanlal Sukhadia University, Udaipur – 313 001, Rajasthan

ABSTRACT

Sixteen pteridophytes have been collected from Kumbhalgarh Wildlife Sanctuary Rajasthan. These are *Actiniopteris radiata* (Swartz) Link., *Adiantum incisum* Forsk., *A. lunulatum* Burm., *A. philippensis* Linn., *Cheilanthes albomarginata* Clarke., *C. farinosa* (Forsk) Kaulf., *Asplenium pumilum* Sw. var. hemenophylloides Fee., *Hypodematium crenatum* (Forsk) Kuhn., *Christella dentata* (Forsk) Holttum., *Pteris vittata* Linn., *Marsilea minuta* L., *M. rajasthanensis* Var. billardii (Gupta), *Salvinia auriculata* Roxb., *Azolla pinnata* RBr., *Equisetum ramosissimum* Desf., *Selaginella repanda* (Desv) Spring.

Keywords: *Pteridophytes, Kumbhalgarh Wildlife Sanctuary, Rajasthan.*

Introduction

Rajasthan is the largest state in India with an area of 3,42,274 sq.km and lies between latitude 23°3′N and 30°12′N and longitude 69°31′E and 78°17′E. The climate of the Rajasthan is one of the great extremes with sudden change in temperature, precipitation and wind, which restricts the pteridophyte flora. The Kumbhalgarh Wildlife Sanctuary is situated in the Aravalli ranges and it has an area of 608.56 sq.km. Geographically it is situated between 73°2′–73°30′ East Longitude and 25°–25°40′ North Latitudes. It is situated in the hilly tracts ofRajsamand, Udaipur and Pali districts.

* Corresponding Author; E-mail: yashpal_khichi@rediffmail.com.

Materials and Methods

The present study was undertaken to identify the pteridophyte flora of the sanctuary. Area was visited several times during different seasons of the year, especially rainy season. Field notes were taken at the time of collection to observe habit, habitat and localities during period under study and brought to the laboratory in the sealed bags and taken photographs of plants and pressed in standard herbarium sheets. The specimens are incorporated in the herbarium of Department of Botany University College of Science for further study.

For the description of genera, standard books and publication by Bir and Verma (1963), Gupta, Bhardwaja and Gena (1958, 1987), Sharma and Bohra (1977) and Khullar (1994, 2000), Advanced Pteridophytes (2002) were consulted. For morphological studies, material was dissected and observed under Olympus Binocular Microscope. Camera Lucida drawings were made under the microscope.

Results and Discussion

The following pteridophytes were identified:

PTERIDALES–ACTINIOPTERIDACEAE

1. *Actiniopteris radiata* (Swartz) Link. (Plate 59.1 Figures A, Specimen no. 2133)

Field Notes

Plants usually grows in dry localities on dilapidated walls or rock crevices in the plants up to 1000m. It is the xerophytic fern.

Locality

Sadri and Bijapur block alt. 100–1100m and extremely common in the sanctuary.

Distribution

Himachal Pradesh, Punjab, Haryana, Uttar Pradesh, Gujarat and Rajasthan.

PTERIDALES–ADIANTACEAE

2. *Adiantum incisum* Forsk. (Plate 59.1 Figure B, Specimen no. 2120)

Field Notes

An abundant low altitude fern growing in rock crevices, slopes and walls.

Locality

Way of Roopnagar Fort and on it walls alt. 675m. Near the valley of Harganga temple alto 450m and Hati Pullya of Ranakepur Ghat alt. 825m.

Distribution

Himachal Pradesh, Punjab, Sikkim, Darjeeling, south central India, Rajasthan.

3. *Adiantum lunulatum* Burm. (Plate 59.1, Figure C, Specimen no. 2129)

Field Notes

This fern is morphologically very variable and ferns of low level flourishing in moist shady situations in the forest and roadside.

Locality

Near Hati Pullya and Jhilwara block alt. 750 to 900 m.

A: ***Actiniopteris radiata*** **(Swartz) Link.**

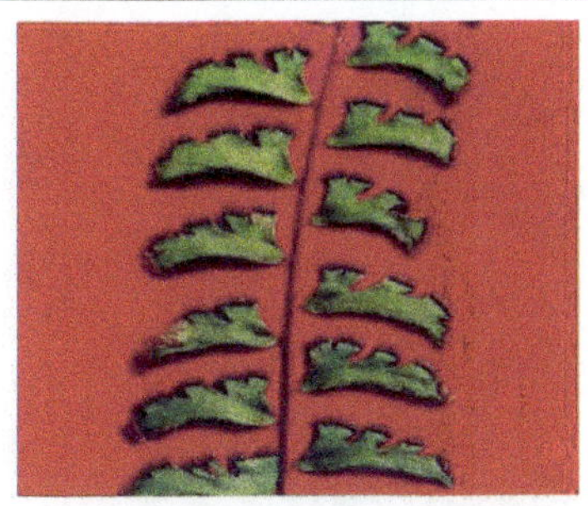

B: ***Adiantum incisum*** **Forsk.**

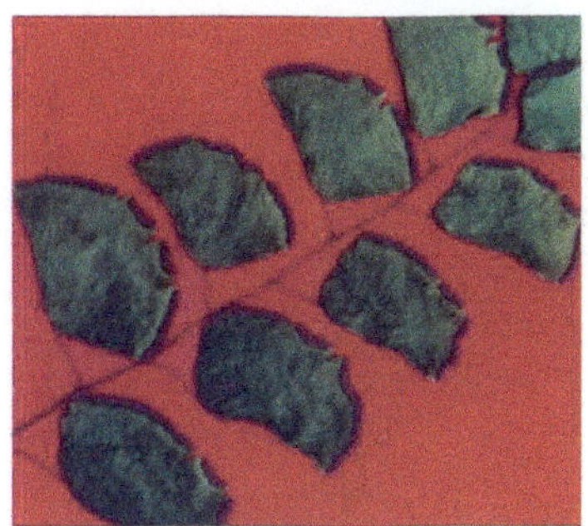

C: ***Adiantum lunulatum*** **Burm.**

D. ***Adiantum philippensis*** **Linn.**

E. ***Cheilanthes albomarginata*** **Clarke.**

F. ***Cheilanthes farinosa*** **(Forsk) Kaulf.**

G. ***Asplenium pumilum*** **Sw. var. hemenophylloides Fee.**

H. ***Hypodematium crenatum*** **(Forsk) Kuhn.**

Plate 59.1

I. ***Christella dentata*** **(Forsk) Holttum**

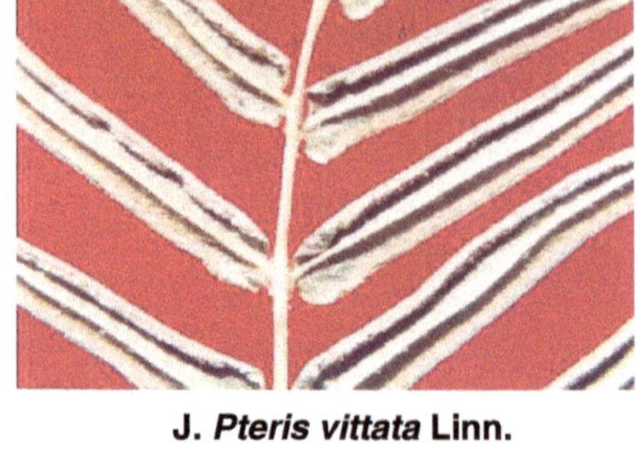

J. ***Pteris vittata*** **Linn.**

K. ***Marsilea minuta*** **L.**

L. ***Marsilea rajasthanensis*** **Var. billardii (Gupta)**

M. ***Salvinia auriculata*** **Roxb.**

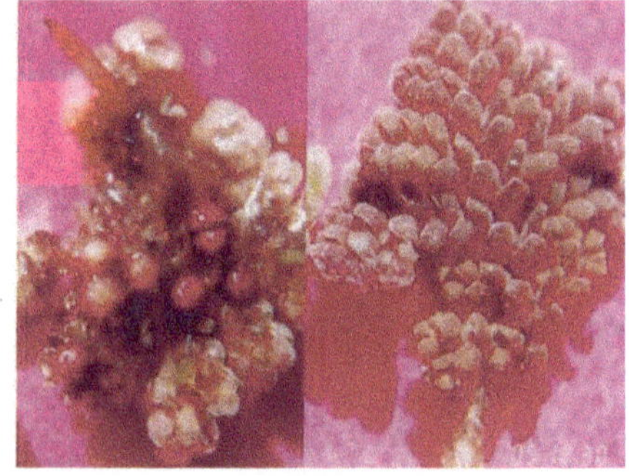

N. ***Azolla pinnata*** **R.Br.**

O. ***Equisetum ramosissimum*** **Desf.**

P. ***Selaginella repanda*** **(Desv). Spring**

Plate 59.2

Distribution

Kashmir, Himachal Pradesh, Dehra Dun, Punjab, Rajasthan and Gujarat.

4. *Adiantum philippensis* Linn. (Plate 59.1, Figure D, Specimen no. 1110)

Field Notes

A very common fern occurring form the plains up to 1000 m altitude throughout the sanctuary common on walls, rocks, banks of streams, water-falls.

Locality

Widely distributed around the Sanctuary alt. 500 to 850 m. Bijapur block alt. 450 m.

Distribution

Throughout the plains of India. Pakisthan, Punjab, Haryana, Uttar Pradesh.

PTERIDALES–CHEILANTHACEAE

5. *Cheilanthes albomarginata* Clarke. (Plate 59.1, Figure E, Specimen no. 2222)

Field Notes

A common fern on Ranakepur Ghat on slopes along rode side.

Locality

Near Ranakepur Ghat Road alto 650–600 m growing on calcareous moist soil.

Distribution

Kashmir, Himachal Pradesh, Uttar Pradesh, Punjab, Haryana, Rajasthan.

6. *Cheilanthes farinosa* (Forsk) Kaulf. (Plate 59.1, Figure F, Specimen no. 2126a)

Field Notes

This is small tufted fern popularly known as the 'Silver Fern' generally grows in dry rock crevices at lower elevations between 300–700 m alt.

Locality

Widely distributed around the Sanctuary alto 300–850 m.

Distribution

Sikkim, Darjeeling hills, Western Himalaya, Maharasthra, South central India.

ASPLENIALES–ASPLENIACEAE

7. *Asplenium pumilum* Sw. var. hemenophylloides Fee. (Plate 59.1, Figure G, Specimen no. 2231)

Field Notes

Distributed rarely in shady and moist rock crevices.

Locality

Way of Ranakaker Rest House alt. 1025 m.

Distribution

Exclusively restricted Mt. Abu.

ASPIDIALES –HYPODEMATIACEAE

8. *Hypodematium crenatum* (Forsk) Kuhn. (Plate 59.1, Figure H, Specimen no. 2230)

Field Notes

Commonly grows in the crevices of dry exposed rocks and embankments and xerophytic fern occur lithophytically on dry rocks in sunny situations in exposed place.

Locality

Way of Ranakaker Forest Rest House alt. 965–1067 m.

Distribution

Western Himalaya, Assam, Maharasthra, Rajasthan, and plains of India.

THELYPTERIALES–THELYPTERIDACEAE

9. *Christella dentata* (Forsk) Holttum. (Plate 59.2, Figure I, Specimen no. 2225)

Field Notes

Plants growing along riverbanks in the edges of the flowing water in shady situations it is extremely rare in sanctuary.

Locality

Jhilwara block alt. 750–900 m. and Deasuri Ghat near Nag temple alt. 615–785 m alt.

Distribution

Kashmir, Himachal Pradesh, Maharasthra, throughout tropics and sub tropics world, Rajasthan and Gujarat.

PTERIDALES–PTERIDACEAE

10. *Pteris vittata* L. (Plate 59.2, Figure J, Specimen no. 2226b)

Field Notes

Common growing in warmer parts in open and shaded locations or at the margins or in the forest and open road side on walls or on banks of rivers and canals in plains.

Locality

Near Ranakepur river alt. 300 to 950 m. and Viro Ka Math alt. 1067 m.

Distribution

Western Himalaya, Kashmir, Gujrat, Uttar Pradesh, Sikkim, and Darjeeling.

MARSILEALES–MARSILEACEAE

11. *Marsilea minuta* L. (Plate 59.2, Figure K, Specimen no. 2250)

Field Notes

Grows in ditches, small ponds, rice-fields, bank of lack and under similar climate and other edaphic condition of soil, manure and water.

Locality

Deasuri Ghat near Nag temple and Deasuri Dam alt. 100–615 m.

Distribution

Assam, Bihar, Orissa, Punjab, Uttar Pradesh, Gujarat and Rajasthan.

12. *Marsilea rajasthanensis* Var. billardii (Gupta) (Plate 59.2, Figure L, Specimen no. 2253)

Field Notes

Plants are xerophytic appear different from the type species in their vegetative growth this small sized aquatic or sub sub-aquatic fern in met with at lower elevations.

Locality

Near Ranakepur river and way of Kumbhalgarh alt 150–500 m.

Distribution

Rajasthan-Ajmer in Ghoogara and Sursura, Kota.

SALVINIALES–SALVINIACEAE

13. *Salvinia auriculata* Roxb. (Plate 59.2, Figure M, Specimen no. 2262)

Field Notes

Slow moving or still lakes, ponds or backwaters. Once introduced into a water body, rapid vegetative growth enables the plant to quite cover most of the free water surface.

Locality

Near Saventri Pond and Maga anicut alt. 200–550 m.

Distribution

Chamba, khajjiar lake, Utter Pradesh, Bengal, Tripura, Maharasthra, Java.

AZOLLALES–AZOLLACEAE

14. *Azolla pinnata* R.Br. (Plate 59.2, Figure N, Specimen no. 2266)

Field Notes

A fairly common floating annual hydrophytes in the plains of northern India. Floating on the surface of small, still ponds or bank waters with not wave action.

Locality

Near Deasuri Dam canal and way of Bijapur block altitude 400–850 m.

Distribution

Uttar Pradesh, Maharasthra, Japan, Africa, Gujarat and Rajasthan.

EQUISETALES–EQUISETACEAE

15. *Equisetum ramosissimum* Desf. (Plate 59.2, Figure O, Specimen no. 2268)

Field Notes

Plant terrestrials or subaquatics of steam banks, beside streams, still water bodies in swamps, beside road cuttings in full sun or partial shape.

Locality

Widely distributed near Hati Pullya alt. 850 m. and Viro Ka Math alt. 1067 m.

Distribution

Uttar Pradesh, Punjab, Gujarat, Rajasthan, Africa, China, Japan and Malaya.

SELAGINALES–SELAGINELLACEAE

16. *Selaginella repanda* (Desv). Spring (Plate 59.2, Figure P, Specimen no. 2239)

Field Notes

Growing on moist soil cover on boulders and rocky slopes in moist and shady depression in open sunny situations.

Locality

Near Parsuram hills and Jhalwara block alt. 950 m.

Distribution

Himachal Pradesh, Uttar Pradesh, Assam, Bihar, Gujarat and Rajasthan.

References

Bhardwaja, T.N., Gena, C.B. and Varma, S., 1987. Status survey of pteridophytic flora of Rajasthan with special reference to endangered ferns and fern allies. *Indian Ferns J.*, 4: 47–50.

Bohra, D.R., Singh, R. and Sharma, B.D., 1980. Survey of the pteridophytic flora of Rajasthan. *Geobios*, 7: 334–336.

Bir, S.S. and Verma, S.C., 1963. Ferns of Mt. Abu. *Res. Bull. (N.S.) Punjab University*, 14: 187–202.

Gupta, K.M. and Bhardwaja, T.N., 1958. Indian *Marsileas:* Their morphology and systematic. III. On the examination of some further collections of *Marsilea* form south India and Rajasthan. *J. Bombay Nat. Hist. Soc.*, 55: 287–292.

Chapter 60

Hair-Black Box for Accumulation of Heavy Metals

R. Gayathiri, A. Bharani and A. Krishnaveni

Department of Environmental Science, Tamil Nadu Agricultural University, Coimbatore – 641 003

Introduction

The hair analysis is an excellent tool to identify specific toxins, and vitamin and mineral deficiencies caused by them. If your hair reveals the presence of heavy metals (arsenic, lead, mercury, aluminum, cadmium, selenium, nickel, copper and iron) this means the tissues and organs of your body contain them too. The hair is a spill over from what is in the body. Generally heavy metals cannot be detected by blood, or urine tests. Heavy metals are toxic, the severity of symptoms being related to the absolutes consumed, and the time factor involved. Large doses consumed can lead to death immediately, smaller amounts over many months or years produce chronic degenerative diseases.

Heavy metals chemically relate to mineral content of the body. For examplex, Aluminum will displace Calcium, magnesium and manganese due to its valence in the atomic table. This produces a deficiency of these three minerals with metabolic dysfunctions. Minerals are the "sparkplugs" of life. They are involved in almost all enzyme reactions within the body. Without enzyme activity, life ceases to exist. A trace mineral analysis is preventative as well as being useful as a screening tool.

The protein in hair fiber holds the composition of the body tissues for a permanent period. By analyzing the hair fiber composition, a trained eye can tell what toxins have accumulated in the body tissues and what vitamins and minerals are depleted or too abundant causing an imbalance in body function. By detoxifying unnatural chemicals like mercury and lead, and by replacing specific vitamins individual to your needs, proper health can be restored using nutrition. The human hair analysis can detail these levels.

Hair Analysis

Analysis of hair fiber, composition which indicates accumulation of toxins, vitamin and mineral deficiency or abundance causing imbalances in body function. Hair is perhaps the best specimen through which to evaluate mineral imbalances and toxicities.

It provides good long-term exposure assessment, is non-invasive, is inexpensive, and allows for investigation of nutrient/toxic interactions which are only beginning to be determined for the other samples mentioned.

Hair Analysis: The Roadmap to Wellness

Hair has the advantage of long-term memory. Hair is a permanent record, like tree rings. A three-inch strand of human hair will give a six-month history of what's going on in the body since head hair grows at a rate of about a half an inch a month. A hair analysis determines exactly what chemicals are inside of you, including radiation, heavy metals like mercury, food chemicals such as aspartame, pesticides, bacteria and more. The hair analysis is the best "roadmap" to view a history of what's going on inside of our body.

Who Needs a Hair Analysis?

Any human or animal that is found ill and no explanation can be found, or treatment is ineffective. Hair analysis opens up a whole new vista for solving the mystery of illness. An imbalance of minerals can be caused by stress, improper diet, taking the wrong vitamins or mineral supplements or excessive amounts, medications and accumulation of toxic metals from the environment.

Below the skin is the hair root, which is enclosed by a sack-like structure called the hair follicle. Tiny blood vessels at the base of the follicle provide nourishment. A nearby gland secretes a mixture of fats called sebum, which keep the hair shiny and waterproof to some extern. Two sets of glands discharge secretions through the skin. Sebaceous glands arise from the walls of hair follicles and sweats glands embed in the subcutaneous layer which in the palms and soles. At the base of the

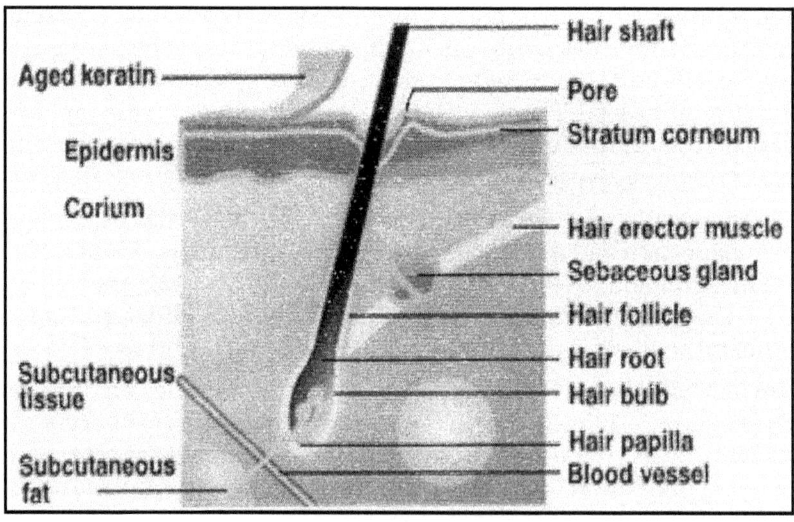

Figure 60.1: Hair Structure

follicle is the papilla; these cells play essential roles in regulating hair growth, hair cycle, and the size of the resultant hair. Surrounding the dermal papilla are epithelial keratinocytes and smaller number of melanocytes.

Hair is structured in three basic layers. Packed dead cells surrounding these structures are the cuticular layers of the hair. In the center of these structures lies the medullar canal, which is actually apart of the excretory system and houses any foreign debris, heavy metals, synthetics and medications that are thrown off by the body and eventually released through the canal. The first layer is the cuticle. A second, thicker layer is called the cortex and sometimes a third, inner layer, called the medulla. The cuticle is the outer layer of protective scales. The cortex provides strength to the hair shaft, and determines the color and texture of hair. The medulla is present only in thick, large hairs. Trace elements are located in hair at different concentrations depending on the distance between the elements and the roots of the hair.

The elements derive from various sources:

1. From the matrix during histogenesis,
2. From deposits of sebum,
3. From the transfer from exocrine sweat glands by absorption or other mechanisms,
4. From incorporation of apocrine sweat glands directly into the hair,
5. From pollution deposits around the hair after extrusion of hair by the skin, and
6. From residues of cosmetic or pharmaceutical products on piliferous skin.

Because of the above, numerous endogenous influences, exogenous interferences, and analytical peculiarities must be taken into account when trace elements are determined in hair. Furthermore, the concentrations of trace dements in hair are influenced considerably by physiological features like age, pigmentation, and sex; 15–18 racial and ethnic characteristics; 19 nutritional ingredients like diet and medication; geochemical conditions; the properties of the location from where the hair was cut; the length of the hair; and the hair color. All of these factors have to be considered before reference ranges can be established.

Hair Chemistry

When the hair is in its normal outstretched state, it is referred to as A of alpha keratin. The original configuration of the hair is held in place by the bonding found in the cortex layers anile hair. There are four types bond.

The Hydrogen Bond

This bond is located between the coils of the alpha helix and is responsible for the ability of the hair to be stretched elasticity and return back to its original shape. These bonds are responsible for approximately 35 per cent of the strength of the hair and 50 per cent of the hairs elasticity.

The Salt Bond

The salt bond is also an ionic (electrolytically controlled) bond formed by the electron transfer from the side chain of a basic amino group to the side chain of an acidic amino acid. It is responsible for approximately 35 per cent of the strength of the hair and 50 per cent of the hair's elasticity.

The Cystine Bond

The cystine bond also known as the disulfide bond, sulfur bond, or just S bond is formed by cross-links between cystine residues of the main polypeptide chains. This bond is perpendicular to the axis of the hair and between the polypeptide chains. It is responsible for the hair's toughness or abrasion resistance.

The Sugar Bond

The sugar bond is formed between the side chain of an amino acid having an OH group and an acidic amino group. It gives the hair toughness but little strength (5 per cent).

Hair Life Cycle

Hair is actually dead material when it leaves its root. The root of a hair fiber sticks in a bag in the skin. The fiber is pushed out of this bag about 0.35 mm per day, making an average growth rate of 1 cm, or half of an inch, per month. The growth rate is however very much related to an individual's age, diet and etc. Healthy hair has an average lifetime of 2–6 years. After a rest period of three months the single hair falls out, and a new fiber starts to grow out of the bag. The lifetime also depends on circumstances and each person. The lifetime of hair is responsible for the maximum of hair length one can have.

Growth Cycle of Hair

There are three phases in the hair life cycle:

Active Growth Phase or Anagen Phase

The hair root produces the cells that form the living part of the hair. This pushes the cells that already exist up and out from the follicle.

Transition Phase or Catagen Phase

New cells are not created at this stage. Instead, the hair follicle actually shrinks about 82 per cent.

Resting Phase or Telogen Phase

The protein hair strand remains connected to the hair follicle, but it doesn't grow. After five or six weeks, the dermal papilla reconnects to the base of the hair follicle and the bloodstream. The hair reenters the active growth

Sources of Trace Elements in Hair

Primary or Endogenous Source

The Matrix

Centre of intense metabolic activity.

Cells along with papilla, circulating blood lymph and extra cellular fluids dissolve trace elements.

Metals react with the Sulphydril groups of follicular proteins forms mercantile or metallo-proteins.

Keratin is formed from the metallo-protein–incorporation of metals into the hair.

The Eccrine Sweat Glands

Sweat contains water, salts of Na and K, lesser amounts of urea, aminoacids, lactic acid and pyruvic acid.

S-S cross linkages in keratin have affinity for Pb, Hg and As.

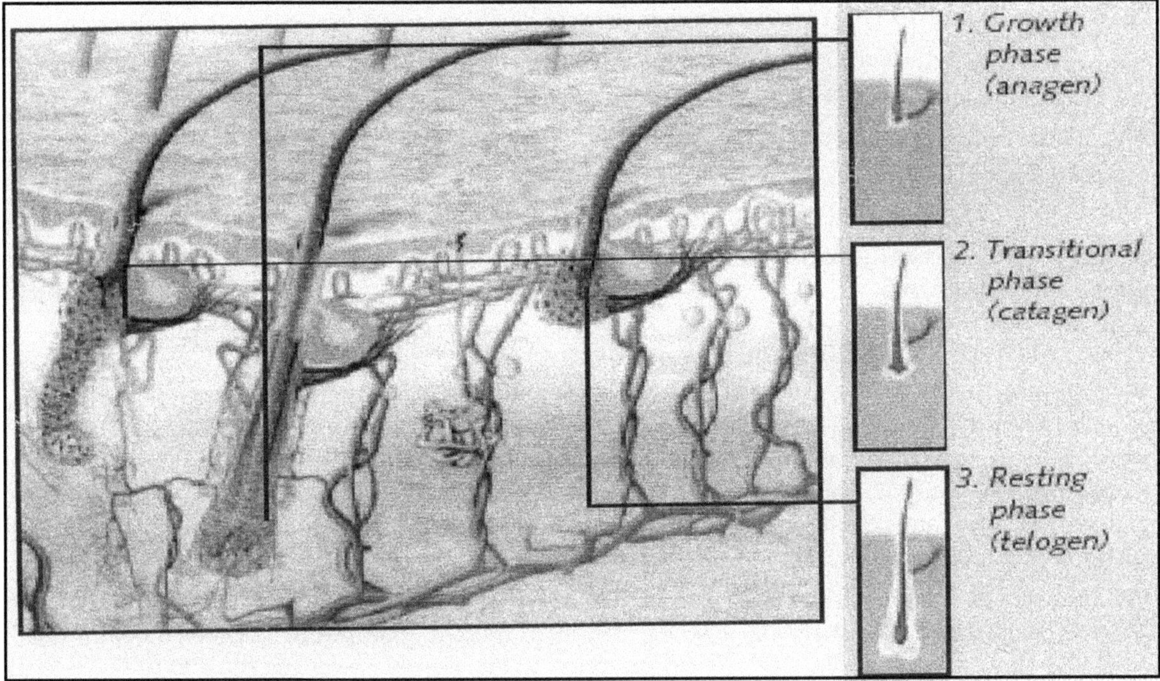

Figure 60.2: Growth Cycle of Hair

The Sebaceous Gland

Consists of mixture of lipids and waxes-sebum.

Provide physical and chemical means of incorporating trace elements into hair. Contrary view sebaceous secretions protect the hair from adsorption of trace elements.

The Epidermis

Minor source of trace elements accumulation in hair.

Exogenous Sources

Particulates in air

Solutes present in water

Hair treatments used for cosmetics, medical or hygienic purposes

Elements directly deposited on hair.

May be removed by washing.

In certain cases the increase in metal concentration from root to the tip of the hair is due to exogenous elements.

Also incorporated in to the hair.

Example

1. Se levels in hair increased by 20 to 40 times due to the use of scalp medications.
2. Use of Arsenic containing shampoos increased the As levels in the hair.

Pathways to Metal into Hair

1. Via the follicle into the hair matrix.
2. Secretion of metals in the sebum on to the hair surface.
3. Secretion of metal in exocrine sweat on to the surface of the hair.
4. Secretion of metals in apocrine sweat on to the surface of the hair.
5. He notes that apocrine sweat may not be important for scalp hair.
6. He discusses the relative merits of head versus pubic hair and concludes that scalp hair is to be preferred.

Mechanism of Trace Element Entering in Hair

Sulphydril (-SH) and disulphide-cystine (-S-S-) bonds in hair follicle can bind heavy metal that enter the body. Considering that sulphides can easily bind with heavy metals, if heavy metals enter the body, then these metals will be bonded by sulphides of the hair follicles. Hair follicle consists of keratin protein. This substance is present in nails, fur and horns of mammals. Hair functions to protect against hot and cold temperature. In hot climate soft and thin fur protects against sun light. In cold climate thick fur can preserve body heat.

The amount of metals in hair correlates with the amount of metals absorbed by the body. Therefore hair maybe used as biopsy material. A study on methyl mercury compound showed that amount of substances in hair is associated with methyl mercury in the area surrounding the hair follicles. With increase of age, the amount of hair on the body surface increases. Hair is thickest in the second decade, and after this period the number of hair follicles decline and during the seventh decade hair follicles are thin as that of infant hair (Saeni, 2000).

Stored Elements in Hair

The hair shaft is a filament formed from the matrix of cells at the bottom of the hair follicle deep in the epidermal epithelium. Each follicle is a miniature organ that contains both muscular and glandular components (Figure 60.3). Human hair is 80 per cent protein, 15 per cent water and small amounts of lipids and inorganic materials. The mineral content of the hair is 0.25 per cent to 0.95 per cent on a dry ash basis of the approximately 100,000 hairs in the average human scalp, 10 per cent are in the resting phase. During the growth phase, the scalp follicles produce hair at a rate of 0.2 to 0.5 mm/day–or about 1 cm each month. The growing hair follicle is richly supplied with blood vessels, and the blood that bathes the follicle is the transport medium for both essential and potentially toxic elements. As these elements reach hair follicles, they are then incorporated into the growing hair protein. As hair approaches the skin surface, it undergoes a hardening process, or keratinization, and the elements accumulated during its formation are sealed into the protein structure of the hair. Because of the exposure of hair follicles to the blood supply during growth, element concentrations of the hair reflect concentrations in other body tissues

Factors Influencing Metal Accumulation in Har

Limic and Vaikovic (1986) considered four groups of factors influencing the trace element levels in human hair: body stores. They pointed out that the concentrations of any element within a single hair represented the contribution of this factor. In view of the application of hair analysis for determining the quantum of population exposure and the numerous factors causing change in the levels of multi

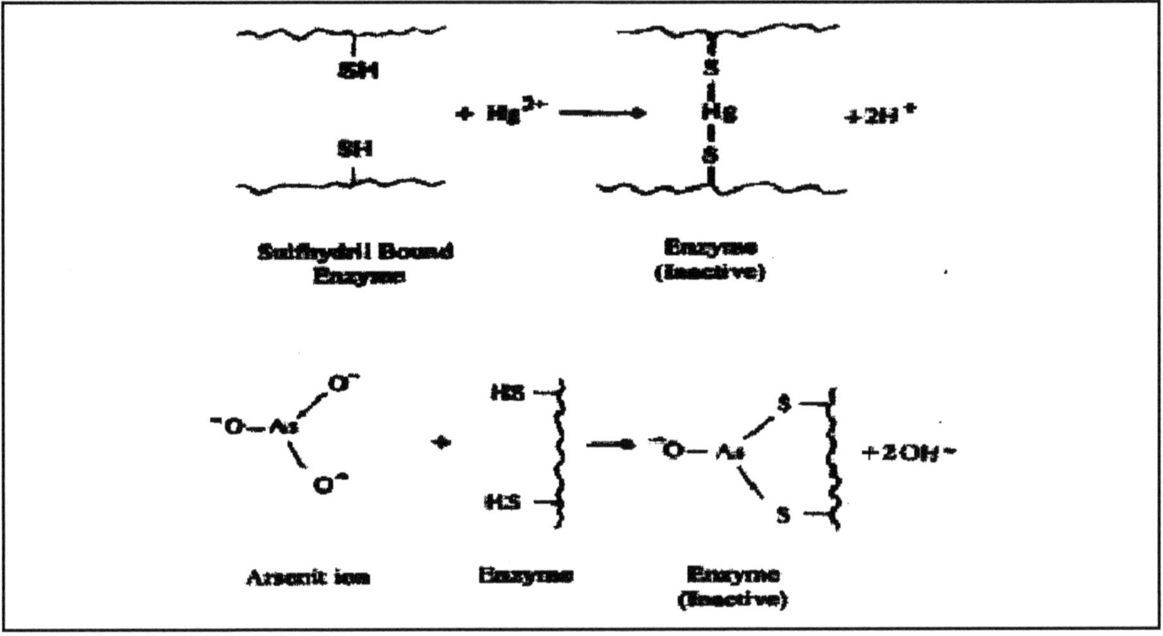

Figure 60.3

elements, the influencing factors reported by others are compiled here and placed under three main categories:

Biological Factors of Hair

1. Type and location of hair on the body
2. Hair color
3. Distance from scalp to tip
4. Growth and time
5. Structure

Demographic (Personal) Factors

1. Age
2. Sex
3. Diet
4. Cooking ware
5. Smoking and drinking habits
6. Health
7. Occupation
8. Duration of service
9. Obesity, weight and height

10. Pregnancy
11. Use of cosmetics
12. Hair care and shampooing
13. Drug use
14. Economic status
15. Customs and culture
16. Race and nationality

Environmental Factors

1. Region and place of residence
2. Rural and urban exposure gradients
3. Distance from the source of pollution
4. Seasonal variation
5. Nature of soil
6. Duration of exposure
7. Time of exposure
8. War and fire

Biological Factors of Hair

A variety of characteristics of the hair have been reported to have a bearing on the quantity of trace clement accumulation in hair.

Type and Location of Hair on the Body

Human hair collected from scalp, facial, chest, pubic regions, and other (auxiliary) of the body has been analyzed for the concentrations of elements, and different values have been reported for such types of hair. Danilovic (1958) showed 2.5 times greater lead levels in the auxiliary hair than scalp hair of Yugoslavian children and reasoned that it might he due to perspiration contamination.

A close correlation was observed between scalp and pubic hair for levels of cadmium, copper, lead, and zinc (Wilhelm *et al.*, 1990: Sikorski *et al.*, 1986): however, no correlation was observed for calcium, manganese. Copper and zinc (De Antonio *et al.*, 1982). A significantly increased level of Pb in scalp hair (Pb-H) compared to pubic hair (Baumslag *et al.*, 1974) resulted mainly from differences in external contamination-apocrine source, growth rate and resting phase. Dang *et al.* (1983) and Petering *et al.* (1973) have suggested that, if there is any question about the significance of high results, pubic and auxiliary hair samples should be employed because scalp hair is mainly contaminated by exogenous sources and samples from the pubic region are not as readily exposed to atmospheric particulate matter, as evidenced by lower Pb values in pubic than scalp hair, Sirkorski *et al.* (1986) considered that pubic hair successfully replaces scalp hair in trace element estimation because of these reasons. Hopps (1977). However, found that pubic hair is less suitable than scalp hair due to its low growth rate, long rest period, and external source of contamination through apocrine gland secretion of lipid-rich fluid.

Similarly, Flynn (1976) believed that use of body hair front regions other than the scalp, such as auxiliary, facial, and pubic regions, is not as useful as head hair. Element levels are evaluated in head hair collected from five distinct regions of the scalp such as frontal. Temporal, vortex (anterior and posterior), and nape; among these, the nape legion has been stated as the best location for collection of hair because this area is considered to be least exposed to external contamination (Jenkins, 1979; Bencze, 1990). Niculescu *et al.* (1983) found that Pb levels varied from one single hair to another in the sante subject and hence it was necessary to calculate the average value for several single hairs. Carvalho *et al.* (1984) studied the effect of hair type and observed that fishermen with straight hair always presented higher Pb-H and Cd-H (Pb and Cd in scalp hair) concentrations than those with curly hair. These variations scent to depend more on biological factors than on exposure level.

Hair Color

A number of investigations demonstrated the influence of color on element concentrations in hair. Different trends of element levels are reported for the influence of hair color. When element levels of hair are compared to the hair color, three patterns of effects have been identified.

Elevated levels were reported in black hair for Zn (Eads and Lambdin, 1973; Dorea *et al.*, 1982), Cu (Anke and Schneider, 1962; Briggs *et al.*, 1972; Bose and Chakravorty, 1964), Mn (Hambidge and Droegemweller, 1974).

1. The enhanced Zn and Cu concentrations observed in black hair may he implicated in the production of the melanin pigments, which are the metal complexes derived from tyrosine and tryptophane (Kikkawa *et al.*, 1955; Eads and Lambdin, 1973). Likewise, high Mn content in hair of Indian Darjeeling subjects is related to its black color (Bose and Chakravorty, 1964). Further, Pb content is the highest for dark hair and decreased with the increase in lightness in hair color of unexposed Marida city residents (Burguera and Rondon 1987).
2. A reversible trend of element levels in hair due to color was predicted. The levels of Cd (Wilhelm *et al.*, 1988) and Pb (Schuhmacher *et al.*, 1991) decreased from red hair to blond, brown, and black hair. Also, Dutcher and Rothman (1951) reported that red hair had greater amounts of Fe compared to hair of any other color.
3. No correlation was found between hair color and level of Cu (Goss and Green, 1955; Dutcher and Rothman, 1951; Lea and Luttrell, 1965; Guillard *et al.*, 1985), Pb and Cd (Eads and Lambdin, 1973; Carvalho *et al.*, 1984), and Ni (Eads and Lambdin, 1973) and Cr (Schroeder and Nason, 1971).

Distance from Scalp to Tip

Variation in concentrations of trace elements along the hair shaft from the scalp outward has been well documented. An increase in the profiles of concentrations from the proximal to the distal parts of scalp hair was reported for Cu (Yukawa, 1984; Gordus 1973; Gangodharan and Sankar Das, 1976; Hambidge, 1973; Eads and Lambdin, 1973), Mn (Gangadharan *et al.*, 1973; Fergusson *et al.*, 1981), Pb (Eads and Lambdin, 1973). The increasing element trend from scalp to distal end of hair is correlated with time of exposure (Jenkins, 1979), the difference in the body source of Ph (Fergusson *et al.*, 1981), and the exogenous Cu contribution for the longer duration to the highest Cu level (Hambidge, 1973). Yukawa (1984) suggested many possible reasons for such a trend. One reason is the absorption and accumulation of elements in hair structure by contamination from the environment, in support; Bate and Dyer (1965) have demonstrated that large amounts of Au, Hg, Zn, Cu, Fe, Co, Mn, Cr, and Ba are absorbed by hair in simulated perspiration at pH 5.5. The second reason could he the movement of elements through the inside of the hair from the scalp to the hair tip. A third reason could he the concentration variation observed from the structural differences of the hair surface, which is weathered by the environment (as by washing and combing) over long periods of time.

Varied patterns of element values have been reported. A decrease in the concentration from the scalp to the tip of the hair shaft for Cl and Br (Yukawa, 1984) and for Cr and Zn (Obrusnik *et al.*, 1973). a constant profile along the hair shaft for Zn (Dybezynik and Boboli, 1976) and for Ni, Mn, and Zn (Eads and Lambdin, 1973), and four different pattern of profiles for Zn (Gangadharan and Sankar Das, 1976) are known. Obrusnik *et al.* (1973) noted an upward profile of Mg, Se and Hg, a slightly decreasing profile for Zn and it generally constant profile for Mn. Many studies showed higher element content at the proximal end than at the distal end of hair, whereas Alder *et al.* (1977) found that this is not true for all elements. Levels are not constant even for Mn, Cu, Fe, Ni, Co and Cr, which are essential elements. However, there appears to be no conspicuous difference in the average element Concentrations in hair from different anatomic sites.

Growth and Time

Growth of hair occurs in four stages: the anagen phase (actively growing period). Catagen (degenerating phase), telogen (resting phase). And an intermediate phase during which a new follicle forms from the remainder of the previous hair cycle. The long anagen phase is important for element analysis (Bencze, 1990) for it provides a relatively uniform sampling of hair and the capacity of time studies (Flynn, 1976) and because uptake or accumulation of elements is higher during the anagen phase than any other (Jaworowski *et al.*, 1966) Hair is formed at a rate of 2–5 mm/d in humans (Hopps, 1977). Thus, sampling of 2–5 mm of hair from the scalp is advantageous, because it has recently grown, showing the element status of 1d earlier accumulation, and has lesser possibility of exogenous contamination. Hence, element analysis of 2–5 mm of proximal scalp hair shows the endogenous sources of an element without exogenous contamination.

Certain factors such as regional and seasonal variations in hair growth rate and cycle must he taken into account when sampling (Flynn, 1976; Strain *et al.*, 1972). Seasonal effects may he a result of stage of hair growth and changes caused by perspiration, surface contamination, and diet. During sampling, however, it is not possible to assess the effect of different growth rates (Wilhelm *et al.* 1990). Thus, external contamination may be the most important factor influencing the level of an element in hair (Flynn, 1976).

Structure

Hopps (1977) and Barrett (1985) reported that metal levels determined in hair depend on hair structure and thickness. No influence of hair structure on Cd, Ct', Pb, and Zn levels in scalp and pubic hair was noted (Wilhelm *et al.* 1990).

Demographic Factors

Age

Age is an important factor in metabolism (Nordberg *et al.*, 1978) and also a relevant factor to be considered when hair is used as an indicator of the body burden of metals (Kostial, 1983). A large number of studies are available on the influence of age on hair element levels.

Cd-H levels decreased with age independently of sex (Bosque *et al.*, 1991). A similar decrease with age was reported for Cd-H from younger females (1–3 yr) to older females (4–70 yr) whereas Cd content was found to be inversely related to age, from 3–7 yr in preschool children (Wilhelm *et al.*, 1988).

Further, Eads and Lambdin (1973) observed a decline in Cd-H levels of women aged 37–72 yr and no difference in Cd-H levels in young and old males. Jones *et al.* (1987) reported that Cd-H appears not to be age dependent. Petering *et al.* (1973) observed that in women Cd-H attained a peak at age 40–50 yr and that in males Cd-H increased with age up to 20 yr and then decreased.

Pb-HI decrease with age in controls and in workers occupationally exposed to petrol lead as well as in young females (1–30 yr) compared to older women (40–79 yr) (Schroeder and Nason, 1969). Eads and Lambdin (1973) found a trend of increasing Pb-H with age in women aged 37–72 yr. Klevay (1973) showed significant decrease with age in Pb-H levels in males but not in females. Petering *et al.* (1973) found a decrease in Pb-H in males and an increase in females up to 35 yr and then a sharp decrease. In contrast, Reeves *et al.* (1975) and Jones *et al.* (1987) indicated that Pb-H levels in humans showed no significant difference with age.

A fairly uniform distribution was noted for Ni-H in both males and females in different age groups (Eads and Lambdin, 1973) and also no increase in Ni-H with age (Schroeder and Nason, 1969). Cu-H decreased from growing females (1–30 yr) to older females (40–70 yr) but did not decrease with age in males (Schroeder and Noson, 1969). Similarly, Eads and Lambdin (1973) reported no change in Cu-H of males with age but reported a decline in females with age. Jones *et al.* (1987) reported the highest Cu-H in young adults, whereas Ohmori *et al.* (1975) observed lesser Cu-H in 20-yr old than in 40-yr-old subjects. No correlation was found between Cu-H level and age (Lebrudo and Marie, 1988). Thus, all patterns of age influence demonstrated for levels of different elements include an increase or decrease of element levels, not any uniform pattern (Wilhelm *et al.*, 1990; Schroeder and Nason, 1969), and no correlation between age and element levels (Randall and Gibson, 1989; Dipietro *et al.*, 1989; Lebrudo and Marie, 1988; Jones *et al.*, 1987).

In the case of infants, age of the mother also influences element levels in the hair of babies. Baumslag *et al.* (1974) reported that Cd-H in the baby decreased as the age of the mother increased. Among different age groups, more pronounced hair levels of As, Cd, and Pb were indicated for 1–5-yr old children compared to adults (Kowal *et al.*, 1979), probably because of high total intake per kilogram body weight. Hence, hair samples from children may be preferable to those of adults for studying element levels in hair.

Sex

Many investigations have indicated significantly higher hair levels in females than in males for Cd (Petering *et al.*, 1973; Bosque *et al.*, 1991), Cu (Schroeder and Nason, 1969; Ohmori *et al.*, 1975; Medeiros and Pellum, 1985), Pb (Lekouch *et al.*, 1999). The higher Zn-H observed in girls than in boys is due to the increased requirement of Zn at puberty in girls, resulting in smaller body reserves of Zn. Additionally, Zn deficiency appears to be more common in females than males.

Other reports also presented higher hair levels in males than in females for Cd (Dipietro *et al.*, 1989: Medeiros and Pellum, 1985: Wilhelm *et al.*, 1988; Leotsenidis and Kondakis, 1990), Fe (Jones *et al.*, 1987), As (Hartwell *et al.*, 1983), and Pb (Hartwell *et al.*, 1983: DiPietro *et al.*, 1989).

Higher levels of Cd (Leotsenidis and Kondakis, 1990) and Cd and Pb (Dipietro *et al.*, 1989) and lower concentrations of Ni, Mg, and Ti in males were shown compared to those of females. Differences, whether higher or lower between males and females could he due to differences in metabolism, ingestion of food, occupation, or simply sampling variation (Bosque *et al.*, 1991).

Diet

Diet is also expected to affect the levels of elements in hair, and especially intake of a high-fiber diet apparently lowers uptake of metals from the intestine and results in a lower metal content in the hair. Although Arunachalam *et al.* (1979) presented no significant difference in the mean hair element levels between non vegetarians and vegetarians, Srikumar *et al.* (1992) found that south Indian vegetarians had significantly higher hair levels of Cu, Mg. Se, Fig, Pb, and Cd, except Zn, than Swedish people on a mixed diet. A correlation exists between dietary intake of trace elements such as Na, Ca, Zn, Mn, Mg, Fe, and Cu (Klevay *et al.*, 987) and of Fe, Cu, Mg, and Mn (Monasterios 1986) and their corresponding concentrations in hair. It is also observed that whatever is consumed may influence element levels in hair.

Those who consume milk and dairy products daily show lower Pb-H than those who consume them occasionally (Hauser *et al.*, 1999). Increased As, Cd, and Pb were assessed in children who ate paint, dust, or clay (Hartwell *at al.* 1983). Herber *et al* (1983) demonstrated that in Caucasian boys as a group for each 1 g intake of Fe, the Fe-H level increases 7.0 mg/kg. The sharp drop in Zn-H levels after weaning or bottle-feeding instead of breast-feeding is of particular interest in showing the influence of Zn-H levels in infants (Van Wouwe and Hammer, 1985). No influence of eating habits for Cd-H levels has been in preschool children Wilhelm *et al.*, 1988).

Cooking Ware

The different types of cooking ware made of various metals are also found to influence element levels in hair. Aluminum and iron cooking ware are used abundantly in Indian cuisine, and perhaps this may he responsible for increased Fe and Al content in hair (Takagi *et al.*, 1986). Similarly, Shrestha and Schrauzer (1989) have pointed out that a possible source of Pb in the Darjeeling people is from their bronze cooking ware, which is widely used and is soldered and repaired with Pb solder; thus, Pb pollution is not only a problem of industrialized nations hut also may he more serious in those communities that maintain a traditional lifestyle of using bronze cooking ware. Subramanian (1991) has also corroborated an enhanced Pb-H level in a particular south Indian people who utilize their traditional lead cooking ware.

Smoking and Drinking Habits

The tobacco smoking habits of male and female rural subjects (near Delhi, India) using the hookah have been associated with an increased level of Cd-H (Subramanian and Sukumar, 1989). Thus, smoking is a contributing factor to the high bioaccumulation of Cd (Chattopadhyay *et al.*, 1990) and of As, Cd, and Pb (Hartwell *et al.*, 1983). Even Cd-El levels in children were affected by their parents' smoking habit (Bosque *et al.*, 1991). No influence of smoking habit was reported, however, for scalp hair Cd levels (Bergomi *et al.*, 1987; Sikorski *et al.*, 1986: Takagi *et al.*, 1986) and for the scalp and pubic hair levels of Cd, Cu, Pb, and Zn (Wilhelm *et al.*, 1990) and Ph (Burguera and Rondon, 1987). The habit of drinking illegally distilled whisky is also correlated with an elevated Pb-H level in the village subjects (Folio *et al.*, 1982).

Health

Changes in hair levels of essential or nonessential elements are ascribed to certain specific cases of health disorders that result as a consequence of altered metabolites and distribution of element levels (Sukumar and Subramanian, 1992a). Thus, the health status of subjects becomes a very important factor, altering hair levels of elements.

Cd-H was related to intelligence tests, motor impairment assessment, and school achievement scores of 149 Maryland public school children aged 5–16 yr (Thatcher *et al.*, 1982). Cd-H levels were higher in adult hypertensive black females and males and hypertensive mothers and their babies than in controls (Medeiros and Pellum, 1984, 1985; Bialkowska *et al.*, 1987; Huel *et al.*, 1981), and Cd-H levels were higher in learning disabled children than in the controls (Phil and Parkers, 1977; Ely *et al.*, 1981). Decreased Cr-H level found in insulin-requiring diabetic children (Hambidge *et al.*, 1977) on the other hand Higher Cd-H levels were measured in patients with cerebral diabetics (Aharoni *et al.*, 1992) than in controls.

Concentrations of Ph-H were found to he elevated in different subjects such as children with chronic lead poisoning (Kopito *et al.*, 1967), learning disabled children (Phil and Parkers, 1977). Children with mild plumbism (Kopito *et al.*, 1967), bipolar patients (Kanofsky *et al.*, 1986), hypertensive black males (Medeiros and Pellum, 1985), and hypertensive South Carolina adolescents (Borgman *et al.*, 1982). children with chronic diarrhoea (Castilloduran, 1985), first-6-month-old breast-fed and bottle-fed infants (Van Wouwe and Hammer, 1985), and infants with diaper rash (Collipp *et al.*, 1985). Antilla *et al.* (1984) observed that in mild chronic Zn deficiency the levels of Zn-H decreased, but in severe deficiency the growth of hair was decreased to such an extent that the measurable Zn concentration remained normal.

Varied patterns of multi element status have been exemplified in hair of the following patients compared to controls:

1. Lower levels of Mg, Cu, and Zn in epileptic patients (Ilhan *et al.*, 1999), Cu and Mn in male hypertensive, Cr and Zn in female hypertensive, and Zn in females with coronary heart disease and diabetics (Subramanian *et al.*, 1989: Sukumar and Subramanian *et al.*, 1992), Zn and Cu in hypothermic children (Baoshan Li, 1989), and Mn, Zn, and Fe in potential mothers.
2. Higher levels of Cd, Ph, and Zn in adult black female hypertensive (Medeiros and Pellum, 1984).
3. Lower Fe, Cu, Mg, higher Ca and insignificant levels of Zn in mentally retarded children (Shrestha and Carrera, 1988).
4. No correlation in levels of Cu and Zn in hypertensive (Vivoli *et al.*, 1987), Ca, Zn, and Cd in hypertensive (Borella *et al.*, 1987), Al and Cu in renal dialysis patients (Hewitt and Day, 1986), Zn and Cu in children with malnutrition (Lebrudo and Marie 988) and Rb, Cr, Co, and Mo in diabetics (Soylak *et al.*, 1995).

Occupation

Variation in hair element levels of human beings is heralded because differences in occupation, which in him causes alterations in exposure level and sources of metals (Majumdar *et al.*, 1999). Hence, hair is employed successfully as an indicator for screening population groups occupationally exposed for longer periods continuously to the polluting element. A large number of studies have reported higher element exposure occurring among industrial workers than controls (Table 60.1). Significantly increased concentrations of Cd, Pb, Ni, Ni, and Cu are measured in the hair of printing, electroplating, and mining workers (Nishiyama and Nordherg 1972; Reeves *et al.*, 1975; Hagedorn-Gotz *et al.*, 1977; Hyoi *et al.*, 1986). These studies ascertained that increase of element concentrations in hair is proximately proportional to exposure in their working environment. Assessment of working environment is essential in evaluating the viability (If element exposure; the possibility of hazard due to direct exposure to high doses of these metals can further be confirmed by observations during sampling (Hyoi *et al.*, 1986). The use of protective procedures such as gloves and masks may alter the availability of elements while working. The parents' occupation influences the Cd-H content in their children (Bosque *et al.*, 1991).

Table 60.1: Significantly Elevated Levels of Elements Reported in the Hair

Sample No.	*Elements*	*Sources/Workers*	*Reference*
1	Pb	Lead workers	Stankovic *et al.*, 1974
2	Pb	Battery factory workers	Fergusson *et al.*, 1981
3	Pb	Male workers	Grandjean, 1979
4	Pb	Gasoline station workers	Burguera and Rondon, 1987
5	Cd	Occupationally exposed mothers and their babies	Huel *et al.*, 1984
6	Cd	Non ferrous smelter workers	Tomza *et al.*, 1983
7	Fe	Steel mill workers	Jamall and Jaffer, 1987
8	Cr	Tannery workers	Randall and Gibson, 1989
9	Zn	Lead processing workers	Ohmori, 1984
10	Ni,Co	Nickels and cobalt workers	Bencko *et al.*, 1986
11	Zn, Cu, MN	Zinc smelters	Junko *et al.*, 1982
12	Mn, Cr	Fire works workers	Sukumar and subramanian, 1992
13	Mn, Cr, Ni, Cu	Welders and smelters	Bencko, 1991
14	Mn, Ni, CD	Match factory workers	Sukumar and Subramanian, 2000
15	Cd, Cr, Cu, Mn	Foundry workers	Sukumar and Subramanian, 2000

In children whose fathers were industrial workers, Cd-H levels were high compared to those whose fathers had other occupations (Wilhelm *et al.*, 1988). The highest Ph-H has been assessed far a boy whose father is a welder, indicating that the occupation of the parents is one of the most relevant factors influencing levels of Ph-H in children (Sehuhmaeher *et al.*, 1991). When compare to Pb-H levels of controls, an intermediate level in the family members and the highest level in battery factory workers were graded, indicating that the Occupation of even one of the family members could alter levels in other member by Pb dust transported home in their working clothes (Fergusson *et al.*, 1981).

Duration of Service

Besides occupation, the duration of service causes change in the element levels in workers' hair. As the duration of Pb exposure increased, elevated concentration of Pb-H were found in occupationally exposed petrol station workers (Burguera and Rondon1987).

Obesity, Weight and Height

Levels of Zn-H were determined to the lower in obese children in Taiwa than in controls (Chen *et al.*, 1988). A few studies have reported no correlation between hair levels of Cu, Mn, Hg and Zn and birth weight of infants (Heinersdoff and Taylor, 1979; Kauf *et al.*, 1984), between Cr-H level and height of workers and controls (Randall and Gibson 1989) and between Zn-H level and height and weight percentile (Van Wouwe *et al.*, 1986).

Pregnancy

Declines in Cu-H and Zn-H concentration noted in women with 30 weeks of gestation were probably caused by the increased demands of the fetus (Dang *et al.*, 1983). Moreover, newborn Cr-H levels exceeded corresponding maternal hair level, suggesting the fetus extracted Cr from the maternal stores (Hambidge and Baum, 1972; Hambidge and Droegemweller, 1974).

Use of Cosmetics

Use of cosmetics and dyes also causes variations in metal content of hair as these contain various metal salts and complexes that contribute to hair levels of metals either through absorption of metal salts on hair roots or by adsorption onto the hair shaft. Various salts or complexes of metal ions allowed for use as coloring agents of hair dyes and cosmetics include Al, As, Ba, Cr(III), Cu, Cu, Fe, Hg, Ti and Zn. Many of these metals used in cosmetics or dyes appear in hair. Males using dandruff shampoo had significantly higher levels of Se and Ti than those using regular shampoo or conditioners.

Further, Ba, Ca, Cu, Mg, Na, and Sr were all elevated in hair of females using permanent wave or color treatments compared to those using only dandruff shampoo, regular shampoo, and conditioners (DiPietro *et al.*, 1989). High Hg-H levels were measured in two women who used a germicidal soap containing 1 per cent (w/w) of HgCl (Gangadharan *et al.*, 1973). Similarly, increased concentrations of Ph-I-I were assessed in Bengali women who used red lead as a cosmetic (Bagchi *et al.*, 1940). Among the 17 elements surveyed, the values of 10 metals were elevated in the hair of U.S. control females who had received hair treatment, whereas levels of Co were the lowest in the non treated hair of males (Vance *et al.*, 1988). Thus, it is evident that shampoos, cold wave lotions, hair sprays, bleaches, and dyes (Abraham, 1982; McKenzie, 1978) and shampoo, water, and other media (Limic and Valkovic, 1986; Leblanc *et al.*, 1999) can affect the results of hair element analysis.

Hair Care and Shampooing

Takagi *et al.* (1986) indicated differences in element levels for hair samples receiving greater frequency of shampooing (3–7/week) against normal shampooing (12/week).

Drug Use

Briggs *et al.* (1972) reported that women taking oral contraceptives had higher Cu-H values than other women. Similarly, Reilly and Harrison (1979) found elevated Cu-H in female students and younger non students and reasoned that it might reflect the use of contraceptive pills. Toshihito *et al.* (1992) stated that long-term anticonvulsant therapy could induce alteration in both the metabolism and distribution of Cu, Zn, and Mg in hair. In contrast, Ilhan *et al.* (1999) suggested no obvious association between patients and anticonvulsant therapy.

Economic Status

Hambidge *et al.* (1979) noted that Zn-H was evidently lower in children of low income families than in children of middle-income families. However, a correlation was not found between the socioeconomic status of the mother and levels of Cu, Zn, kin, and V in the hair of children (Gibson and Dc Wolfe, 1979).

Customs and Culture

In the population of UK Tedi region, Papua New Guinea, the cultural custom of using ochre clay daubs as facial hair and body adornment by men and boys is related to enhanced concentrations of Fe-H as a result of external adsorption of Fe to the hair (Jones *et al.*, 1987).

Race and Nationality

Race is a factor that expresses itself among other things, in the color, shape, and thickness of hair, and affects the concentrations of Cu-H and Zn-H. Ethnic origin influences Zn-H with Asians having the highest levels and blacks the lowest (Briggs *et al.*, 1972). Higher Mn-H was observed in Indians than in French persons (Umarji and Bellare, 1966). Blacks had higher concentrations of Ph-H and Cd-H than whites (Medeiros and Pellum, 1985). Similarly, the Pb-H level was higher in blacks than in whites due to higher intake or exposure (Baumslag *et al.*, 1974).

Takagi *et al.* (1986) made an international correlation of 21 element levels between Japan, India. Canada. the US and Poland and found significant differences in levels of trace elements corresponding to the independent nature of living habitat. Sadiq *et al.* (1992), however, reported that nationality has no explicit effect on element levels in the scalp hair. Batzevich (1995) determined that concentrations of elements (Zn, Cu, Se, Mn, Hg, Fe etc.) in hair are highly variable because of local factors (geochemical condition or nutritional factors) but that racial or ethnic identification is impossible using trace element analysis.

Environmental Factors

The wide variation of element levels in the hair in the general population living in the same environment possibly indicates an influence by the source of environmental pollutants. The environmental gradients of elemental pollution result from the production of high concentrations of one or more toxic trace elements from a single source or combined sources (Jenkins 1979). Hair levels of elements vary because of environmental gradients resulting from a place of residence nearer to heavy traffic or highways, or to mines or smelters of Pb, C, Cu, Ni, or Zn, or to urban industrialized areas, or as a result of regional topography, local morphology, or population density (Caprio *et al.*, 1974; Dissanayake *et al.*, 1984).

Region and Place of Residence

The influence of place of residence is reported for the high Pb-H levels of subjects from the Republic of Panama (Klevay and Forks, 1973). Further, increased Pb levels in hair of women of child-bearing age are related to enhance Pb exposure in the city of Karachi (Jamall and Allen, 1990). Compared to people from a clean region. Cu and Mn levels were high in the hair of 170 men and children living in the region of a Ferro metallurgical plant (Horst *et al.*, 1992).

Carvalho *et al.* (1981) reasoned that elevated Cd and Pb in the hair of fishermen from Sabae river basin, Brazil, were due to their close proximity to the smelter. Hammer *et al.* (1971) related the Cd, Cu, Ph, and Zn concentrations in hair of school boys living in four industrial cities with a known environmental exposure gradient and concluded that hair levels of Cd and Pb accurately reflected community exposure. Jones *et al.* (1987) reported that Cd-H levels in the population of ok Tedi region. Papua New Guinea, were associated with the local naturally occurring sources of highly soluble and particulate Cd in groundwater, the observed Pb-K. Higher in the Port Arthur subjects than in the Hanover subject was attributed to the high airborne Pb content in the gasoline production center (Eads and Lambdin, 1973).

The increasing Pb-H in schoolchildren from Jeddah to Makkah, Tabuk, and Riyadh indicated an increasing trend in environmental Pb pollution due to rapid modernization with increased use of leaded gasoline (Ahmad *et al.*, 1988). Jervis *et al.* (1977) reported values of Ph and Cd in hair of residents living in the neighborhood of a lead refinery were higher but Zn was lower than normal.

Bioaccumulation of Pb, Cd, Ni, and Cr in the hair of Fort McKay residents may be related to exposure to extraction plant effluents (Moon *et al.*, 1988).

Rural and Urban Exposure Gradients

Levels of Pb-H were higher in subjects from Metro Manila and Camerrene Sur than in those from rural areas, Palawan and Mindara, where there was very limited vehicular traffic (Kapauan *et al.*, 1982). Similarly, the values of Pb-H and Cd-H were higher in urban Delhi subjects than in village farmers and were related to exposure to automobile exhaust and industrial pollution (Sukumar and Suhramanian, 1992c).

Folio *et al.* (1982) showed that levels of Pb-H and Cd-K were significant ally higher in the urban group from Middle Tennesse than in a rural group. They reasoned that several urban subjects living near major roadways were exposed to Pb from automobile exhaust.

Moreover, the elevated levels in rural subjects were related to emission from machinery run on leaded gasoline and to the use of insecticides and illegally distilled whisky. The higher Cd-H concentration observed in urban children from Duisburg than in rural children from Westfalen, Germany, was associated with the major route of Cd uptake through ingestion via food, drinking water, and contaminated soil and dust (Wilhelm *et al.*, 1988).

Distance from the Source of Pollution

Many investigations have established a relationship between element levels in hair and the distance from the source of pollution. Stankovic *et al.* (1974) graded population into three types based on Pb exposure. Pb-H levels were high in peasants living far from the source of pollution, intermediate in urbanites, and the highest in subjects living near a lead smelter.

Likewise, Chaltopadhvay *et al.* (1977) classified the nature of exposure in rural and urban subjects based on distance from the source of pollution and hair levels: 20 ppm in the rural subjects, above 20 ppm in urban controls, and 60 ppm in urbanites living near the smelter. Hartwell *et al.* (1983) also emphasized that among blood, hair, and urine samples hair was the most useful in determining the relationship between levels of Ph and Cd and distance from the source of pollution and body burden

Seasonal Variation

An influence of season on the levels of elements in hair was reported. Strain *et al.* (1966) stated that Zn content was higher in hair collected during summer than at any other time of the year. Moreover, Wilhelm *et al.* (1988) observed that the levels of Cd-H were higher in the summer than in the winter. Seasonal effects may be due to the stage of growth of hair, changes caused by perspiration, surface contamination and diet (Combs *et al.*, 1982) and to the difference in water supply (Wilhelm *et al.*, 1990).

Nature of Soil

Barltrop *et al.* (1975) reported that hair levels of Pb were high in subjects living in a high-lead soil area of the U.K., when compared to that of subjects living in a low-lead soil area. Similarly, Stauber and Florence (1989) indicated an elevated Mn-H level in Aborigines living in Mn-rich soil in Groote Eylandt in the Northern Territory of Australia but not in Aborigines of a non manganese area.

Duration of Exposure

Individuals who were living in the vicinity of Zn and Cu smelters and spent more time out-of-doors had higher concentrations of As, Cd and Pb than those who spent less time outside (Hartwell *et al.*, 1983). Wilhelm *et al.* (1990) showed that Pb-H levels were high in preschool children with more outdoor activities in summer in comparison to those of children who spent less time outdoors.

Time of Exposure

When Weiss *et al.* (1972) investigated Pb levels in antique K (1871–1923) hair samples of children and adults and a contemporary (1971) population of children and adults, they found that Pb content of human hair was significantly higher in both children and adults of the antique period than in the present-day population, indicating greater ingestion of Pb by the earlier population.

War and Fire

The Gulf War and Kuwaiti fire might be contributing factors to the elevated concentrations of Pb, Ni,and V in the hair of people in Deharan in Saudi Arabia (Sadiq *et al.*, 1992). So far, high hair levels of elements (mostly Pb and Cd) have to be related to environmental pollution, hut the concentrations of Co and Zn are not connected to environmental pollution because these are nutritional elements Hammer *et al.*, 1971). Low values of elements are also observed and related to low automobile use, low air Pb levels, and environmental pollution (Eltayeb and Van Grieken, 1990; Schumacher *et al.*, 1991; Reilly and Harrison, 1979).

Factors Influencing the Framework of Hairs

One has to distinguish between endogenous influences that represent uninfluenced biological variations in vivo, and exogenous disturbing factors that depend on effects in vitro during the pre analytical period.

Endogenous Influences

Various endogenous factors influence the composition and framework of the hair, which delimit hair from other biological materials in the human body. Amino acids in the hair alter heterogeneously by genetic disposition, nutrition, status of health and disease, season, weather, cosmetics, and environmental conditions. The water content very much depends on the relative atmospheric humidity. The lipids change up to 10 per cent following free fatty acids; mono di-, and triglycerides; esters; alcohols; and hydrocarbon compounds. A considerable amount of minerals and trace elements are lost during the dry incineration of hair in the pre analytical procedure.

Exogenous Influences

Shampoos, colorants, body lotions, cosmetics, and other environmental influences are important considerations. From the environmental health point of view, information on the home and workplace, and leisure activities is essential. The outer epidermis of a hair is not an impermeable layer. Very small fissures and orifices facilitate immigration and deposition of chemicals. The same applies for the protein structure of hair, in which components of shampoos, cosmetics, *or* tap water may be absorbed and deposited. Therefore, hair contains elements of both endogenous and exogenous origin. Solvents are used for pre analytical hair sample preparation.

These solvents, however, cannot differentiate between the body's own deposits and potential environmental contaminants. With solvent extraction, relevant information can be obtained. Elements that are primarily loosely bound to the hair surface may become affixed to the hair structure by the pre analytical decontamination procedures. Hair may be regarded as a type of ion exchanger with its small, superficial holes and fissures operating as ports of entry.

Symptoms Linked to Toxic Element Exposure

Arsenic

Fatigue, headaches, dermatitis, increased salivation, muscular weakness, loss of hair and nails, hypo pigmentation of skin, anemia, and skin rashes.

Cadmium

Loss of sense of smell, anemia, dried scaly skin, hair loss, hypertension, kidney problems.

Lead

Children: delayed mental development, hyperactivity, delayed learning, behavioral problems. Children and adults: fatigue, anemia, metallic taste, loss of appetite, weight loss and headaches, insomnia, nervousness, decreased nerve conduction, possibly motor neuron disorders.

Mercury

Reduced sensory abilities (taste, touch, vision and hearing), metallic taste with increased salivation, fatigue, anorexia, irritability and excitability, psychoses, mania, anemia, parenthesis, tremors, in coordination, cardiovascular disease, hypertension with renal dysfunction.

Why Blood Samples are not Reliable?

Hair is used as one of the tissues of choice by the Environmental Protection Agency in determining toxic metal exposure. A 1980 report from the E.P.A. stated that human hair can be effectively used for biological monitoring of the highest priority toxic metals. This report confirmed the findings of other studies in the U.S. and abroad, which concluded that human hair may be a more appropriate tissue than blood or urine for studying community exposure to some trace elements. The U.S. Environmental Protection Agency* provides a concise, well-researched answer to this question with the following summary:

One of the most unsatisfactory criteria for determining the relevance of trace element analysis is the choice of specimen as an indicator of the differences between physiological and pathological events. Usually, serum/plasma and whole blood are reliable body samples, and they are in the clinical chemistry laboratory. Unfortunately, however, these samples are not reliable for trace elements in the following instances:

1. Missing organ specificity and deficit of organ storage,
2. Variable oxidation status and changing chemical bonds to transport vehicles and other proteins (speciation); and
3. Actual demands of essential elements from biochemical and metabolic processes of various organs and, in consequence, a short-term, reversible efflux from the blood.

   The milk, urine, saliva and sweat measure the component that is absorbed but excreted.

   The blood measures the component absorbed and temporarily in circulation before excretion and/or storage.

   The hair nails and teeth are tissues in which trace minerals are sequestered and/or stored" (Ananth Rao, 2005).

"...Human hair has been selected as one of the important monitoring materials for worldwide biological monitoring in the Global Environmental Monitoring System (GEMS) of the United Nations Environmental

Program." (Environmental Protection Agency, Biological Monitoring of Trace Metals, EPA-600/3-80-089).

Why Use the Hair? Why not Use the Blood?

A hair is ideal tissue for sampling and testing. First, it can be cut easily and painlessly and can be sent to the lab without special handling requirements. Second, clinical results have shown that a properly obtained sample can give an indication of mineral status and toxic metal accumulation following long term or even acute exposure.

A HTMA reveals a unique metabolic world: intracellular activity, which cannot be seen through most other tests. This provides a blueprint of the biochemistry occurring during the period of hair growth and development.

Examples

1. Thirty to 40 days following an acute exposure, elevated serum levels of lead may be undetectable. This is due to the body removing the lead from the serum as a protective measure and depositing the metal into such tissues as the liver, bones, teeth and hair.
2. Calcium loss from the body can become so advanced that severe osteoporosis can develop without any appreciable changes noted in the calcium levels in a blood test.
3. Symptoms of iron deficiency can be present long before low iron levels can be detected in the serum.

Hair, Blood or Urine Analysis

For more than 30 years, the significance of measuring element concentrations in scalp hair, blood, and urine has been studied. These biological samples reflect the body's dynamic equilibrium. Hair acts as a depot and indicates element storage over time. Studies correlate elements in hair with exposure to smelters and mines and with disease and physiologic or pathologic effects of nutritional excesses or deficiencies. Additionally, geographic variation and historical trends in hair element levels have been published, and hair analysis is used in forensic medicine. Consequently, hair analysis provides a long-term record that reflects normal and abnormal metabolism, assimilation and exposure. Authors concluded that hair is a "meaningful and representative tissue for biological monitoring of most of the toxic metals." Hair analysis is also useful as a prognostic tool to ascertain whether an individual has a specific biochemical uniqueness, which can then be addressed in a therapeutic or prophylactic program. Hair element testing is best viewed as a means to monitor element imbalances and environmental toxicity. Follow-up blood testing or provocative urine testing is useful to confirm hair element findings.

Extensive work is underway in the realm of urine element testing. Hamilton, Poulsen, Sabbioni, Van der Venne, and others have undertaken the Herculean task of compiling reference range data for urine and blood elements in various European populations. Urine is an appropriate sample to assess the excretion of potentially toxic elements, providing a window on levels retained in the body and indicating duration of exposure–especially significant diagnostically because of the prevailing patterns of exposure.

In most developed countries, the situation parallels that in Denmark, where "...high-dose environmental or occupational trace element exposure rarely occurs and health risk assessment is mainly pertained to the health effects of long-term low dose exposure." Under certain conditions, urine samples are optimal to gauge the Effects of this long-term, low-dose type deposition of toxic elements. Thus, urine analysis can provide important information to the clinician that may not be readily available with blood analysis. Minerals can be stored in various tissues where they may cause damage or metabolic interference in the depot structures (kidney, bone, nerve tissue) without causing particularly elevated blood levels. Toxic elements are often cleared rapidly from the blood, leaving only a relatively brief time window in which, it is essential that hair be properly handled and analyzed for accurate results. Elements are deposited in the hair in two ways. One way is by ingestion, whereby the element level is a result of systemic processes.

The second way is when toxic or nutrient elements permeate the hair's surface, known as external or exogenous contamination. In the past, the biggest problem with hair analysis was the lack of standard procedures for collecting, preparing and analyzing samples. Fortunately, hair is easy to collect and relatively free of biological deterioration. Unfortunately, hair is easily contaminated, often because of perms or dyes and swimming pool use. Therefore, proper collection and sample preparation is critical for reliable results. Great care must be exercised to prevent environmental contamination in the lab. GSDL uses the most advanced techniques, including the industry's only "clean room," which dramatically reduces external mineral dusts from affecting the analytic results.

Urine Elemental Analysis

This source shows what elements the body is currently excreting. It provides good qualitative information if a person has been recently exposed to a toxic element (days-weeks), and it gives quantitative information of excreted elements before, during, and after provocative challenge.

Blood Elemental Analysis

Blood provides information about what the body has recently (hours to days–in some cases weeks) absorbed. Blood levels are largely independent of tissue deposition. Blood levels vary according to the actual component analyzed (plasma, serum, RBCs). They can be transient in nature, and are subject to the body's homeostatic mechanisms to maintain levels within narrow ranges.

Is the Hair Analysis Really Accurate?

The hair analysis is as credible as a standard blood or urine test–maybe more. The key to a good hair analysis is the capability of a credentialed laboratory and the expertise of the practitioner interpreting the results. The American government and law enforcement agencies have depended upon hair analyses for decades to provide conclusive evidence in crime investigations and for drug testing for prison parolees. The cause of death by poisoning can only be determined conclusively via a hair analysis, as in the case of President Andrew Jackson. A hair analysis is not understood in traditional medicine, but is highly respected in the world of forensics, scientific research, and drug testing. A hair analysis works the same as a soil or water sample to a geologist and engineer–it leaves a permanent imprint of the body's 'environment' within the protein in the hair follicles. If you want to determine if your well or tap water is polluted, you have a water test performed, right? If you're drilling for oil, you have a soil test done. So, it is with the human body. If you want to know if toxins are within your tissues, have a hair analysis performed.

How Accurate is Hair Testing?

When performed correctly, which includes not washing the hair at the laboratory; mineral analysis is as accurate as or more so than standard blood tests. Some hair testing laboratories wash the hair with detergents or acetone for three to ten minutes. Research indicates this is unwise. Washing erratically removes up to 50 per cent of the water-soluble minerals. Commercial hair testing laboratories are inspected by the government, and must adhere to the same standards as blood laboratories. Often they are under greater scrutiny because the test is less conventional.

Who are Exposed to Heavy Metals?

1. Traffic police
2. Taxi drivers
3. Car station workers
4. Petrol station workers
5. Industrial workers like coal, paint, acid production, etc.
6. People living near the industries.
7. Smokers, alcoholics and drug users.
8. Drinking of tea and coffee that contains Cd.
9. Uptake through contaminated food and water.

Table 60.2: Maximum Permissible Limit of Heavy Metal in Hair

Metal	*Limit (ppm)*
Al	< 10–15
Ar	< 7–10
Cd	<2
Cr	< 10
Hg	< 3–5
Pb	< 10–15

Table 60.3: Ranges of Concentration of Trace Elements Found in Human Hair

Sl.No.	*Element*	*Range (g/g)*	
		Minimum	*Maximum*
1.	Aluminium	3.7	9.0
2.	Antimony	0.064	0.212
3.	Arsenic	0.06	1.2
4.	Barium	0.6	5.6
5.	Bromine	0.17	2.5
6.	Cadmium	0.34	1.6
7.	Calcium	135	1150
8.	Chlorine	158	5190
9.	Chromium	0.65	33
10.	Cobalt	0.24	0.34
11.	Copper	14.5	1705
12.	Gallium	0.02	0.14

Contd...

Table 60.3–Contd...

Sl.No.	*Element*	*Range (g/g)*	
		Minimum	*Maximum*
13.	Germanium	0.90	307
14.	Gold	1.55	505
15.	Iodine	2.7	12.0
16.	Iron	73	110
17.	Lanthanum	0.49	0.83
18.	Lead	14.5	52
19.	Magnesium	7.3	21.8
20.	Manganese	1.7	808
21.	Mercury	1.08	1.55
22.	Molybdenum	0.13	3.1
23.	Nickel	20	28
24.	Phosphorus	120	209
25.	Potassium	72.6	900
26.	Rubidium	0.2	0.23
27.	Selenium	0.48	0.79
28.	Silicon	1.05	27.9
29.	Silver	10.9	2.6
30.	Sodium	0.4	1250
31.	Strontium	516	12
32.	Sulphur	0.75	841
33.	Tin	0.39	1.6
34.	Titanium	3.9	24.0
35.	Tungsten	0.3	604
36.	Zinc	122	246
37.	Zirconium	1.0	2.3

Table 60.4: Analytical Methods

Sl.No.	*Analytical Method*	*Metals*
1.	Graphite furnace atomic absorption	Hg-methyl and inorganics
2.	Inductively coupled argon plasma mass spectrometry	Al, B, Ba, Ca, Cd, Co, Cr, Cu, Fe, Mg, Mn, Mo, Ni, Pb, Se, Zn
3.	Inductively coupled plasma optical emission spectrometry	Ar, Cd, Cr, Cu, Mn, Mg, Li, Be, Ni, Pb, Se, Zn
4.	Proton induced X-ray emission (PIXE) spectrometry	Se, Mo, Pb, Thallium
5.	X-ray analysis	
6.	Neutron activation analysis	

Benefits of Hair Analysis

1. Accurately indicates past ingestion or exposure.
2. Element remains permanently in the hair matrix
3. Concentrations higher in hair than blood or urine
4. Longer window of detection-months instead of hours
5. Cost effective
6. Time saving
7. Sensitive than urine and blood analysis
8. Simple to collect, store and transport
9. Eliminates the need for random testing
10. Differentiates between one time usage or addictive behaviour.

Case Studies

Table 60.5: Total Mercury Concentrations in Human Hair from 13 Countries in Relation to Fish Consumption and Location

Sl.No.	*Country*	*Mean Hair Mercury (ppm)*	*Range (ppm)*
1.	Australia	1.7	1.3–3.6
2.	Canada	1.8	1.8–2.2
3.	China	2.7	2.7–8.8
4.	W. Germany	0.7	0.1–1.3
5.	Hong Kong	4.0	1.6–6.2
6.	Italy	1.6	0.5–7.5
7.	Japan	5.0	1.6–22.0
8.	Monaco	1.7	0.2–11.4
9.	New Zealand	2.0	1.8–2.2
10.	New Guinea	1.4	0.8–7.8
11.	South Africa	1.8	1.4–6.3
12.	U.K.	5.3	15–8.8
13.	U.S.A.	3.0	1.1–7.6

Source: Airey, 1983.

Differences Between Countries

The data for this experiment were collected to examine inter country variability in hair mercury concentrations for 3 groups of people when each group ate the same quantity of fish, Fish consumption has been shown to have an effect but what other reasons could account for the difference in means of each group (*e.g.* group A). I suggest that some or all the following factors could account for the observed trends, although this data set alone, statistical proof is not possible: (*i*) latitude of the country, average yearly temperature of the land or the amount of rainfall, (*ii*) industrial density, (*iii*) size of fish eaten on average (hence fish mercury concentration) and the amount of fish in a fish meal, (*iv*) the position donors live relative to the sea, (*v*) population density. People living on highly populated islands or in places with little potential for agricultural development might turn to the sea for their protein. Besides eating larger amounts of local fish, the mercury containing industrial and domestic effluent from such populations may tend to accumulate in fish from local fishing grounds. It has been shown that, using the data from China as an example, the closeness of the donors' homes to fishing ports is also a factor to be considered.

When the data from this survey and a combination of all data available from the literature are grouped according to latitude an interesting trend occurs, although it cannot be proven statistically. When either data set is grouped in latitude bands mercury concentrations from people in the latitudes >400 are significantly less, at the 0.005 level, than those from other latitudes (*i.e.* 2.62 ppm compared to 3.10 and 3.32 ppm). Data from this study although not collected randomly are given for comparison (1.5 ppm compared to 2.35 and 2.21 ppm). One reason for these lower levels could be a temperature effect. At high temperatures it has been shown that more mercury volatilizes from the ground (Airey, 1982) so at persistently low temperatures even in industrial areas, less mercury will be volatilized from surfaces and get into the pathways that eventually reach man. Such information is important to people involved in global mercury modeling (Millward, 1982) who at present use a single value for fluxes from the earth at all latitudes.

As more fish is eaten, the body burden of mercury increases, as shown by hair mercury concentrations. Groups of people who ate fish every day, every week, twice a month and once a month had mean hair mercury concentrations of 11.6 ppm, 2.5 ppm, 1.9 ppm and 1.4 ppm respectively.

Table 60.6: Trace Elements in Hair–Mumbai Industrial Workers

Element	*Mean (ppm)*	
	Root Section	*Distal Section*
Na	107	90
Cr	1.03	0.94
Mn	3.9	6.5
Fe	110	91
Cu	22	22.7
Zn	137	166
Cd	0.9	0.62
As	0.13	0.26
Se	1.32	1.26
K	56	33

Source: Bhat *et al.*, 1982.

Elevated values were particularly conspicuous in the hair piece samples, most of which were cut in the period 1880–1929 and are considered to have been contaminated from exogenous sources by application of hair oil and cosmetics on every occasion of use. According to the factor analysis, a similar pattern of loading of factors 1 and 2 was found for these elements. As already discussed in the previous report, the very high levels of inorganic mercury such as those found in the present samples must have resulted from exogenous contamination, and organic mercury has its origin in the intake of methyl mercury via food (fish) consumption. (Bhat *et al.*, 1982)

Therefore, the positive loading of factor may be relevant to exogenous contamination. Our findings are similar to the previous report that hair specimens from historical populations (1871–1923) contained a significantly greater amount of lead (93.4 for adults and 164.2pgg' for children) than does contemporary hair from the U.S.A. The authors who reported these data simply interpreted the elevated level as reflecting a greater ingestion of lead in historical populations.

Therefore, contamination from an exogenous source is less likely to have occurred for these elements. Non-significant loadings of factor in the factor analysis also suggest little possibility of exogenous contamination. As for Zn, the elevated levels in the older groups and moderate loading of factor are to be noted. These facts suggest that there has been some contamination of the specimens. Moderate loading of factor 1 was also found for Na, the level of which was very high in unwashed samples from older groups and was comparable to the level found in contemporary groups after washing. Thus, the possibility of exogenous contamination in older samples cannot be neglected. Levels of K were lower in unwashed hair samples than have been reported in the literature and comparison with the contemporary level revealed significantly lower levels in washed samples. The meaning of this finding is not clear at present. Not only exogenous contamination of Na but high salt intake in older groups might have contributed to elevation of the Na levels and the Na/K ratio.

Table 60.7: Element Concentrations in Women's Hair from Old and Contemporary Populations (Japan) (Suzuki and Hongo, 1984)

Elements	*Historical Hair Samples*		*Contemporary*
	1911–29	*1930–68*	*1981–83*
Sr	7.4	5.0	5.9
Mn	33.5	4.2	0.9
Fe	1284	103	15.4
Cu	67	25.8	14.5
Zn	380	236	184
Pb	100	20	3.4
Na	19.7	17.9	15.1
Hg	7.7	3.1	1.1

A comparison of the trace elements in hair (TEH) of persons living in "industrial" locations with those in "normal-industrial" areas leads to the following observations: (a) elevated levels of Mn, by a factor of two, in people working in/living near a fossil fuel based power station compared to the "normal" population. Trace element accumulation very high in Historical hair samples than Contemporary. Because due to various reason application of hair oil and cosmetics in old period and high level of organic mercury due to fish consumption.

The mean concentrations of seven elements have been shown for the subjects of four groups namely, fishermen (Table 60.1), students (Table 60.2), businessmen (Table 60.3) and non-mining workers (Table 60.4). Further, the values of elements have also been given for the sub-groups of subjects classified based on their characters such as age, diet, smoking habit etc. Significantly, few elemental levels were found different due to certain characters. Mainly, Cu (fishermen), Mn (businessmen) and Cd (workers) were different due to age whereas, Cu (students), Cd, Ni and Zn (businessmen) were varied between the non-vegetarians and vegetarians. Moreover, Cd, and Ni (fishermen) and Zn (students) levels were influenced by the smoking habit. Similarly, Cd and Cu were affected by the tobacco chewing habit of fishermen. The Cu and Cd concentrations differed respectively owing to income of businessmen and year of service of workers. When the element levels in the hair of each group were compared with those of other groups, certain elemental levels were significantly different. Remarkably, high Cd, low Mn and Zn levels in the non mining workers, low Ni and high Pb levels in the students and businessmen and low Zn level in the fishermen were observed.

Table 60.8: Levels of Elements in the Hair of Fishermen from Pondicherry (ppm)

Sl.No	Subjects	Samples	Cd	Cr	Cu	Mn	Ni	Pb	Zn
1.	Total subjects	40	0.2	1.2	21.3	3.8	1.8	6.3	140.6
2.	Age: 31–45 year	20	0.2	1.4	35.3	6.1	2.6	4.6	149.2
	46–60 year	16	0.1	1.1	15.7	7.3	1.5	6.5	137.5
	61–75 year	4	0.2	1.3	18.8	6.2	1.7	8.1	138.7
3.	Smoking habit Non Smoking	13	0.1	1.6	27.0	2.8	3.2	4.0	124.7
	Smoking	13	0.3	1.0	22.1	6.6	1.5	7.5	149.9
	Tobacco chewing	14	0.4	1.7	14.3	5.7	1.9	4.2	122.4
4.	Income: Medium Income	12	0.3	1.3	18.8	6.2	1.7	8.1	138.7
	Low Income	28	0.1	1.3	21.4	3.9	1.6	6.6	144.1
5.	Hair colour Black colour	19	0.1	1.3	21.7	3.7	1.9	6.9	141.0
	White colour	1	0.4	1.3	6.1	3.1	1.6	3.4	139.9
	Mixed colour	20	0.1	1.2	22.4	4.6	1.7	6.1	140.7

Source: Sugumar and Subramanian, 2002.

Table 60.9: Levels of Elements in the Hair of Students from Chennai (ppm)

Sl.No	Subjects	Samples	Cd	Cr	Cu	Mn	Ni	Pb	Zn
1.	Subjects Total normal	35	0.2	1.5	18.8	2.7	0.4	22.4	175.8
2.	Age: 15 year age group	5	0.3	0.9	13.2	2.6	0.7	23.9	199.8
	16-30 year age group	30	0.2	1.6	19.7	2.7	0.3	19.8	169.8
3.	Diet: Vegetarian	19	0.2	1.6	12.5	2.9	0.4	18.1	18.7
	Non Vegetarian	16	0.2	1.3	26.3	2.4	0.4	27.6	168.1
4.	Smoking habit: Non Smoking	27	0.2	1.6	15.6	2.6	0.4	20.4	191.5
	Smoking	8	0.1	1.0	29.3	2.9	0.2	29.1	139.4
5.	Income: High Income	1	0.7	1.7	21.2	2.4	0.1	12	127.8
	Medium Income	32	0.2	1.5	19.1	2.7	0.4	23.2	18102
	Low Income	2	0.2	1.8	12.2	3.8	0.2	15.0	143.8
6.	Hair colour: Black colour	32	0.2	1.4	19.1	2.8	0.4	23.2	181.2
	Brown colour	3	0.05	1.7	15.2	1.3	0.04	14.0	127.8

Source: Sugumar and Subramanian, 2002.

Table 60.10: Levels of Elements in the Hair of Business from Madras

Sl.No.	Subjects	No. of Samples	Mean Concentrations (ppm±SE) in Har						
			Cadmium (Cd)	Chromium (Cr)	Copper (Cu)	Manganese (Mn)	Nickel (Ni)	Lead (Pb)	Zinc (Zn)
1.	Businessmen total (Normal)	37	0.4±0.07	1.2±0.2	16.3±3.5	3.2±0.7	1.0±0.2	30.2±13.8	212.3±14.8
2.	Age:								
	16–30 year age group	11	0.4±0.1	1.2±0.2	15.1±1.2	2.2±0.4	1.3±0.5	11.7±4.3	201.8±26.2
	31–45 year age group	14	0.4±0.1	1.0±0.3	20.8±3.0	5.7±0.4	0.6±0.2	33.3±6.7	2541.1±21.9
	46–60 year age group	10	0.4±0.2	1.5±0.4*	13.1±1.6	1.6±1.7	1.2±0.8	37.2±18.0	189.1±19.5
	61–75 year age group	2	0.3±0.2	1.0±0.1	10.5±0.5	1.6±1.5	0.9±0.8	10.2±2.6	214.0±39.5
3.	Diet								
	Vegetarian	17	0.2±0.06	1.1±0.2	16.5±2.0	3.4±1.0	0.6±0.3	37.6±2.2	162.1±17.0
	Non-vegetarian	20	0.7±0.01**	1.2±0.2	15.7±1.1	2.8±0.5	1.8±0.4**	27.5±4.7	245.7±20.3**
4.	Smoking habit								
	Non-smoking	24	0.4±0.07	1.1±0.2	15.9±1.6	2.2±0.7	2.3±2.3	20.4±2.3	213.7±16.3
	Smoking	13	0.4±0.4	1.2±0.2	18.7±1.2	2.5±0.7	0.9±0.02	17.3±6.1	192.1±32.0
5.	Income								
	High income	12	0.1±0.1	2.7±0.3	11.7±3.7	2.2±0.7	2.3±2.3	16.4±2.3	183.0±22.3
	Medium income	25	0.4±0.08	1.0±0.1***	16.0±1.5	3.1±0.7	0.9±0.2	31.3±14.9	221.5±14.2
6.	Hair colour								
	Black colour	17	0.3±0.09	1.1±0.2	17.5±1.7	1.6±1.5	0.9±0.8	10.2±2.6	214.0±39.5
	White colour	1	0.03	0.8	23.8	3.8	2.7	8.0	248.9
	Mixed colour	19	0.5±0.1	1.8±0.3	13.9±1.9	4.2±1.6	1.2±0.5	18.2±39.0	202.6±18.4

Source: Sugumar and Subramanian, 2002.

Table 60.11: Levels of Elements in the Hair of Male Workers of Neyveli Lignite Corporation, Tamil Nadu

Sl.No.	Subjects	No. of Samples	Mean Concentrations (ppm±SE) in Hair						
			Cadmium (Cd)	Chromium (Cr)	Copper (Cu)	Manganese (Mn)	Nickel (Ni)	Lead (Pb)	Zinc (Zn)
1.	Workers total	50	1.3±0.2	0.9±0.2	22.5±2.7	1.2±0.2	1.2±0.3	6.8±0.8	9.8±6.6
2.	Age:								
	16–30 year age group	21	0.9±0.1	1.0±0.2	21.9±3.4	1.0±0.2	1.5±0.5	6.7±0.8	90.3±9.5
	31–45 year age group	17	1.5±0.3	1.0±0.6	27.8±7.5	1.3±0.3	0.8±0.2	8.9±1.9	87.0±1.9
	46–60 year age group	11	1.9±0.5*	0.4±0.2	17.2±4.2	1.4±0.6	1.0±0.5	4.5±2.8	112.8±25.3
3.	Diet:								
	Vegetarian	18	1.9±0.3	0.7±0.3	23.8±6.0	0.91±0.5	2.1±0.6	8.4±3.4	121.9±26
	Non-vegetarian	32	1.2±0.03	0.9±0.2	22.3±3.0	1.2±0.2	1.1±0.4	6.5±0.8	95.3±6.5
4.	Smoking habit								
	Non-smoking	29	1.4±0.2	0.9±0.2	22.5±19	0.9±0.2	1.2±0.5	4.8±1.2	97.3±8.4
	Smoking	21	1.2±0.2	0.8±0.4	21.9±2.6	1.5±0.4	1.3±0.4	4.7±0.6	102.7±11.2
5.	Year of service:								
	1–10 year service	23	1.1±0.1	0.9±03	23.5±3.4	1.1±0.1	1.4±0.5	1.3±0.9	95.9±6.4
	11–20 year service	17	0.9±0.3	0.9±0.5	11.1±3.5	2.2±1.5	0.9±0.3	5.5±2.7	95.3±10.9
	21–30 year service	10	1.9±0.4*	0.5±0.2	22.5±4.9	1.0±0.3	0.8±0.4	5.8±0.5	108.5±22.0
6.	Hair colour								
	Black colour	28	1.3±0.2	0.8±0.2	17.9±2.9	0.9±0.1	0.9±0.03	6.4±1.0	96.9±12.6
	Brown colour	4	1.5±0.4	1.3±1.0	19.8±2.3	1.6±0.7	2.7±2.3	6.6±3.0	100.1±11.1
	White colour	2	1.2±0.4	0.8±0.3	9.5±0.9	2.1±1.0	1.1±0.1	10.5±5.5	89.4±3.6
	Mixed colour	16	1.3±0.6	0.8±0.4	23.5±1.2	1.1±1.4	1.1±01	6.7±0.8	102.3±5.1

Significance: 16–30 Vs 46–60 year age groups and 1–10 Vs 21–30 year and service.

*: $P < 0.05$

Source: Sugumar and Subramanian, 2002.

The lower level of Zn found in the fishermen (non-vegetarian) in comparison to other groups, might be ascribed to their monotonous life style of food habit mainly including fish as their stable food. The levels of elements, except for Cd and Ni levels, in the hair of fishermen and students representing the general Population of Madras, were within the normal ranges reported by Schroeder and Nason). But the higher Pb level observed in the students might be related to the variations in exposure to sources of traffic and industrial pollution, as reported elsewhere). The high levels of Cd in the hair of non-mining workers including clerks, watchmen, security personnel and drivers of the lignite open mine, may indicate the possible exposure to Cd from the open mining and the nearby lignite thermal power plant. The low levels of Mn and Zn observed in the non mining workers were ascertained to low nutritional intake and the rapidly changing life style of dietary habits. Jones *et al.* pointed out high levels of Cd (3.83 ppm) anti normal levels of Cu (10.29), Pb (23.79) and Zn (113.1) in the hair of residents of gold and copper mining area of Ok Tedi. When compared to these values, our values of Cd and Pb were lower and Cu level was higher. It has long been known that mining activity can markedly change the status and distribution of certain heavy metals in the adjacent environment. A few parameters such as age, diet, smoking habit, year of service and income of family may account for change in the concentrations of elements in the hair of certain subjects. The Cu level in particular, was higher in the middle age group (31–45 yr) than in the older age group (46–60 yr) of fishermen. But the Cu and Cd levels were lower in the younger age group (16–30 yr) than in the middle age group (31-45 yr) of businessmen and the older age group (46–60 yr) of non-mining workers. Thus, both the higher and lower concentrations of elements were observed in the younger age group of subjects when compared to the element concentrations found in the older age groups.

Lower levels of Pb in the hair of males and higher levels in the females up to the age of 35 yr and then a sharp decrease. In the case of vegetarians, the high phytate intake may result in reduced availability of Zn for intestinal absorption. It may be the reason that significantly lower levels of lements were observed in the hair of vegetarians than in the non vegetarians; the Cu level in the students and the Cd, Ni and Zn in the businessmen from Madras was significantly lower. But insignificant difference in the hair contents of elements between vegetarians and non-vegetarians. Tobacco chewing and smoking may be correlated with the change in hair contents of elements. Higher Cd concentration was observed in the fishermen smoking and chewing tobacco as well as lower levels of Ni (fishermen with smoking habit), Cu (fishermen with tobacco chewing habit) and Zn (student-smokers) was observed, pointed out higher levels of Cd and Pb in the smokers than in the non-smokers while others found out insignificant difference between smokers and non-smokers. Further, the concentrations of Cd were higher in the workers with 21–30 years of service than in the workers with 1–10 years of service. The level of Cr was higher in the hair of businessmen with high income than in the businessmen with low income. Conclusively, it was observed in the present study that the concentrations of elements in the four groups of subjects from three different places were varied due to the difference in the place of residences, working places, different food habits (vegetarianism and no vegetarianism), age, smoking and tobacco chewing habits, income of family and year of service but not due to colour of hair (Sugumar and Subramanian, 2002).

The average concentration and the comparative situations of all the trace metals in the washed and unwashed scalp hair of the donors are summarized. Of all the analyzed trace metals, the order of accumulation was Zn > Cr > Pb > Cu > Ni > Cd with minimum value. The ratio between the samples were found to be 1.54 for Ni and 1.35 for Pb, and relatively lower ratio of 1.27 for Cu. 1.19 for Cd. 1.18 for Cr and 1.07 for Zn. These values revealed a positive exogenous contribution of all the industrial metals examined in the hair of occupationally exposed workers.

Table 60.12: Comparative Average Trace Metal Concentration in Unwashed and Washed Hair Samples–Coimbatore

Metal	*Sample*	*Cases*	*Mean*	*S.D.*	*Ratio*
Chromium	25	Washed	29.88	4.228	1.178
		Unwashed	35.200		
Lead	25	Washed	18.400	5.987	1.350
		Unwashed	24.840		
Zinc	25	Washed	247.963	17.543	1.069.1
		Unwashed	265.186		
Copper	25	Washed	7.648	1.987	1.269
		Unwashed	9.704		
Nickel	25	Washed	4.328	1.024	1.543
		Unwashed	6.480		
Cadmium	25	Washed	1.756	0.418	1.194
		Unwashed	2.096		

Source: Vishwanathan *et al.*, 2002.

The occupationally exposed auto drivers showed a typical exposure to industrial/vehicular pollution and the results were in good agreement with the similar reports for the hair metal concentration in the occupationally exposed workers of the industrial cities (Ashraf *et al.*, 1994; Sen *et al.*, 1996). The estimated level of lead was found to be higher than those reported earlier (Qureshi *et al.*, 1982). The lead and zinc levels of the school children in Ghana (Golow *et al.*, 1994) are comparable with the present study.

The drivers irrespective of time, spend their whole day in the city polluted with traffic fumes and hence could have high vehicular lead and zinc in their hair similar to that found in San Diego (Chow *et al.*, 1970). Lead concentrations of more than 20 $\mu g\ g^{-1}$ is known to be an indicative of significant lead exposure (Krause *et al.*, 1987), which showed a positive trend in the present investigation. The correlation coefficient observed between the trace metals showed significance at 5 per cent level for Ni-Pb and Ni-Cd. Many researchers (Baumslag *et al.*, 1974; Leotsinidis *et al.*, 1990; Sen *et al.*, 1996) have reported correlation between different metal concentrations in human scalp hair, but no possible reasons for the correlation were advanced. Presently it is difficult to ascertain the reasons because further studies need to be carried on the elution and absorption of different elements and the nature of binding of elements in hair. By and large, the subjects in the present study are under threat to automobile exhausts and industrial pollution in urban area.

The overall observations made in the present study with respect to the estimated trace metals possess prime importance in the evaluation of environmental and occupational effects on the professional drivers of urban/industrial areas. Further detailed work in relation to comparative evaluation of environmental, occupational/non-occupational exposure levels, large sample size with distinct exposure periods will be more informative to ensure a better understanding of trace metal accumulation in hair of occupationally exposed drivers.

Conclusion

In conclusion, biomedical trace element research will continue to thrive in future. This field is still wide open, practically inexhaustible and is limited only by the imagination of the investigators. This is a multidisciplinary field and concerted efforts through team work and mandatory. Although there are many techniques for analysis of heavy metal accumulation in the human body, there is no question that hair analysis can be used as a biomedical and environmental research tool in the near future.

References

Abraham, J.L., 1982. Trace Elements in hair. *Lancet*, 2: 554–555.

Aharoni, A., Tester, A., Paltieni, Y., Tal, J., Dori, Z. and Sharf, M., 1992. Hair chromium content of women with gestational diabetes compared with non-diabetic pregnant women. *Am. J. Clin. Nutr.*, 55(1): 104–107.

Airey, D., 1982. Contribution from coal and industrial materials to mercury in air, rain water and snow. *Sci. Total Environ.*, 25: 19–40

Airey, D., 1983. Total mercury concentrations in human hair from 13 countries in relation to fish consumption and location. *Sci. Total Environ.*, 31: 157–180.

Alder, J.F., Samuel, A.J. and West, T.S., 1977. The anatomical and longitudinal variation of trace element concentration in human hair. *Anal Chim Acta*, 92: 217–221.

Anke, H. and Schneider, H.J., 1962. Mineral stoffgehalt de frauen and mannerhaare. *Dtsch Z Verdau Stoffnechaclkr*, 22: 31–35.

Arunachalam, J., Gangadharan, S. and Yegnasubramanian, S., 1979. Elemental data on human hair sampled from Indian student population and their interpretation for studies on environmental exposure. In: *IAEA Symposium, Nuclear Activation Techniques in the Life Sciences*, Vienna, 1978, pp. 499–513.

Ashraf, W., Jaffar, M. and Mohammad, D., 1994. Trace metal contamination study on scalp hair of occupationally exposed workers. *Bull. Environ. Contam. Toxicol.*, 53: 516–523

Bagchi, K.N., Ganguli, H.D. and Sirdar, J.N., 1940. Lead content of human hair. *Indian J. Med. Res.*, 27: 777–791.

Banumslag, N., Yeager, D. and Levin, L., 1974. Trace element content of maternal and neonate hair. *Arch Environ. Health*, 29: 186–191.

Baoshan, Li, 1989. Primary inquiry of relationship between zinc and copper in the hair and unknown cause of children hyporprenia in cretinism area. *J. Xi Med. Univ.*, 10(2): 170–172.

Barltrop, D., Strehlow, C.D., Thornton, I. and Webb, J.S., 1975. Absorption of lead from dust and soil. *Postgrad Med J.*, 51: 801–804.

Barrett, S., 1985. Commercial hair analysis: Science or scam. *JAMA*, 254(8): 1041–1045.

Bate, L.C. and Dyer, F.F., 1965. Trace elements in human hairs. *Nucleonics*, 23: 72–76.

Batzevich, V.A., 1995. Hair trace element analysis in human ecology studies. *Sci. Total Environ.*, 164(2): 89–98.

Baumslag, N., Yeager, D., Levin, L. and Petering, H.G., 1974. Trace metal content of maternal and neonate hair-zinc, copper, iron and lead. *Arch. Environ. Health*, 29: 186.

Bencko, V., Wagner, V., Wagnerova, M. and Zavazal, V., 1986. Human exposure to nickel and cobalt: Biological monitoring and immuno-biochemical response. *Environ Res.*, 40: 399–410.

Bencze, K., 1990. What contribution can be made to biological monitoring by hair analysis? Part I. Fresenius. *J. Anal. Chem.*, 337(8): 867–876.

Bergomi, M., Borella, P., Fantuzzi, G., Tartoni, P.C. and Vivoli, G., 1987. Cadmium levels in tooth, hair, blood of children. In: *Proceedings of International Conference on Heavy Metals in the Environment*, New Orleans, CEP Consultants, UK, 2: 92–94.

Bhat, K., Arunachalam, I., Yegnasubramanian, S. and Gangadharan, 1982. Trace element in hair and environmental exposure. *Sci. Total Environ.*, 22: 169–178.

Bialkowska, M., Hoser, A., Szostak, W.B., Dybczynski, R., Sterlinksi, S., Nowicka, G., Majchrzak, J., Kaczorowski, I. and Danko, B., 1987. Hair zinc and copper concentration in survivors of myocardial infraction. *Ann. Nutr. Metab.*, 31: 327–332.

Borella, P., Bergomi, M., Fantuzzi, G., Rovesti, S. and Vivoli, G., 1987. Cadmium, zinc and copper in the hypertension. In: *Proceedings of International Conference on Heavy Metals in the Environment*, New Orleans, CEP Consultants, UK, 2: 89–91.

Bose, A.K. and Chakravorty, M.D., 1964. Manganese concentration in hair. *Indian J. Ind. Med.*, 10: 110–114.

Bosque, M.A., Domingo, J.L., Lobert, T.M. and Corbella, J., 1991. Cadmium in hair of school children living in Tarragona Province, Spain: relationship to age, sex and environmental factors. *Sci. Total Environ.*, 28(2): 147–156.

Briggs, M.H., Briggs, M. and Wakatama, A., 1972. Trace elements in human hair. *Experientia (Basel)*, 23: 406–407.

Burguera, J.L. and Rondon, C., 1987. Lead content in hair, as related to petrol-lead exposure in Merida City, Venezuela. In: *Proceedings of International Conference on Heavy Metals in the Environment*, (Eds.) S.E. Lindberg and T.C. Hutchinson, New Orleans, CEP Consultants UK, 2: 272–276.

Caprio, R.J., Margulis, H.L. and Jaselow, M.M., 1974. Lead adsorption in children and its relation to urban traffic densities. *Arch. Environ. Health*, 28: 195–197.

Carvalho, F., Tavares, T.M., Souza, S.P. and Linhares, P.S., 1984. Lead and cadmium concentrations in the hair of fisherman from the Subae River basin, Brazil. *Environ. Res.*, 33: 300–306.

Chattopadhyay, A., Roberts, T.M. and Jervis, R.E., 1977. Scalp hair as a monitor of community exposure to lead. *Arch. Environ. Health*, 31: 226–236.

Chattopadhyay, P.K., Joshi, H.C. and Samaddar, K.R., 1990. Hair cadmium level of smokers and non-smokers human volunteers in and around Calcutta city. *Bull. Environ. Contam. Toxicol.*, 45: 177–180.

Chen, M.D., Lin, D.Y., Lin, W.H. and Cheng, V., 1988. Zinc in hair and serum of obese individuals in Taiwan. *Am. J. Clin. Nutr.*, 48: 1307–1309.

Chlopicka, J., Zagrodzki, P., Zachweija, Z., Krosniak, M. and Folta, M., 1995. Use of pattern recognition methods in the interpretation of heavy metal content (lead and cadmium) in children's scalp hair. *Analyst*, 120(3): 943–945.

Chow, J.J., 1970. Lead accumulation in roadside soil and grass. *Nature*, 225:295–297.

Collipp, P.J., Kuo, B., Castromagana, M., Chen, S.Y. and Salvatore, S., 1985. Hair zinc, scalp hair quantity and diaper rash in normal infants. *Cutis*, 35(1): 66–70.

Dang, H.S., Jaiswal, D.D., Mehta, U. and Deshpande, A., 1983. Trace element changes during pregnancy: Preliminary study. *Sci. Total Environ.*, 31: 187–192.

Danilovic, V., 1958. Chronic nephritis due to ingestion of lead contaminated flour. *Br. Med. J.*, 1: 127–128.

De Antonio, S.A., Katz, S.A., Scheiner, D.M. and Wood, J.D., 1982. Anatomically related variations in trace metal concentrations in hair. *Clin. Chem.*, 28: 2411–2413.

DiPietro, E.S., Donald, L.P., Daniel, C.P. and Jane, W.N., 1989. Determination of trace elements in human hair: Reference intervals for 38 elements in non-occupationally exposed adults in the US and effects of hair treatments. *Biol. Trace Elem. Res.*, 22: 83–100.

Dissanayake, C.B., Sanarathe, A. and Weerasooriyan, S.V.R., 1984. Environmental significance of trace elements in human hair: A case study from Sri Lanka. *Int. J. Environ. Stud.*, 23: 41–48.

Dorea, J.G., Homer, M.R., Bezara, V.L., Pereira, M.G. and Salomon, J.B., 1982. Hair zinc levels and nutritional status in urban children from Ilheurs, Bahia, Brazil. *Hum. Nutr. Appl. Nutr.*, 36A: 63–67.

Dutcher, T.F. and Rothman, S., 1951. Iron, copper and ash content of human hair of different colors. *J. Invest. Dermatol.*, 17: 65–68.

Dybezynik, R. and Boboli, K., 1976. Forensic and environmental aspects of neutron activation analysis of single hair. *J. Radioanal. Chem.*, 31: 267–289.

Eads, E.A. and Lambdin, C.E., 1973. A survey of trace metals in human hair. *Environ. Res.*, 6: 247–252.

Eltayeb, M.A.H. and Van Grieken, R.E., 1990. Iron, copper, zinc and lead in hair from Sudanese population of different age groups. *Sci. Total Environ.*, 95: 157–166.

Ely, D.L., Mostardi, R.A., Woebkenberg, N. and Worsteli, D., 1981. Aromatic and hair trace metal content in learning disabled children. *Environ. Res.*, 25: 325–339.

Fergusson, J.E., Hibbard, K.A. and Ting, R.L.H., 1981. Lead in human hair: General survey, battery factory employees and their families. *Environ. Pollute*, B2: 235–248.

Flynn, A., 1976. Hair element analysis as a measure of mineral status. *J. Appl. Nutr.*, 29: 51–57.

Folio, M.R., Hennigan, C. and Errera, J., 1982. A comparison of five toxic metals among rural and urban children. *Environ. Pollute.* A29(4): 261–269.

Gangadharan, S. and Das, M. Sankar, 1976. Hair as an indicator of environmental exposure. In: *Report to AGM on Application of Nuclear Methods in Environmental Research*. International Atomic Energy Agency, Vienna.

Gibson, R.S. and Wolfe, M.S.De, 1979. Copper, zinc, manganese, vanadium and iodine concentrations in the hair of Canadian low birth weight neonates. *Am. J. Clin. Nutr.*, 32: 1728–1733.

Golow, A.A. and Kwaansa-Ansah, E., 1994. Comparison of lead and zinc levels in the hair of pupils from four towns in the Kumasi municipal area of Ghana. *Environ. Contam. Toxicol.*, 53: 325–331.

Gordus, A., 1973. Factors affecting the trace metal content of human hair. *J. Radioanal. Chem.*, 15: 229–243.

Goss, H. and Green, M.M., 1955. Copper in hair. *Science*, 112: 112–330.

Grandjean, P., 1979. Lead content of scalp hairs as indicator of occupational exposure. In: *Toxicology and Occupational Medicine*, (Ed.) W.D. Deichman. Elsevier, Amsterdam, pp. 311–318.

Grant, E.C.G., Howard, J.M., Davis, S., Chasty, H., Hornby, B. and Galbraith, J., 1988. Zinc deficiency in children with dyslexia: concentration of zinc and other minerals in sweat and hair. *Br. Med. J.*, 296(662): 607.

Hagedorn-Gotz, H., Kuppers, G. and Stoepler, M., 1977. On the nickel contents in urine and hair in case of exposure to nickel carbonyl. *Arch. Toxicol.*, 38: 275–285.

Hambidge, K.M. and Baum, J.D., 1972. Hair chromium concentrations of human newborn and changes during infancy. *Am. J. Clin. Nutr.*, 25: 376.

Hambidge, K.M., Chavez, M.N., Brown, R.M. and Walravens, P.A., 1979. Zinc nutritional status of young middle-young middle-income children and effects of consuming zinc-fortified breakfast cereals. *Am. J. Clin. Nutr.*, 32: 2532–2539.

Hambidge, K.M., Rodgerson, D.O. and O'Brein, D., 1968. Concentration of chromium in the hair of normal children and children with juvenile diabetes mellitus. *Diabetes*, 17: 517–519.

Hammer, D.I., Finkles, J., Hendrick, R.H., Shy, C.M. and Horton, R.J.M., 1971. Hair trace metal levels and environmental exposure. *Am. J. Epideminol.*, 93: 84–92.

Hartwell, T.D., Handy, P.W., Haris, D.S., Williams, S.K. and Genlbach, S.H., 1983. Heavy metal exposure in populations living aground zinc and copper smelters. *Arch. Environ. Health*, 38: 284–295.

Hauser, G., Vienna, A., Wofsperger, M. and Goessler, W., 1999. Milk consumption, smoking and lead concentration in human hair. *Coli. Anthropol.*, 23(2): 433–436.

Heinersdoff, N. and Taylor, T., 1979. Concentration of zinc in the hair of school children. *Arch. Dis. Child.*, 45: 958–960.

Hewitt, C.D. and Day, J.P., 1986. Aluminium and copper concentrations in hair and serum are unrelated in renal patients. *Acta. Pharmacol. Toxicol.*, 59(87): 442–445.

Hopps, H.C., 1977. The biological base of using hair and nail for analysis of trace elements. *Sci. Total Environ.*, 7: 71–89.

Horst, G., Marquardt, D., Wlbrandt, B. and Leppins, S., 1992. Biological monitoring in environments of ferro-metallurgical industry. *Zentbl. Hyg. Umweltmed*, 192(6): 509–521.

Huang, B., Lin, S., Chen, S., Zhou, G., Yin, F., Lou, Z. and Bi, M., 1991. Hair chromium levels in patients with vascular diseases. *Biol. Trace Elem. Res.*, 29(2): 133–138.

Huel, B., Boudene, C. and Ibrahim, M.A., 1981. Cadmium and lead content of maternal and newborn hair: Relationship toparity, birth weight and hypertension. *Arch. Environ. Health*, 36: 221–227.

Hyoi, N., Imahori, A., Shiobara, S. and Fukashima, I., 1986. Study on trace element concentrations in hair and blood of electroplanting workers. *Jpn. J. Ind. Health*, 28: 200–201.

Ilhan, Am, Uz, E., Kali, S., Var, A. and Akyol, O., 1999. Serum and hair trace element levels in patients with epilepsy and healthy subjects: Does the antiepileptic therapy affect the element concentration of hair? *Eur. J. Neurol.*, 6(6): 705–709.

Jamall, I.S. and Allen, P.V., 1990. Use of hair as an indicator of environmental lead pollution in women of child-bearing age in Karachi, Pakistan and Bangladesh. *Bull. Environ. Contam. Toxicol.*, 44(3): 350–356.

Jamall, I.S. and Jaffer, R.A., 1987. Elevated iron levels in hair from steel mill workers in Karachi, Pakistan. *Bull. Environ. Contam. Toxicol.*, 39(4): 608–613.

Jaworowski, Z., Bilkiewiez, J. and Kostanecki, W., 1966. The uptake of 210Pb by resting and growing hair. *Int. J. Radiat. Biol.*, 11: 563–566.

Jeejeebhoy, K.N., Chu, R.C., Marless, E.B., Greenberg, G.R. and Roberston, A.B., 1977. Chromium deficiency, glucose intolerance and neuropathy reversed by chromium supplementation in a parent nutrition. *Am. J. Clin. Nutr.*, 30: 531–538.

Jenkins, D.W., 1979. Toxic trace metals in mammalian hair and nails. EPA 600/04–79–049. U.S. Environmental Protection Agency, Washington, DC.

Jervis, R.E., Tiefanbach, B. and Chattopadhyay, A., 1977. Scalp hair as a monitor of population exposure to environmental pollutants. *J. Radioanal. Chem.*, 37: 751–760.

Jones, G.L., Wily, D., Lumsden, B., Taufa, T. and Lourie, J., 1987. Trace metals in the hair of inhabitants of the Ok Tedi region, Papua New Guinea. *Environ. Pollut.*, 48(2): 101–115.

Junko, A., Ichrio, F. and Akira, I., 1982. Multi-element hair analysis of mining industry workers. *J. Radioanal. Chem.*, 68(1–2): 59–65.

Kanofsky, J.D., Rosen, W.A., Ryan, P.B.B., Decina, P.L., Fiove, R.R. and Kanofsky, P.B., 1986. Lead levels in the hair of bipolar patients and normal controls. *Med. Hypotheses*, 20(2): 151–156.

Kapauan, P.A. and Beltran, I.L. and Cruz, C.C., 1982. Trace metal pollutants in Filipino human head hair. *Philipp. Sci. III* (3–4): 145–155.

Kikkawa, H., Ogita, Z. and Fujito, S., 1955. Nature of pigments derived from tyrosine and tryptophan in animals. *Science*, 121: 43–47.

Klevay, L.M., Bistrain, B.R., Fleming, C.R. and Neumann, C.G., 1987. Hair analysis in clinical and experimental medicine. *Am. J. Clin. Nutr.*, 46(2): 233–236.

Klevay, L.M. and Forks, H.G., 1973. Hair as a biopsy material. *Arch. Environ. Health*, 26: 169–172.

Kopito, L., Byers, R. and Schwachman, H., 1967. Lead in hair of children with chronic lead poisoning. *N. Engl. J. Med.*, 276: 949–953.

Kostial, K., 1983. The absorption of heavy metals by the growing organism. In: *Health Evaluation of Heavy Metals in Infants: Formula and Junior Food*, (Eds.) E.H.F. Schmidt and A.G. Hildebrandt. Springer, Berlin, pp. 99–104.

Kowal, M.E., Johnson, D.E., Kraemer, D.F. and Pahren, D.F., 1979. Normal levels of cadmium in diet, urine, blood and tissues of inhabitants of the United States. *J. Toxicol. Environ. Health*, 5: 995–1014.

Krause, C. and Chutsch, M., 1987. Haare als Indicator fur die Erfassung von pb and Cd Belastungen, Schriftner. *Ver Wasser Boden. Lufthyg*, 71: 101–109.

Lea, C.M. and Luttrell, V.A.S., 1965. Copper content of hair in kwarshiokor. *Nature (Lond)*, 206: 413.

Lebrudo, C.J. and Marie, T.M.C., 1988. Hair zinc and copper levels in healthy and malnourished Filipino Children. *Acta Mineral*, 37: 93–98.

Leostenidis, M. and Kondakis, X., 1990. Trace metals in scalp hair of Greek agricultural workers. *Sci. Total Environ.*, 95: 149–156.

Limic, N. and Valkovic, V., 1986. Incorporation of trace elements from environment into the hair structure. *Biol. Trace Elem. Res.*, 12: 363–374.

Majumdar, S., Chatterjee, J. and Chaudhuri, K., 1999. Ultrastructural and trace metal studies on radiographers hair and nails. *Biol. Trace Elem. Res.*, 67(2): 127–138.

McKenzie, J.M., 1978. Alteration of the zinc and copper concentration of hair. *Am. J. Clin. Nutr.*, 31: 470–476.

Medeiros, D.M. and Pellum, L.K., 1984. Elevation of cadmium, lead and zinc in the hair of adult black female hypertensives. *Bull. Environ. Contam. Toxicol.*, 32: 525–532.

Millward, G. E., 1982. Non steady state simulation of the global mercury cycle. *Int. J. Geophys. Res.*, 87: 88–91.

Moon, J., Davidson, A.J., Smith, T.J. and Fodl, S., 1988. Correlation clusters in the accumulation of metals in human scalp hair: Effects of age, community of residence and abundance of metals in air and water. *Sci. Total Environ.*, 72(5): 87–112.

Niculescu, T., Dumitru, R. and Bolha, V., 1983. Relationship between the lead concentration in hair and occupational exposure. *Br. J. Ind. Med.*, 40: 67–70.

Nishiyama, K. and Nordberg, G.F., 1972. Adsorption and elution of Cd on hair. *Arch. Environ. Health*, 25: 92–96.

Nordberg, G.F., Fowler, B.A., Friberg, L., Jernelov, A., Nelson, M., Piscator, M., Sandetead, H., Yostal, J. and Vou, B.V., 1978. Factors influencing metabolism and toxicity of metals: A consensus report. *Environ. Health Perspect.*, 25: 3–41.

Obrusnik, I., Gislason, J., Macs, D., Mcmillan, D.K., Diauria, J. and Pate, B.G., 1973. The variation of trace element concentrations in single human head hair. *J. Radioanal. Chem.*, 15: 115–134.

Ohmore, S., Mitura, T., Kusaka, Y., Tsuji, H., Sagawa, T., Fumya, S. and Tamari, Y., 1975. Nondestructive multi-elementary analysis of human hairy by neutron activation. *Radioisotopes*, 24: 396–402.

Petering, H.G., Yeager, D.W. and Wilhemp, S.O., 1973. Trace metal content of hair. II. Cadmium and lead of human hair in relation to age and sex. *Arch. Environ. Health*, 27: 327–330.

Phil, R.O. and Parkers, M., 1977. Hair element content in learning disabled children. *Science*, 198: 204–206.

Qureshi, I.H., Chaudry, M.S. and Ahmed, S., 1982, Trace element concentration in head hair of the inhabitants of the Rawalpindi-Islamabad area. *J. Radioanal. Nucl. Chem.*, 1–2: 209.

Randall, J.A. and Gibson, R.S., 1989. Hair chromium as an index of chromium exposure of tannery workers. *Br. J. Ind. Med.*, 46(3): 171–175.

Reeves, R.D., Jolley, K.W. and Buckley, P.D., 1975. Lead in human hair: Relation to age, sex and environmental factors. *Bull. Environ. Contam. Toxicol.*, 14: 579.

Reilly, C. and Harrison, F., 1979. Zinc, copper, iron and lead in scalp hair of students and non student adults in Oxford. *J. Hum. Ntr.*, 33: 248–252.

Sadiq, M., Milan, A.A. and Al Thagafi, K.M., 1992. Intercity comparison of metals in scalp hair collected after the Gulf war 1991. *J. Environ. Sci. Health*, A27(6): 1415–1431.

Saeni, M.S., 2000. Determination of heavy metal pollution by hair analysis. *Elec. J. Indo. Med. Assoc.*, 1(6): 1–10.

Schroeder, H.A. and Nason, A.P., 1969. Trace metals in human hair. *J. Invest. Dermatol.*, 53: 71–78.

Schroeder, H.A. and Nason, A.P., 1971. Trace element analysis in clinical chemistry. *Clin. Chem.*, 17(6): 461–474.

Schuhmacher, M., Domings, J.L., Llobet, J.M. and Corbehla, J., 1991. Lead in children's hair as related to exposure to Tarregona Province, Spain. *Sci. Total Environ.*, 104(3): 167–174.

Sen, J. and Das Chaudhuri, A.B., 1996. Hair trace element concentrations in the bengalee population. *J. Ecotoxicol. Environ. Monit.*, 4: 237–342.

Shrestha, K.P. and Carrera, A.E., 1988. Hair trace elements and mental retardation among children. *Arch. Environ. Health*, 43(6): 396–398.

Shrestha, K.P. and Scharuzer, G.N., 1989. Trace elements in hair: A study of residents in Darjeeling (India) and San Diego, California (USA). *Sci. Total Environ.*, 79: 171–177.

Sikorski, R., Juszkiewicz, T., Paszkowski, T., Randomanski, T., Szkoda, J. and Milart, P., 1986. Hair trace metal concentration of pregnant women at term in comparison with blood and milk levels. *Eur. J. Obstet. Gynecol. Reprod. Biol.*, 23: 349–357.

Soyalk, M., Saraymen, R. and Dogan, M., 1995. Investigation of lead, chromium, cobalt and molybdenum concentrations in hair samples collected from diabetic patients. *Fresenius. Environ. Bell.*, 4(8): 485–490.

Srikumar, T.S., Ockerman, P.A. and Akesson, B., 1992. Trace element status in vegetarians from South India. *Nutr. Res.*, 12(2): 187–198.

Stauber, J.L. and Florence, T.M., 1989. Manganese in scalp hair: problem of exogenous manganese and explications for manganese monitoring in Grook Eylandt Aborigines. *Sci. Total Environ.*, 83: 85–98.

Strain, W.H., Pories, W.J., Flynn, A. and Hill, O.A., 1972. Trace element nutrient and metabolism through head hair analysis. In: *Trace Substances in Environmental Health*, (Ed.) V.D.D. Hemphill. University of Missouri, Columbia, pp. 383–397.

Stunkovic, B., Stankovic, M., Milies, S., Koricanac, Z., Milovanovic, L.J. and Dugadzic, M., 1974. Trace metal content in hair and the levels of S-HI AA in the urine of population exposed to lead. In: *International Symposium Proceedings*, CEC, WHO, EPA, ECE, Luxembourg, 4: 2285–2291.

Subramanian, R., 1991. Metals in hair as a indicator for metal burden of the body. In: *Biological Monitoring of Exposure to Chemicals: Metals*, (Eds.) H.K. Dillon and M.H. Ho. Wiley, New York, p. 255.

Subramanian, R. and Sukumar, A., 1989. Elements in scalp hair indicating body burden of metal pollution. *J. Trace Elem. Exp. Med.*, 2(2–3): 88.

Sukumar, A. and Subramanian, R., 1992a. Elements in hair and nails of urban residents of New Delhi: CHD, hypertensive and diabetic cases. *Biol. Trace Elem. Res.*, 34: 89–98.

Sukumar, A. and Subramanian, R., 1992b. Trace elements in scalp hair of manufactures of fireworks from Sivakasi, Tamil Nadu. *Sci. Total Environ.*, 114: 161–168.

Sukumar, A. and Subramanian, R., 1992c. Elements in hair and nails of residents from a village adjacent to New Delhi: Influence of place of occupation and smoking habits. *Biol. Trace Elem. Res.*, 34: 99–105.

Sukumar, A. and Subramanian, R., 2003. Elements in the hair of non-mining workers of a lignite open mine in Neyveli. *Ind. Health*, 41: 63–68.

Sukumar, A., 2002. Factors influencing levels of trace elements in human hair. *Rev. Environ. Contam. Toxicol.*, 175: 47–78.

Suzuki, T. and Hongo, T., 1984. Element concentrations of Japanish women's hair from historical samples. *Sci. Total Environ.*, 39: 81–91.

Takagi, Y., Matsuda, S., Imai, S., Ohmore, Y., Masuda, T., Vinson, J.A., Mehra, M.C., Puri, B.K. and Kaniewski, K., 1986. Trace elements in human hair: an international comparison. *Bull. Environ. Contam. Toxicol.*, 36: 793–800.

Thatcher, R.W., Lester, M.I., Alaster, M.C. and Horst, R., 1982. Effects of low levels of cadmium and lead on cognitive functioning in children. *Arch. Environ. Health*, 37(3): 159–166.

Tomza, U., Janicki, T. and Kossman, S., 1983. Instrumental neutron activation analysis of elements in hair a study of occupational exposure to a non-ferrous smelter. *Radiochem. Radioanal. Lett.*, 58(4): 209–220.

Umarji, G.M. and Bellare, R.A., 1966. Hair manganese levels in normal subjects. *Indian J. Exp. Biol.*, 14: 212–217.

Van Wouwe, J.P. and Hamer, V.D., 1985. Hari zinc in infancy and childhood. *Sci. Total Environ.*, 42: 149.

Vance, D.E., Ehmann, W.D. and Markesbery, W.R., 1988. Trace element imbalances in hair and nails of Alzheimer's disease patients. *Neutrotoxieology*, 9(2): 197–208.

Viswanathan, H., Hema, A. Edwin, Deepa and Usha Rani, M.V., 2002. Trace metal concentration in scalp hair of occupationally exposed autodrivers. *Environ. Monit. Assess.*, 77: 149–154.

Vivoli, G., Borella, P., Bergomi, M. and Fantuzzi, G., 1987. Zinc and copper levels in serum, urine and hair of humans in relation to blood pressure. *Sci. Total Environ.*, 66: 54–64.

Weiss, D., Whitten, B. and Leddy, D., 1972. Lead content of human hair, 1871–1971. *Science*, 178: 69–70.

Wihelm, M., Ohnesorge, F.K. and Hotzel, D., 1990. Cadmium, copper, lead and zinc concentration in human scalp and public hair. *Sci. Total Environ.*, 92: 199–206.

Wilhelm, M., Hafner, D., Lombeck, I. and Ohnesorge, F.K., 1988. Variables influencing cadmium concentrations in hair of pre-school children living in different areas of the Federal Republic of Germany. *Int. Arch Occup. Environ. Health*, 60: 43–50.

Yukawa, M., 1984. The variation of trace element concentration in human hair: The trace element profile in human long hair by sectional analysis using neutron activation analysis. *Sci. Total Environ.*, 38: 41–54.

Chapter 61

Physico-chemical Characteristics of a Pond Ecosystem Polluted by Tannery Effluents and Hospital Sewage Outlets

T. Prabhu[1], *S. Ramalingam*[1] *and S.M. Nagarajan*[2]

[1]*Department of Physics,* [2]*Department of Botany*
A.V.C. College (Autonomous), Mannampandal – 609 305, Tamil Nadu

ABSTRACT

The present work has been carried out to study the effect of tannery effluent and hospital sewage outlet on the soil and groundwater of a fresh water body, in Mayiladuthurai city. The samples were collected and analyzed for the presence of micro and macronutrients and heavy metals. From the results, it has been observed that the pond and the surrounding area were polluted severely.

Keywords: *Tannery effluent, Hospital sewage outlet, Pond ecosystem, Micro and macronutrients.*

Introduction

Life depends on water and natural water can be described in terms of their physical, chemical and biological conditions. Physically, the water must be clear with an ambient temperature and free from suspended solids, colorations, obnoxious odour and taste. The soil is a natural medium for the production of living organisms and it supplies nutrients to the plants. Nowadays water, soil and air are being contaminated by many factors. Contamination here refers to pollutions that exists in many forms and affects many different aspect of the earth's environment. Soil and water pollutions are the build up of toxic chemical compounds, Salt and pathogens (disease-causing organism) that affects plants, human being and animal life. The wastewaters from the industries and domestic areas affect the water quality by their colored nature, high organic content, widely varying pH, presence of heavy metals and other pollutants. (Bhuvaneswari and Devika, 2005).

The toxic materials are generally discharged from solid wastes, which are unwanted solid materials such as garbage, paper, metals and wood. This solid waste is the primary pollutant in major cities.

All these above facts for a thorough monitoring of dangerous concentrations of hazardous materials, such as heavy metals and toxic compounds in the heart of the city, with reference to water and soil. While doing so, attention may be taken to various physico-chemical parameters of water and soil due to the effluents of leather washing factory and hospital sewage outlet (Sujatha and Gupta, 1986). Therefore, in this present study, an attempt has been made to analyze the heavy metals distribution in soil and water of a fresh water body polluted by tannery effluents and hospitals sewage outlets (Mariappan and Vasudevan, 2002).

Materials and Methods

The samples of tannery effluents and the hospital sewage outlets were collected aseptically in sterile glass bottles from the pond, where they discharged and stored in Mayiladuthurai city, Nagapattinam district, Tamil Nadu. The samples were subjected to various physico-chemical analyses as per standard methods (APHA, 1989) and tabulated in Table 61.1 and 61.2. The concentrations of micro and macronutrients and heavy metals were studies and compared with standard methods.

Table 61.1: Soil Analytical Report of Mixture of Tannery Effluents and Hospital Sewage Outlets

Sl.No.	*Parameter Analysed*	*Analytical Value*
1.	pH	8.30
2.	EC (dms^{-1})	0.64
3.	Colour	Brownish Black
4.	Texture	Sandy Clay Loam
5.	Lime Status	Present
6.	Oraganic Matter	1.06
	Available Macronutrients (Kg/ac)	
7.	Available Nitrogen	152.4
8.	Available Phosphorus	9.5
9.	Available Potassium	31.5
	Available Micronutrients (in ppm)	
10.	Available Zinc	2.18
11.	Available Copper	2.19
12.	Available Iron	13.49
13.	Available Manganese	4.29
	Soil Fractions (in per cent)	
14.	Fine Sand	45.19
15.	Coarse Sand	28.29
16.	Silt	12.19
17.	Clay	14.33

Contd...

Table 61.1–Contd...

Sl.No.	*Parameter Analysed*	*Analytical Value*
18.	**Cat ion Exchange Capacity** (C. Mole Proton$^+$/kg)	23.6
	Exchangeable Bases (C. Mole Proton$^+$/kg)	
19.	Calcium	10.9
20.	Magnesium	9.5
21.	Sodium	3.19
22.	Potassium	0.51
	Heavye metals (in ppm)	
23.	Boron	0.12
24.	Molybdenum	0.22
25.	Cobalt	0.13
26.	Chromium	1.29
27.	Lead	0.22
28.	Mercury	0.02
29.	Cadmium	0.10

Results and Discussion

In the present investigation, the physico-chemicals characteristics of the tannery effluent and hospital sewage outlet from the area have been analyzed and the results shown in Tables 61.1 and 61.2. The color of the effluents was brownish black and oduor was unpleasant. It may be due to the presence of various chemicals being added for washing leather and disposed medicines from hospitals.

The pH of the effluents was in the range of basic level. The values of pH effluents discharged indicates that it is not well with in the permissible limits of CPCB (1995).

The temperature was 31°C and turbidity was 100 NTU. It will affect the oxygen combining capacity there by indirectly affecting the dissolved oxygen in water (Goel, 2000).

The result of organic matter, indicates tannery effluent and hospital sewage outlet has higher values compared to the standard (300 mg/l) prescribed by CPCB 1995. Most of the values of inorganic substances and salt that show good conductivity (Manivasakam, 1984). In the present study, high levels of BOD in tannery effluent and hospital sewage outlet was recorded which may be due to the presence of considerable amount of the organic and inorganic matter.

The levels of BOD were considerably higher than the (CPCB 1995) standard pretrial for BOD (30–100 mg/l) discharge of effluent into land surface waters and for irrigations purpose. Here the value of DO was 1.8 and high value of BOD and COD (Naga-Prupurna and Shrikanth 2002), they suggested that there is a need to decrease the DO content for increasing the turbidity, BOD as well as total suspended solids of lake water. Mariappan and Vasudevan (2002) studied the relation ship of varies physico-chemical parameter of pond in eastern part of sivagangai district.

The present investigation revealed that high values of COD in the effluent samples from tannery effluent and hospital sewage outlet crossed the limits of standard prescribed by (CPCB 1995) for effluents discharge in to land surface water (COD levels of 250 mg/l). The high COD values may be due to significant amount of biologically resistant organic matter (Nagarajan and Ramachandramoorthy, 2002).

Table 61.2: Water Analytical Report of the Mixture of Tannery Effluents and Hospital Sewage Outlets

Sl.No.	*Parameter Studied*	*Samples Details*
	Physical Parameter	
1.	Colour	>3 hue
2.	Odour	Unpleasant
3.	Turbidity	100 NTU
4.	Temperature	31 degree C
	Chemical Parameter	
1.	pH	8.08
2.	EC (dms^{-1})	39
3.	Total Dissolved Solids (TDS) (mg/l)	24960
4.	Suspended Solids (mg/l)	1478
5.	BOD (mg/l)	13400
6.	COD (mg/l)	10480
7.	DO (mg/l)	1.80
8.	Phosphorus (mg/l)	0.12
9.	Potassium (mg/l)	28.08
10.	Sodium (mg/l)	7360
11.	Calcium (mg/l)	1600
12.	Magnesium (mg/l)	720
13.	Carbonatee (mg/l)	160
14.	Bi-Carbonatee (mg/l)	3843
15.	Chloride (mg/l)	10675
16.	Sulphate (mg/l)	1824
17.	Nitrate (mg/l)	10.13
18.	Nitrite (mg/l)	0.02
19.	Fluoride (mg/l)	3.49
20.	Zinc (mg/l)	0.59
21.	Copper (mg/l)	0.22
22.	Iron (mg/l)	4.19
23.	Manganese (mg/l)	2.19
24.	Chromium (mg/l)	7.89
25.	Nickel (mg/l)	0.11
26.	Mercury	BDL
27.	Cadmium	0.19

The maximum level of TDS was present in effluents which could be due to high salt content. The TDS values of chemical effluents in diversity were found to be beyond the limits prescribed by ISI (1979).The level of suspended solids was drastically higher in the effluents where compared to the permissible level of 30 mg/l prescribed by the ISI (1979) for effluents discharge. The effluents sample showed high values of TSS, The ions especially calcium, sulphate, sodium, potassium impart hardness to water.

In the case of heavy metals, the tannery effluent and hospital sewage outlet showed the tremendous amounts of Boron, Molybedumn, Cobalt, Chromium, Lead, Mercury and candium. They crossed the limits of standard (CPCB 1995).

From the result of the present study it can be inferred that the physicochemical parameters such as pH, BOD, COD TDS, TSS, of tannery effluent and hospital sewage outlet were found to be higher and surpassed the permissible limits of CPBP (1995). Release of such effluents into the water course will cause water and soil pollution and affects aquatic organisms drastically. Hence the effluents should be treated properly before discharge.

References

APHA (American Public Health Association), 1989. *Standard Methods for the Examination of Water and Wastewater*. AWWA. Wat. Pollut. Control Fed., Washington, DC.

Ashok Kumar, C., 1984. Toxicity and bioaccumulation of heavy metals in a estuarine bivalve, *Anadra rhombea* (Born).

Bhuvaneswari, K. and Devika, R., 2005. Studies on the physico-chemical and biological characteristics of Coovum river. *Asian Journal of Microbiology, Biotechnology and Environmental Sciences*, 7(3): 449–451.

CPCB, 1995. *Pollution Control: Acts, Rules, and Modifications*, issued there under Central Pollutions Control Board, New Delhi.

Goel, P.K., 2000. *Chemical and Biological Methods for Water Pollution Standards*. Environmental Publications, Karad, India.

ISI, 1979. Guide for treatment and disposal of effluents of cane sugar industry IS: 4903 (1st revision) ISI, New Delhi.

Manivasakam, N., 1984. *Physico-chemical Examination of Water, Sewage and Industrial Effluents*. Pragati Prakashan, Meerut, p. 42.

Mariappan, P. and Vasudevan, T., 2002. Corelation coefficient of some physico-chemical parameters of drinking water ponds in eastern part of Sivagangai District, Tamil Nadu. *Poll. Res.*, 21(4): 403–437

Nagarajan, P. and Ramachandramoorthy, T.R., 2002. Oil and grease removal from steel industry wastewater by chemical treatment. *J. Ecotoxicol. Environ. Monit.*, 12(2): 181–184.

Naga Prapuma and Srikanth, 2002. Pollution level in Husain Sagar lake of Hyderabad: A case study. *Poll. Res.*, 21(2): 187–190.

Sujatha, P. and Gupta, A., 1996. Tannery effluent characteristics and its effects on agriculture. *J. Ecotoxicol. Environ. Monit.*, 6(1): 45–48

Chapter 62

Stratified Scavenging System with Exhaust Gas Recirculation for Two Stroke Spark Ignition Engine

P. Srinivasa Rao[1] *and K. Raja Gopal*[2]

[1]*Department of Mechanical Engineering, Chaitanya Bharathi Institute of Technology, Gandipet, Hyderabad – 500 075, Andhra Pradesh*

[2]*Jawaharlal Nehru Technological University, Hyderabad – 500 072, Andhra Pradesh*

ABSTRACT

From the beginning, two stroke engines have suffered from high emissions and poor fuel economy compared to the larger, heavier but more efficient four stroke engines. Over the last century, many innovative ideas have been designed and patented with the intent or improving these two critical short comings. Because of stringent emission regulations introduced in the last two decades, designers have now given impetus to revive, revise and apply technologies to meet these requirements.

Among the most successful technological improvements, stratified scavenging and charging is the most important one which reduced the two stroke engine emissions by as close as 40 per cent. However the major research is only concentrated either on low capacity engines as 60cc, 65cc and 75cc or as high as 250cc and 350cc engines. But in Indian subcontinent, major part or the two wheelers is of either 125cc or 150cc engines, which emit dangerous pollutants causing health hazards to human beings apart from environmental disorders. The suitability of technology as stratified charging or scavenging to such engines is not validated and substantiated so far. This chapter presents the results and conclusions from a conceptual design study for a stratified charging concept for a higher cubic capacity engines.

Introduction

The civilization of any county depends on the number of vehicles used by the public. For heavy duties, diesel engines are preferred, while for individual transport, a small duty, two-stroke petrol engines are being employed. However, from the beginning, two stroke engines have suffered from high emissions and poor fuel economy compared to the larger, heavier but more efficient four stroke engines. The major pollutants emitted from these two-stroke spark ignition engines are carbon monoxide and un-burnt hydro carbons. Breathing of these emissions causes detrimental effects on human, animal and plant life besides environmental disorders (Sharma, 1996; Fulekar, 1999). Hence globally, stringent regulations are made fix permissible levels of pollutants in the exhaust of 2 and 4 stroke spark ignition engines. A two-stroke engine is recently receiving a renewed interest in the automotive industry due to its lower weight and volume resulting in a more compact body and twice in power compared to four stroke engines of same cubic capacity. The short circuiting loss of fresh charge has been known ever since two stroke engines were first made more than a century ago by Sir Dugald Clerk in 1879 equally old are all the innovative technologies and concepts that have been proposed or attempted in order to circumvent the short circuiting loss of fresh charge, as stratified scavenging through air head, exhaust gas recirculation, air assisted fuel injection, Compressed Wave Injection (CWI), direct or indirect fuel injection etc. The charge stratified engine discovered by Heiko Rosskemp (2003), G.P. Blair (1981) and B.W. Hill (1983) are in another interesting design disclosed by Huber (1943) stratification occurs in three phases air, air fuel mixture and air which improves trapping of fresh charge. In the year 1901 patent (Huber, 1943) Woolf used poppet valve in a unitlow crankcase scavenged and carbureted two stroke engine. In another interesting design more or less similar to today's air head scavenging system is the invention by Stephenson (1911), in which he fitted poppet valve which is fitted at the top of the transfer passage to fill the transfer passage with air during the upward stroke or the piston. Poppet valve closes when the piston approaches TDC and fuel and air mixture is admitted into the crankcase through piston controlled intake port. These are some or the improvements suggested and successfully implemented but only suitable for the small capacity engines *viz.*, 65cc to 75cc engines only.

An automatic Ball Valve admitting air from outside to the top of transfer passage for a 240cc engine is also attempted (Woolf, 1901; Stephenson, 1911) the number of patents issued or applied for indicates the research activities pertaining to the design and application of such engines. Engine manufacturers have commercialized some of the technological advancements as Air-Head scavenging particularly Komatsu-Zenoah of Japan.

Experimental Programme

The Table 62.1 shows the specifications of the single cylinder engine used for the experimentation.

The Figure 62.1 shows the modifications done to conventional spark ignition engine used for the experimentation. The two stroke engine is altered in the basic design as another port is created in the transfer port in the vicinity of the port entrance into the cylinder and is provided with a reed valve to have unidirectional flow into the cylinder along with the charge from the crankcase and a part of exhaust gas is recirculated into the cylinder. As the exhaust gas and the fresh air are also accommodated into the cylinder the quality of the charge that is supplied to the cylinder is to changed to have a proper and better combustion in the chamber.

Table 62.1: Specifications of the Experimental Engine

Cylinder capacity	152cc
Engine Power	1.1 Kw
Rated Speed	4500rpm
Port Size	12mm × 14.5mm
Bore	58.24mm
Stroke	58.40mm
Compression Ratio	8.5
Inlet Port Opening	146° (ATDC)
Inlet port Closing	146° (ABDC)
Exhaust Port Opening	128° ATDC
Type Cooling method	Air cooled

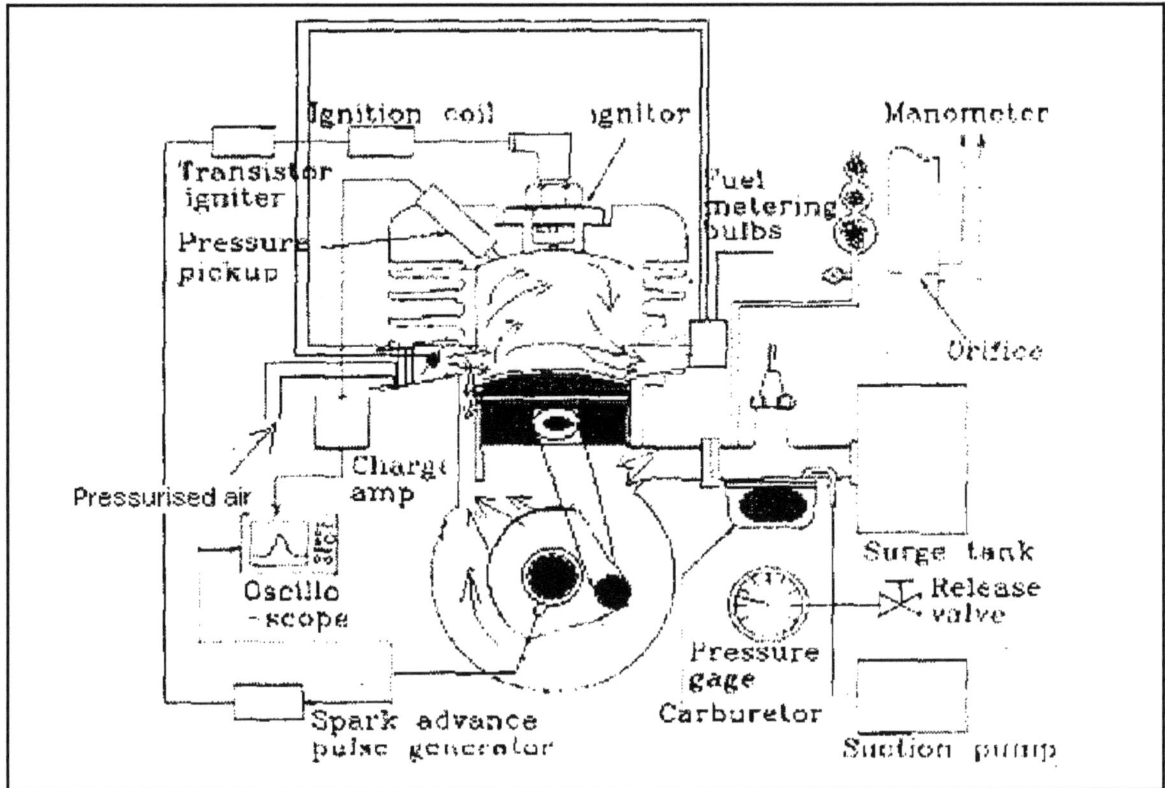

Figure 62.1: Stratified Scavenged Two Stroke Engine with EGR

As the fresh air and exhaust gases are being pumped into the cylinder the fresh charge that is entering into the cylinder through the crankcase may be diluted. The carburetor is provided with the suction pump to have a rich mixture when it is needed to overcome the above problem. A filter is provided to filter the particulate mater which may reenter the cylinder. The pollutants of CO and UBHC emissions are recorded with Netal Chromotograph CO/HC analyzer.

The fluid flow (air and charge) through the Engine under working conditions is shown in the Figure 62.2.

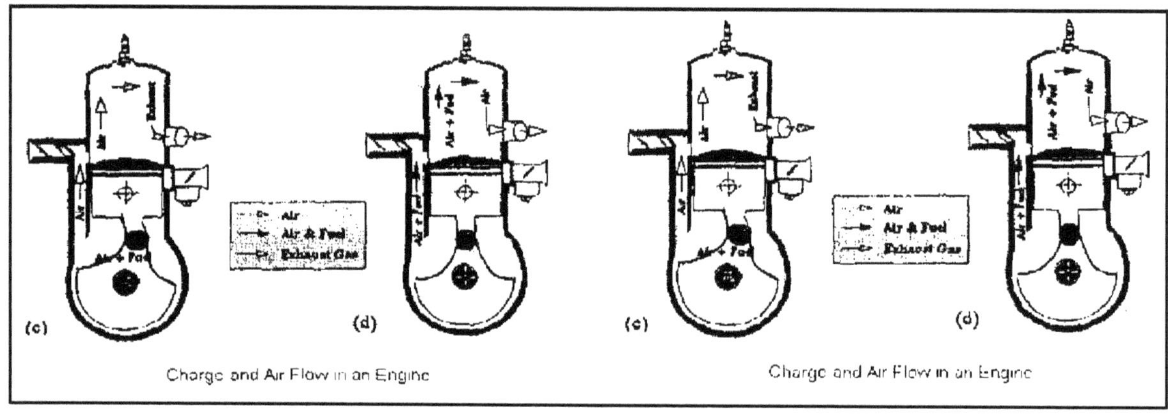

Figure 62.2: Charge and Air Flow in an Engine

Results and Discussions

Figures 62.3 and 62.4 shows the variation of UBHC and CO with fresh charge delivery ratio at different equivalence ratios respectively.

It is observed from the Figure 62.3 as fresh loss ratio increases Hydrocarbons decrease. The trends are observed to be same for both residual and non residual cases. As the re-circulated gas which is sand witched between fresh air and fresh charge as the scavenging processes begins the fresh air pushes the burned gases out from the chamber and before the fresh charge enters the cylinder the re-circulated gas acts as an artificial piston to throw away the flue gases out completely leading to reduction in Hydrocarbon emissions.

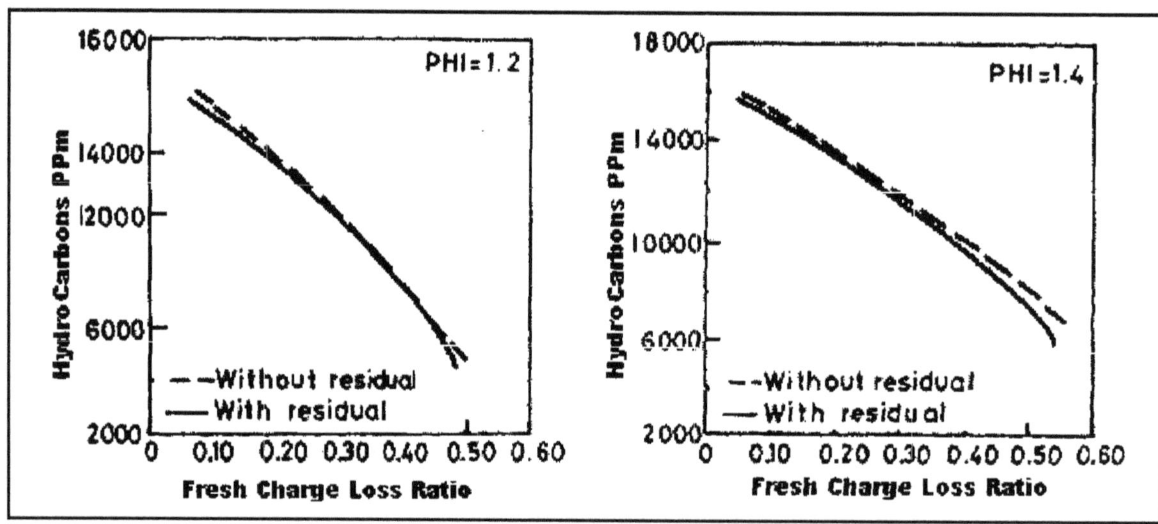

Figure 62.3: Effect of Fresh Charge Loss Ratio on HC Concentration at Different Equivalence Ratio

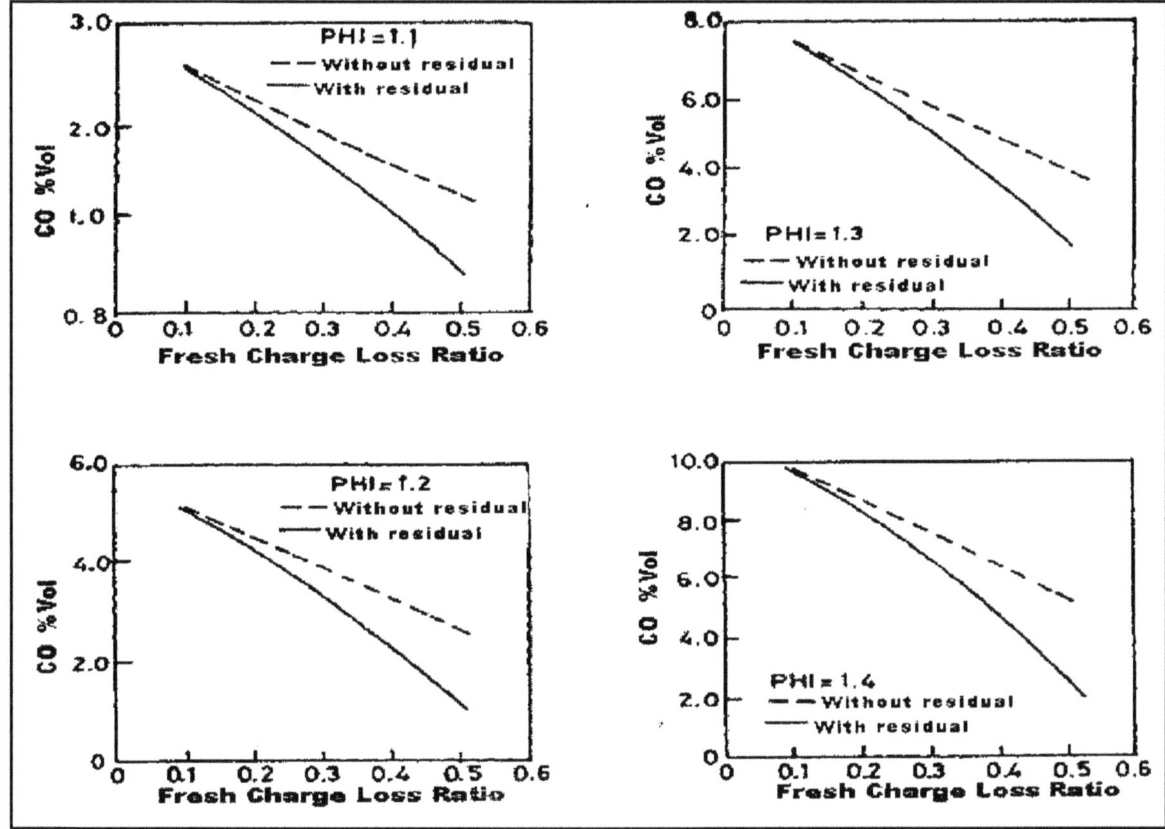

Figure 62.4: Effect of Fresh Charge Loss Ratio on Carbon Monoxide Concentration at Different Equivalence Ratio

From the Figure 62.4 it is observed that CO emissions decrease as the fresh charge loss ratio increase at all equivalence ratios and it is more pronounced at lower equivalence ratios. The CO emissions decrease due to the better oxidation and combustion reactions because of the admission of air into the combustion chamber. Similar trends are observed for both residual and non residual cases.

Conclusions

1. Based on the experimental investigations on a two stroke stratified scavenged engine with part-exhaust gas recirculation and fitted with extra port to provide the pressurized air through transfer port the pre an post combustion gases are clearly analyzed and found that the scavenging and trapping efficiencies are strong functions of delivery ratio.
2. The reduction of Hydrocarbon emissions is found to be 40 per cent when compared with conventional engine.
3. The CO emissions are decreased by 22 per cent to 25 per cent with stratified engine compared with non-stratified engine.

References

Blair, G.P. and Hill, B.W. Reduction of fuel consumption of a spark-ignition two-stroke cycle engine. *SAE*, 830093.

Blair, Gordon P., 1981. US Patent No. 4, 253, 433 March 3.

Fulekar, M.H., 1999. Chemical pollution: A threat to human life. *Indian J. of Environ. Prot.*, 1(3): 353–359.

Hill, B.W. and Blair, G.P., 1983. Further tests on reducing fuel consumption two stroke cycle engine. *SAE* Paper No. 831303.

Huber, Fritz and Lenz, Anton, 1943. US Patent No. 2. 317, 772, April 27.

Mavinahalli, Nagesh S. and Neelakantappa, B.P., 1991–92. Performance improvement or a two stroke engine using non-return valves in transfer passages. Student project report. Malnad College of Engineering, Hasan, India.

Rosskamp, Heiko and Hahnadorff, Markus, 2003. US Patent No. 6, 598, 568, July 29.

Sharma, B.K., 1996. *Engineering Chemistry*. Pragathi Prakashan (P) Ltd., Meerut.

Stephenson, William, 1911. US Patent No. 1, 012, 288, December 19.

Woolf, Ellis J., 1901. US Patent No. 683, 886, January 29.

www.Komatsu–Zenoah, Japan.

Yoshida, Y., 1999. Development of Stratified Tow Stroke Cycle Engines for Emission Reduction. *SAE*, 0103269.

Chapter 63

Effect of Different Animal Manures on Freshwater Fish of *Catla catla*

R. Sivakami, R.S. Grace Sims, P.P. Shimna and G. Prem Kishore

P.G. and Research Department of Microbiology,
Kurinji College of Arts and Science, Tiruchirappalli – 620 002, Tamil Nadu

ABSTRACT

India with exploding population has to increase its food production through less expensive different modes. One such mode is fish culture, however 40–60 per cent of the production cost in aquaculture goes for fish food. India has large amount of organic manure. Hence an attempt was made to use a combination of these manures to enhance fish production. Result indicates that a combination of pig manure and biogas slurry recorded highest growth when compared to cow dung + biogas slurry and poultry dropping + biogas slurry. These results suggested that these manure can be utilized for reducing the cost of fish production besides the cleaning of environment.

Keywords: *Organic manure, Catla catla, Pond productivity.*

Introduction

India will have to look for alternative sources of food to meet its exploding human population. One way is to use fish as a source of protein. Today, India has made rapid studies in pisciculture. Supplementary diet, in aquaculture contributes to about 40–60 per cent of the total production cost (Sooch *et al.*, 2005). A wide variety of manures have been attempted to fertilize the pond to increase the food production. Literature reveals that animal manure has different effects on fishes. Kaur *et al.* (1987) reported 100 per cent survival of *Cyprinus caripio* when fertilized with cow dung and biogas slurry while the same species showed above 90 per cent survival using poultry manure (Kaur and Kaur, 1987). Dhawan and Toor (1989) who recorded maximum growth in ponds fertilized with poultry droppings followed by cow dung. Sooch *et al.*, (2005) however, reported maximum growth of *C. Carpio* when fertilized with poultry droppings than with biogas slurry and cow dung. Hence an attempt was made in the present study the efficacy of three combinations *viz.*, pig manure + biogas slurry, cow dung + biogas slurry and poultry droppings + biogas slurry mixed in ratio of 1 : 1 on the growth of the fresh water fish *Catla catla.*

Materials and Methods

Experiments were conducted in the Department of Microbiology, Kurinji College of Arts and Science, Tiruchirappalli, in Central Tamil Nadu. Various physico-chemical variables of water were also analysed. These were done by using standard methods (APHA, 1985). In addition, the total plankton production was also done by using a secchi disc rafter counting cell, the plankton counting cell, the plankton count included both phyto and zooplankton. This was done during the sampling of water for physico-chemical analysis. Seven tanks measuring 5 × 2.5 m with a depth of 1.0 m were selected. Of the seven, two tanks each had the same combination and concentration of nutrients, while the seventh tank was used a control. Manuring was done with pig manure + biogas slurry (Tanks 1 and 2) cow dung + biogas slurry (Tanks 3 and 4) and poultry droppings + biogas slurry (Tanks 5 and 6). All the tanks had manure combination in the ratio of 1 : 1. Fingerlings obtained from private fish pond at Gundoor, Tiruchirappalli District were used for the present study. The study was done for a period of 60 days.

Results and Discussion

The effect of different manures and its combinations and fish growth are presented in Tables 63.1–63.3.

Table 63.1: Effect of Different Manures on Water Quality Parameters

Sl.No.	Parameters	Days	Manure			Optimum Range of Water Quality Parameters
			Pig Manure + Biogas Slurry	Cow Dung + Biogas Slurry	Poultry Dropping Biogas Slurry	
1.	Water Temperature (°C)	0	27	27	27	15–37.5
		20	28	28	28	
		40	29	29	29	
		60	30	30	30	
2.	Secchi Disc Transparency (cm)	0	26.0	26.0	26.0	–
		20	23.0	22.0	23.0	
		40	21.0	20.0	21.0	
		60	16.0	17.0	16.0	
3.	Hydrogen Ion concentration (pH)	0	7.5	7.5	7.5	6.5–9.6
		20	7.6	7.8	8.2	
		40	7.8	7.9	8.3	
		60	8.2	8.2	8.4	
4.	Dissolved oxygen (mg/l)	0	6.6	6.6	6.6	7.8–9.0
		20	6.9	6.8	6.4	
		40	7.0	7.1	6.9	
		60	7.8	7.4	7.5	

Contd...

Table 63.1–Contd...

Sl.No.	Parameters	Days	Manure			Optimum Range of Water Quality Parameters
			Pig Manure + Biogas Slurry	Cow Dung + Biogas Slurry	Poultry Dropping Biogas Slurry	
5.	Alkalinity (mg/l)	0	146	114	150	137–230
		20	186	164	156	
		40	196	176	160	
		60	216	190	196	
6.	Free CO_2 (mg/l)	0	Nil	Nil	Nil	15.0
		20	Nil	Nil	Nil	
		40	Nil	Nil	Nil	
		60	Nil	Nil	Nil	
7.	Nitrogen (mg/l)	0	0.055	0.045	0.040	–
		20	0.065	0.055	0.050	
		40	0.075	0.065	0.060	
		60	0.085	0.075	0.070	
8.	Phosphorous (mg/l)	0	0.0030	0.0025	0.0025	–
		20	0.0035	0.0030	0.0030	
		40	0.0040	0.0035	0.0035	
		60	0.0045	0.0040	0.0040	

In the present study, phytoplankton was maximum observed in the tank manure with pig manure and biogas slurry. The same was true with zooplankton also. This was followed by the tank containing cow dung + biogas slurry (Tank 3 and 4) and poultry dropping + biogas slurry (Tank 5 and 6). Literature reveals that biogas slurry is a potential manure enhancing the growth of both phyto and zooplankton (Rai 1981; Kalayan and Shetty, 1987; Sooch *et al.*, 2005). The high planktonic count in tank containing pig manure and biogas slurry was maximum because of the high nutrient content which enhanced plankton production (Tables 63.1–63.2).

Table 63.2: Effect of Different Manures on Pond Productivity

Pond Productivity	Days	Pig manure + Biogas slurry (1 : 1)	Cow dung + Biogas slurry (1 : 1)	Poultry dropping + Biogas slurry (1 : 1)
Phytoplankton (ml/l)	0	14.2	15.2	15.4
	20	20.4	19.4	18.2
	40	28.6	22.3	24.6
	60	36.4	31.4	29.3
Mean		24.9	22.07	21.87
Zooplankton (ml/l)	0	20.6	30.2	20.4
	20	43.7	35.3	30.8
	40	48.8	46.4	38.8
	60	50.6	48.3	43.3
Mean		40.92	37.55	33.32

Table 63.3: Effect of Different Manures on Fish Growth of *Catla catla*

Days	*Fish Growth*	*Manures*		
		Pig Manure + Biogas Slurry (1 : 1)	*Cow Dung + Biogas Slurry (1 : 1)*	*Poultry Dropping + Biogas Slurry (1 : 1)*
Initial (0 days)	Length (cm)	4.3	4.3	4.3
Final (60 days)		9.8	8.6	7.3
Length gain (per cent)		20	12	10
Initial (0 days)	Weight (gm)	2.0	2.0	2.0
Final (60 days)		40	32	30
Weight gain (per cent)		34	30	28
Per cent of survival		96	95	95

Fish growth in terms of weight gain was again observed to be maximum in the tank manured with pig manure + biogas slurry followed by cow dung + biogas slurry and poultry dropping + biogas slurry. Sooch *et al.* (2005) and Dhawan and Toor (1989) observed maximum fish growth in the tank manure with poultry droppings. In the present study the maximum growth was observed in the tank containing manure of pig manure + biogas slurry could be due to their synergistic effect.

Water quality parameters (Table 63.1) remains favourable through out the period of study. Dissolved oxygen remained optimum for all the three combinations. Thus their was conducive environment for the fish to grow through out the study period.

The result indicates that the organic waste present in large quantity in India can be used as a supplementary food for increasing fish food production. The present study also indicates that the combination of these manure rather than alone appears to be more beneficial for fish production. In addition to this system also helps in cleaning the environment and reaping economic benefits thus serving as dual purpose.

References

APHA, 1985. *Standard Method of Examination of Water and Wastewater.*

Dhawan, A. and Toor, H.S., 1989. Impact of organic manure and supplementary diet on plankton production and fish growth and feeding of an Indian major carp, *Cirrhina mrigala* (Ham.) in fish ponds. *Biol. Waste,* 29: 289–298.

Kalyani and Shetty, 1987. In: Sooch, S.S., Dhewan, A. and Singh, D.P., 2005. Strategies for utilization of biogas slurry in fish pond manuring. *Indian J. Environ. and Ecoplan.,* 10(3): 757–760.

Kaur, K., Sehgal, G.K. and Sehgal, H.S., 1987. Efficiency of biogas slurry in carp. *Cyprinus carpio* var. *communis* (Linn.) culture. Effects on survival and growth. *Wastes*, 22: 139–146.

Rai, 1981. In: Sooch, S.S., Dhewan, A. and Singh, D.P., 2005. Strategies for utilization of biogas slurry in fish pond manuring. *Indian J. Environ. and Ecoplan.*, 10(3): 757–760.

Sooch, S.S., Dhewan, A. and Singh, D.P., 2005. Strategies for utilization of biogas slurry in fish pond manuring. *Indian J. Environ. and Ecoplan.*, 10(3): 757–760.

Chapter 64

Analysis of Cyanobacteria in Two Freshwater Lentic Systems of Central Tamil Nadu

R. Sivakami, X. Casino, A. Anthonisamy and G. Prem Kishore

P.G. and Research Department of Microbiology,
Kurinji College of Arts and Science, Tiruchirappalli – 620 002, Tamil Nadu

ABSTRACT

Cyanobacteria is an essentially important group as it is used for a variety of activities like food, feed, biofertilizers, biodegradation of waste and fuel production. Hence it was thought worthwhile to assess the Cyanobacteria population of two freshwater bodies of varied nature located in Tiruchirappalli, Central Tamil Nadu with in distance about 10 kms from each other in Kulumani region. Results show that temple tank recorded only 11 species belonging to 9 genera, while the river fed pond recorded 18 species belonging to 10 genera. The temple tank however recorded 2 unique species while the river fed pond, recorded 9 unique species. The presence of economically important Cyanobacteria, like *Anabaena* (biofertilizer) *Spirulina* (feed) in these water bodies indicates that these water bodies can be utilized for the benefit of man.

Keywords: *Cyanobacteria, Freshwater lentic systems, Species diversity.*

Introduction

Freshwater habitats occupy relatively a small portion of the earths surface as compared to marine and terrestrial habitats, but their importance to man is for greater than their area (Odum, 1971).

Because of its several unique properties, fresh water supports a variety of life. Investigations on lakes over the world have shown that they possess a great variety of combination of properties in many types. These water bodies are natural habitat for a wide variety of microorganisms especially Cyanobacteria. Cyanobacteria today is an economically important group as it is used for a variety of activities like food, feed, biofertilizers, biodegradation of waste and fuel production (Sivakami *et al.*, 2006). Hence it was thought worthwhile to assess the Cyanobacterial population of two fresh water bodies of varied nature located in Tiruchirappalli, Central Tamil Nadu within distance about 10 kms from each other at Kulumani region.

Materials and Methods

The city of Tiruchirappalli, located in Central Tamil Nadu (Lat 10°50′N Long 78°36′E) has innumerable temple tanks besides other water bodies of varying sizes. The present study was undertaken on two water bodies one an artificial temple tank of about 0.2 hectare made of rock stone and the other, a pond with a muddy bottom. While the temple tank was rain-fed, the other was river fed getting water from Canal of the River Cauvery. However both the water bodies were perennial. Surface water samples were collected with the help of a plankton net from the water bodies between 8.00 am–9.00 am for a year (January–December 2006) on a monthly basis. According to Kanungo *et al.* (2005) samples were immediately transferred to 4 per cent formal in for later microscopic analysis. Care was also taken to observe fresh samples also. The sign (+) was used to express the presence of Cyanobacteria.

Results and Discussion

The Cyanobacteria present in both the ponds are presented in Tables 64.1 and 64.2. While temple tank recorded only 11 sps. belonging to 9 genera, the river fed pond recorded 18 species belonging to 10 genera with both ponds recording 9 common species. The temple pond however recorded two unique species *Aphanocapsa roseana* and *Synechocystis pevalekii* which were not found in river fed pond while this pond, 9 unique species. Thus two ponds, though perennial, showed differences with, the river fed pond recording greater Cyanobacteria diversity. Further in the river fed pond, the largest genera was *Aphanbocapsa* recording 5 species. The temple tank however did not record any dominant genera. Instead the genera *Aphanocapsa, Oscillatoria* and *Synechocystis* recorded two species each respectively. In the river fed pond, *Anabaena circinalis, Chroococcus giganteus* and *Synechocystis aquatilis* were found to be perennial but in temple tank only two species (*Anabaena circinalis* and *Microcyctis aeruginosa*) were perennial. However, none of the species recorded blooms during the period of study. In general, the maximum number of species was recorded during the October and November season (rainy season) in both the ponds. But the river fed ponds had 10 species while temple tank, only 6 species during this period. The higher diversity recorded in river fed pond is probably due to increased amount of nutrients in this pond has because it receives water from the River Cauvery. The higher amount of diversity noticed during October–November period merges with the rainy season as both the water bodies would have received increased amount of nutrients from the run off water. Literature reveals that similar results were also observed by (Hosmani and Bharati, 1980; Seenayya, 1980; Pearsall, 1932; Smith, 1983; Sinha and Chandrawati, 1996; Mendhe, 1996; Bose and Gorai, 1993; Kanungo *et al.*, 2005). The results of the present study show that economically important Cyanobacteria like *Anabaena* (biofertilizer) *Spirulina* (feed) occurs in these water bodies which can be harnessed for use by man.

Table 64.1: Seasonal Abundance of Cyanobacteria in River Fed Pond (Muddy pond) of Tiruchirappalli District, Tamil Nadu

Sl. No.	*Name of Cyanobacteria*	*Months*											
		Jan.	*Feb.*	*Mar.*	*Apr.*	*May*	*Jun.*	*Jul.*	*Aug.*	*Sep.*	*Oct.*	*Nov.*	*Dec.*
1.	*Anabaena circinalis*	+	+	+	+	+	+	+	+	+	+	+	+
2.	*A. utermohii*									+	+	+	
3.	*Aphanocapsa elachista*										+		
4.	*A. grevillei*										+	+	
5.	*A. montana*		+										
6.	*A. muscicola*											+	
7.	*A pulchra*	+			+	+						+	
8.	*Chroococcus giganteus*	+	+	+	+	+	+	+	+	+	+	+	+
9.	*C. turgidus*				+	+					+		
10.	*Dactylococcopsis raphidiodes*	+	+										
11.	*Gloeocapsa atrata*			+						+	+	+	+
12.	*G. punctata*												+
13.	*Lyngbya limnetica*									+	+		
14.	*Microcystis aeruginosa*			+	+	+	+	+	+			+	
15.	*Oscillatoria acuta*	+	+	+									
16.	*O. peromateta*										+	+	
17.	*Phormidium tenue*	+	+	+		+	+	+					
18.	*Synechocystis aquatilis*	+	+	+	+	+	+	+	+	+	+	+	+
19.	*Spirulina major*	+									+	+	+

Note: +, ++, +++ are degree of abundance of Cyanobacteria in river fed pond.

Table 64.2: Seasonal Abundance of Cyanobacteria in Temple Tank (Rocky pond) of Tiruchirappalli District, Tamil Nadu

Sl. No.	*Name of Cyanobacteria*	*Months*											
		Jan.	*Feb.*	*Mar.*	*Apr.*	*May*	*Jun.*	*Jul.*	*Aug.*	*Sep.*	*Oct.*	*Nov.*	*Dec.*
1.	*Anabaena circinalis*	+	+	+	+	+	+	+	+	+	+	+	+
2.	*Aphanocapsa elachista*						+	+					
3.	*Aphanocapsa roseana*				+								
4.	*Chroococcus giganteus*	+										+	+
5.	*Gloeocapsa atrata*												
6.	*Microcystis aeruginosa*	+	+	+	+	+	+	+	+	+	+	+	
7.	*Oscillatoria acuta*	+	+	+	+	+							
8.	*Oscillatora peromata*										+	+	+
9.	*Phormidium tenue*	+	+	+									
10.	*Synechocystis aquatilis*						+	+	+	+		+	
11.	*Synechocystis pevalekii*										+	+	+

Note: +, ++, +++ are degree of abundance of Cyanobacteria in temple tank.

References

Bose, S.K. and Gorai, A.C., 1993. Seasonal fluctuation of plankton in relation to physico-chemical parameters of a freshwater tank of Dhanbad, India. *J. Freshwater Biol.*, 5: 133–140.

Hosmani, S.P. and Bharati, S.G., 1980. Limnological studies in ponds and lakes of Dharwar-comparative phytoplankton ecology of four water bodies. *Phykos.*, 19: 27–43.

Kanungo, V.K., Verma, J.N. and Patel, D.K., 2005. Cyanobacteria of a temple tank at Raipur Chhattisgarh. *Indian J. Environ. and Ecoplan.*, 10(3): 743–745.

Mendhe, K., 1996. Study of phytoplanktonic population of Bilwali Talab Indore. *Indian J. Appl. Pure Biol.*, 11: 83–88.

Odum, E.P., 1971. *Fundamentals of Ecology*. W.B. Saunders, Philadelphia, p. 1–574.

Pearsall, W.H., 1932. Phytoplankton in the English lakes. II. The composition of the phytoplankton in relation to dissolved substances. *J. Ecol.*, 20: 241–262.

Seenayya, G., 1980. Water pollution and algal blooms in some of the ponds of Hyderabad. In: *Progress in Ecology*, (Eds.) Agrawal, V.P. and V.K. Sharma, 4: 37–52.

Sinha, R. and Chandrawati, J., 1996. Physico-chemical studies and phytoplankton population in four ponds of Patna. *Biol. Journal*, 8: 83–89.

Sivakami, R., Kumar, Jijendra, Rani, Sorna and Kishore, G. Prem, 2006. Cyanobacteria of two different water bodies of Tiruchirappalli, Tamil Nadu. *Indian J. Environ. and Ecoplan.* (accepted).

Smith, V.H., 1983. Low nitrogen to phosphorus ratios favour dominance by blue-green algae in lake phytoplankton. *Science*, 221: 669–671.

Chapter 65

Effect of *Kirganelia reticulata* (Euphorbiaceae) Plant Extract against the Growth and Development of the Human Filarial Vector *Culex quinquefasciatus* Say (Diptera : Culicidae)

M. Manimegalai, S. Umavathi and A. Naresh Kumar

P.G. and Research Department of Zoology,
Kongunadu Arts and Science College, Coimbatore – 641 029, Tamil Nadu

ABSTRACT

Methanolic extract of *Kirganelia reticulata* leaves were evaluated for the larvicidal and adult emergence inhibition activity against human filarial vector *Culex quinquefasciatus.* The rate of mortality was observed after 24 hours using the concentration of 0.1–0.8 per cent and the values of LC_{50} and LC_{90} were tabulated by the method of Fenny. Qualitative analysis of the phytochemicals of methanol extracts revealed the presence of alkaloids, flavonoids, glycocides, steroids, phenols, tannins, saponins carbohydrate and protein. In this research showed that the plant *Kirganelia reticulata* having the larvicidal properties and it can be used as a natural biocide against the filarial vector *Culex quinquefasciatus.*

Keywords: *Kirganelia reticulata, Euphorbiaceae, Phytochemicals, Culex quinquefasciatus,* Filariasis.

Introduction

In India mosquito fauna include 225 species grouped under 16 genera, 58 species belonging to the family *Anopheles*, 57 species of *Culex*, 111 species of *Aedes* and 7 species of *Mansonia* (Nagpol and Sharma, 1995). Mosquitoes constitute a major public health menance, serves as a vector for transmitting diseases like malaria, dengue, filariasis and Japanese encephalitis (JE) to humans and these mosquitoes have the potential and lethal capacity to kill more than a million victims a year around the world (Vatandsoon and Hoin vaizin, 2001). Human filariasis is a major public health hazard and remains a challenging socio-economic problem in many of the tropical countries.

Lymphatic filariasis is caused by *Wuchereria bancrofti* and transmitted by mosquito *Culex quinquefasciatus* is found to be more endemic in the Indian subcontinent and it is reported that it infects more than one million individuals world wide annually. Control of such mosquito born diseases is becoming more and more difficult because of increasing resistance to pesticides, lack of effective vaccines and drugs against disease causing mosquitoes. Hence an alternative approach for mosquito control is the use of extracts of plant origin. Search for natural insecticides that do not have any ill effects on the non-target population and are easily degradable remains to be one of the top priority issues for the tropical countries (Redwane *et al.*, 2005; Rajkumar and Jabanesan, 2005). Many plants in the environment having the mosquitocidal effect.

Most of *Euphorbiaceae* is used as fencing plants and their availability is plenty at the backyards and they are resistance to the water stress and they can easily regenerated. As a result of growing these kinds of fence plants, the environment also purified with the fresh air and free from pollution. Hence in the present study the plant extract of *Kirganelia reticulata* was tested against the egg and various developmental stages of mosquito *Culex quinquefasciatus* in a search for affordable natural products to be used in the control of *Filariasis.*

Materials and Methods

Rearing and Maintenance of Larvae

Egg rafts were collected in the early morning around 5 to 6 a.m from the stagnant water bodies in a plastic container with unchlorinated tap water and allow hatching in the laboratory conditions (29±2°C). After 24 hours freshly hatched larvae were maintained in separate containers with tap water and fish food was given as the source of food (Renugadevi and Thangaraj, 2006).

Preparation of Plant Extract

The leaves of *Kirganelia reticulata* was collected from the field were brought to the laboratory they were washed with tap water and followed by distilled water, the leaves were dried under the shade and powdered using an electrical blender and kept it in the airtight container for further use. 100 gms of *Kirganellia reticulata* powdered leaves was extracted with 300 ml of methanol by-using Soxhlet apparatus for 8 hours (Vogel, 1978). The extract was concentrated in vacuum evaporator to yield a dark greenish, gummy extract. The residue was made into 1 per cent stock solution with respective solvent and taken for further bioassay test.

Phytochemical Screening

The extracts were used for phytochemical analysis for the presence of alkaloids, flavonoids, saponins, glycocides, steroids, phenols, tannins, resins, carbohydrate and protein by using standard procedures to identify the constituents as described by Harborne (1973) and Gopiesh and Kannabiran (2007).

Bioassay Test

To obtain different concentration of the test medium the crude extract 1 to 10 ml of stock solution were dissolved in the water and then mix thoroughly and with the dry ingredients of the diet suggested by Miller and Chamberlain (1985). Freshly laid eggs, newly emerged 1st instar to IV instar and freshly molted pupae of *Culex quinquefasciatus* were exposed to different percent test concentration (0.1 per cent–0.8 per cent) of *Kirganellia reticulata* over a period of 24 hours for both treated as well as control and the mortality was observed. To test the effect of extract each trail contained minimum 20 eggs, larvae and pupae. LC_{50} and LC_{90} values were calculated by the method of Finney (1971). Controls were maintained by using methanol and it is negligible for calculation.

Measurement of Oocytes

To find out the methanolic extract of *Kirganelia reticulata* on the development of ovarian follicle in *Culex quinquefasciatus*. The mosquito larvae at I instar stage were exposed to different sub lethal concentration of leaf extract (0.05, 0.01 and 1 per cent) till pupation. The surviving pupae were sexed and put into separate cages with respect to the concentration of the toxicant. The emerged adult males were fed with 10 per cent sucrose solution soaked in cotton wick (Al-Sharook *et al.*, 1991) and the females were fed with human blood as the source of food as detailed by Meola and Readio (1980). The follicle length was measured in adults at 24, 48, 60 and 72 hours, after blood meal. Ovaries were dissected into *Aedes* saline (Hayes, 1953) and the follicle length along the longitudinal axis was measured using Oculometer (Masler *et al.*, 1980).

Results and Discussion

The effect of the extract of *Kirganelia reticulata* extract sources was tested against the aquatic stages of the human filarial vector *Culex quinquefasciatus*. This plant extract showed good result against the *Culex quinquefasciatus* mosquito and the results were given in the Table 65.1. The 90 per cent of 1st larvae were killed in the concentration of 0.5 per cent concentration and the LC_{50} and LC_{90} values are 0.11328 and 0.52927. In II, III, IV instars larvae 80 per cent mortality at 0.5, 0.6, and 0.7 per cent respectively and the LC_{50} and LC_{90} values are 0.1993, 0.2993, 0.3993 and 0.67325, 0.77325, 0.87325. The lowest mortality was observed in lower concentration. 20 pupae were introduced in different concentration of the plant extract for about 24 hours and after 24 hours the treated pupal forms were treated to normal condition and allow for adult emergence were observed. 70 per cent adult emergence was noticed at 0.4 per cent concentration. The pupal forms showed more showed more resistance than eggs and larval forms. As the exposure concentration of the pupal forms to leaf extract increased, percentage of adult emergence was to be decreased. Similar findings of adult inhibition activity of *Melia azederach* and various neem extracts against *Culex pipens* (Al-Sharook *et al.*, 1991, Muthukrishnan *et al.*, 1999).

Table 65.1: Effect of Methanol Extract of *Kirganelia reticulata* Against the Growth and Development of *Culex quinquefasciatus*

Stages of Mosquito	LC_{50}	*LFL*	*UFL*	LC_{90}	*LFL*	*UFL*	X^2
1st instar	0.11328	0.03422	0.16352	0.52927	0.46274	0.64135	0.626
2nd instar	0.19930	0.13635	0.24376	0.67325	0.57610	0.84924	0.096
3rd instar	0.29930	0.23635	0.34376	0.77325	0.69325	0.94924	0.96
4th instar	0.39930	0.33635	0.44376	0.87325	0.77670	1.04924	0.96
Pupa	0.60000	0.55450	0.64550	1.09230	0.96945	1.32075	0.008

The normal ovary showed a gradual increase in the length and the maximum length of 442.75µ at 72 hours after blood meal whereas the length of the treated ovary did not exhibit a similar increase suggested retarded development (Table 65.2). It was emphasized that growth and maturation of oocytes were largely depend on reserve material available in pupae (Engleman, 1971). Accordingly the delayed effect of follicle development may be due to nutritional deficiencies caused by the constituents present in the *Kirganelia reticulata*. Similar reduction in the length of the ovaries of *Culex quinquefasciatus* and *Aedes aegypti* was reported after treatment with neem products (Shyamala *et al.*, 2003 and Mohanraj *et al.*, 2005).

Table 65.2: Length of Ovary Follicles of *Kirganelia reticulata* Treated and Normal Females of *Culex quinquefasciatus* at Different Time Intervals

	24 hrs	*48 hrs*	*60 hrs*	*72 hrs*
Control	204.25±0.1	286.73±0.15	436.8±0.14	442.75±0.5
0.05 per cent	95.46±0.05	272.4±0.2	394.43±0.22	395.25±0.1
0.1 per cent	84.75±0.25	253.56±0.29	372.63±0.35	374.28±0.15
1 per cent	73.85±0.024	244.93±0.15	361.38±0.38	362.3±2.20

* All the values are expressed in the unit of .

The qualitative studies of phytochemicals were carried out in the methanolic extract of *Kirganelia reticulata*. The presence of medically active constituents such as alkaloids, flavanoids, glycosides, steroids, phenols, resins, carbohydrate and protein, saponin and tannins are present in this extract (Table 65.3). Tetracyclic phenols and saponins are reported to be effective in controlling of mosquito *Aedes aegypti*, the vector of yellow fever (Marston *et al.*, 2001). Aluminium chloride known for its phenolic complexing activity obtained from aldes leaf also reported to have the larvicidal activity against *Aedes aegypti* (David *et al.*, 2001). Isoflavonoids from tubers of Neorautanenia mitis had larvicidal effect against malarial and filarial transmitting vector *Anopheles gambiae and Culex quinquefasciatus* respectively (Joseph *et al.*, 2004). A piperdine alkaloid from *piper longum* fruit was found to be active against mosquito larvae of *C.pipiens* (Lee, 2000).

Table 65.3: Results of Phytochemicals Screening of *Kirganelia reticulata* Leaves Using Methanolic Extract

Alkaloids	*Flavanoids*	*Glycocides*	*Steroids*	*Phenols*	*Saponins*	*Tannins*	*Resins*	*Carbohydrate*	*Protein*
+	+	+	+	+	+	+	–	+	+

+: Presence of the compound; –: Absence of the compound.

Weisman and Chapagain (2006) reported that saponins alkaloids extracted from the fruit of *Balanites aegyptica* showed 100 per cent larval activity against *Aedes aegypti* mosquito larvae. Osbourn (1999) have suggested that the saponin molecules interact with cuticle membrane could be the most probable reasons for the larval death. Hostettman and Marston (1995) were reported saponins containing plants are used as folk medicines, especially in Asia and are intensly used in food, veterinary and medical industries. Saponin and glycocides are very toxic to cold-blooded animals, but apparently not to mammals. Plant extract containing a high percentage of saponins are commonly used in Africa to treat water supplies and wells contaminated with disease causing vectors; after treatment, the water is safe for human drinking (Hall and Walker, 1991). However, much study is required to find out the mechanism by which saponins kills the larvae. The findings of the present investigation revealed that the leaf extract of *Kirganelia reticulata* also possess might be due to the presence of saponins and alkaloid cause larvicidal and adult emergence inhibition activity against *Culex quinquefasciatus*. So this plant can be used as an environment friendly and sustainable insecticide to control the mosquito *Culex quinquefasciatus*.

References

Al-sharook, Z., Balan, K., Jianing, Y. and Rembold, H., 1991. Insect growth inhibitors from two tropical *Meliacea*: Effect of crude seed extracts on mosquito larvae. *J. Appl. Entomol.*, 111: 425–430.

Chapagain, B. and Wiesman, Z., 2005. Larvicidal effects of aqueous extracts of *Balanities aegyptica* (Desert date) against the larvae of *Culex pipiens* mosquito. *Afr. J. Biotechnol.*, 4(11): 1351–1354.

David, J.P., Rey, D., Meyran, J.C. and Marigo, G., 2000. Involvement of lignin like compounds in toxicity of dietary alder leaf letter against mosquito larvae. *J. Chem. Ecol.*, 27(1): 161–174.

Englemann, F., 1971. Juvenile hormone controlled synthesis of female specific protein in cockroach *Leucophaea maderae*. *Arch. Biochem. Biophy.*, 145: 439–447.

Finney, D.G., 1971. *Probit Analysis*, 3rd edition. Cambridge University Press, Cambridge, pp. 245.

Gopiesh Khannan, V. and Kannabira, K., 2007. Larvicidal effect of *Hemidesmus indicus*, *Gymnema sylvestre* and *Eclipta prostrata* against *Culex quinquefasciatus*. *Afric. J. Biotech.*, 6(3): 307–311.

Hall, J.B. and Walker, D.H., 1991. *Balanites aegyptiaca Del.: A Monograph*. School of Agriculture and Forest Science, University of Wales, Banger, UK.

Harborne, J.B., 1973. *Phytochemical Methods*. Chapman and Hall Ltd., London, pp. 49–188.

Joseph, C.C., Ndoile, M.M., Malima, R.C. and Nkunya, M.H., 2004. Larvicidal and mosquitocidal extracts, a coumarin, isoflavonoids and pterocarpans from *Neorautanenia mitis*. *Transactions of the Royal Society of Tropical Medicine and Hygiene*, 98(8): 451–455.

Hayes, R.O., 1953. Determination of physiological saline solution for *Aedes Aegypti* (L). *J. Eco. Entomol.*, 46: 624–627.

Hostettman, K. and Marston, A., 1995. *Saponins: Chemistry and Pharmacology of Natural Products*. University Press, Cambridge.

Lee, S.E., 2000. Mosquito larvicidal activity of pipernonaline, a piperidine alkaloid derived from long pepper, *Piper longum*. *J. Amer Mosq. Control Assoc.*, 16(3): 245–247.

Marston, A., Maillard, M. and Hostettmann, K., 1993. Search for antifungal, mooluscicidal and larvicidal compounds from African medicinal plants. *J. Ethanopharmacol.*, 38(2–3): 215–223.

Masler, E.P., Fuchs, M.S., Sage, B. and O'Connor, J.D., 1980. Endocrine development ovarian development in the autogenous mosquito, *Aedes atropalpus*. *J. Gen. Comp. Endocrinol.*, 41: 250–259.

Meola, R. and Readio, D., 1987. Juvenile hormone regulation of the second biting cycle in *Culex pipens*. *J. Agric. Entomol.*, 10: 9–19.

Miller, M. and Chamberlin, 1986. Evolution of traps and parasitoid muscidifurax reptor Girault and sanders to manage houseflies and stable flies on dairy forms. *J. Agric. Entomol.*, 10: 9–19

Mohanraj, R.S. and Dhanakkodi, B., 2005. Effect of neem extracts on ovarian length and protein of *Aedes aegypti*. *J. Exp. Zool., India*, 7: 125–128.

Muthukrisnan, J., Pushpalatha, E. and Kasthuribhai, A., 1997. Biological effect of four plant extracts on *Culex quinquefasciatus* Say larval stages. *Insect Science Applic.*, 17: 389–394.

Nagpal, B.N. and Sharma, V.P., 1995. *Indian Anophelenes.* Oxford and IBH Publishing Co. Pvt. Ltd., Delhi, India.

Rajkumar, S. and Jebanesan, A., 2005. Larvicidal and adult emergence inhibition effect of *Centella asiatica Brahmi* (Umblliferae) against mosquito *Culex quinquefasciatus* Say (Diptera : Culicidae). *African Journal of Biomedical Research*, 8(1): 31–33.

Redwane, A., Lazrek, H.B., Bouallam, S., Markouk, M., Amarouch, H. and Jana, M., 2002. Larvicidal activity of extracts from *Querus lusitania* var infectoria galls (oliva). *J. Ethanopharmacol.*, 79: 261–263.

Renugadevi Arasappan and Thangaraj, T., 2006. Mosquitocidal effect of the plant extract against the yellow fever mosquito, *Aedes aegypti* L. *Ind. J. Ecoplan.*, 12(2): 389–394.

Shyamala, V., Manimegalai, M. and Dhanokkodi, B., 2003. Effect of neem derivatives on the egg hatchability and growth of ovarian follicle in *Culex quinquefasciatus. Pestology*, 2: 19–22.

Vatandoost, H., Moin vaziri, 2001. Larvicidal activity of neem extract (*Azadirachta indica*) against mosquito larvae, in Iron, 2000. *Pestol.*, 25(1): 69–72.

Vogel, A.I., 1978. In: *Textbook of Practical Organic Chemistry*. The English language Book Society and Longman, London, pp. 1368.

Chapter 66

Occurrence of Chlorococcales of Uyyakkondan Channel of the Ayilapettai Region, Tiruchirappalli, Tamil Nadu

R. Sivakami, V. Guru, P. Sugi and G. Prem Kishore

P.G. and Research Department of Microbiology, Kurinji College of Arts and Science, Tiruchirappalli – 620 002, Tamil Nadu

ABSTRACT

In general the luxuriant growth of Chlorococcales indicates an eutrophic condition of the water body (Krishnan and Suresh, 2005). Further very few studies has been done on Chlorococcales in India especially in Tamil Nadu. This prompted the present study in Uyyakkondan Channel of Ayilapettai region. Results show that a total of 9 sps occurred in 3 sites of the Channel. Among this *Chlorella* and *Gloeotaenium loitlesbergerianum* was found to occur in all the 3 sites. Positive correlation between Chlorococcales and pH, BOD, COD and TSS could be observed.

Keywords: *Chlorococcales, Physico-chemical parameters, Uyyakkondan channel.*

Introduction

The majority of Chlorococcales are free floating and planktonic, growing mostly in shallow confined waters (Philipose, 1959 and 1967). Gonzalves and Joshi (1946) had also observed that this order abundance in shallow waters. In general, the luxuriant growth of Chorococcales indicates an eutrophic condition of the water body (Krishnan and Suresh, 2005). Hence the present study was undertaken to see whether the Uyyakkondan Channel of Ayilapettai region in Tiruchirappalli, Central Tamil Nadu has been subjected to eutrophication.

Study Area, Materials and Methods

The Uyyakkondan Channel is an outlet of the River Cauvery which runs through the Ayilapettai region (Lat 10°50′N Long 78°36′E) in Tiruchirappalli, Central Tamil Nadu and finally rejoins the river Cauvery at Grand Anaicut. Water flows through this channel almost throughout the year except during the dry season (June–July) when water discharge to the channel stops. The maximum depth of the channel is 240 cm, the bottom being of sandy clay type. The banks on either side of the channel are lined with a number of herbs and shrubs belonging to various families. This channel was chosen for the present investigation because of easy accessibility and the absence of any industrial or construction work. However the channel is surrounded by agricultural held on both the sides of the bank.

Monthly collections were made from three selected sampling points of Uyyakkondan channel from January–December 2006. Each site was approximately half a kilometer away from one another. Algal samples were collected in polyethylene bottles and studied fresh and also after preservation in 4 per cent formalin. For identification of Chlorococalean taxa, the monograph on Chlorococcales by Philipose (1967) was used. In addition certain physico-chemical parameters of water samples were also analyzed. This was done by following the standard methods (APHA, 1985).

Table 66.1: Chlorococcale Population of 3 Different Sites of Uyyakkondan Channel at Ayilapettai Region

Sl.No.	*Species*	*Site–1*	*Site–2*	*Site–3*
1.	*Chlorococcum infusionum*	–	+	+
2.	*Tetraedron trilobulatum*	+	+	+
3.	*T. trigonum* variety	–	+	+
4.	*Chlorella vulgaris*	+	+	+
5.	*Gloeotaenium loitlesbergerianum*	+	+	+
6.	*Oocystis borgei*	+	–	+
7.	*Nephrocytium agardhianum*	+	–	+
8.	*Coelastrum microporum*	–	+	+
9.	*Scenedesmus bijuga* variety	–	+	+

Note: +, ++, +++, are the abundance of Chlorococcale population in 3 different sites.

Results and Discussion

The Chlorococcales that occurred in the 3 sites of the channel are presented in Table 66.1. A total of 9 species belonging to 8 genera could be observed from all the three sites. While site one recorded 5 species, site 2 recorded 6 species where site 3, 9 species.

Among these, only *Chlorella* and *Gloeotaenium loitlesbergerianum* was found to occur throughout the period of study in all the 3 sites. The taxa collected were unicellular as well as colonial and near to the edges of the Uyyakkondan Channel. In general, the maximum population was found to occur during the January and February months. A perusal of physico-chemical variables along with Chlorococcales population reveals that pH during January and February was 7.2–7.5. This shows that Chlorococcales population refer slightly alkaline media for growth. Similarly the maximum COD and BOD were recorded during January and February period (COD: 75–110 mg/l, BOD: 25–30 mg/l). Krishnan and Suresh (2005) also recorded maximum population when COD and BOD were highest, probably showing that they prefer high COD and BOD levels. DO levels however did not show much variation and no specific correlation could be observed with Chlorococcales population. Similar was the case with total suspended solids also. This is in correction with the finding of Krishnan and Suresh (2005) who also recorded no significance correlation between DO, TSS and Chlorococcales. The presence of *Chlorella vulgaris* throughout the year shows that it can grow both in polluted as well as clear waters (Tiwari *et al.*, 2002). Luxuriant growth of Chlorococcales indicates the Eutrophic condition of the water body. This might be the reason for higher values of COD and BOD and TSS in the months of January and February.

Table 66.2: Chlorococcale Population Identified in the Uyyakkondan Channel at Ayilapettai Region, Tiruchirappalli

Sl.No.	*Species*	*Sites*	*Months*											
			Jan	*Feb*	*Mar*	*Apr*	*May*	*Jun*	*Jul*	*Aug*	*Sep*	*Oct*	*Nov*	*Dec*
1.	*Chlorococcum*	1	–	–	–	–	–	Dry	Dry	–	–	–	–	–
	infusionum	2	+	+	+	–	–	Period	Period	–	–	–	–	+
		3	+	+	+	–	–			–	–	–	–	+
2.	*Tetraedron*	1	+	+	+	+	+	Dry	Dry	–	–	–	–	–
	trilobulatum	2	+	+	+	+	+	Period	Period	–	–	–	–	–
		3	+	+	+	+	+			–	–	–	–	–
3.	*T. trigonum* variety	1	–	–	–	–	–	Dry	Dry	–	–	–	–	–
		2	–	–	–	–	–	Period	Period	+	+	+	–	–
		3	–	–	–	–	–			+	+	+	–	–
4.	*Chlorella vulgaris*	1	+	+	+	+	+	Dry	Dry	+	+	+	+	+
		2	+	+	+	+	+	Period	Period	+	+	+	+	+
		3	+	+	+	+	+			+	+	+	+	+
5.	*Gloeotaenium*	1	+	+	+	+	+	Dry	Dry	–	+	+	+	+
	loitlesbergerianum	2	+	+	+	+	+	Period	Period	+	+	+	+	+
		3	+	+	+	+	+			+	+	+	+	+
6.	*Oocystis borgei*	1	+	+	+	+	+	Dry	Dry	+	–	–	–	–
		2	–	–	–	–	–	Period	Period	–	–	–	–	–
		3	+	–	–	–	–			–	+	+	+	+
7.	*Nephrocytium*	1	+	+	–	–	–	Dry	Dry	–	–	–	–	–
	agardhianum	2	–	–	–	–	–	Period	Period	–	–	—	–	–
		3	+	+	+	+	+			+	+	+	+	+
8.	*Coelastrum*	1	+	+	+	+	+	Dry	Dry	+	+	+	+	+
	microporum	2	–	–	–	–	–	Period	Period	–	–	–	–	–
		3	+	+	+	+	+			+	+	+	+	+
9.	*Scenedesmus*	1	+	+	+	+	+	Dry	Dry	+	+	+	+	+
	bijuga variety	2	–	–	–	–	–	Period	Period	–	–	–	–	–
		3	+	+	+	+	+			+	+	+	+	+

Table 66.3: Physico-chemical Parameters of Uyyakkondan Channel at Ayilapettai Region

Parameter Details	*Unit*	*Range*
Temperature	°C	22°C–32
pH	–	7.0–7.8
DO	mg/l	6.8–8.0
CO_2	mg/l	Nil–2
BOD	mg/l	8–30
COD	mg/l	18–110
Total Alkalinity	mg/l	50–150
TSS	mg/l	15–60
Oil and grease	mg/l	Nil–5

References

APHA, 1985. *Standard Method of Examination of Water and Wastewater.*

Gonzalves, F.A. and Jooshi, D.B., 1946. Freshwater algae near Bombaly. *J. Bombay Nat. Hist. Soc.*, 46: 154–176.

Krishnan, M. and Suresh, G., 2005. Studies on the ecology of chlorococcales of Sitarampur Dam, Jharkhand, India. *Indian J. Environ. and Ecoplan.*, 10(3): 859–864.

Philipose, M.T., 1959. Freshwater phytoplankton of inland fisheries. In: *Proc. Symp. Algol.* (Ed.) P. Kachrooo. I.C.A.R., New Delhi, pp. 272–291.

Philipose, M.T., 1967. *Chlorococcales.* ICAR, Krishi Bhawan, New Delhi.

Tiwari, D., Singh, S. and Kaur, P., 2002. Chlorophyceae of the River Pandu. *Phykos*, 41: 115–120.

Chapter 67

Impact of Monocrotophos on Behavioural, Respiratory and Change of Biochemical Constituents of *Channa punctatus* (Bloch)

***P. Kennadi*[1]*, *S. Raveendran*[2] *and C. Sivasubramanian*[2]**

[1]*Department of Zoology, Government Arts College, Ariyalur*
[2]*P.G. and Research Department of Zoology, Khadir Mohideen College, Adirampattinam, Thanjavur District*

ABSTRACT

India is rich in Inland fishery resources, but the indiscriminate use of the pesticides causes serious threat to such resources. Pesticides are stable compound and they enter into the aquatic ecosystem through the agricultural run off. The pesticides, which enter the body tissues of the fish, affect the physiological activities and the nature of main food stored like glycogen and protein. When the fish were treated with various concentrations of the monocrotophos the amount of protein and glycogen contents in the muscle of the fish decreases slowly sometimes rapidly on prolonged exposure even in the same concentrations. There was clear and steady increased in the rate of opercular movements and follows decreased rate of oxygen consumption was observed with increasing concentrations of the pesticide monocrotophos.

* Corresponding Author.

Introduction

Pesticides are stable compound. Indiscriminate liberal and injudicious use of pesticides by man to control the crop pests and diseases for higher agricultural productivity has led to a slow but steady deterioration of the aquatic ecosystem, since *water* as the ultimate sink. These pollutants also destroy the quality of the aquatic media and render it unfit for various aquatic organisms particularly fishes. A variety of organophosphate, organchlorine and carbamide pesticides are extensively used in agriculture for the control of pests. It can be seen that the toxic action is specific for a particular animal in a particular toxicant. The toxicity levels were influenced by the sex (Victoriamma and Radhakrishanaiah, 1982) by the nutrient supply (Arunachalam *et al.*, 1980) by the pH (Eisler, 1970). The effects of several insecticides and pesticides on various physiological responses of fishes were reported by Anderson, 1971; Brett, 1973; Monoharan and Subbiah, 1982; Devi Sewtheranyan, 2000; Prashanth, 2003 and Mathivanan, 2004. The present investigation has been takes up to elucidate the effects of monocrotophos at sublethal concentration on certain biochemical constituents of the fish muscle and also to examine the sublethal effect of the behaviour and respiratory activity of the fish *Channa punctatus.*

Materials and Methods

The fish for the experimental purpose health live *Channa punctatus* were collected from the local ponds. The experiments were carried out with help of small circular trough of 15 litre capacity in which a iron wire gauge covered the trough at the surface level of water to avoid for aerial respiration. The LC_0, LC_{50} and LC_{100} values were calculated and tabulated. Before the actual starting of the experiments, the test fishes were divided into 6 groups of the same weight wee selected from the stock tank and transferred them into the test chamber with test solution of various concentrations of monocrotophos *i.e.* 0.0000 ppm, 0.5 ppm, 1.0 ppm, 1.5 ppm, 2.0 ppm, 2.5 ppm. Each group is consisting of 10 fish per trough. These were actually subjected to both short and long term exposure periods, the former lasting for about 24 to 48 hrs and the latter lasting for 72 to 96 hrs. Both the long term and the short-term exposures were given to all the five groups involving the different concentrations on monocrotophos. A control (pesticide free water) was also maintained in the above said manner.

After the exposure period was over one of the fish was taken out and scarified for the analysis of selected biochemical constituents *viz.*, protein (Lowry *et al.*, 1951 method) and glycogen (Antheron method). Behavioural effects were also studied in normal and treated fishes, which were maintained at room temperature.

Results and Discussion

Pesticides are one of the major xenobiotic substances that have been used in India for a longer period for management of pests in agriculture fields and control of vectors in public health operations. Most of the insecticides are hydrophobic that they can easily be absorbed by soil particles and can migrate to natural water systems such as rivers, lakes and ponds through the run off causing severe aquatic pollutions. The indiscriminate use of the pesticides, however, resulted in the pollution of our ecosystem causing hazards to non target organism including fishes.

Glycogen and proteins are the chief nutrients of the animals. They have a variety of functions. The glycogen supplies energy in the form of ATP molecules which are formed during TCA cycle. Glycogen is autoketogenic and aids in the utilization of body fats. Similarly proteins are made up of amino acids which form the building blocks of the body like essential constituents of protoplasm of all the cells. The proteins in different tissues differ in composition and properties.

In the present study the results obtained clearly indicate that there was a decreased amount of protein and glycogen content to resist the effects of pesticides. That is to provide immediate energy to the fighting elements of the body and protect all systems of the body from the harmful effect of the pesticides. As the fish was constantly kept in the medium of different concentration of pesticides dissolved in water there was no way for the animal to move away from the toxic medium. The effect of pesticides dominates. The organ system and the glycogen content are slowly depleted due to the utilization of the already stored glucose contents of the body. At particular level poison overtakes the organisms so that the level of the carbohydrates content is very low in the muscles because of the utilization.

Table 67.1: Effect of Different Concentration of Pesticide Monocrotophos on Mortality Percentage of *C. punctatus* as a Function of 24, 48, 72, 96, hrs Exposure

Concentrations (ppm)	*Percentage of Mortality and Hours of Exposure*				*Remarks*
	24	*48*	*72*	*96*	
0.5	0	0	0	0	Lc 0/96 hr
1.0	0	0	0	0	Lc 0/96 hr
1.5	0	0	0	0	Lc 0/96 hr
2.0	0	0	0	0	Lc 0/96 hr
2.5	0	0	0	0	Lc 0/96 hr
3.0	0	0	0	10	Lc 10/96 hr
3.5	10	20	40	50	Lc 50/96 hr
4.0	10	20	40	70	Lc 70/96 hr
4.5	10	20	50	80	Lc 80/96 hr
5.0	40	80	90	100	Lc 100/96 hr

Table 67.2: Sub-lethal Effects of Monocrotophos on Oxygen Consumption and Opercular Beats of *Channa punctatus*

Concentration (ppm)	*Opercular beats (No.Imin)*	*Oxygen consumption (ml/g/h)*
0	125±3.85	0.98±0.05
0.5	130±3.25	0.98±0.03
1.0	160±4.65	0.75±0.08
1.5	180±6.25	0.60±0.18
2.0	195±3.50	0.40±0.25
2.5	225±4.50	0.25±0.45

With regard to the exposure of fishes to different concentrations of Monocrotophos pesticide there was no much change within 24 hours of treatment with mild concentrations of 0.5 ppm and 1.0 ppm but the same on prolonged exposure upto 96 hours the glycogen content was observed in decreased amount (Table 67.3) when the concentration was increased to 1.5 ppm, 2.0 ppm and 2.5 ppm. The glycogen content was observed in the decreasing order with increasing concentrations and with more exposure periods the glycogen content was found more and more in decreased. Because of the stress the fish makes suitable adjustments for which the stored energy is utilized. This may be the reason for the decreased amount of glycogen content.

Table 67.3: The Total Protein Content and Total Glycogen Content of Muscles of *C. punctatus* Exposed to Pesticide Monocrotophos (Each value represents the mean±SD of five values.)

Concentration ppm of Monocrotophos	*Exposure Period (hours)*	*Amount of Protein Content (mg/g)*	*Amount of Glycogen Content (mg/g)*
Control	24	23.50±0.4325	3.25±0.1550
0 ppm	48	22.10±0.2535	3.10±0.1662
	72	21.10±0.1725	2.90±0.1775
	96	19.75±0.2202	2.50±0.2250
Treated	24	22.25±0.5155	3.10±0.2250
0.5 ppm	48	21.75±0.6250	2.95±0.1350
	72	20.25±0.2550	2.35±0.1825
	96	19.25±0.1550	1.90±0.1450
Treated	24	22.00±0.2505	2.85±0.1100
1.0 ppm	48	21.25±0.1505	2.50±0.1250
	72	18.50±0.1608	2.25±0.1360
	96	18.25±0.1725	1.90±0.1450
Treated	24	21.50±0.1325	2.65±0.2250
1.5 ppm	48	20.40±0.1225	2.35±0.5500
	72	1810±0.1450	1.70±0.5125
	96	15.25±0.1150	1.50±0.6225
Treated	24	20.20±0.1158	2.50±0.5525
2.0 ppm	48	19.50±0.1255	1.95±0.6250
	72	16.50±0.1150	1.15±0.6350
	96	15.25±0.1225	0.50±0.2550
Treated	24	19.76±0.01010	2.25±0.9250
2.5 ppm	48	17.50±0.2050	1.70±0.7250
	72	16.00±0.2355	1.10±0.6250
	96	15.25±0.5250	0.40±0.5050

Such reduction in stored glycogen content has been reported in *Tilapia mossambica* exposed to methyl parathion (Rao and Rao, 1929) effect of endosulfan reported by Vasanthi and Ramasamy (1987) in saratherodon mossambicas and in *Channa punctatus* following malathion intoxication reported by Shah and Duable (1983) and in *O. mossambicus* to metacid by Baskaran (1991).

The protein content in the muscle of C. *punctatus* is decreased with increasing concentrations of pesticide monocrotophos. Even with the same concentrations longer exposure resulted in decreased amount of protein content (Table 67.3) which indicates that the tissue protein undergoes proteolysis. This results in the production of free amino acids, which are used in the TCA cycle for energy production under stresses (Kabeer Ahamed, 1979). There are similar reports of the effects of toxicants on total protein in other fishes by Rath and Mishra (1980), Shah and Dubale (1983), Palanisamy *et al.* (1989). In the present study, the opercular beats increased with increasing concentrations of pesticides. The initial increase in opercular movement in the less concentrated medium as a primary response to sudden stress was reported by Anbu and Ramaswamy (1991) as in *Channa striatus* exposed to carbomate pesticide sevien. Baskaran and Palanichamy (1990) reported that the opercular beats increased with increasing concentrations of fertilizer in *A. scandens.*

Oxygen consumption of pesticide treated fish showed an initial increase with lower concentration of pecticides and decreased with increasing concentrations (Table 67.2). The reduced oxygen consumption could be attributed to gill damage or to hypochronic mirocytic anemia as suggested Palanichamy and Baskaran (1995). Similar decrease in oxygen consumption was observed by Larcolivin (1994) in *Percu fluiatiels.*

The decrease in oxygen consumption appears to be a protective measure to ensure that there is a low intake of the toxic substance. Reduced oxygen consumption at higher concentrations of pecticides could also arise as a result of respiratory inhibitory factors that come into play as suggested Rafia sultana and Umadevi (1995) in *Mystus gulio* under heavy metal pollution.

Oxygen consumption and opercular activity of the fish were concentration dependent. The initials increase oxygen uptake probably to meet energy demand during early periods of exposure. Which was followed by gradual decrease in oxygen uptake in later period of exposure, which may be due to onset of toxicity as suggested by Fry (1971). In order to meet the increased demand for oxygen by the tissues opercular activity is enhanced but this too has a limit beyond which the activity stops resulting in the death of the animal.

The pesticide monocrotophos dissolved in water brings about extensive changes in the physical parameters of water such as pH, salinity, alkalinity and severe depletion in the dissolved oxygen content. So the aquatic systems are drastically altered due to this new stress in the environmental medium which may alter the biochemical contents and respiratory parameters to cope with those situations.

References

Anbu, R.B. and Ramasamy, M., 1991. Adaptive changes in respiratory movement of an air breathing fish *Channa striatus* (Bloch) exposed to Carbonate pesticide seiven. *J. Ecobiol.*, 3(1): 11–16.

Anderson, J.M., 1971. Sublethal effects of changes in ecosystems: Assessments of pollutants of physiology and behaviour. *Proc. R. Soc.*, London, 177: 307–320.

Arunachalam, S., 1980 Toxic and sub-lethal effects of carboryal on the fish water catfish *Mystus vittatus* (Bi). *Arch. Environ. Contain. Toxicol.*, 9: 307–316.

Baskaran, 1991. Use of biochemical parameters in biomonitoring of pesticide pollution in some freshwater fishes. *Excotoxical, Environ. Monit.*, 2: 103–109.

Baskaran and Palanichamy, S., 1990. Sublethal effects of ammonium chloride on feeding energetics and protein metabolism in the fish *Orechromis mossambicus. J. Ecobiol.*, 2: 97–106.

Sewtharanyam, Devi, 2000. The effects of endosulfan on oxygen consumption of the fish *Oreochromis missambicus. J. Ecotoxicol. Environ. Monit.*, 10(1): 21–24.

Eisler, K., 1970. Factors affecting pesticide induced to toxicity in estuarine fish. Technical papers of the behaviour of sports for the fish and wildlife.

Ahamed, S.I. Kabeer, 1979. Studies on some aspects of protein metabolism and associated enzymes in freshwater teleost, *Sarotherodon mossambicus* subjected to malathion exposure. *Ph.D. Thesis,* S.V. University, Thirupuathi, India.

Collivin, Lars, 1984. The effects of copper on maximum respiration rate and growth rate of perch, *Perca fluviatilis* L. *Water Res.,* 18: 139–144.

Lowry, O.H., Rosebrough, N.J., Farr, A.I. and Randhall, R.J., 1951. In protein measurement with folin phenol regent. *J. Biol. Chem.,* 195: 265–273.

Mathivanan, R., 2004. Effects of sublethal concentration of Quinolphos on selected respiratory and biochemical parameters in the freshwater fish *Oreochromis mossambicus. J. Ecotoxicol. Environ. Monit.,* 14(1): 57–64.

Monoharan, T. and Subbiah, G.W., 1982. Toxic and sublethal effects of endosulfan on *Barabar stigma. Proc. Ind. Acad. Sci (Animal Science),* p. 523–532.

Palanichamy, S., Arunachalam, S. and Baskaran, P., 1989. Impact of pesticides on protein metabolism in the freshwater cat fish *Mystus vittatus. J. Econobio.,* 1: 90–97.

Palanichamy, S. and Baskaran, P., 1995. Selected biochemical and physiological responses of the fish *Channa striatus* as biomonitor to assess heavy metal pollution in freshwater environment. *J. Excotoxical. Environ. Monit.,* 5(2): 131–138.

Prashanth, M.S., David, M. and Kuri, Riveendra C., 2003. Effects of cypermethrinon toxicity and oxygen consumption in the freshwater fish *Cirrhinus mrigala. J. Ecotoxicol. Environ. Monit.,* 13(4): 271–277.

Sultana, Rafia and Devi, Uma, 1995. Oxygen consumption in a catfish *Mystus gulio* (Ham) exposed to heavy metals. *J. Environ. Biol.,* 16(3): 207–210.

Rao, K.S.P. and Rao, K.V.R., 1929. Effect of sublethal concentration of methyl parathion on selected oxidative enzymes and organic constituents in the tissues of freshwater fish *Tilapia mossambica. Current Science,* p. 46.

Rath, S. and Mishara, B.N., 1980. Changes in nucleic acid and protein content of *Tilapia mossambica* exposed to Dichlorovas (DDVP). *Indian Journal of Fisheries,* 27: 76–81.

Shah, P.H. and Double, M.S., 1983. Biochemical changes induced by malathion in the body organ of *Channa punctatus. Journal of Animal Morphology and Physiology,* 30: 107–118.

Vasanthi, M. and Ramasami, M., 1987. A shift in metabolic pathway of *Sarotherodon mossmbicus* (Peters) exposed to Thiodon (Endosulfan). *Proc. Indian Acad. Sci., (Anm. Sci.),* 96: 56–6.

Victorimma, D.C. and Radhkrishnaiah. 1982. Mercury tolerance of fresh water field crab *Ozinokelphus senex. Symp. Physiol Resp. Animal Pollution.*

Index

www.ingramcontent.com/pod-product-compliance
Ingram Content Group UK Ltd.
Pitfield, Milton Keynes, MK11 3LW, UK
UKHW052224270726
14059UKWH00003B/108